Studio Recording Procedures

How to Record Any Instrument

Mike Shea

McGraw-Hill

New York   Chicago   San Francisco   Lisbon
London   Madrid   Mexico City   Milan   New Delhi
San Juan   Seoul   Singapore   Sydney   Toronto

The McGraw·Hill Companies

Library of Congress Cataloging-in-Publication Data

Shea, Mike
    Studio recording procedures: how to record any instrument / Mike Shea.
        p.   cm.
      ISBN 0-07-142272-2
        1. Sound—Recording and reproducing.   2. Sound studios.   I. Title.

    TK7881.4.S47   2004
    781.49—dc22

                                                                2004058156

1 2 3 4 5 6 7 8 9 0   DOC/DOC   0 10 9 8 7 6 5

ISBN 0-07-142272-2

The sponsoring editor for this book was Steve Chapman and the production supervisor was Sherri Souffrance. It was set in New Century Schoolbook by MacAllister Publishing Services.

Printed and bound by RR Donnelley.

 This book was printed on recycled, acid-free paper containing a minimum of 50% recycled, de-inked fiber.

McGraw-Hill books are available at special quantity discounts to use as premiums and sales promotions, or for use in corporate training programs. For more information, please write to the Director of Special Sales, Professional Publishing, McGraw-Hill, Two Penn Plaza, New York, NY 10121-2298. Or contact your local bookstore.

This book is dedicated to my mom, Mrs. Mary Antionette (Giorgio) Shea, who always told me, "If you're going to do something, do it right."

CONTENTS

Contents

Contents

PREFACE

First, take it from me, if you're trying to improve the quality of your current recording project, your best bet is to rip out the section of this book that covers recording procedures and stuff it in your back pocket when you head out to do your next recording session. That's where you need it, because correct microphone selection and placement is essential to getting a good sound.

I'm not talking about the typical "Place a directional mic at the snare's rim" instructions. Instead, in this book you'll find explicit directions such as "Use mics X, Y, or Z positioned so that their 120-degree point is facing the high hat when the mic is aimed at the most worn area of the drum's head. Then use processors A, B, and C to achieve these specific sounds."

This right-to-the-point approach will be fully appreciated by those faced with the realities of getting the job done (especially the beginning recordist) as they often find themselves pinched between a rock (such as an impatient client watching the clock with one hand on his wallet) and a hard place (like an irate studio owner who will not tolerate customer complaints and to whom the term "down time" results in volcanic eruptions).

The rest of this book was meant to be a reference or textbook that gives an overview of the scientific and theoretical aspects of professional audio recording.

What I wanted to accomplish with the section that covers basic audio technology was for readers to achieve an understanding of decibel references and to comprehend the values displayed on every type of audio meter. This would enable them to become skillful at managing and controlling the levels of audio signals as they passed through various gain stages—in other words, to be able to handle the fundamental requirement of audio engineering. Obviously the sound will improve when the recording engineer understands the ins and outs of signal routing and level setup.

In my opinion this book provides only a fraction of the total information needed by today's professional recording engineer. Yet, you do not have to know everything in each section to accomplish any particular recording task, and attempting to memorize any of it would be a waste of time.

I wanted to make this a handy reference tool that will always serve the reader after he or she gives it a good once over. After you've read it once, you'll have a good bead on what it entails and how it all fits together, so when you need to acquire a better understanding of a specific subject area, you will know exactly where to look.

For example, I disclosed a method of getting an automatically updated recording of the mix bus not only during mix-down, but while you're recording basics and overdubs so that at the end of every session, you have a mix that's ready to copy or refer to at the start of the next session. While this is very helpful during the recording and mix-down stages, it also involves syncing up the two-mix recorder, as I will explain in the section on synchronization.

I could not possibly get into every recording console and describe its functioning, so I've presented the basic types, explaining the input section, the output section, and the monitoring section, along with the use of computer-assisted recording methods. Consoles can be configured in a variety of ways, but by understanding what comprises these basic sections you can work your way through any console's signal routing.

In writing this book, time became extremely critical. Due to health problems (you start to hurry things up when they tell you that you have only six months left), it became more important to get as much factual information in the book as possible. Therefore, I did not get to cover everything that I wanted to cover, such as using measurement mics in recording applications. But the manufacturers who produce these provide reams of information on various applications, and by using the methods that I have outlined, you will know how to put it all to use. I also wanted to compile an extensive mic closet (or directory) section and go over a great many mics, their specs, and their characteristics. But by using the information provided, you can do the same thing yourself and you will be better off for it.

I did not see the practicality of getting into basic digital audio electronics or an expanded section on recorders. This is because these technologies frequently change and anything described here may have been outdated almost as soon as it was printed.

Learning how to develop your hearing perception is very much a part of every recordist's training. Listen and pay attention to the ambience and wide dynamics that take place on old, big band recordings. These engineers (your forebears) often only used a couple of mics, and they recorded these tracks "live to two" at most. But the sound they got was in-your-face real, with big drums, loud horns, and tons of room ambience. While all of it was captured using distant micing techniques, the vocalist always sounds great and comes through clear as a bell. It is not a mystery as to how this was accomplished. Once you understand the fundamentals, all you will ever need to do is just listen and you will be able to figure out *all* the answers. *No lie!*

As far as learning to follow along with music charts, you can pick up a book on the subject, but do yourself a favor and get the recording along with the sheet music to follow along with, as this will be more in keeping with the task at hand. I also strongly suggest you get a how-to book on using oscilloscopes because these tools provide the audio engineer with the bottom line truth about sound quality. You've got a lot of work to do, and outside of a little bit of math, it's all fun. So, have a ball! I did.

INTRODUCTION

This book was written to serve as a textbook and reference source that provides an overview of the theory, science, and physics involved in professional audio recording. It's the book I wish I had when I was first starting out. Secondly, because it was also meant to be used like a road map or guide to get good sounds rapidly, its full of hands-on practical information that was designed to get even a beginning sound recordist up, running, and achieving professional quality recordings quickly. As you will see, this is not only generalized information. I tried to be more specific because in today's recording environment, there's a need to be factual and direct. This is particularly essential for the interns, apprentices, assistants, and beginning recordists who need to increase their recording skills, pronto.

Anyone who reads trade magazines understands that there is something new on the scene every day. What I have done is to give you the tools to go out and continue learning on your own. My goal was to have the readers know how to interpret information about audio equipment and understand how to use this knowledge to help with the work they are doing.

My approach was "This is the information you need to understand; here's where you find it; here's how you interpret it; and this is how you make it work for you." But in order for you to be able to do any of this, you must first understand the language involved. What are the units of voltage, power, and level that are used in measurement saying about amplifier gain and recorder or mixer distortion?

After we discuss that, we can get into specifications and find out what intermodulation distortion and signal-to-noise ratio really means, as well as what things like equivalent input noise are all about. You have to have a firm grip on operating levels, be they +4 dBu, 0 dBm, or −10 dBV. Once you comprehend these areas, you will be able to interpret information about audio equipment for yourself.

You'll no longer have to wonder what AC meters are indicating or how RMS, peak, PPM, and peak-to-peak relate to VU. If you do not understand what decibels are all about, you will not be able to understand audio specifications.

So who needs specs? You do. They clue you in on how things will sound when using one device as opposed to another. For example, a microphone's sensitivity rating will warn you when it is not safe to put a $3,000.00 mic right in close on that bass speaker because it may result in distortion and possible damage to its irreplaceable diaphragm!

Back in the late 1960s and early 1970s when I was trying to learn the technology involved in audio recording, very little was offered in schools of higher education. This was partly due to the fact that most recording studio engineers received their education via on-the-job training. This often entailed, as it later did for me, helping to build the studio's mixing board (which provided a basic education in electronics) as well as helping to construct the studio rooms themselves (which taught about controlling noise transmission, vibration isolation, and acoustics). In the past, studios could afford the kind of clean-up/go-for/tape-op/apprentice/assistant engineer/staff recording engineer in-house training progression, which had been *the* main way to learn this craft.

The training that actually related to recording came from one's move up the ladder, starting out as a coffee go-for and then a tape deck operator ("Tape Op"). This was before remote controls became common, and it afforded the opportunity to get acquainted with following music charts as well as understanding how to prepare for and react to production cues. From here, you were moved to the studio side of the glass where you became a second set of eyes, ears, and hands for both the recording engineer and the producer. Aside from the obvious lessons learned—such as the correct microphone placement for recording various instruments along with the unobtrusive layout of cabling, and the positioning of acoustic screens and baffles—the subtleties of studio etiquette started to really become ingrained. Keeping quiet during recording "takes" may seem like a given, as does making sure a vocalist has a glass of water handy, but one also becomes a quick study in keeping one's subjective opinions and any other comments to oneself.

Obtaining steady employment of this type in the late 1960s was nearly impossible (I would have to wait until the mid-1970s). So I took courses in electronic music at The New School in New York City and attended seminars such as those given by famous acousticians like Leo Branneck at institutions such as Harvard and MIT. I also ended up taking classes at Queens College (part of The City University of New York), which offered a B.A. in Communications Arts and Science, but they lacked a true music recording program, as theater and broadcast were the school's main areas of focus. I ended up working in a music rehearsal studio, which, in time, developed into a full-fledged recording facility.

A decade later, in the mid-1980s when I was both writing about and teaching professional audio recording, the audio education scene had changed. There was already a trickling of university-level audio recording education programs being offered, predominantly in conjunction with a degree in music, but, to my knowledge, there were no associate degree programs. There were a few accredited programs (similar to trade schools), such as those at The Institute of Audio Research, which offered 441 hours of course work that fully covered all the basics of audio recording technology. But most audio recording education was still to be had in recording studios. Unfortunately, this on-the-job training was no longer structured to the point of achieving a truly competent staff as the old stepping-stone method of training was no longer cost effective.

Today, virtually no structured on-the-job training remains, but there are now at least 100 schools in the United States and Canada offering some audio as part of their curriculum. Many of these are short programs (under one year) of the trade school and private studio type, but a large number of associate degree programs are now offered. Additionally, there are dozens of university-level, four-year bachelor degree programs as well as a couple of master degree programs available. Yet it's very difficult to teach everything there is to know about recording in a year or two because there's simply so much material to learn. It is not unusual for graduates to have trouble with direction in terms of continuing their development. For example, I can't count how many times I've been asked, "Do you need an assistant? I sure could use a mentor."

I was never in a position to hire any permanent help, and the few beginners I offered to teach mil-spec soldering were not interested. Therefore, I always thought it might be worthwhile to write a textbook covering audio recording that would be beneficial to audio recording students, especially those attending schools whose programs are attached to music degrees, as these can often be lacking on the technological side.

But maybe you are not at the beginning stages of your recording career, or you think that pros do not need to know any more about audio basics.

I once heard a saying that, while meant to be humorous, comes across as a warning. It went something like this: Engineers are people who have a vast amount of knowledge about a limited subject area. Throughout their career, they learn more and more about a continually narrowing and more-defined area of interest. Soon they end up knowing just about everything about practically nothing.

This is not the situation for the audio engineer/recordist. Your chosen area of interest is too multidisciplined and ever-changing for there to be a comfort zone. So even seasoned recordists must continually expand their knowledge base.

I guess I'm just partial to facts. I don't even like the term "stainless steel" because it is an advertising ploy. Since it dates back to the 1930s, people take it for granted yet it is a bold-faced lie. There are steels that are stain resistant, but there is no such thing as a stainless steel. Give me a piece of steel, even with the highest chromium content you can find, and I guarantee I will be able to stain it.

Sticking with the metallurgy analogy (I know about this stuff because machining replacement parts is often part of microphone repair), I've heard it said that titanium is stronger than steel and lighter than aluminum, when in fact it is stronger than aluminum and lighter than steel. It bugs me when audio recording is subjected to falsehoods like these.

Fact: Solutions to complex problems emerge from having a detailed understanding of audio recording's underlying technology.

I understand that some people are afraid of math. But hey, nobody enjoys working through difficult equations, and even the brightest among us are leery of logarithms, especially in audio, where the same numbers can mean many different things.

So who needs all this complicated technological stuff? Music is supposed to be fun right? Well, it is actually not much fun when it sounds bad and all the technology is about making things sound good.

But don't worry. You will be able to understand *all* the technology in this book. How do I know that? Because I was able to understand it. *What?* Well, in February of the year 2000, my brain was cut in half.

Later, they told me that I had had a stroke nine or ten months earlier and that at the time, a tiny capillary had burst inside my brain. I didn't feel a thing for months, and when I finally went to the hospital with massive head pain, it took four doctors 11 hours of emergency surgery to remove a blood clot the size of an orange from my brain. They saved my life and I even had the ability to speak and answer questions right after the operation. If you had asked me to explain dBm and dBu, I could have done it without a hitch. However, if you asked me anything about how they related to each other, I'd have drawn a blank. Some things were no longer connected! You see, most of my neurological connections had been severed.

At first it was much more apparent on the physical level, as I was like a newborn baby. I had to re-learn how to move a finger. Crawling came much later, and it took me three and one-half years until I was able to walk confidently without canes and braces.

If you think I'm telling you all this so that you'll feel sorry for me, think again. Many of the stroke victims in the hospital with me had lost control of half of their body and would never regain it. I lost everything but got it *all* back, so I have nothing to complain about. I only bring it up because this is the reason why I know the technology covered in this book will be easy for you to understand. I *had* to write it that way so that *I* could understand it. I just could not trust my brain. For two and a half years I went over everything repeatedly until it was all laid out logically and explained in terms that were simple enough for me to follow and hold together without getting confused. You see, because I wasn't confident in my mental capacity, I triple-checked and quadruple-checked everything so that it now seems like it is all just common sense.

The fact is, I became *the* perfect person to write this book. I actually knew all this stuff inside-out and had done the job for 35 years. I also didn't presuppose any prior knowledge on the reader's part and made absolutely sure that all information flowed in a logical order and could be easily understood.

Oh, you'll still have to work at it. For example, even though I wrote out and solved all the mathematical equations (you're welcome), the information won't sink in until you follow along using a calculator. I would also suggest getting your hands on a dozen or so microphone specification sheets and practicing up on your new-found skill at deciphering them to figure out possible applications they may serve in making your job easier while providing a superior sound.

While I have almost 40 years of extensive experience in this field, I do not consider myself *the* top expert on audio recording. No one person can, as the field is just too vast and ever-changing for anyone to claim that title.

Although I had a complete test-bench set up for checking the specifications of equipment I reviewed for professional audio publications, the series of recording procedure articles that part of this book is based on was far from a one-man-subjective-opinion show. These were written in the early 1980s while I was a freelance recording engineer handling a lot of substitute work (mostly overdubs) for other over-booked engineers. At the same time, I was also a freelance maintenance technician for a number of facilities, instructing graduate courses at the Institute of Audio Research, and serving as Technical Editor of *Recording World* magazine. So my getting studio time to check out equipment or microphone techniques was never a problem. More important is the fact that I was definitely not self-consumed with my ego. You see, I knew that any statement concerning the art of microphone selection and placement was basically going to be subjective and that *every* engineer has his or her own special tricks and techniques. Therefore, I always called upon musicians, other recording engineers and techs, as well as studio owners and layman/consumers for their input and opinions on all my tutorial and review articles.

I deliberated over the viability of including so many reviews of older equipment, but I was urged by other audio professionals to go ahead with my outline exactly as I had laid it out. Covering many types of control parameters would not only familiarize the reader with most of the subtle differences, but when using software-based controls, the reader would better understand the functions and ways to utilize certain adjustments.

While some other segments of this book may now seem dated, music instruments and microphones (no matter what the advertising) have not really changed all that much. In fact, some of the principles covered go back to the 1930s and were taught to me in the 1960s by electronics/recording engineers who had been WWII radio operators and later CBers and engineers in commercial radio. These "old-timers" started me off with shunt resistors and never let me get away with any "slip-short" (superficial) methodologies.

I found myself thinking about how there just doesn't seem to be any of those old-timers around any more to spread the word to the next generation of audio recordists when I realized that "them old-timers" is me! I may only be in my fifties, but hey, that's about how old those "geezers" were!

So now it's my turn to pay back into the "spreading the audio information" fund because I feel I owe all the audio recordists who came before me a debt. And so do you! Sharing the knowledge is now your responsibility, too.

When I started teaching, I found that the test of whether you really understood the basic principles was if you could explain them so that they were easily understood. It also doubles as a kind of payback to yourself. If you take the time to explain something to someone, it increases your knowledge.

It's all about the code you choose to live by. Hey, if you think my writing this book is about money, let me clue you in. What you're holding isn't a romance novel. I've done this writing thing before, and it's very rare for royalties to exceed $500 a year. Even with a shelf-life of ten years ($5,000 total), you're still talking about a wage that is less than two dollars an hour! There is nothing original here. Everything has been said before. The only credit due me is that I made the effort to figure all this out at a time when there were no schools teaching audio recording. Now there are plenty of schools, but many leave out the basics. For them it is an economic fact of life because potential students don't like math! Hey, who does?

Take it from me, audio "magic" doesn't happen on its own. Recording engineers need to have a firm grip on the technology they're using, and I've laid it all out nice and simple for you. Make sure you do the same.

1

The Basics

Why the Big Mystery About Decibels?

Your ability to comprehend just about anything in audio requires that you understand what decibels are. This may seem fairly obvious, yet some of those involved in professional audio lack a complete understanding of the term *decibels* (dB) or at least are not capable of fully explaining it to others. This shouldn't be surprising; just try and find a thorough, easily understood explanation. Most of the time, if discussed at all, the meaning of decibels is barely touched upon or is glossed over with open-ended statements such as the following, which end up leading the novice in circles:

- The rules simply state that for voltage, dB = 20 log 10 × (E divided by the reference voltage value), while for power, dB = 10 log 10 × (P divided by the reference power value).

- A 1-milliwatt sine wave into a 600-ohm load has an RMS value of 0.775 volts and is the reference level, equal to 0 dB, when decibels are expressed as dBm.

- "dBA" refers to a decibel measurement taken while the signal was being modified by an "A"-weighted filter.

- To convert power output (*P*) in watts to decibels, dB = 10 log 10 × (the power level divided by the reference power level).

- To convert power output in decibels to power in watts, *P* in watts = antilog of (the decibels divided by 10 and multiplied by the reference level).

- You cannot add or subtract logarithmic values on their own, but graphs have been computed to aid in finding the results of adding or removing an amount of decibels to or from the whole.

All these statements, while true, can leave you hanging, waiting for another clue. Even statements that end in conclusions often have a condition of some kind attached, such as the following:

When an amp is said to put out 12 dB when fed an input of 6 dB, the ratio is plainly understood because it is a change in the rate of gain. It does not matter what the decibels in question are referenced to as long as they are both referenced to the same value. However, many times the decibel number alone is a relative value based on the relationship between two levels, as opposed to its being an individual quantity or some kind of an independent value.

Power (watts) in decibels is referenced to dBm, or 1 milliwatt into a 600-ohm load. But audio test equipment, including common meters, derive their reading from voltage instead of wattage.

Because 1 milliwatt into 600 ohms is equal to 0.775 volts, that is the input level required for most audio meters to read "0 dB," excluding most of those used in recording studios that have a resistor added that shunts (converts) their 0 VU level to +4 dB, or 1.23 volts.

These kind of statements leave the novice struggling to comprehend their meaning. Not only is this worthless information, but it subverts a process that should be relatively easy.

How About Some Answers?

Just so you understand it, here's how that 0.775 volts was calculated. The current (I) is 1 milliwatt, or 0.001 watts. The resistance (R) is 600 ohms. Power (I) = voltage (E) squared and then divided by R. Take out your calculators please, and follow along.

Simplifying this equation we come up with

$$\sqrt{I \times R} = E$$

Here, the square root of IR (0.001 × 600) = E, or the square root of 0.6 = E. So we end up with E = 0.77545967, or 0.775 volts.

If you've somehow managed in this day and age to acquire a calculator that lacks log and square root function keys, it's time for an updated version. You are going to need it in your chosen field.

It should be obvious that we need standardized specifications and reference points to make comparisons between pieces of equipment and to judge a signal's strength and quality.

As an example: A *signal-to-noise* (S/N) ratio of "−85 dBm" simply means that background noise, tape hisses, and line and other noises have been measured to be 85 dB lower than the signal's 0 dB level. That is when (and only when) the signal's 0 dB level is equal to 1 milliwatt into 600 ohms of resistance, because the dBm designation means it is being referenced, or compared, to 1 milliwatt into 600 ohms, or 0.775 volts. So, that little letter m is critical.

"What's with 600 ohms?" you may ask. "Nothing I know of has a 600-ohm input or output impedance." Well, radio, then television, and finally recording studios adopted the standard of 600 ohms for all inputs, outputs, and lines back in the tube era, when a matched input and output impedance of 600 ohms was found to be optimal for signal power transfer because it allowed tube electronics to operate at peak performance without being overloaded.

Enter progress, in the form of the transistor, and input and output impedance requirements became the simpler "bridging" type of impedance matching. Bridging impedance is simpler in terms of design and manufacturing implementation because it only calls for a fairly low output impedance and a high imput or load impedance (so as not to load down the signal). But somewhere along the way things got a little goofy. Most audio meters do not determine level according to wattage, and 600 ohms is no longer the de facto standard for audio impedance interface. But "0 dB" can still be referenced to that same 1 milliwatt into 600 ohms, which yields 0.775 volts, and some meters are still calibrated to indicate "0 dB" at this level.

Next thing you know people are saying it's time to chuck "dBm" and instead use "dBv." But wait a second, the folks in Japan are already using "dBV" and it's referenced to 1 volt, not 0.775, and since capital V and lowercase v could get confusing, we better use "dBu" for the 0 dB = 0.775 volt reference.

I think it's best to stop at this point to distinguish dBu and dBm. Simply put, 0 dBm is referenced to a level of 0.775 volts into a 600-ohm load, while dBu is referenced to a 0 dB level of 0.775 volts into a load that is higher than 600 ohms.

Yet the fact remains, dBm and dBu are based on the same mathematical equations, and both the 0 dB levels are equal to 0.775 volts. That letter after the "dB" is important because it gives us our reference point. It would be nice if there was also some indication when a meter's 0 dB reading was referenced to +4 dBu, or 1.23 volts (I've seen it), but usually there is none.

Saying that something puts out X dB without stating what it's referenced to is like a fortune teller's giving someone the score of a game without stating which game. If a reference is not given when stating a value in decibels, it is useless information.

Decibels, Dumbbells, and Jezebels

Decibels can easily turn into *dumb-bels* or *jeze-bels* in the wrong hands. I once listened to someone representing a company that manufactured a noise reduction product for the automotive industry rattle off several pages of decibel equations. First he stated that "if" a preinstallation noise level is reduced by 6, 12, or 18 decibels, it would equate to the noise remaining being one-half, one-fourth, and one-eighth of what it was. At best a truism. But he followed this by stating that his product reduced noise levels by 32 dB. Then he went on very quietly to say "suppose" you have a noise level of 110 dB that is reduced to 85.6 dB by the inclusion of our product, and you wish to know the percentage of the eliminated noise. He then justified the subtraction of decibels, as in 110 dB − 85.6 dB = 24.4 dB, with some fancy mathematical concepts, such as "the subtraction of logs is equivalent to dividing the antilogs." He went on to "clarify" this by giving another page of complex equations referenced to a "ratios of pressure" table that *he* had devised. He concluded with the statement, "Therefore, the effect of using this product, which reduced [no "suppose" here] the noise from 110 dB to 85.6 dB is equivalent to saying that the product reduced the noise by 94 percent."

Let's face it; don't you think that if everything he proposed was factual, there would have been a way to state it in easily understood layman's language?

As you can see, the decibel can be a pain, but it doesn't have to be.

If you want to understand the decibel, then grab your pocket calculator and take a short walk with me. I promise that there will be no tests, no difficult equations, and nothing to memorize, and you can even forget everything you've read as soon as you're done reading it.

You see, by following along *and* doing the calculations, you'll end up knowing how and why everything works out. Then if you ever need to find any values in terms of their decibel relationships, you can look for the equation that answers what you need, be it resistance, voltage, current, or power in wattage or decibels. Once you locate an equation that works with the variables you have, simply substitute your values and do the calculating (which you'll already have experience doing), and you'll have the answer you need. Then you can forget all the math until you need it again.

Some Historical Background

I have found it makes good sense to start the explanation of decibels with a bit of history.

The decibel actually dates back to the measurement of radio signals, when they were still being transmitted through telephone lines. Those early engineers had to predict the loss in level due to those lines, so to test signal levels they set up a 2-mile "loop" of wire, which was actually the total of 1 mile out and 1 mile back through the same two conductor-19 telephone cable. Believe it or not the resulting unit of measurement was called a "MOL" (miles of loss), such as 4.15 MOL. The range of values was so vast they had to express it logarithmically. This ended up changing the numbers, so the new unit of measurement was termed a "transmission unit." Then, in honor of Alexander Graham Bell, the name was changed to a "Bel." The Bel also proved too great an amount for ease of use, so it was subdivided into 10 units called "decibels," as in "dB."

A decibel, as in dBm or dBu, is simply a designation for a quantity or an amount or level that is actually based on a ratio of two levels. This means the quantity of the one in question is being compared with a reference amount. The ratio between the two levels is then converted to "Bels" by taking the common logarithm of that ratio. Therefore, the level difference of the ratio in "Bels" is the log of that ratio, and for decibels (or one-tenth of a Bel) the level of the ratio equals 10 times the log of the ratio. Which is where that crazy

$$dB = 10 \log 10 \times (P \text{ divided by reference power value})$$

equation comes from.

Logarithms

Since the decibel is a logarithmic value, it's a good idea to understand what logarithms are all about. A logarithm's "exponent" number, which is positioned to the right and slightly above it's "base" number, tells you to what power that base number must be raised to so as to equal another number, called the "antilog." This doesn't come close to clearing anything up, so let's go at it another way:

$$2^3 = 8$$

Here, 2 (the first power) times 2 (the second power) = 4, which is then multiplied by 2 (the third power) = 8, where 2 is the base number, 3 is the exponent, and 8 is the antilog.

This becomes a little less difficult once you realize that the exponent number minus 1 is the number of times the base number is multiplied by itself. Having a log function on a calculator is a dream come true for anyone who had to deal with this bear of a process in the precalculator days.

Decibels are used to express three basic types of ratios: the *difference* in levels, the *gain* in level, and the level as *referenced to* a standard level. For a difference in level between two power values (P1 and P2) the ratio of the two will be P2 divided by P1; therefore, the difference in level (in decibels) = $10 \times \log (P2/P1)$.

You Still Gotta Problem with Decibels?

Most people have a problem with decibels because they are based on logarithms. Understanding wattage ratings is easy, even though those are based on a square, simply because they act in a linear manner. For example, 12 watts is twice as large as 6 watts and four times 3 watts. No problem, right? But logarithms, and therefore decibels, are not nice and linear at all: 12 dB is twice the level of 9 dB, four times that of 6 dB, eight times the value of 3 dB, and 16 times the level of 0 dB. And 12 dB is also 32 times the value of −3 dB! That is if we're talking about power; if it's voltage we're discussing, then 12 dB is twice the level of 6 dB, four times the level of 0 dB, and eight times the level of -12 dB. This is because a 3 dB increase represents a doubling of power, while a 6 dB increase is a doubling of voltage.

Why should everyone involved in audio subject themselves to this kind of lunacy? Well, that's because we need to use logarithms to express our human hearing's nonlinear response to variations in level. Be thankful for this nonlinear response because if your response were linear, not only would your dynamic range allow you to hear a whisper across a room, but the sound of a car horn might knock you unconscious.

According to standard definitions, humans can hear levels as low as 0.0002 dynes per square centimeter. A dyne is a measure of force. What's being stated is that if $^{2}/_{10,000}$ of a dyne of force is put on an object a square centimeter in size, that power would equal the force required to make a sound you would just begin to acknowledge. Pretty useless information unless you design microphones, test hearing, or manufacture hearing aids.

Now let's discuss that same amount of power in terms of good old linear watts. In watts per square centimeter, our "threshold of hearing" begins, or kicks in, around 0.00000000001 watts (that's 1^{-11})! The maximum level that humans can withstand before the onset of pain is approximately the sound a jet aircraft creates when taking off. That sound level referenced to watts equals about 10 watts of force on that same square centimeter.

A range of hearing from 1^{-11} to 10 is quite large. No wonder we were able to survive the dinosaur age. If you divide 10 watts by 1^{-11} (or 10×-12, as it is more commonly stated), you'll have the dynamic range of human hearing, which ends up being from 1 to 10,000,000,000,000. No lie! Imagine a producer asking you to "Turn up the bass a thousand million units or so."

Now stick with me a bit and you'll see I'm not talking complete nonsense. If I turned that bass line up until the level was 10 times higher, the producer wouldn't hear it as being any more than twice as loud. If I bumped it up another 10 times in level (bringing it to 100 times its original level), it would seem to be only four times as loud. This is because our hearing system perceives a gain of 10 times in level as only a doubling of the loudness, which is good for us, because, as already stated, with the kind of dynamic range our ears are capable of, we'd end up being distracted by the conversation of people across the street or knocked unconscious by a horn blast from a passing car.

So logarithms, and therefore decibels, do a good job of illustrating and providing us a workable relationship that matches the range of our hearing. You can check this out for yourself by eyeballing an amplifier with a readout in watts or one connected to a wattmeter (easily accomplished with a visit to your local electronics store). Set the level to 1 watt, listen, and then turn it up to 2 watts. It won't sound twice as loud. You'll have to crank that amp up to around 10 watts before it sounds twice as loud. That's just the way human hearing works. So blame the need to utilize logarithms in audio measurement on our not-so-shabby human hearing capabilities.

Levels measured simply as voltage can't compare with the usefulness of the decibel. Just imagine a manufacturer trying to make a big splash about their latest amplifier offering by advertising that the amp puts out a full 16 volts. That statement wouldn't mean anything. In fact, if the input fed to that amp were 24 volts, instead of amplifying it would actually be attenuating the signal! Voltage measurements simply do not define the actual amount of energy or work that's being performed. The load impedance must be known and used along with the measured amount of voltage in order to calculate the more meaningful values of energy, such as power, current, and watts. So why do we bother using measured voltage or for that matter dBu values at all?

Standards are not only constantly being updated, they also go in and out of vogue. As an example, amplification in vacuum tubes is controlled by a voltage, while with transistors the control is via current. Yet in the days of tubes the favored measurement of level was in dBm, which is an expression of wattage and more of an indication of current. With the widespread changeover to transistor electronics, dBu, which is related to voltage, was championed as the preferred measurement standard. As you'll come to understand, this is all of no concern, since although different, "dBm" as a value of power and the voltage value in "dBu" are very much interrelated, and both will always be used.

Who Really Cares?

For designers, standards writers, theorists, and academicians, it doesn't matter because they don't seem to have much problem sorting any of this out. In fact I think they kind of enjoy playing around with the stuff, and I know they feel it's their responsibility to help try to keep it all straightened out and correct.

For those involved in calibration, repair, and specification checks and who use pure sine waves set at exacting input levels, it makes no difference what standard is used as long as it's the same for both of the values being dealt with (or they are able to use conversion tables or calculations to match up the readings).

For those involved in audio recording, the need to understand exactly what their reference meters are stating about levels with regard to the limitations of the equipment being used is obviously important. This requirement is complicated by the deficiency of our human eye/brain combination in "keeping up with" sound-level changes as they occur in real time. Yet we still require some sort of humanly visible "warning indication" to aid us in supervising and controlling audio levels.

Your being able to read level meters so as to set gain stages properly is what this whole discussion is really all about, and I know you've followed along and understood everything so far, right? Just kidding! You see we've gotten way ahead of ourselves here, because having a little understanding of basic electronics is needed to help clear all this up.

That's the problem with most explanations of the decibel; they leave out the "scary" math and electronics parts, so they end up being incomplete. I know because I had to refigure out this decibel thing several times from different angles before it all made sense. That's why I know that you'll not only understand it all but be able to explain it to anyone else.

Basic Electronics

Positively charged protons and neutrons (which have no charge) make up the nucleis of atoms. Orbiting around this nucleus are electrons, which have a negative charge. The electrons and protons usually equal each other in number, keeping things all stable and unexciting as far as the electrical charge goes. But some atoms can give off a couple of electrons from their outermost orbits, leaving behind more protons.

Now things start to get interesting, because the atoms that receive those extra electrons end up having a negative charge, while those that are left deficient in electrons end up with more protons and thus have a positive charge.

Electrons are able to move about and make up a "current flow" only when three requirements are met. There has to be some negative atoms with excess electrons that are just itching to travel, and there has to be a path, or a conductor, for them to move through so they can get to the third requirement: some atoms with excess protons and thus a positive charge. Now you got a party, and the conductor handles all the introductions, thank you very much.

Conductors are made up of atoms that will pass on electrons through their outer orbits from one atom to another. The number of electrons that can pass through a conductor is determined by how thick that conductor is (the number of atoms in the conductor's cross section). Gold, copper, and silver make good conductors, while rubber, glass, air, and plastics, like Bakelite, are poor conductors and are therefore called *insulators*. However, those insulators are helpful for corralling and directing electrons to travel where you want them to without any dangerous wandering about. But enough electrons *can* break down an insulator, just the way lightning breaks down air's insulation properties and uses it as a conductor to ground. Understanding only this much information allows for the conceptualization of a simple circuit.

How Simple a Circuit?

How about a battery that's powering a lightbulb? Okay, batteries use a chemical reaction to produce positive and negative atoms, hence their "+" and "−" terminal designations. But remember, we're visualizing electron flow as only the negative atoms actually moving about. So in our circuit we hook up a conductor that's inside an insulator, a configuration that's commonly referred to as a *wire,* to each of the battery's terminals and let those electrons FLOW, BABY, FLOW.

But we must remember the conductor's "cross section" limitation as far as the number of electrons it will be able to pass on. Too many electrons trying to pass through a thin conductor will cause heat. If the conductors are attached to an even thinner conductor, the heat will concentrate there. If that thin area is inside an air-free vacuum so that it doesn't immediately burn out, it will get red hot and glow. Light up that party!

All of this may be fine for the average Joe, but for us super brainiac audio engineer types, we demand exacting definitions to describe our subjective opinions on how good things sound. So we need to have a bead on terms like volts, amps, ohms, and watts, which are an everyday part of our language. But honestly, it's not any more complicated than that battery circuit, and the proof will be your ability to explain it to anyone so that they readily understand it too.

Please feel free to skip ahead to the section entitled "Alternating Current" and come back to this section another time, but things will go easier if you get it over with now.

If an atom has an excess of electrons and is negatively charged or lacks electrons and is positively charged, it has the "potential" to produce a force that attracts other atoms of the opposite charge or repels those of the same charge. The extra electrons, moving bucket brigade style from one atom to another, are measured as *voltage*, or "*E*," sometimes referred to as electromotive force, which is rated in coulombs. Now here's a bit of useless information: 6,280,000,000,000,000,000 electrons = 1 coulomb.

The flow itself is called *current*, and 1 coulomb of electrons passing by a single point in 1 second is equal to 1 ampere (A) or "I" (for intensity of electron flow). The force required to accomplish this movement is 1 volt — 1 volt, 1 coulomb, and 1 ampere. Convenient, no?

To get a flow of current you must have potential and a conductor. No conductor works perfectly because they all present some opposition to the flow of electrons. This opposition, called *resistance,* is measured in ohms, symbolized by the capital Greek letter omega (Ω).

We've made a big step here because now we have an understanding of voltage, current, and resistance, and the interrelationship of these three is the foundation of all electronic measurement and terminology.

Prefixes

Before we go any further let's get prefixes out of the way. To handle the vast range of electronic values, these prefixes are placed in front of (*pre*-) terms so as to set (fix) their value range. It is critical for correct measurements and comparisons that you know what these prefixes are saying about the numbers you're dealing with.

Prefixes that multiply a value	Prefixes that divide a value
kilo- = × 1,000	milli- = ÷ 1,000 (or × 0.001)
mega- = × 1 million	micro- = ÷ 1 million (or × 0.000001)
giga- = × 1 billion	nano- = ÷ 1 billion (or × 0.000000001)

There Ought to Be a Law: Ohm's Law

Enter George Ohm with a mathematical relationship between voltage (E), current (intensity, I), and resistance (R). At first I had a hard time remembering if voltage was the E or the I in the equations until I related E with the V of voltage by thinking of the microphone manufacturer Electro Voice's nickname of "EV." Hey, it worked for me.

George Ohm figured out that the result of multiplying current I by the resistance R equaled the value of the voltage E ($E = I \times R$). Therefore, if a conductor presents 250 ohms of resistance to a current flow of 2 amps, the result will be 500 volts.

Playing algebra with this $E = I \times R$ relationship results in formulas for current $I = E/R$ and resistance $R = E/I$. So now you know that if you have 100 volts flowing through a conductor with 50 ohms of resistance, the current should be 100/50, or 2 amps. With a current of 2.5 amps at 50 volts you'd expect the resistance of the conductor to be 50 ÷ 2.5, or 20 ohms. Ohm's law covers $E = I \times R$, $I = E/R$, and $R = E/I$.

But never forget those prefixes. If the current flow of the last example was 2.5 mA (milliamps, or thousands of an amp), the equation would have been 50 ÷ 0.0025, and the result would be 20,000 ohms, or 20 kilo-ohms ($k\Omega$). A big difference!

Yet it's all still easy, just slightly more time consuming. If E is 100 volts and R is 10 K-ohms, then I is 100 ÷ 10,000, which equals 0.01 amp, or 100 milliamps. If R is 200 ohms and I is 1.5 milliamps, then the voltage E would be 200 × 0.0015 amps, equaling 0.3 volts, or 300 millivolts.

Normally, in electronics courses there would now be a whole column of these equations for you to sit down and work your way through. Playing around with these equations would make a lasting impression But since you have this book to refer to, you can skip it because as far as I'm concerned, you'd be better off, as an audio recordist playing around with some scales on the piano so as to augment your lasting impression of relative pitch.

Watts It All About: Watt's Law

Power in watts is what it's all about, and it rates the amount of work done over a set period of time. Measuring this work in terms of horsepower goes back to a Mr. James Watt, who was not only a scientist and inventor, but also a salesman. You see, Mr. Watt invented a steam engine that pumped water, and he wanted English coal miners to pump water out of their mines with his engine instead of the horses they had been using. Therefore, he had to come up with a way to compare and thereby show his steam engine's superiority over horses. He was able to demonstrate that his machine outperformed horses by calculating the work done by both in "foot-pounds." As a side note, 746 watts was equal to 0.7376 foot-pounds, or 1 horsepower, meaning that if you've got an amp that's putting out 750 watts, it's delivering a little more than 1 horsepower. Big deal, right?

In audio electronics, we don't care about horses or water pumps, but watts and power count big time. It's important for us to be able to relate the work being done to voltage, current, and resistance. Power, in watts, is the end result of the way voltage (E), current (I), and resistance (R) interrelate while doing work.

When current flows, it produces heat (lamps and heaters take direct advantage of this). When current flows through a conductor it also sets up a magnetic field (used in motor operation and wireless communication, as in radio waves).

NOTE: *The magnetic field "associated" with the movement of electrons is present along the full length of the conductors they are flowing through. A hint as to why and where you position your wiring can cause hums and other noise in the system. With direct current a magnetic field can only be set up during the times that the current values change (as when it's being switched on and off). So if there is a constant DC current flowing through a conductor, it does not produce a magnetic field.*

As already stated, the power being generated is the outcome of the way I, E, and R interrelate, and Watt's formula for power is: power (P) is equal to current (I) multiplied by voltage (E), or $P = I \times E$. Starting with this and then adding in the equations from Ohm's law, we end up with equations that cover many variables.

From Ohm's law we know that voltage (E) is equal to current (I) multiplied by resistance (R). This means that power, in watts, is equal to ($I \times I$) $\times$ R, or $I^2 \times R$. Ohm's law also states that current (I) equals voltage (E) divided by resistance (R). So power, in watts, equals voltage (E) multiplied by itself (E^2) divided by resistance (R). To these three, you can add the equations derived strictly from Ohm's law.

$$P = \qquad \frac{E^2}{R} \qquad I^2 \times R \qquad I \times E$$

$$E = \qquad I \times R \qquad \frac{P}{I} \qquad \sqrt{P} \times R$$

$$I = \qquad \frac{E}{R} \qquad \frac{P}{E} \qquad \frac{\sqrt{P}}{R}$$

$$R = \qquad \frac{E}{I} \qquad \frac{P}{I^2} \qquad \frac{E^2}{P}$$

Now if you use your pocket calculator to figure along with the following few examples of the total watts produced, it should help your understanding of the equations.

I = 4 amps and E = 600 volts (4 × 600 = 2400 watts)

R = 335 ohms and I = 2 milliamps (0.002 × 0.002 = 0.000004 × 335

= 0.00134 watts or 1.34 milliwatts)

I = 500 milliamps and R = 1 k–ohms (0.5 × 0.5 = 0.25 × 1000 = 250 watts)

That wasn't too painful, now was it?

Series and Parallel Interconnections

Series and parallel interconnections are most easily understood when related to hooking batteries up together because each method increases the amount of either voltage or current available.

A *series* connection, as with the batteries in Figure 1-1, has terminals hooked up in a positive-to-negative formation. The total positive charge ends up located at the first battery's positive output terminal, while the total negative charge is located at the last battery's negative terminal. The resulting *voltage* would be the *sum* of all four. If all the cells used were AA (1.5-volt) batteries, the four would total 6 volts. The resulting current would remain the same as that of a single battery cell.

A *parallel* connection, as in Figure 1-2, has all the battery cells' positive terminals hooked together and all the cells negative terminals also joined. In this configuration the total voltage potential of all four AA batteries would remain set at 1.5 volts. However, the total *current*

Figure 1-1
With series circuits, the connections are end to end, or daisy-chained.

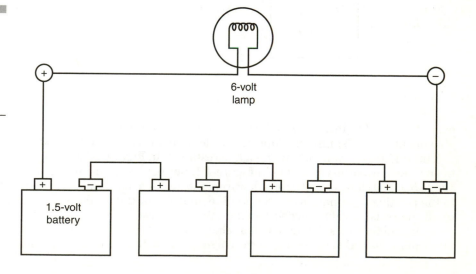

6-volt lamp

1.5-volt battery

Figure 1-2
With parallel
circuits, the
connections
are side by
side, or
ganged
together.

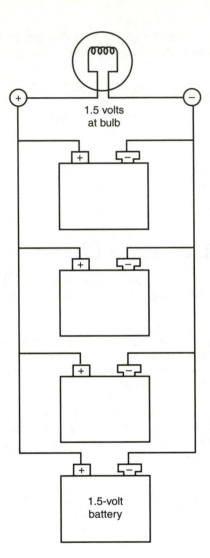

1.5 volts
at bulb

1.5-volt
battery

available would be increased to the *sum* of the currents of all four cells. So if each battery was able to put out 0.25 amps, the four in parallel would produce 1 full amp of current.

  This difference is important enough to restate it. In Figure 1-3 a series circuit is set up end to end like our battery example in Figure 1-1, but now we're using resistors in series with a single battery. Here, the electrons flow from the negative terminal through the three resistors to the positive terminal. In a series (end-to-end) circuit the total resistance, or *load*, is equal to all the resistor values added together. We know from Ohm's law ($I = E/R$) that the total resistance determines the current available in a circuit. In a series circuit the sum of the resistors equals the total resistance, so the current will be the same throughout the circuit.

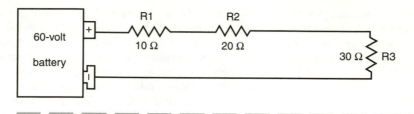

Figure 1-3
A series circuit set up end to end like our battery example in Figure 1-1, but here we have three resistors in series with a single battery.

On the other hand, the voltage will vary as an amount of potential is used up in getting the electrons past the resistance of each component. In this example, the battery's 60 volts is the total potential. So an *E* of 60 divided by an *R* of 60 will equal a total current of 1 amp.

Kirchoff's Law

Now we get to Kirchoff's law, which states that *all* the voltage will be expended in the circuit. Because of this we can follow and "plan on" the voltage drops throughout the circuit by using Ohm's law for voltage, which states that $E = IR$. Therefore, the voltages at the three resistors are:

$$R1 = 1 \times 10 \text{ ohms} = 10 \text{ volts}$$
$$R2 = 1 \times 20 \text{ ohms} = 20 \text{ volts}$$
$$R3 = 1 \times 30 \text{ ohms} = 30 \text{ volts}$$

Adding all these voltages up we get 60 volts, which when subtracted from the total voltage potential of 60 volts means we've used up all the voltage, per Kirchoff's Law, which actually is "All voltage drops must add up to the total voltage applied, or the sum of the added voltage drops when subtracted from the source will *always* equal zero."

Thank you, Mr. Ohm. Now let's bring out Jim Watt for power drops, or consumption, as determined by Watt's law, stating that P = *IE*.

$$I = E \text{ (60 V) divided by R (60 ohms)} = 1 \text{ A}$$

Back to total power: P = *IE*, or 1 amp × 60 volts = 60 watts. So we can now find our individual power drops across each resistor:

$$P1 = 1 \times 10 = 10 \text{ watts}$$
$$P2 = 1 \times 20 = 20 \text{ watts}$$
$$P3 = 1 \times 30 = 30 \text{ watts}$$

All the individual power drops add up to 60 watts, meaning the total power potential is being dissipated, so our man Kirchoff is quite happy to be proven right again.

Now, remember that phrase "plan on" a few paragraphs back? Well, that referred to the use of voltage dividers (or series resistors), as in providing different voltages from a single power supply. Take the same 60-volt battery potential, but say you need 9-, 36-, and 15-volt supplies to power three different effects boxes. With Ohm's law you can now figure out the value in ohms of the required voltage dividers to do the job (I = E/R, or 1 amp = 60 volts/60 ohms). So here 9, 15, and 36 ohms will do quite nicely. This may be too simple an example, but it's not really all that difficult to solve if you measure total voltage with a multimeter and note the battery or power supply's maximum current rating.

Let's say you measure a power supply rated at 500 milliamps and find that it puts out 52 volts. This seems like it's not enough power to fulfill our hypothetical three-device requirement. Yet if the devices you're trying to power can operate (even if less efficiently) at a lower voltage, you may give it a shot in a pinch.

According to Ohm's law, total $R = E/I$, or 52 volts divided by 0.5 amps, which equals 104 total ohms. Now to keep this thought process going we'll say that the device that wants 36 volts can operate down to 28 volts. So, using equivalents, the needed voltage (28 volts) divided by the total voltage of 52 volts = the required resistance X divided by the total resistance of 104 ohms. Or 28 volts divided by the total voltage of 52 volts is equivalent (equal to) the ratio of needed resistance divided by 104 ohms:

$$28 \text{ volts} \ \times \ 104 \text{ ohms} \ = \ 2912$$

which can then be divided by 52, which results in 56 ohms:

$$15 \text{ volts} \ \times \ 104 \text{ ohms} \ = \ 1560$$

which can then be divided by 52, which results in 30 ohms:

$$9 \text{ volts} \ \times \ 104 \text{ ohms} \ = \ 936$$

which can then be divided by 52, which results in 18 ohms.

Kirchoff's happy because all three add up correctly to 104. But what about the power rating of those three resistors? Hello, Mr. Watt and $I^2 \times R = P$. So:

$$0.5 \text{ amps} \ \times \ 0.5 \text{ amps} \ = \ 0.25 \ \times \ 56 \text{ ohms results in 14 watts}$$
$$0.5 \text{ amps} \ \times \ 0.5 \text{ amps} \ = \ 0.25 \ \times \ 30 \text{ ohms results in 7.5 watts}$$

and

$$0.5 \text{ amps} \ \times \ 0.5 \text{ amps} \ = \ 0.25 \ \times \ 18 \text{ ohms results in 4.5 watts}$$

"Hey," says Mr. Kirchoff, "that seems like a lot of watts from a little 52-volt, 0.5-amp power supply." You got to pay to play, so okay, total $P = I^2R$, or $(0.5 \times 0.5 = 0.25) \times 104$ total ohms = 26 total watts, and $14 + 7.5 + 4.5 = 26$ watts. Seems like we pulled it off. But where can you find a 56-ohm resistor that's rated at 14 watts (or 30 ohms at 7.5 watts and 18 ohms at 4.5 watts for that matter)? How about using a Hugh variable resistor, as in a 25-watt rheostat? A crazy proposition? I agree. But if it's a show-must-go-on pinch and anything goes, it may be worth smoking a power supply or two.

Isn't circuit analysis grand? I agree; it stinks, which is why I'm not an electronics design engineer.

Personally, I don't expect the average audio recordist to tear apart and add onto, or "build out," a power supply so as to drive multiple devices. No, that's the job of a tech. But I do expect you to know it is possible and to understand that you now have this reference guide should you need to figure something like this if you find yourself in a show-must-go-on pinch.

You see, I just can't imagine anyone in this field not being willing to take the time to achieve at least a basic understanding of electronics. To my way of thinking it's the same as an audio recording engineer not being able to follow along (albeit slowly) with a written music score. Hey, shouldn't a racing car driver at least know where the gas goes in? At the very least, now you won't be stupefied when a tech accomplishes something "magical" like this.

At this point our basic course in electronics moves on to parallel and combined series/parallel circuits. Parallel circuits are important, if only to clue us in on the problems of wiring multiple speakers in parallel. You can't get by with just series circuits because in order to reach the goal of understanding the decibel, meter readings, and optimum gain stage setting, the others *must* be dealt with too. I think it's called paying your dues. You can attack this subject and series/parallel combination circuits and all the equations with your pocket calculator on a long, boring ride or whenever you want to impress someone with your high IQ.

Parallel Circuits

An example of parallel circuitry is shown in Figure 1-4. Here, where the ends of the resistors are joined together closest to the battery's negative output terminal (on the bottom at point Y), the current flow will divide up, with part of it going through each resistor. These individual current flow routes will then recombine at the point where the resistors are joined together closest to the battery's positive output terminal (on the top at point X). Unlike series resistance, where the voltage varies, with parallel resistance the current varies. In a parallel circuit the potential voltage across points x and y will equal that of the full input, and so the voltage remains the same on both sides of the joined-together resistors. The voltage

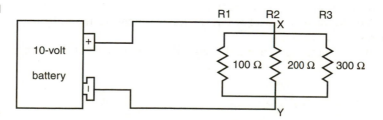

Figure 1-4
An example of parallel circuitry: The three resistors are wired in parallel.

across each of the individual resistors in parallel configurations will also be equal to the full input voltage, which in this case is 10 V.

In parallel circuits the total current equals the sum of each of the resistors.

Ohm's law states

$$I = E \text{ divided by } R; \text{ or } I \times \frac{E}{R}$$

At R1,

$$I = 10 \text{ volts divided by } 100 \text{ ohms} = 0.1 \text{ amperes; or } I = \frac{10v}{100\Omega} = 0.1$$

At R2,

$$I = 10 \text{ volts divided by } 200 \text{ ohms} = 0.05 \text{ amperes; or } I = \frac{10v}{200\Omega} = 0.05$$

and at R3,

$$I = 10 \text{ volts divided by } 500 \text{ ohms} = 0.02 \text{ amperes; or } I = \frac{10v}{500\Omega} = 0.02$$

The total current will equal 0.17 amperes.

Total power (Pt) equals the sum of the power expended in going through each of the three resistors:

$$P = EI, \text{ at R1, } 10 \text{ volts} \times 0.1 \text{ amps} = 1 \text{ watts}$$

At R2,

$$10 \text{ volts} \times 0.05 \text{ amps} = 0.5 \text{ watts}$$

At R3,

$$10 \text{ volts} \times 0.02 \text{ amps} = 0.2 \text{ watts}$$

The total power in this circuit is 1.7 watts.

You can calculate the total power when you know the total current by using $P = EI$: 10 volts $\times$ 0.17 amperes = 1.7 watts.

The equivalent resistance in a parallel circuit is equal to the input voltage divided by the total current:

$$10 \text{ volts divided by } 0.17 \text{ amps} = 59 \text{ ohms}$$

Watt? I Mean What?

The parallel resistance rule states (and I couldn't make this up) "The combined resistance of a parallel circuit is equal to the reciprocal of the sum of the reciprocals of the individual resistors." You don't have to think about this. You know I didn't make it up, and it can be figured out mathematically, so forget about the theory, okay?

But just in case you *have* to know: Finding the total or equivalent resistance starts with finding the total current, I (t):

$$I (t) = I (a) + I (b) + I (c)$$

or

$$I (t) = E (t) \text{ divided by } R (t)$$

$$I (a) = E (a) \text{ divided by } R (a)$$

$$I (b) = E (b) \text{ divided by } R (b)$$

$$I (c) = E (c) \text{ divided by } R (c)$$

By not dealing with the current values, the equations change into

$$E (t) \text{ divided by } R (t) = E (a) \text{ divided by } R (a) + E (b)$$

$$\text{divided by } R (b) + E (c) \text{ divided by } R (c)$$

We divide both sides of the equation by E (t) and it becomes

$$1 \text{ divided by } R (t) = 1 \text{ divided by } R (a) + 1 \text{ divided by } R (b) + 1 \text{ divided by } R (c)$$

Dividing both sides of the equation by 1 makes it

$$(Rt) = \cfrac{1}{\dfrac{1}{Ra} + \dfrac{1}{Rb} + \dfrac{1}{Rc}}$$

Checking it out, we get
$$(Rt) = \cfrac{1}{\dfrac{1}{100} + \dfrac{1}{200} + \dfrac{1}{500}}$$

$$(Rt) = \frac{1}{0.01 + 0.002 + 0.005} = \frac{1}{0.017} = 59 \text{ ohms}$$

See, this stuff really works! Now just forget about it until you need it, and then look it up. But to set your mind totally at ease it simply boils down to this: Using two resistors in parallel (R1 × R2) ÷ (R1 + R2) gives you the total resistance. Big help, right? Hold onto your seat because starting with this you then have to take this total and move on to figuring in the next resistor in parallel, then the next, and so forth. All this means: R (1) × R (2), or 100 × 200, = 20,000, divided by R (1) + R (2), or 100 + 200 = 300 = 66.666666, and 66.666666 × R (3), or 500 = 33333.333, divided by 66.666666 + R (3), or 566.666666 = 59 ohms.

For the audio recordist it is only important to understand that when two equal resistors are used in parallel, the *combined* resistance is one-half the value of one resistor. With three resistors of equal value this will result in one-third the resistance of one resistor. This should give you a clue as to the reason for the high current drain in parallel circuits, as in *red hot and smoking amplifiers* trying to feed multiple speakers wired in parallel.

Take a break before reading on, because next we tackle the fun, fun, fun topic of series/parallel circuits.

Series/Parallel Circuits

Series/parallel wiring combinations require more in-depth analysis, so it calls for a comparison of the two.

Series	Parallel
E = the sum of the individual voltage values	Voltage is the same throughout the circuit
I is the same throughout the circuit	Current = the sum of the individual current values
R is the sum of the individual resistors	Resistance is the reciprocal of, or 1 divided by, the resistance of the individual resistors
Total power is the sum of individual powers at each component	Power = the sum of the individual powers of each component

Just one look at the series/parallel circuit of Figure 1-5 shows you the need to simplify, or divide up, the sections of combination circuits, as in Branches A and B in the figure.

Resistance R:

Branch A consists of two resistors in series, R1 and R2, totaling 120 ohms.

Branch B has two resistors in parallel (R3 × R4 ÷ R3 + R4) = 2400 ÷ 100 = 24 ohms.

This is in series with R5, which at 36 ohms brings the total to 60 ohms.

Total R:

Branches A and B (120 ohms and 60 ohms) are in parallel, so the combination and total resistance of the whole circuit is 120 × 60 = 7200 ÷ (120 + 60), or 180, which equals 40 ohms.

Figure 1-5
A series/parallel
circuit, with the
A and B
branches
designated

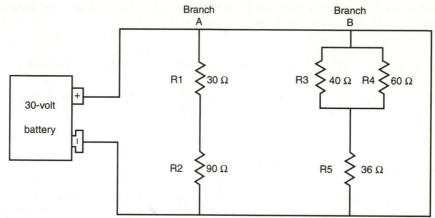

Stick with me now as we figure current values.

Current I: With this information we can now find the total current values: I total = voltage E, 30 volts, divided by the resistance.

Branch A: 30 volts/120 ohms = 0.25 amperes (or 250 milliamps)

Branch B: 30 volts/60 ohms = 0.5 amperes (or 500 milliamps)

Total I: 0.25 amperes + 0.5 amperes, or 0.75 amperes (or 750 milliamps)

Voltage $E = I \times R$:

Branch A:

at R1: 30 ohms × 0.25 amperes = 7.5 volts

at R2: 90 ohms × 0.25 amperes = 22.5 volts

Kirchoff's law checks out at 30 volts total voltage E.

Branch B:

at the combination of R3 and R4: 24 ohms × 0.5 amperes = 12 volts

at R5: 36 ohms × 0.5 amperes = 18 volts

Mr. Kirchoff is smiling ear to ear because we have a total of 30 volts.

Now that we have the values for the total amount of voltage and current we can find the total power:

Total P, in watts = total E × total I, or 30 volts × 0.75 amperes = 22.5 watts

The power levels at each component equal the voltage E squared and then divided by the resistance R:

R1 = (7.5 volts)^2, or 56.25, ÷ 30 ohms = 1.875 watts

R2 = (22.5 volts)^2, or 506.25, ÷ 90 ohms = 5.625 watts

Even though R3 and R4 have the same voltage drop, because they have different resistive values their power values will differ.

$$R3 \ = \ (12 \text{ volts})^2, \text{ or } 144, \ \div \ 40 \text{ ohms} \ = \ 3.6 \text{ watts}$$

$$R4 \ = \ (12 \text{ volts})^2, \text{ or } 144 \ \div \ 60 \text{ ohms} \ = \ 2.4 \text{ watts}$$

$$R5 \ = \ (18 \text{ volts})^2, \text{ or } 324 \ \div \ 36 \text{ ohms} \ = \ 9 \text{ watts}$$

$$\text{Total} \ = \ 22.5 \text{ watts}$$

Now you at least know that although solving mathematical equations is far from a pleasant way to spend your time, it still isn't anything to be afraid of. It may bore you to tears, but not to death. And guess what? That's it. The hard part is over.

Alternating Current

So far we've dealt with currents that are constant (as from a battery) and that move in a single direction from negative to positive. This is *direct current,* as in DC. Yet most of the currents and voltages we work with in audio recording are not DC but are continuously changing signals. These often have waveforms like those of the sine wave or much more complex shapes.

It is possible to have a current that periodically reverses, or alternates, in its direction of flow. This is called *alternating current*, or AC. These reversals in direction of current flow can happen infrequently, or they can occur very rapidly. Whether they take place a couple of times a second or billions of times a second, two changes in direction are considered one cycle. During this single cycle the current flows first in one direction and then in the other, and then it returns to its original position, ready for the start of the next cycle. The number of cycles that take place in the period of 1 second is called the *frequency* of that alternating current, and this measured unit is termed *hertz* (Hz).

Some Help from Our Graphic Friends

This is one of the many times when a graph makes everything as clear as a freshly polished bell (see cymbal cleaning). Such graphs are made up of a horizontal axis and a vertical axis. Here the horizontal axis represents an increase in time as you move from left to right, while the line itself represents zero, ground, or no potential. The area above this horizontal axis represents positive voltage, or amplitude, and the strength or value can be determined by its position relative to the level designations on the vertical axis. The area below the horizontal line represents negative potential, and like the direction of flow the level, or amplitude, designations as marked on the vertical axis increase negatively with a downward movement away from the zero point. See Figure 1-6, graph A. Graph B shows a power "turn on" point, at which time the current moves to the amplitude level indicated by its height above the horizontal axis. As we follow the passage of time from left to right along the horizontal axis we see that

The Basics

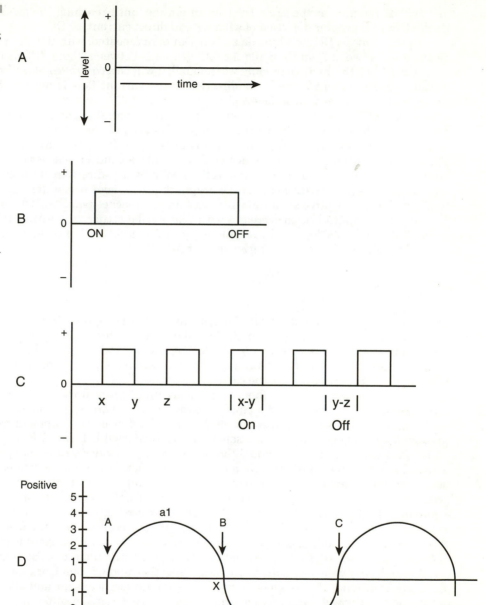

Figure 1-6
Graph A shows a representation of level versus time. Graph B depicts direct current, or DC. Graph C depicts pulsating DC. Graph D shows a sine wave used to depict alternating current, or AC.

the current remains at the same level, or amplitude, until reaching the switched-off point. Here we have a graphic depiction of what we call direct current, or DC.

Graph C shows a DC switching on and off sequence repeated, with the "on" period depicted by the area between X and Y, while the "off" period is between Y and Z. This can be termed *intermittent* DC. But more commonly, when found, it will be a *controlled,* or *pulsating* DC. This is still a direct current because the direction of the current flow is always the same, here always in the positive area of the graph.

In graph D we see the current beginning at zero potential (right on the horizontal axis). The current then moves, or flows, into the positive area of the graph, where its amplitude increases with time (left-to-right movement) until it reaches point "a1" (its positive peak). Then as time continues the current decreases in amplitude until it reaches the zero point once more and completes a "half cycle." Now at time point "X" the direction of the current reverses and moves into the negative part of the graph; that is, it goes below the horizontal axis. As time goes on the negative amplitude increases until it reaches point "a2" (the negative peak). The negative amplitude then decreases with time until it returns to zero and the current flow reverses direction, as depicted by its crossing over the horizontal axis again, and completes a full cycle. This is called *alternating current*, or AC.

Ground

So what does *ground* actually mean? Well, it can refer to a couple of things. First, there is utility (power line) *earth* (*neutral*) ground. When lightning strikes it can travel along power lines, destroying equipment, starting fires, and killing people (and animals). By connecting one of the power lines to earth, the lightning is provided with a fast, short, and low-resistance path to exactly where it wants to go, the earth.

There is also *safety* ground. Should some insulating material within a device break down and cause a short between the hot power line conductor and an exposed metal part, and someone (say, a very important superstar client) touch this piece of equipment and a normally grounded piece of equipment at the same time, it could spell E-L-E-C-T-R-O-C-U-T-I-O-N.

In order to avoid the "ugly dance," the safety ground provides a third wire, which connects exposed metal to the third prong in an AC power outlet. This prong is connected to either a separate green wire or the metal conduit that travels all the way to the main breaker or fuse panel, where it is connected to the neutral. This sets up a "fault current" path. Now, should an equipment fault occur, the hot-to-metal short will cause high current to run along the fault current path to the breaker panel, where it trips the breaker or blows the fuse, disconnecting that line and thus ending the danger. This is why it's considered dangerous (and makes you liable) to use "ground lifter" three-prong-to-two-prong outlet adapters. (See direct boxes.)

Although they will eventually be connected together, audio ground, or common, is different from power ground. This common serves as a signal reference point and also functions as a return path for power supply leakage and noise. There is a major difference between a common as used in balanced lines and one as used in unbalanced lines. In balanced circuitry the positive and negative sides of the signal run on two conductors, and the zero, or common, ground reference conductor and shield are not used to transfer the audio signal. With unbalanced operation the common, or ground, conductor must be used to transfer the audio signal because this type of cable generally has only two conductors. (See balanced and unbalanced operation.)

When we dealt with DC amps we were measuring steady-state current, as in graph B of Figure 1-6. But AC is continually changing in amplitude and periodically changing in direction of current flow. To discuss both on an even basis, an AC ampere is defined as the current that causes the same amount of heat generation as a DC ampere. For a sine wave (and *only* a sine wave) this value is derived by taking its maximum, or peak, amplitude (such as points "a1" and "a2" of Figure 1-6D) and multiplying it by 0.707. This is then termed its *effective value*. The value measured at any other point is called an *instantaneous value*. We'll get further into measuring and reading AC values and levels in the later section on "Audio Recording Meters." The alternating current shown by graph D is called a sine wave, short for *sinusoidal*. (*Never* use the term *sinusoidal*, because doing so can cause a severe depreciation in your cool!)

The variations in most audio AC waveforms are not as smooth, and those half cycles are not so exactly shaped. Yet many of the more complex waveforms are actually made up of the sum of two or more sine waves whose frequencies are multiples of some lower frequency. The base frequency is called the *fundamental,* and the multiples of the fundamental are called *harmonics*.

NOTE: *If a waveform has a frequency of two cycles per second it lasts a period of one-half second. So with a frequency of 100 cycles per second a waveform has a period of $^{1}/_{100}$, or 0.01, seconds. Thus, we end up with the two equations:* $t = 1/f$ *and* $f = 1/t$, *or time* (T) = 1 divided by (or the inverse of) the frequency (F), as well as frequency (F) = 1 divided by (or the inverse of) time (T).

Sine waves actually turn up just about everywhere: in generators of all kinds, motors, springs, pendulums, resonant (harmonic) elements, and even backyard swings. Its predictability makes it very important to audio for measurement, calibration, and specification checks. It is common practice to feed a 1 kHz (1000 cycles per second) sine wave at 0 dB into equipment to check the operating levels. When the output is viewed on an oscilloscope, any sine wave with even a slight amount of distortion will just scream out at you from those smooth-rolling, hill-like waves. It is also the only waveform that can pass through a circuit made up of passive components that consist of resistance, inductance, and capacitance and still retain its basic shape, which is a nice transition into a discussion of passive electronic components.

I know it's the last thing you want to deal with. Can there be anything more boring than a discussion of the properties of electronic components? But like everything else that you've had to bear with so far, it's another step on the road to gaining an understanding of, and therefore having a proficiency in, setting up gain stages properly throughout the recording system. However, just to make it a little more palatable I've thrown in some fun "messin' with sound" stuff you can do when you understand how components operate.

How about a fuzz-tone box like nobody ever heard the likes of, or jack the volume up all the way to distortion pickups that you can place on or close to any kind of instrument whose sound is based on a vibrating metal part and will even make a thumb piano an electric monster metal instrument. I'm talking about mucho sustain and lots of distortion.

Hey, all they gave me was more difficult equations, a soldering iron, and a list of stuff to pick up for lunch.

Passive Electronic Components

Resistors

We have already seen that resistors can be used to limit current and to set voltages to specific levels. Resistors are considered passive electronic components because their value, like that of all passive components, is set during the manufacturing process and is not supposed to change when used within its rated operating parameters. As you know, resistance values can be calculated using Ohm's law: $R = E/I$.

The amount of current flowing through a resistor multiplied by that resistor's value in ohms results in a set voltage ($E = IR$). The relationship between voltage and current as influenced by resistance also occurs with power ratings, in watts. Here either voltage or current squared and then divided by resistance yields the wattage.

Resistance impedes current flow by converting the energy of electron movement into heat. The amount of current or heat that resistors (of the same value) can dissipate is pretty much a function of their size. Power ratings above 2 watts generally call for the use of *wire-wound* resistors, which are composed of wire wrapped around a core of insulating material, such as porcelain.

Older resistors rated at under 2 watts are generally of the carbon type. These are a mixture of carbon and nonconductive, insulation materials that have been combined to yield a specific resistance. Carbon resistors have a tenancy to age poorly (change value) and are commonly replaced with newer, film-type resistors (which have identifiable end bulges). Film resistors do not crack as easily and are also less likely to produce amplifier noise than those older, carbon types, whose values drift with age, humidity, temperature, and so on.

Capacitors

Capacitors are capable of holding, or storing, electrical energy, somewhat like a battery but on a more temporary basis. They can be fixed, polarized, or variable. They are made up of two conductive plates separated by an insulating material called a *dielectric*. The reference value that dielectric materials are compared to is that of air, which has been given the rating of 1. Mica is rated 6, and ceramic materials can reach 1000. However, any dielectric can break down from an excessive charge that causes arcing across the two plates. You can witness this phenomenon yourself during a thunderstorm when lightning jumps from the clouds to the earth when the electrical charge building up between the two exceeds the insulating capabilities of air.

Typical fixed capacitors can be made with mica between the plates or even paper (which is rolled around tin foil). Newer varieties with ceramic dielectrics have the plates (often made up of silver) *fired*, or sputtered, directly onto their outside surfaces. Electrolytic capacitors are polarized, so they are useful in smoothing out ripples in DC power supplies.

Variable capacitors like those that were used to tune station frequencies in older radios are made up of two sets of plates, with the air between them acting as the dielectric. Here one set rotates, changing its distance from a stationary set, thereby varying the amount of stored charge. Using a very lightweight material, such as gold foil or gold actually sputtered onto Mylar, as a plate, makes it possible for sound pressure waves that impinge upon the plate to cause a change in its position relative to another fixed plate. Thus, the sound pressure waves cause a change in the stored or polarized electrical charge in that capacitor. Now, to help understand how capacitor or electret microphones work, consider that the sound pressure not only causes a change in electrical charge, but that change would also be relative to that sound pressure's level, frequency, and wave shape.

Capacitors also have the ability to stop DC from passing through them. Yet AC and pulsating DC, because of their varying charge, will induce current to flow through them. Furthermore, how much current flows is directly proportional to the frequency or the rate of change. Here as the frequency increases, the amount of current moving across the capacitor's plates increases. This relationship is termed *capacitive reactance*. It is measured in ohms, and as the rate of varying increases, the capacitive reactance, in ohms, decreases. Stated another way, as the frequency of the voltage increases, the capacitor's opposition to current flow decreases.

Since current flow is dependent on the rate of change of the voltage, current and voltage cannot occur at the same time. Current is said to *follow* voltage, and thus the two are not in phase with each other. With inductors, as you'll see, the rate of current change causes variations in the opposition to the flow of voltage. It is easy to see how these two components lend themselves to use in filters and crossovers.

Capacitors are also used to stop unwanted DC signals from entering audio lines. In fact some very inexpensive high-frequency crossovers consist of nothing more than a single capacitor.

Lists of "high pass" capacitor values in relation to the frequencies blocked are available in electronics handbooks. You can use these to find the value of a capacitor that will get rid of some low-frequency hum. Better yet, try this trick: "Y" a bass guitar's output and route one path through a cap that passes only the top end. Now you can have a field day putting all kinds of effects on that high-end signal and then mix a bit of it back in with the normal bass sound to make it jump right out in the mix, especially when the player has a percussive style.

Here are a couple of examples: A Mylar or, better yet, a tantalum capacitor (wired in series) rated at 27 microfarads is commonly used to block DC from getting into the audio signal. Reducing the cap's value to 0.2 microfarads causes the attenuation to start at 200 Hz and to yield a 10 dB loss in level at 20 Hz. Reducing it even further, to 0.05 microfarads, causes the low-frequency roll-off to start at 700 Hz, and it renders a 10 dB loss at 60 cycles. Further reduction of the rating will, well, you get the picture. That's why they're called high-"*pass*" filters.

Inductors (Coils)

It's already been stated that when current travels through a conductor it sets up a magnetic field that is present along the full length of that conductor. What happens is that as the electrons flow from negative to positive, a rotating magnetic field is produced around the conductor.

NOTE: *With direct current this can happen only during the times that the current values change (as when it is switched on and off). Unlike AC, which has varying current polarity, a constant DC current flowing through a conductor cannot set up a magnetic field.*

Inductance is a measure of a conductor's ability to convert a magnetic field that is near it into a voltage. The symbol for inductance is L, and the value is rated in henries, after the scientist Joseph Henry.

When the current flowing in a conductor varies, its magnetic field varies. This varying makes it possible to transfer energy from one conductor to another. Here voltage is induced into a conductor that travels through or near a magnetic field created by another conductor.

In order to measure inductance values, the amount of induced voltage is divided by the change in the current value, as in "amperes per second." Therefore, both inductance and the voltage that's induced are directly related to the rapidity or frequency of the current changes. Lenz's law states that induced voltage opposes change in the current that produced it. This means that induced current will flow in the opposite direction, or polarity, and thus oppose the potential that created it.

Think about these two facts for a moment and you'll begin to understand how the inductor can be frequency selective and why its opposition or resistance to current make this component another good candidate for use in frequency-selective filtering (equalizers and crossovers). The same kind of charts for inductive "low-pass" filters can be found in electronics reference books, as are "high-pass" capacitive filters.

A magnetic field can be condensed by winding the insulated (coated) conductor into a coil, making it appear stronger, and the magnetic flux can be concentrated around the coil windings if an iron-based magnetic material is in the center of the windings. Circuits with many windings of wire develop large amounts of inductance. These coils also oppose any change in current, and the voltage induced in them flows in the opposite direction. The amount of inductance a coil can produce is determined by the length of wire in its windings. Doubling the number of turns yields four times the inductance.

NOTE: *Hopefully, this will explain why it is not a good idea to coil up excess power or audio cabling, especially bundling the two together, as in the back of a rack.*

A larger turn circumference (bigger core) will also increase the inductance if more wire length is used for the same number of turns. But if a thinner wire is used, the inductance decreases, because the magnetic field becomes less concentrated.

Inductors can be interconnected in an aiding (in phase) or in an opposing manner. In an aiding manner the coil's magnetic fields and the voltage they induce will increase. In opposing, the magnetic field will be the difference of the sum of the two, meaning that by wiring them out of phase with each other, any signal that appears equally in both will be canceled out, while the part of the signal that's different (the differential signal) will not. Taken individually these facts do not seem very exciting, but when you combine them, as in humbucking pickups, the results are, as you'll see, nothing short of extraordinary.

Transformers

When two coils are placed together, they can form a transformer. Here the first coil, called the *primary*, is energized by an AC input, and the second coil, called the *secondary*, has voltage induced into it. The primary is not necessarily the input coil because transformers are two-way streets. However, transformers can have more than one secondary, wherein the secondary coil is *tapped*, or split up, into several shorter coils.

The ratio of the voltages and currents in the primary and the secondary of a transformer is set by the number of windings in each, and this is called the *turns ratio*. If the secondary has twice the number of turns as the primary, then twice the amount of voltage will appear at its output.

However, nothing is free, and this *step-up* transformer's current output would be half the current of the input. Reducing the number of secondary turns would reduce the voltage and also increase the current proportionally.

How proportional? Well, to the point of the number of turns T in the (p) primary divided by the number of turns in the (s) secondary should equal the voltage E in the primary divided by the E in the secondary, and they both should equal the current I in the secondary divided by the I in the primary, or $Tp/Ts = Ep/Es = Is/Ip$. That's about as proportional and as predictable as it gets. In all cases the *power* in the primary and the secondary will always be equal because $P = EI$.

NOTE: *Nothing is perfect, so expect losses due to stray capacitance, wire resistance, and insulation variables to cause less than ideal transformer operation.*

Remember an inductor's opposition to current change is called inductive reactance and is rated in ohms.

Although measured in ohms, inductive reactance is very different from resistance. Here (much like with capacitance) the reactance values are proportional to the frequency of the current change. Additionally since the induced voltage depends on a current change they also cannot occur at the same time. Current must lead voltage. Designers put this phase shift to use when inductors are used in equalizers and filters.

What's that got to do with transformers? It simply means that when choosing a transformer for isolation or impedance matching you want to use one with an inductive reactance that changes as little as possible throughout the audio frequency spectrum. This means that, as always but even more so here, it is important to check frequency response specifications.

Transformers make good ground isolators because there is no need for a physical connection between the primary and secondary sections of a transformer because magnetic coupling is utilized to pass signals across the coils. So noise will not be added to the signal from the noise flowing through any shield or ground conductor that is not physically connected across a transformer. Now, even though it should be obvious, it must be pointed out that "ground isolators" cannot remove any noise that's already part of the signal, but they can work wonders when positioned between a noisy line and the signal-carrying conductor. But some "hum-eliminator" ground isolation devices use unshielded low-power-handling transformers. So you want to check the low-frequency response, power-handling capability, distortion, and transient response specifications of ground isolators to avoid what could be serious signal degradation.

Impedance matching and level attenuation can be accomplished by using an attenuator network, called a *minimum loss* pad, composed of resistors or by using a *broadband* transformer. Besides providing a specific amount of attenuation to the signal, a resistor network can match up source and load impedances of unequal values. There are many computer-generated lists of resistor values to accomplish just about any attenuation or impedance matchup.

Altec's Technical Letter No. 133, part number CP-176-1K, dates back quite a few years now, but it still provides a wealth of information on this subject: "Minimum loss networks can provide the desired impedance match with only a slight attenuation to the signal but when a large amount of impedance mismatch has to be dealt with, resistor pads can have the disadvantage of causing an unacceptable amount of attenuation to the signal."

In this case a transformer may be more appropriate. To specify the proper transformer parameters, three factors must be known: the impedance levels to be matched, the lowest frequency that will be passed, and the power level that must be handled. With this information the transformer's core can be chosen and the number of turns and the desired wire thickness specified. It's simply a trade-off. For instance, power supply transformers are designed to operate over a very narrow range of frequencies to help avoid some of the trade-offs in power handling and size.

Even with low-power audio line transformers there's always a compromise between efficiency (in terms of size and number of windings), power handling, and frequency response. A high turns ratio will limit the high-frequency response. A low power-handling capability could cause core saturation, in which the magnetic field loses strength, particularly at low frequencies.

Humbucking Coils

Guitar pickups are often hot-rodded with additional coil windings (*rewound pickups*), for added distortion, and sustain provided by their hotter output, but they can also gain the ability to cancel radio-type airborne frequency buzzes and electromagnetic interference hums too, hence the term *humbucking* coils. You can also get an improvement in power transfer efficiency. Ready for this?

A coil wrapped around a magnetized rod sets up a magnetic field, which becomes weaker with distance (per the inverse square law). If an object made of ferrous material (such as a steel guitar string) moves across this field it causes electricity to be generated and flow through the coil. We're converting mechanical energy into electrical energy, and, simply stated, that's what pickups do. The resulting current that flows through the coil will be alternating, that is, flowing back and forth, mimicking the string motion, and eventually the speaker cone that's driven by the amplifier the guitar is plugged into will do the same.

Rewinding a pickup and adding more turns to it will make the signal stronger, as will using a rod with stronger magnetism. But if you add too much winding, you'll lose some high-frequency content. Additionally if you use too strong a magnet, you'll lose some of the sustain because the stronger magnetic field will oppose the string movement to the extent where until it actually slows it down more rapidly. Again there is always some trade-off, here between the most power, the widest frequency response, and the longest sustain.

But let's see what we can get away with using a noise-cancellation humbucking pickup scheme. First off, you must understand your enemy. The noises we are dealing with are "room" noises, like fluorescent light fixtures, in-wall AC wiring, radio waves, and (motor-) electro-mechanical-induced noises that do not show up until we amplify our guitars, microphones, and, yes, those old-time record players. These noise sources have the ability to induce voltage into coils without regard to the magnetism of the center poles. The noise doesn't even know those magnets are there. We can utilize the fact that noise is ignorant by adding another magnet-coil configuration to each of the string's pickups and wire it out of phase with the first and simply cancel out the noise. But wait a minute, what about the audio signal?

NOTE: *The point is that in order to fully comprehend Humbucking coils, you must first understand the affects that series and parallel wiring, reversals in magnetic polarity, and in and out of phase interconnection have on the flow of both the audio signal and any induced noise. This leads to a better understanding of* common-mode rejection ratio *(CMRR) and differential inputs as well as speaker wiring and countless other facets of audio technology. So, it's very important that you understand this stuff. A tall order, I know, but that's why I used nine illustrations just to lay everything out in the open. See Figure 1-7, a through i.*

Picture it this way (see Figure 1-7): the two ends of the wire that make up the coil are opposite in polarity. The interference will at any point in time cause an *identical* current flow in both coils. If we simply wire our two coils out of phase we would cancel out both the interference *and* the audio signal. Someone cleverly figured out that if you reversed the polarity of one of the magnetized rods, the audio signal would be out of phase, but the interference would not, because this voltage is induced directly into the coils regardless of the polarity of the magnetized poles those coils are wrapped around. Now the outputs of the two coils when wired out of phase cancel the in-phase interference and at the same time correct the out-of-phase audio signal. Very nicely done, you may think, but we've got a one–two punch going for us here because we can choose to wire those two out-of-phase coils in either a parallel or series configuration.

If we choose parallel it will cause the combined two coils' voltage output to be the same as for a single coil. On the other hand, with series interconnection the voltage yielded by the combined two coils will be double the amount of a single coil.

Now hold onto your hats because

- By using coils with half the number of turns you end up with less inductance, meaning a broader frequency response.

- The use of series connection results in the voltage output of two coils being added, so there is no loss of output.

Battery Representations of Parallel Connections of Coil Winding

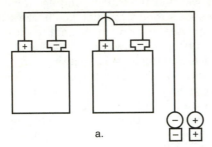

a.

Two batteries wired in parallel and in phase.

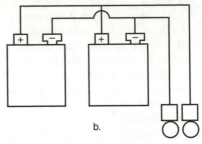

b.

This is probably the worse case of all wiring scenarios. Coils wired in this way output no voltage or current because their power is shorted out. Batteries wired this way get hot, which could cause their casings to split, and the escaping acid could contaminate electronics and skin. (Watch your eyes!) Wiring a pair of amplifier outputs this way will result in no output and two blown amps!

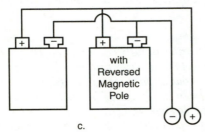

with Reversed Magnetic Pole

c.

The wiring of these two batteries depicts out-of-phase parallel wiring as it affects the audio output, when one of them has a reversed magnetic polarity.

Figure 1-7
Humbucking Coils: The leads from each of the coil outputs end in circles "○," which give the polarity of the audio signal, and in squares "□," which give the polarity of the noise signal. With a + inside, the output is positive; with a − inside, the output is negative; and when it is empty, there is no output. Battery representations of parallel connections of coil winding (a., b., c.). Battery representations of series connection of coil winding (d., e., f.). Wiring guitar pickup coils (g., h., i.). (continued)

- Since the two magnets are reversed in polarity, there is virtually no loss of string sustain.

- Any noised induced into the coils will be canceled because they are wired out of phase.

In other words, humbucking pickups give you the frequency response, high output, sustain, and noise cancellation you're looking for *all at the same time*. This is an audio rarity. I have no idea who came up with this conceptualization, but the achievement is duly noted in my book and has been for quite some time.

If you have any difficulty understanding any of this, do not pass "GO" but go directly back to series and parallel circuitry. See I told you this stuff is important (and it can be fun too).

Battery Representations of Series Connections of Coil Winding

Wiring Guitar Pickup Coils

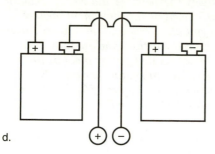

d.

Two batteries wired in series and in phase.

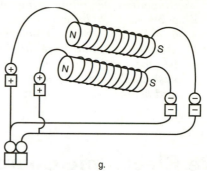

g.

Two coils with the same magnetic polarity, wired in parallel and out of phase. As can be seen from the lack of + and − signal outputs, this configuration results in no noise, no audio signal, and, as if it mattered, less string sustain.

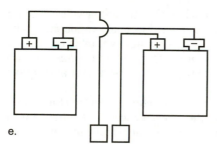

e.

The interconnection of these two batteries depicts out-of-phase series wiring and the way it affects the noise signal.

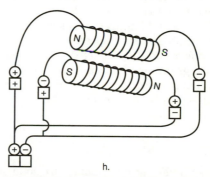

h.

Two coils with opposite magnetic polarity wired in parallel and out of phase result in no noise, twice the audio current, audio voltage equaling that of a single coil, full string sustain, but reduced high-frequency bandwidth.

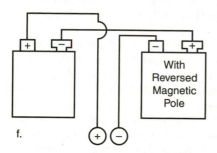

f.

Two batteries, one with reversed magnetic polarity, depicting out-of-phase series wiring and its effect on the audio signal.

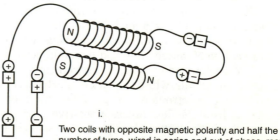

i.

Two coils with opposite magnetic polarity and half the number of turns, wired in series and out of phase, result in no noise, full audio signal voltage, full audio signal current, full string sustain, and full audio bandwidth. "BINGO, You win!"

Figure 1-7 (continued)

NOTE: *Acoustic guitar pickups are relatively inexpensive and easy to rewind and can be used on anything with a tone produced by the movement of steel (Rhoades tynes, piano strings, metal bells, cymbals, vibes, and on and on), any of which might benefit from some added power, distortion, and sustain. I once attached (with beeswax) one and a half fully hot-rodded guitar pickups to a thumb piano, and the results were impressive. The musician very quickly became a big fan of stomp boxes, Marshal stacks, and the use of feedback-type sustain. Who knew he was heavy metal at heart? The thing wailed big time and he tore the house apart.*

Active Electronic Components

Active components have parameters, or values, that change when the current or voltage going through or across them varies. Here the impedance, or resistance value, of a component might be variable, depending on the amount of current applied to it. This new value of resistance might also return to its previous level when current change ends. The first active components were mechanical, and many were used for communication. They included switches and relays for the telegraph and even speakers.

In fact, one of the earliest active components was a microphone. The carbon "button transmitter" used granules of carbon that reacted by changing their resistance value when sound waves applied pressure to them. This in turn proportionally varied the output of the electrical current across them, producing the sound.

Diodes

Far from being mechanical, diodes are solid-state semiconductor devices used to control the flow of current. Also known as *rectifiers,* they act upon current flow somewhat like a one-way street, impeding or resisting the flow of current in one direction while allowing more freedom to current flow in the other.

Crystal semiconductors such as silicon and germanium are manipulated during the manufacturing process to contain either an excess of electrons (arsenic is added) or a deficiency of electrons (boron is added). A deficiency of electrons creates a positively charged material known as a P material, while materials with an excess of electrons have a negative charge and are called N material. N and P should be easy to remember.

What you end up with is electron-carrying currents moving across the N materials toward the P materials, while the lack of electrons, or positive "holes," in P materials, acting as acceptors, take electrons out of the flow. One theoretical explanation is that as the electrons are taken from the flow, holes, or positive charges, are left behind. As electrons move in one direction (negative to positive), the holes move in the opposite direction (positive to negative).

So what's all this got to do with audio recording? For starters we need DC for power, and power utility companies only provide AC. But stick with me here a bit because those little diodes can also create a really nice and raunchy "We don't care what you think, we like it to sound nasty!" effect as well.

Diodes are made up of both N and P materials. This creates a *junction*, and all the extra electrons gather together on the N side while all the positive "holes" gather on the P side, just waiting to party. Depending on how things are *biased*, our adding a little voltage to the mix will either cause a nice electron flow across the junction or we'll end up with stubborn resistance to the current flow. Biasing simply refers to the alignment, polarity-wise, of the way we connect the voltage through the diode. Negative to the N side (or *cathode*) and positive to P side (or *anode*) makes for a *forward bias* and results in current flow almost free of resistance.

However, a *reverse bias*, or positive connected to the N side and negative connected to the P side, pulls the electrons and holes away from the junction. The party possibilities are now facing a lot of resistance, which limits current flow.

So when dealing with an alternating current, our diode will alternate between letting the current flow and resisting the current flow, because when AC is connected to a diode it's being fed a signal that constantly alternates between forward and reverse biasing.

An easy way to remember the polarity of a diode in terms of its biasing is to picture the vertical line in the schematic symbol or as drawn on the actual component as a tennis net. The side of the net where the player (either the triangle or the larger of the two sides of the component) is missing is the negative side, or where the negative connection is made for forward biasing and for less impedance to current flow (see Figure 1-8).

So What's Rectification?

Since a forward-biased diode results in current flowing almost free of resistance and a reverse-biased diode will result in high resistance, when faced with an alternating current a diode will alternate between letting the current flow and resisting the current flow because of the constant reversing (alternation) between forward and reverse biasing.

When used with a meter that reads only DC, a diode will rectify, or resist, the reverse-biased half waves of an AC signal passing through a conductor, and the result will be pulsating DC (refer to Figure 1-6C).

Figure 1-8
The schematic symbol for a diode, with a depiction of the actual component

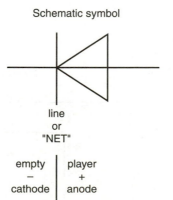

Schematic symbol

line
or
"NET"

empty	player
−	+
cathode	anode

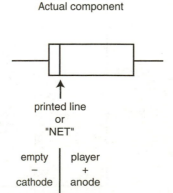

Actual component

printed line
or
"NET"

empty	player
−	+
cathode	anode

Full-wave rectification calls for dual diode usage. In this case *both* conductors are rectified, which doubles the number of pulses. If one of the diodes is reversed biased there'll be one positive and one negative pulse instead of just one positive pulse per cycle, which makes for a more stable DC meter reading.

Many power supplies use a total of four diodes, or a *bridge rectifier*, which places two diodes on each conductor, one negatively biased and one positively biased (see Figure 1-9). Now there's not only a negative and a positive pulse output during each cycle, but one conductor carries only positive and the other only negative pulses. So you end up with two identical outputs of reversed polarity.

To help understand this, think of common 60 Hz AC. With a DC meter, an incoming AC signal at 60 cycles going through half-wave rectification will produce 60 DC pulses per second, while full-wave rectification will yield 120 pulses per second (60 positive and 60 negative). By using a bridge rectifier the full wave is rectified and all the negative pulses will be on one conductor and all the positive pulses will be on the other conductor. The addition of capacitors, filters, and regulators smoothes out the DC even further, which can result in close to a pure DC output from the power supply. However, those two extra diodes result in the doubling of the voltage drop through the diodes, so this is not something for low-voltage applications.

You want fuzz? Okay, here comes the fun part. With a music instrument's AC signal, a single diode will render that 1960s "fuzz box" distortion that ruled the airwaves for a very short time (until it seemed everybody had used it). Connecting a negative lead to the cathode, or "N," side of the diode (again notated by the line on the schematic symbol and the printed line on the component) is called *forward biasing*. This allows for close to an unrestricted negative flow while heavily impeding the positive flow. Here an audio signal's alternating current will end up with the positive side of its waveform clipped (distorted). When connected in reverse-bias mode, the negative part of the waveform will be clipped.

The four hookup possibilities provided by the two diodes' orientations render different distortion sounds with both diodes being configured in forward- or reverse-bias mode, providing the height of raunch.

Few people ever used this raunchy guitar or synth sound, created by the audio signal's trying to pass through one of the four possible configurations (see Figure 1-10) of a diode across

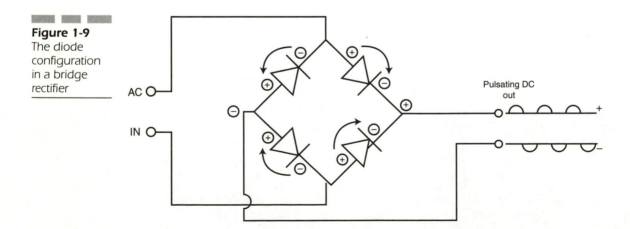

Figure 1-9
The diode configuration in a bridge rectifier

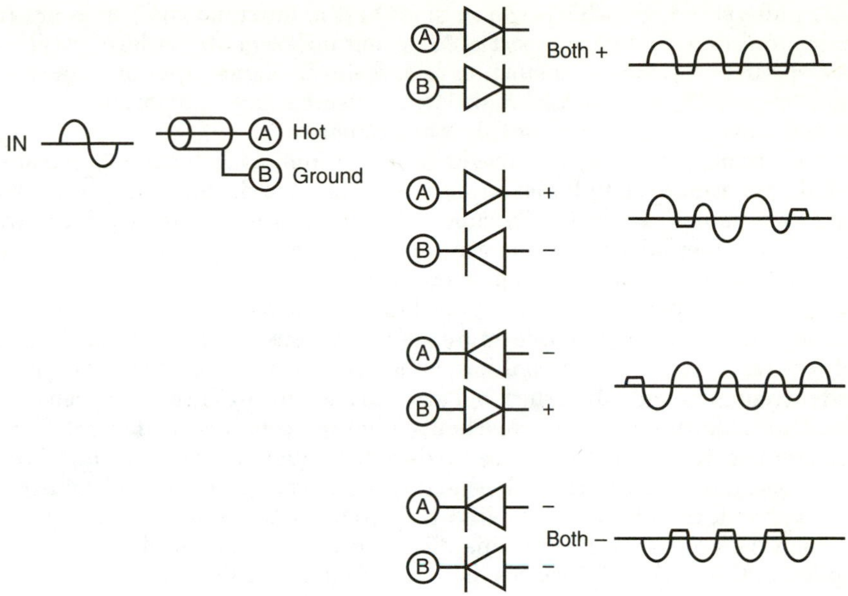

Figure 1-10
The four possible diode hookup configurations of a "nasty fuzz box"

both conductors. This is cheap and easy to try out yourself, and once you decide on the configuration, the two diodes will easily fit inside a quarter-inch guitar connector.

Tubes and Transistors

A basic electronics course would now continue by discussing additional information on electronic components, and you can readily find a wealth of information on this subject should you wish to research it further. This is not in fact a book on electronics, so we will not discuss op-amps and integrated circuits, nor will we delve into transistors and tubes to any depth, but I do want to make the following points.

We have seen that in a diode, current flows across the junction between N and P materials more freely in one direction than in the other. Transistors have two junctions between N and P materials. An overly simplistic description of a transistor's functioning is that they are like dual diodes, with one controlling the other so that base current controls collector current.

While the following is slanted by my personal tastes, it cannot be considered a purely personal opinion. The operation of a transistor is very different from that of a vacuum tube, even though both can be used for the same types of functions, such as amplifying and switching. Tubes are controlled by voltage; that is, the plate current, or the current flowing through the tube, is controlled by the voltage applied to the grid. Transistors are current controlled, in that an input current controls the output, or the current that flows through the transistor. This single difference may seen slight, but with audio signals it is significant in terms of both harmonic content (the addition of warm, even-order or harsher-sounding odd-order harmonics) and the way peak levels are handled (either heavily compressed to the point of being almost squashed before distortion or more readily reaching the point of distortion).

To understand why this is so you must first understand that tubes are inherently slower acting, and this results in a sort of "covering up" of jumps in level, which might cause the faster-acting transistor to distort. In fact, with very large input overloads, a semiconductor's collector-emitter junction can break down, causing a large current flow without an increase in the base current. (The term for this is *avalanche*.)

In a nutshell, tubes (at least triodes) are less prone to distortion than transistors, and when both do distort, the sound from a tube will be less harsh, since the negative feedback used in transistor amplification heavily clips the output, as in making a square wave out of a sine wave. This is actually one of the ways square waves are produced. On the other hand, transistors display little or no microphonics (noise caused by physical vibration) and require only low voltages to operate. Tubes are impedance sensitive, in that they require a more consistent load or they deliver a nonuniform frequency response. Too low a load impedance causes not only overheating but additional harmonic distortion. Transistors when faced with incorrect load impedances not only distort, but also burn up due to the heat caused by excessive current flow. Tubes like analog tape have a built-in natural compression feature. This results from the fact that the signal on the plate can reduce the gain and limit the amplification.

You see, the plate current is dependant on grid voltage. If the grid voltage drops, so does the plate current, thus limiting amplification. In older tubes, which didn't have much shielding between the grid and the plate, this effect was more pronounced. Let's say a positive signal is applied to the grid, which causes more plate current to flow, causing more negative signal at the plate. This increase in negative potential would reduce the grid's positive signal, decreasing the plate potential. The higher the gain, the more the plate overcomes the effect of the grid. In these old-time tubes, when the grid increased the current, the plate decreased it, and when the grid decreased the current, the plate increased it.

This nonlinearity is one of the tube's most endearing qualities as far as an electric guitar amp's sustain is concerned. In short, many people, including myself, find tube distortion to be more pleasant sounding than transistor distortion. This doesn't mean that tubes are better than transistors, just different. In fact I use transistor distortion all the time.

Interface impedances affects both transistor and tube operation. With vacuum tube electronics, matched impedances produced peak power transfer. The introduction and widespread use of transistors in audio electronics imposed different limitations on the output and input impedance variables because semiconductors cannot operate even near the peak of their power output capabilities, as this could cause too much internally generated heat. Here "seeing" too low an input impedance will cause a transistor to draw too much current. This not only will distort the signal, but could possibly generate enough heat to destroy the component itself. In order to truly understand how impedances affect audio signals, it helps first to understand the differences between the two main decibel references.

dBu vs. dBm

I think it's best to start off the discussion of dBu and dBm with an explanatory statement. dBm is referenced to 0 dB and equals a level of 0.775 volts when applied across a 600-ohm load. dBu is referenced to 0 dB and equals a level of 0.775 volts when applied across a load that is higher than 600 ohms.

Everyone understands that we need standardized specifications in order to make equipment comparisons and to judge both signal strength and quality. Dealing with ratios based on a change in a level is not a big challenge; you may have to play around with logarithmic values a bit, but no massive computational undertakings are required. If the level is 3 volts going into a device and 7.5 volts coming out, the level was increased by 2.5 times. But a decibel is a ratio, or a relative relationship between two values, as opposed to one stand-alone quantity, so things do get a little more complicated, but still not terribly difficult.

To reiterate: Power, in decibels, when referred to as "dBm" is referenced to 1 milliwatts into a 600-ohm load. Yet most test equipment and the common VU meter derive their readings from voltage levels instead of wattage. The use of power ratings, as in "dBm" and matching 600-ohm input and output impedances, goes back to those early telephone line-loss problems and remained predominant until transistors became popular. Matched input and output impedances of 600 ohms were used for tube electronics to operate without overloading. With transistors the requirement became a much easier "bridging" type of impedance matching. Bridging is more economical in terms of design and manufacturing implementation because it only calls for fairly low output impedances and very high input impedances so as not to load down the signal.

Tube electronics transfer power, while transistors transfer voltage. Power transfer is most efficient with matching impedance interfaces, while voltage transfer only requires a nonloading interface. When a transistor "is looking at" increasingly higher input impedances, its power output level will increase, up to a point. Its current gain will at the same time be diminished, while its voltage gain could be increased to very high and possibly a disastrous (transistor burnout) extent—the old trade-off routine again. So assuming that the lowest possible output impedance along with the highest input impedance results in the best matchup is not correct. Hey, doesn't it sound a bit too easy? But you'll never buy that line if you have a firm understanding of the principles involved in the relationship between level and impedance.

Take a simple circuit such as the one in Figure 1-11. We'll set the output impedance R1 at 600 ohms and the output voltage E1 at 1.55 volts. The only variable we'll be changing is the load impedance. Why we start with at an output voltage of 1.55 volts is no accident, because

$$E2 \ = \ 1.55 \ \times \ 600 \text{ divided by } (600 \ + \ 600) \ = \ 1.55 \ \times \ 0.5 \ = \ 0.775 \text{ volts}$$

Isn't it nice and neat having 1 milliwatt feeding into a 600-ohm load for a 0.775-volt output as our 0 dB reference?

Now we can vary just the load impedance values in our little circuit and hopefully discover something about level variation when interfacing audio signals.

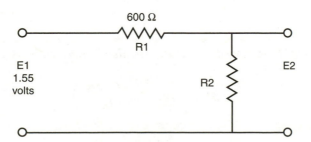

Figure 1-11
A simple circuit used to show the relationship between level and impedance

The Values

E1 = the source voltage, here set at 1.55 volts

R1 = the source output impedance, here set at 600 ohms

R2 = the load impedance, which is the only variable and its value
is given at the start of each set of equations

The equations are used to find the following five values:

E2 = the voltage at the load (across R2)

P1 = the total available power in watts

P2 = the power in watts delivered to the load

dBm = the logarithmic ratio, in decibels, of the power,
in wattage, at the load to the total power available

dBu = the logarithmic ratio in decibels of the voltage
delivered to the load to the total voltage

The Equations

$$E2 = E1 \times \frac{R2}{R1 + R2}$$

$$P1 = \frac{E1^2}{R1 \ (600 \ ohms) + R2 \ (in \ ohms)}$$

$$P2 = \frac{E2^2}{R2}$$

$$dBm = 10 \times \log\left(\frac{E2}{0.001}\right)$$

NOTE: *0.001 watts, or 1 milliwatt, can also be referred to as 10^{-3} watts.*

$$dBu = 20 \times log\left(\frac{E2}{0.775}\right)$$

We will list the results of all five equations later and also graphically plot both the voltage (E2) and the power (P2) levels as they appear at the load. This will hopefully make it nice and easy to compare the voltage and the power levels that result from varying just the input impedance of the load.

The following seemingly endless number of mathematical equations are *not* here for you to read. The equations involved in completing this exercise have been laid out for you to check your results should you go astray while *you* do the calculations. Yes, I actually expect you to follow along *and* do all the equations, and yes, it could take you a couple of several-hour sessions to complete the whole process. But it will be much quicker if you've already done the parallel and series/parallel equations.

However, before you start to moan about how crummy math is and how it doesn't have anything to do with being cool and hanging out with big-time recording stars making great music and all, we're talking about comprehending decibels here. You know, that "dB" term that gets thrown about so much. Face it, you really need to understand this stuff in order to set up gain stages properly.

My intention when I did all this arduous work was to give anyone serious about his or her involvement in the art and science of recording engineering the ability to comprehend the concept of decibels and to use these equations in the comparative measurement of audio levels. To accomplish this I decided to lay it all out so as to make it nice and easy. Don't get me wrong; I do not expect anyone to memorize or even remember any of these equations once the task is completed. Yet by doing the math, you will end up with a gut, commonsense understanding of the relationship between two levels as defined by a decibel value.

I know this for a fact, because I once spent almost a whole semester plowing through a textbook on acoustics. The drill was straight ahead, to do several lines of complex mathematical equations, read a few lines of text that varied some of the values (such as room dimension, frequenty content, power levels, temperature changes), and then do another four or five lines of more complex equations. This went on for over 300 pages. The minute I finished that book I forgot every equation and formula I had dealt with. A waste of time? Not a chance. A subject like acoustics when truly comprehended becomes almost like commonsense knowledge. Yes, I could say that I may have understood it all before, but going through that rote, step-by-step method, as mundane as it might have seemed, caused the full conceptualization of this subject to fully manifest itself very clearly, right there in my own little noodle.

You'll see that you may not remember any of the specifics either. But hey, you have this book as a reference to look anything up anytime you need to. More importantly, you'll instantly know when someone is using a decibel specification incorrectly or in a calculating manner, so "dB double-talk" will never fool you. That's why I spent several days laying out all this math in long form, just so you'd be able to get this info that you really need down pat. Take heart, the math is very easy when you use an inexpensive pocket calculator.

When I first did the equations, I had to use a slide rule.

NOTE: It is best not to round off any numbers until all the equations are completed.

We'll start with what will be our reference, the load impedance, or R2, having a value of 600 ohms:

$$E2 = 1.55 \text{ volts} \times \left(\frac{600}{1200} \right)$$

$$1.55 \times 0.5 = 0.775 \text{ volts}$$

$$P1 = \frac{2.4025}{1200} = 0.002 \text{ watts}$$

$$P2 = \frac{0.60025}{600} = 0.001$$

$$dBm = \frac{0.001}{0.001} = 1 \quad \text{the log of } 1 = 0$$

$$10 \times 0 = 0 \text{ dBm}$$

$$dBu = \frac{0.775}{0.775} = 1 \quad \text{the log of } 1 = 0$$

$$20 \times 0 = 0 \text{ dBu}$$

With R2 having a value of 300 ohms:

$$E2 = 1.55 \times \left(\frac{300}{600 + 300} \right)$$

$$1.55 \times 0.3333333 = 0.5166667 \text{ volts}$$

$$P1 = \frac{2.4025}{900} = 0.0026694 \text{ watts}$$

$$P2 = \frac{0.2669444}{300} = 0.0008898 \text{ watts}$$

$$dBm = \frac{0.0008898}{0.001} = 0.8898$$

$$\text{the log of } 8.8898 = -0.0507076 \times 10 = 0.507076 \text{ dBm}$$

$$dBu = \frac{0.5166667}{0.775} = 0.6666667$$

$$\text{the log of } 0.6666667 = -0.1760.912 \times 20 = -3.5218246 \text{ dBu}$$

With R2 having a value of 150 ohms:

$$E2 = 1.55 \times \frac{150}{750}$$

$$1.55 \times 0.2 = 0.31 \text{ volts}$$

$$P1 = \frac{24025}{750} = 0.0032033 \text{ watts}$$

$$P2 = \frac{0.0961}{150} = 0.0006407 \text{ watts}$$

$$dBm = \frac{0.0006407}{0.001} = 0.6406667$$

the log of $0.64066667 = -0.19336787 \times 10 = -1.933679$ dBm

$$dBu = \frac{0.31}{0.775} = 0.4 \quad \text{the log of } 0.4 = -0.39794$$

$$\times 20 = -7.9588002 \text{ dBu}$$

With R2 having a value of 60 ohms:

$$E2 = 1.55 \times \left(\frac{60}{660}\right)$$

$$1.55 \times 0.0909091 = 0.1409091 \text{ volts}$$

$$P1 = \frac{2.4025}{660} = 0.0036402 \text{ watts}$$

$$P2 = \frac{0.0198554}{60} = 0.0003309 \text{ watts}$$

$$dBm = \frac{0.0003309}{0.001} = 0.3309229$$

the log of $0.3309229 = -0.4802732$
$\times 10 = -4.8027322$ dBm

$$dBu = \frac{0.1409091}{0.775} = 0.1818182$$

the log of $0.1818182 = -0.7403627$
$\times 20 = -14.807254$ dBu

With R2 having a value of 1000 ohms:

$$E2 = 1.55 \times \left(\frac{1000}{1600} \right)$$

$$1.55 \times 0.625 = 0.96875 \text{ volts}$$

$$P1 = \frac{2.4025}{1600} = 0.0015016 \text{ watts}$$

$$P2 = \frac{0.9384766}{1000} = 0.0009385 \text{ watts}$$

$$dBm = \frac{0.0009385}{0.001} = 0.9384766$$

$$\text{the log of } 0.9384766 = -0.0275766$$
$$\times 10 = -0.2757657 \text{ dBm}$$

$$dBu = \frac{0.96875}{0.775} = 1.25$$

$$\text{the log of } 1.125 = 0.09691$$
$$\times 20 = 1.9382003 \text{ dBu}$$

With R2 having a value of 2000 ohms:

$$E2 = 1.55 \times \left(\frac{2000}{2600} \right)$$

$$1.55 \times 0.7692308 = 1.1923077 \text{ volts}$$

$$P1 = \frac{2.4025}{2600} = 0.000924 \text{ watts}$$

$$P2 = \frac{1.4215976}{2000} = 0.0007108 \text{ watts}$$

$$dBm = \frac{0.0007108}{0.001} = 0.7107988$$

$$\text{the log of } 0.7107988 = -0.1482533$$
$$\times 10 = -1.482533 \text{ dBm}$$

$$dBu = \frac{1.1923077}{0.775} = 1.5384615$$

$$\text{the log of } 1.5384615 = 0$$
$$\times 20 = 0 \text{ dBu}$$

With R2 having a value of 3000 ohms:

$$E2 = 1.55 \times \left(\frac{3000}{3600} \right)$$

$$1.55 \times 0.8333333 = 1.2916667 \text{ volts}$$

$$P1 = \frac{2.4025}{3600} = 0.0006674 \text{ watts}$$

$$P2 = \frac{1.6684028}{3000} = 0.0005561 \text{ watts}$$

$$dBm = \frac{0.0005561}{0.001} = 0.5561343$$

the log of $0.5561343 = -0.2548204$
$\times 10 = -2.5482035$ dBm

$$dBu = \frac{1.2916667}{0.775} = 1.6666667$$

the log of $1.6666667 = 0.2218488$
$\times 20 = 4.436975$ dBu

With R2 having a value of 6000 ohms:

$$E2 = 1.55 \times \left(\frac{6000}{6600} \right)$$

$$1.55 \times 0.9090909 = 1.4090909 \text{ volts}$$

$$P1 = \frac{2.4025}{6600} = 0.000364 \text{ watts}$$

$$P2 = \frac{1.9855372}{6000} = 0.0003309 \text{ watts}$$

$$dBm = \frac{0.0003309}{0.001} = 0.3309229$$

the log of $0.3309229 = -0.4482732$
$\times 10 = -4.8027322$ dBm

$$dBu = \frac{1.4090909}{0.775} = 1.8181818$$

the log of $1.8181818 = 0.2596373$
$\times 20 = 5.1927462$ dBu

With R2 having a value of 10,000 ohms:

$$E2 = 1.55 \times \left(\frac{10,000}{10,600} \right)$$

$$1.55 \times 0.9433962 = 1.4622642 \text{ volts}$$

$$P1 = \frac{2.4025}{10,000} = 0.0002267 \text{ watts}$$

$$P2 = \frac{2.1382164}{10,600} = 0.0002138 \text{ watts}$$

$$dBm = \frac{0.0002138}{0.001} = 0.2138216$$

the log of $0.2138216 = -0.6699483$
$\times 10 = -6.6994833$ dBm

$$dBu = \frac{1.4622642}{0.775} = 1.8867925$$

the log of $1.8867925 = 0.2757241$
$\times 20 = 5.5144826$ dBu

With R2 having a value of 20,000 ohms;

$$E2 = 1.55 \times \left(\frac{20,000}{20,600} \right)$$

$$1.55 \times 0.9708738 = 1.5048544 \text{ volts}$$

$$P1 = \frac{2.4025}{20,600} = 0.0001166 \text{ watts}$$

$$P2 = \frac{2.2645867}{20,000} = 0.0001132 \text{ watts}$$

$$dBm = \frac{0.0001132}{0.001} = 0.1132293$$

the log of $0.1132293 = -0.9460411$
$\times 10 = -9.4604105$ dBm

$$dBu = \frac{1.5048544}{0.775} = 1.9417476$$

the log of $1.9417476 = 0.2881928$
$\times 20 = 5.7638556$ dBu

Table 1-1
dBm vs. dBu Equation Results

R2 Load Impedance	E2 Voltage at the Load	P1 Total Watts Available	P2 Watts at the Load	dBm Wattage Ratio	dBu Voltage Ratio
600 ohms	0.775	0.002	0.001	0 dBm	0 dBu
300 ohms	0.5166667	0.0027	0.0009	−0.5	−3.5
150 ohms	0.31	0.003	0.0006	−1.9	−8
60 ohms	0.14	0.0036	0.0003	−4.8	−14.8
1,000 ohms	0.96875	0.0015	0.0009	−0.275	1.938
2,000 ohms	1.19	0.0009	0.0007	−1.48	3.74
3,000 ohms	1.29	0.0006	0.0005	−2.548	4.43
6,000 ohms	1.409	0.00036	0.00033	−4.8	5.19
10,000 ohms	1.46	0.00022	0.0002	−6.699	5.5
20,000 ohms	1.5	0.00011	0.0001	−9.46	5.76
100,000 ohms	1.54	0.0000239	0.0000237	−16.2	5.96
2,300 ohms	1.23	0.008	0.000657	−1.8	4

Table 1-1 and the two graphs in Figure 1-12 show the results of all our equations. Looking at the level values of the two vertical layouts you'll notice that we're dealing with two differing sets of quantities, one for power and one for voltage. It may appear we are giving voltage twice the value of power, but remember that Power = *IE* and *E = IR,* so power = (I × I) × *R.* Power also equals voltage squared (or *E* × E) × R, but we've already dealt with the fact that the voltage is squared within the equations. However, in order to compare the power and voltage results, we still have to take into consideration the fact that dBu = 20 × the log while dBm = 10 × the log. That's why it takes 6 dB to double the voltage while only 3 dB doubles the power. Hence, twice the voltage value is equal to the power value. This does not mean that power is twice voltage. It was simply a way to take the equation differences of 20 × the log and 10 × the log into consideration so that the two could be more easily compared.

The top of the bell-shaped curve in Figure 1-12 shows the highest current value, which "happens" to correspond to the matched 600-ohm load impedance reference example. We see by the slope of the voltage curve that as load impedance increases, the level continues to rise, but not indefinitely, for at some point we begin to see diminishing returns with added impedance. The crossing point of the two curves is also interesting because here neither the voltage nor the current levels seem to be depleted to any great extent. But just to make sure we have this situation well in hand, let's do a couple more equations.

First let's check out that crossing point, which appears to be right at 2,300 ohms.

Figure 1-12
The graphic
representation
of all the load
impedance
equation
results

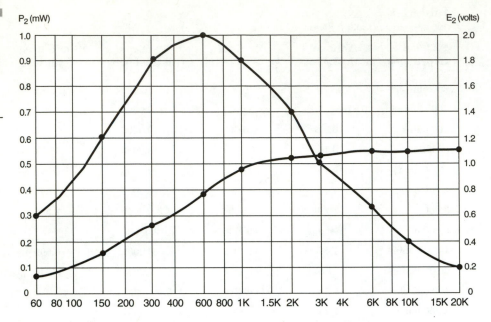

With R2 having a value of 2,300 ohms:

$$E2 = 1.55 \times \left(\frac{2,300}{2,900} \right)$$

$$1.55 \times 0.7931035 = 1.2293103 \text{ volts}$$

$$P1 = \frac{2.4025}{2,900} = 0.0008284 \text{ watts}$$

$$P2 = \frac{1.5112039}{2,300} = 0.000657 \text{ watts}$$

$$dBm = \frac{0.000657}{0.001} = 0.6570452$$

the log of $0.6570452 = -0.1824048$
$\times 10 = -1.8240476$ dBm

$$dBu = \frac{1.2293103}{0.775} = 1.5862069$$

the log of $1.5862069 = 0.2003598$
$\times 20 = 4.0071967$ dBu

So it seems possible that a design engineer who requires an input level of +4 dBu could set up and run a computer program that would point right to a value of R2 (here being 2,300 ohms). Probably worth the program writing effort, no?

Now let's check for any diminishing voltage returns with an increase in the load resistance beyond that which appears to be an optimum value for R2.

With R2 having a value of 100,000 ohms:

$$E2 = 1.55 \times \left(\frac{100,000}{100,600} \right)$$

$$1.55 \times 0.9940358 = 1.5407555 \text{ volts}$$

$$P1 = \frac{2.4025}{100,600} = 0.0000239 \text{ watts}$$

$$P2 = \frac{12.3739274}{100,000} = 0.0000237 \text{ watts}$$

$$dBm = \frac{0.0000237}{0.001} = 0.0237393$$

$$\text{the log of } 0.0237393 = -1.6245326$$
$$\times 10 = -16.245326 \text{ dBm}$$

$$dBu = \frac{1.5407555}{0.775} = 1.9880716$$

$$\text{the log of } 1.9880716 = 0.298432$$
$$\times 20 = 5.9686403 \text{ dBu}$$

So we see that excessive load impedance is no bargain because we'd end up losing a great deal of current potential for virtually no voltage gain.

Now you know for a fact that having the lowest possible output impedance along with the highest possible load impedance is not the answer to every audio interface situation. And you now know that while dBm and dBu are interrelated, they are *not* the same.

How about a postulation that states that all that is needed is to match all input and output impedances for the best audio interface? Let's check out what happens with a couple of matched R1 and R2 impedances at a value other than 600 ohms.

With R1 and R2 both equaling 10,000 ohms:

$$E2 = 1.55 \times \left(\frac{10,000}{20,000} \right)$$

$$1.55 \times 0.5 = 0.775 \text{ volts}$$

$$P1 = \frac{0.775}{20,000} = 0.0000388 \text{ watts}$$

$$P2 = \frac{0.600625}{10,000} = 0.0000601 \text{ watts}$$

$$dBm = \frac{0.0000601}{0.001} = 0.0600625$$

the log of $0.0600625 = -1.2213966$
$\times 10 = -12.213966$ dBm

$$dBu = \frac{0.775}{0.775} = 1 \quad \text{the log of } 1 = 0 \times 20 = 0 \text{ dBu}$$

With R1 and R2 both equaling 100 ohms:

$$E2 = 1.55 \times \left(\frac{100}{200}\right)$$

$$1.55 \times 0.5 = 0.775 \text{ volts}$$

$$P1 = \frac{0.775}{200} = 0.003875 \text{ watts}$$

$$P2 = \frac{0.600625}{100} = 0.0060063 \text{ watts}$$

$$dBm = \frac{0.0060063}{0.001} = 6.00625$$

the log of $6.00625 = 0.7786034$
$\times 10 = 7.7860341$ dBm

$$dBu = \frac{0.775}{0.775} = 1 \quad \text{the log of } 1 = 0 \times 20 = 0 \text{ dBu}$$

Well, that tells us a lot about impedance matching. As with most measurements in audio, the results depend on both the components and other variables involved. Here the dBm level ranges between −12 and over +7 dB! It's the same as saying that the lowest possible input and the highest possible output impedances are best for every audio interface. Forget it, because as far as I know no absolutes exist in the world of audio. In my experience, if something sounds too straight ahead and right down the center without any deviation so as to make everything seem mindlessly simple, look out, because something's about to bite you in your rear.

And covering your rear is part of what doing a good job is all about, especially in the field of audio recording, where you could find out a week after the fact that a session you engineered during which an internationally famous superstar made a once-in-a-lifetime surprise guest appearance ended up having an excessive amount of S/N ratio, overly flat dynamics, or less than adequate headroom or was just straight out distorted. This actually happened to someone I know!

If you do not set your signal levels high enough, then further down the audio chain the gain might have to be raised to the point where the background noise (there's always some noise) will become apparent during quieter passages. Noise is usually lower in level than the program material, so when the program material is loud it easily "masks" this component of the signal. However, if during quiet passages the noise level is close to that of the average signal level, *no* fix is going to do the trick.

At the same time, you obviously don't want to cause the gain stage you're adjusting to distort. So preserving your reserve of headroom is critical, especially with percussive or other signals that have a wide dynamic range. Remember the average can be 10 to 15 dB below peak levels, so it would seem that the capability for handling +15-dB levels is all that's required. Guess again! Some signals can have peaks that are more than 30 dB higher than their average output. Yet even a signal with a more restricted or mellower dynamic range could distort if the level you've set it at is too great for a gain stage it will pass through further down the signal chain. It can be a delicate balancing act, to be sure, and it's important to understand how to handle it.

Simply put, if you do not know how to set up your signal path gain structures, you're nothing more than a glorified elevator operator, which, in the field of audio recording, means you're cooked.

The signal path itself can often be a key ingredient, for the way your cables and interconnections behave can radically affect a signal's quality. Luckily, you do not have to worry about actually "matching" up the impedances between equipment, at least as far as the design stage is concerned, because that's the job of an electronics design engineer. So why bother going through all the preceding text and *all that math*?

Well, now that you have an understanding of the basics of impedance matching, you'll be able to comprehend balanced lines, differential input stages, CMRR, and, after that, gain structure setup, unity (gain) level, headroom, and S/N ratio. A tough go, but as you'll also see, at least you have a great ally in your meters.

Impedance Explained

While it's important to understand what impedance is all about, it's always been hard to nail down a complete definition that has you walking away with confidence in your knowledge. Well, I've got my hammer out and I'm going to have a go at it starting right here with line impedance and the impedance of interconnecting cables. While the exact meaning of *balanced* or *unbalanced* operation refers to the inputs and outputs the cables are connected to, the choice of balanced operation is often due to the imperfections inherent in those cables.

All audio cabling can be thought of as a grouping of passive electronic components (resistors R, capacitors C, and inductors L). First off, all conductors present some resistance to current flow, and this deficiency is rated as ohms per foot, mile, meters, or whatever. The insulation between the conductors inside a cable is also less than perfect. But here, because the insulator's job is in fact to present resistance to any current flow, any deficiency in this area is thought of in terms of its *conductance*, or leakage. This is actually the reciprocal of resistance (1/resistance) and is rated in "mhos" per foot, mile, or whatever. I'm not kidding; they really do call it mhos, pronounced "Moe's" (as in Curly's, Larry's, and, . . .). The wires within the cables also have inductance along their entire length and capacitance between them. Because of this R, C, and L content, some audio cabling can act just like a filter.

Facts to Forget

Once you've read the following facts, for your own sanity please forget them.

- If you divide the resistance by the conductance and take the square root of the result, you'll have the impedance of the cabling as it applies to DC and low frequencies.

- If you divide the inductance by the capacitance and take the square root of that result, you'll have the impedance as it applies to higher frequencies.

- If the impedances derived from both calculations are the same, then alert the press because you've gotten hold of the perfect cable and the result will mean distortionless transmission, minimum loss, and your extreme wealth.

- If they are not the same, and the result from the resistance and conductance calculation is used to "match" impedances, then you'll have minimum loss at DC and low frequencies, but the transmission will not be stable at higher frequencies due to what's called *refection (load bounce-back) losses* caused by a load mismatch.

- If the result from the inductance and capacitance calculation is used to "match" impedances, then you'll have uniform transmission but less than minimum loss at DC and lower frequencies due to the added resistance.

- If the impedances of *both* of the devices the cable is connected to are the same as the cable impedance value derived from dividing the resistance by the conductance, then there will be minimum loss of signal at DC and low frequencies.

- If the impedances of *both* of the devices the cable is connected to are the same as the cable impedance values derived from dividing the inductance by the capacitance, then there will be minimum loss of signal and uniform transmission at higher frequencies.

- If the impedances of the devices are much higher than the calculated impedance of the cable connecting them, then the insulation resistance or the conductance, in mhos, will cause losses at DC and low frequencies and the capacitance will cause losses at higher frequencies.

- If the impedances of the devices are much lower than the calculated impedance of the cable connecting them, then the resistance will cause losses at DC and low frequencies and the inductance will cause losses at higher frequencies.

There they are, just the facts.

Getting involved in this subject area to this depth may seem a little overdone, but it can get way more ridiculous when the discussions and calculations start to encompass bizarre qualities such as an insulating material's differing capacitance ratings caused by the amount and color of its pigment content! Some folks really do talk about this kind of stuff. The only reason I brought this up was so you'd see that, as usual, there are no sure-fire simple answers in audio, just a world of trade-offs to contend with. If someone tells you different, best to ask lots of questions.

Noise

Most of the 60-cycle hum that's picked up by cables comes from the AC power lines that pass throughout a building. I once tested a spring reverb device that, though it came with wall-mounting attachments, could not be mounted on a wall anywhere near AC power cabling because it had *no* shielding and thus no isolation from electromagnetic interference. This thing was an almost perfect receptor. On the other side, certain devices, such as fluorescent lighting and triac light dimmers, cause distortion in the power line's 60-cycle waveform, making their radiated noise stronger, more easily picked up, and more physically audible as well. Electronic equipment or devices that readily produce or pick up electromagnetic interference are not acceptable in any recording situation.

Just in case you're interested: Fluorescent lights use an inductor to limit current flow. This component starts and stops the electrical conductance very quickly, which in turn creates complex waveforms that are more spiky and radiate stronger electromagnetic fields than the normal 60 Hz sine wave AC power ones. With Triac dimmers the voltage builds up gradually and then quickly gets cut off, which causes noise to radiate. You can easily hear these because both sources create those annoyingly audible humming and buzzing sounds. It should be obvious that audio recording equipment and interconnecting cables require extra protection from noises such as electromagnetic interference. I don't know what the manufacturer of that spring reverb was thinking of. Luckily we get a good deal of help by using balanced lines.

Balanced Lines

Audio cables, whether balanced or unbalanced, require only two conductors to pass audio signals. With balanced circuits it is critical that these two conductors have equal impedances between themselves and the ground (or common) conductor. To understand this requirement let's suppose a balanced line is subjected to a 60 Hz magnetic field because it runs across or alongside a power cord. How much of the interference noise, here 60-cycle hum, each of the two conductors picks up will depend directly on how close they are to the source of the noise. The advantage of using balanced cable is that *if* the noise is picked up by the *same* amount on *both* conductors, it will be canceled out when it passes through a "differential stage." So making those two conductors equally distant from all noise sources is important.

Achieving this seemingly impossible task is as simple as twisting the two conductors around each other. This yields two internal wires that have the same average distance from any source of interference they happen to run across or alongside of in this noise- and interference-prone modern world we live in.

Some people have the misconception that a cable's shield is enough to protect the signal-carrying conductors from electromagnetic interference. While properly connected shielding is very helpful, especially with radio interference, the ground noise current they carry can, because of the shield's proximity to the signal-carrying conductors, actually induce noise into them. This may be how one theory of audio wiring, which extols the use of a twisted pair without any shield to avoid this type of leakage, got its start. Another audio truism.

With low-impedance lines, simply twisting the two conductors around each other places both the same distance from an interfering magnetic signal, and using "star-quad" cabling with its dual twisted pairs in parallel reinforces this property. Going along with the no-shield theory, all you have to do is add some insulation around the outside of the conductors and you're pretty much set. That is until you end up with CBers, ham radio operators, and walkie-talkie users chattering all over your recorded tracks.

Radio frequency (RF) interference pickup will be reduced substantially by using cabling with a properly grounded metallic sheath surrounding a twisted pair of conductors. Twisted pairs do help with inductive cross-talk *and* capacitive cross-talk, *but* they should be balanced and end up at a fully (meaning true differential) balanced input. Additionally, the ground-return/shield conductor should be common to both lines and have equal impedance between itself and the two signal lines.

But the shield's drain wire might be connected to neither, both, or either one of the sender's or receiver's common, or ground, and you'll often find all four schemes in a single rack of equipment.

On the other hand, noise and cross talk (signal bleed) from the ground reference will have no effect on the cabling if an input transformer with a "floating" primary is used, because there is no ground connection across that type of transformer's windings. No truism there! So I think it's time for a side trip in our journey.

Transformers and Line Noise

Additionally, since with transformers the only coupling between the primary and the secondary is magnetic in nature, all noise, such as pops and hums from a ground connection, or clicks, buzzes, and other cross talk from capacitive coupling will be greatly attenuated if not completely eliminated by using a metal shield placed between the primary and secondary windings. The best results are achieved when both the primary and the secondary windings have separate shields. In fact a 100 dB rejection of common-mode noise is not unusual.

You'll find this especially true with low-level input transformers, which are used to couple low-output transducers like microphones to preamps or with transformers that are used in coupling lines to step up voltage ratios or to give a slight amount of attenuation so as to provide unity gain.

The common passive "direct box" may include attenuation in the form of a resistor pad and a ground "lift" switch, but mainly it consists of a transformer with a turns ratio designed to couple higher-level sources like music instrument outputs to the lower-level inputs of microphone preamps. Keeping the high-impedance cables that connect the instrument's outputs to these devices as short as possible is important because those cable's capacitance can adversely affect high-frequency response.

Transformers can also be supplemented with an R/C network (resistor capacitor filter) across the secondary to suppress any transient distortion, such as in ringing from overshoot. Terminating resistors can be added so that a controlled defined load is presented to, or "seen by," the source. All this adds to the cost, but it's worth it.

Back to Balanced Lines

In balanced lines, while the two signal conductors are opposite each other in polarity, they should carry an equal amount of voltage and hopefully the impedance between both of them and ground, as well as all other conductors, will be the same. This means that any noise leakage that's picked up will be picked up equally on both audio conductors. At first glance this doesn't seem like such a great idea, until you take into consideration that the receiver of this signal will be a "differential input," which, whether it's made up of active (op-amp) or passive (transformer) circuitry, inherently cancels out noise. Additionally, this noise cancellation will have no effect on the audio traveling through them, because the cancellation caused by differential inputs affects only those signals that are common to both conductors. This is more than likely why it's called common-mode rejection.

Common-Mode Rejection Ratio (CMRR)

CMRR is simple to understand, just look it up. You find definitions like "the ratio of the longitudinal induced *electromotive force* (EMF) to the voltage (normal or differential-mode) caused by this EMF in the load resistor." Glad we got that cleared up.

The level of the original noise as compared to that of the noise that remains after passing through a differential stage is called the CMRR. This specification is given as a decibel amount that tells how much the noise level will be suppressed, so the larger the decibel number, the better. But in order to really understand common-mode rejection (and it's very important that you do), you first need to understand what differential stages are all about.

No matter what, deferential stages are never perfect. This is simply because the components and materials that make them up are not. Component tolerances as tight as ±5% will still degrade the differential stage's common-mode nulling capabilities. One way to lessen this problem is to make the common-mode conductor impedances low at the output and high at the input. Another way is to use a high-quality (meaning expensive) shielded input transformer with specifications that exceed the bandwidth and power-handling requirements of the situation.

Therefore, the best CMRR results occur with differential stages that utilize high-quality transformers or have carefully matched bridging impedance ratios (of at least 7:1). With that we move on to differential input stages.

Differential Input Stages

It's fairly easy to grasp the fact that any noise signal that appears equally on both of the input conductors connected to a transformer's primary cannot set up any kind of a magnetic field, so they will not "be seen" by the transformer's secondary coil. The audio signal passes through to the secondary, while the noise is canceled. Differential stages take advantage of this capability to cancel out noises that are common to both conductors.

Active differential circuitry is a little more difficult to comprehend. To help simplify the explanation, forget about any actual amplification and consider the op-amps as set up to neither amplify nor attenuate.

Operational amplifiers have two input connections: one for positive and one for negative (these are out of phase with each other). The input signal itself will actually be the difference between these two input connections. An op-amp's single output connection is designed to be in phase with one of the inputs and out of phase with the other input. A positive signal connected to the negative input will produce negative output, and a negative signal connected to the positive input will produce positive output. This is called *inverting* op-amp operation. When the input signal and op-amp connection polarities are matched (positive to positive and negative to negative), positive inputs result in positive outputs and negative inputs result in negative outputs. This is called *noninverting* op-amp operation. So op-amps have the capability to operate as inverting or noninverting amplifiers, depending on how the input signal is connected to them. That's all you need to know, so don't bother yourself about the theory any further.

To reiterate, when a negative signal is connected to an inverting op-amp's positive input, it ends up being positive at the amp's output. When a positive signal is connected to an inverting op-amp's negative input, it ends up being negative at the amp's output. Connecting the positive to positive and negative to negative provides noninverting operation, which leaves all the input's polarities intact. Active differential stages take advantage of this.

The following is a further explanation of differential stages. Read it slowly so you can get on with your life.

Although there are many active differential op-amp combinations, I'm using a fairly uncommon quad (4) op-amp *integrated circuit* (IC) type in this example so that everything will be separated and all out there, making it easier to grasp (see Figure 1-13).

Op-amp A: The positive conductor that is connected to noninverting op-amp A's positive input will include the noise, such as RF or *electromagnetic interference* (EMI), as well as any ground or common noise leakage that's been picked up by that conductor, and this signal combination will pass through the op-amp, with the audio signal remaining positive with the noise noninverted.

Figure 1-13
A differential stage made up of four op-amps

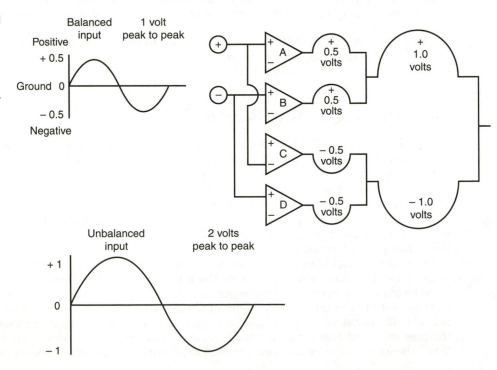

Op-amp B: The negative signal conductor that is connected to inverting operation op-amp B's positive input will include the same noise as the positive conductor. At the output of this inverting op-amp you will have a positive audio signal and an inverted noise signal.

Op-amp C: The positive signal conductor, including any noise that it's picked up, is also connected to inverting op-amp C's negative input. This inverting stage will output a negative audio signal and an inverted noise signal.

Op-amp D: The negative conductor is also connected to noninverting op-amp D's negative input. Again this signal will include the noise that's been picked up by that conductor, and this signal combination passes through the op-amp, with the audio signal remaining negative and the noise noninverted.

The two audio signals are then combined at the output of the differential stage, and the negative and positive audio signals will add together, yielding a gain of 2, or a doubling of the voltage. The two noise signals, on the other hand, cancel each other out.

Wait a minute! How come one gets doubled and the other gets wiped out when they have both been through the same ordeal? Good question. Don't worry, I wouldn't let this discussion end with just some superficial explanation. The main point to remember is that with balanced lines one of the audio conductors is positive and the other is negative when referenced to the ground or common line. This is not the case with the noise signals on those two conductors. They will both have the same polarity when referenced to ground or common. Let's follow along with our signals on their journey, starting with the audio signal.

We are dealing with alternating current, so we'll use a sine wave as an example since it's easy to picture. Because the audio signal is an AC sine wave, one-half of the wave is positive and the other half-wave is negative. Flipping each out of phase thus makes the negative half-positive and the positive half-negative. Yet when you recombine them you'll still have the full sine wave. If both of those half-waves measured 0.5 volts, when combined as an unbalanced signal the peak-to-peak level would be 1 volt, or twice the level of each half-wave, resulting in +6 dB of gain. You really didn't double anything; you just added the two half-waves together. The two "balanced" signal conductors are now combined to one (peak-to-peak) conductor, which is now an "unbalanced" signal. But that +6 dBu of gain does appear in your levels, and when not expected it can give you quite a problem, as you'll see, especially when dealing with certain outboard equipment.

Meanwhile, things are very different for the noise signals. We'll start with the false assumption that the noise signals flowing through each of the two signal conductors are identical. We know this is close to impossible, but like sine waves it helps with the illustration. These two noise signals are not positive or negative with respect to each other, since they are identical. Furthermore, each of the noise signals in both of the conductors when referenced to ground or common will both have the same positive or negative polarity at any given point in time. Yet the noises will vary in their positive and negative orientation according to the waveform of the noise and the way in which it is referenced to the ground or common conductor. Therefore, at any given point in time the noise signal in one of the two conductors will be inverted while the other is noninverted.

Stick with me now, because the lightbulb is about to go on. Here's how it works out: When the noise signals are positive, the one connected to the inverting op-amp's negative input will be inverted and made out of phase with the noise signal on the second conductor, which is connected to the noninverting op-amp's positive input and not inverted. When the noise signals are negative, the one connected to the noninverting op-amp's negative input remains negative, while the one connected to the inverting op-amp's positive input is inverted and made positive and is thus out of phase with the noise signal in the first conductor. At the output of the differential stage, the noise signals are combined together right along with the audio signals. However, if the two noise signals in both of the signal conductors are identical, they will *always* end up being out of phase with each other when they are passed through a differential stage. Therefore, they will, in theory, cancel each other out when they are combined.

That's twice the words "in theory" were used. This was no accident, because for this all to work according to plan there must be equal impedance between the two signal-carrying conductors and between each of them and the ground or common conductor for the full length of the cable so that any noise picked up will be equal on both conductors. If not, there will not be 100 percent CMRR. I have yet to see a CMRR specification of 100 percent, although with double-shielded-quality transformers, 90 percent is not uncommon.

Remember that 6 dBu jump in level when going through differential stages? A gain of +6 dBu does appear in the level and can be a major problem when using outboard equipment with both balanced and unbalanced inputs and outputs. But at least you've got a great ally in your meters.

Audio Recording Meters

Recording studio meters have only two seemingly simple functions, to measure steady-state signals and to give the recording engineer an indication of the relative strength of audio signals.

With measurement, because a steady-state signal is used, the meter's response time is unimportant, unless using an overly sensitive meter results in continually changing readings, which can happen at very low frequencies.

The signals most often used in testing are sine waves. These are rectified fairly easily, but that rectification should still encompass the full wave and induce no errors into the waveform's symmetry. The common practice of judging signal quality by giving a meter a quick glance to augment your listening evaluations demands that the meter have the kind of accurate response you can be absolutely certain of, particularly in respect to its overshoot, rise time, and fall time movement ballistics.

Measurements are taken to define the level in question. But how do you measure something that's constantly changing? Audio signals seem to take change to the extreme. If you ever get a chance to look at an oscilloscope display of a violin being played or, even more bizarre looking, a percussionist's key loop being jingle-jangled in front of a microphone, you'll come away with a whole new outlook on dynamics and transient response. You will also quickly come to the conclusion that scrutinizing signal quality by observing its (peak reading) performance on a scope is not the preferred method, at least for humans. "Real time" is just too fast for our eye/brain combination to handle. Some instantaneous-reading meters can give the value in voltage or current every instant so precisely that if that particular instant happens to coincide with a point in time when an AC signal's waveform is alternating polarity and crossing over the zero point, the reading will be zero.

Average (VU) reading meter movements would normally respond only to a DC input. The input force acting on the magnetic field and thus on the coil-needle combination at any instant will be equal to the current flow at that instant. Therefore, without rectification, alternating current, which has an average current of zero, will cause a DC reading meter to have no discernable pointer deflection. What's rectification?

Again using a "forward-biased" diode results in a negative-to-positive electron flow that's almost free of resistance. However, a reverse-biased diode will result in high resistance. So when faced with an alternating current, a diode will alternate between letting the current flow and resisting the current flow, because AC is constantly reversing, or alternating between, forward and reverse biasing. Used with a meter that reads DC, a diode will rectify, or suppress, the reverse-biased half-waves of an AC input and the result will be pulsating DC. Full-wave rectification calls for dual diode usage. In this case *both* sides of the waveform are rectified, which doubles the number of pulses, making for a more stable meter reading.

VU-type meters take the *average* level of all the positive excursions of an AC waveform that have been "rectified," or converted to pulsating DC. This value is then "scaled" to correspond with a required reference value. It's a meter's rectification and scaling factors that cause its readout to be that of *root mean square* (RMS), or average.

RMS is related to power measurement, in that it takes the square root of the sum of many individual voltage readings that have been squared. Although RMS is short for root mean square, we're actually talking about the square root of the mean (or average) square. This measurement takes a number of samples (S) of the waveform (Sa, Sb, Sc, Sd, to Sz), squares each, and then adds the results. This sum is then divided by the number of samples taken so as to give an average (or *mean*) amount or level. And the square root of this average equals the RMS value.

Average-reading meters can provide RMS indication by using simple diode rectification and *scaling* (multiplying the average value by 1.111).

Just in case you ever need to know the difference, here are some conversions:

$$\text{RMS} = 0.707 \times \text{peak}$$

$$\text{RMS} = 1.111 \times \text{the average}$$

$$\text{Peak} = 1.414 \times \text{RMS}$$

$$\text{Peak} = 1.571 \times \text{the average}$$

$$\text{Average} = 0.637 \times \text{peak}$$

$$\text{Average} = 0.901 \times \text{RMS}$$

NOTE: *When the input signal is a square waveform, peak, RMS, and average levels will all be the same. This may be meaningless as far as a meter's actual day-to-day use, but it does provide the easiest method I know of to simultaneously check for the correct calibration setup of almost all the meters found in an audio recording facility.*

The majority of meters are calibrated, or set to "equivalent" RMS sine wave values, but most of the signal detection circuitry involved, such as with *light-emitting diode* (LED) and D'Arsonval (VU) meters, is peak or average sensing. Yet digital meters can be "true" RMS, average, or peak sensing or some combination. The only inherent RMS-sensing meters are electrodynamometers and thermocouple meters, but they are rarely used today.

Or you may be looking at peak-to-peak meters whose reading equals the input waveform's maximum fluctuation into the positive *and* negative. Peak value is the largest amplitude of the signal's waveform every instant.

Peak-to-peak value is the sum of the signal waveform's peak positive and the peak negative amplitude amounts at any given instant. As crazy as it sounds, peak-to-peak levels are twice as high as peak levels!

For a spoken sentence with an RMS reading of 0 dB, an average-reading meter could show a level of −3 dB while a peak meter could read +12 dB. Since studio VU meters are average reading and most peak meters read peak to peak, the difference in the readings between the two could be 15 dB for the same *spoken* sentence. Depending on the signal content, the difference can easily be 10 dB greater than that of speech. Furthermore, even if you take the time to calibrate your VU meters to 0 dB, what will the *actual* level be when the pointer hits the right-hand stop? Is it +7 VU or +17 peak, +10 VU or +20 peak, or would that needle have reached +16 VU or +26 dB peak? If you are not able to listen to the actual *end result* signal (as you can with a three-head analog tape deck) or if the signal is being shipped out of the facility before it reaches the point in the chain where actual audible monitoring takes place, you may be in very deep trouble.

You see, today some meters are still calibrated to indicate "0 dB" with an input level of 0.775 volts. However, most recording studio meters are set up to read 0 dB referenced to 1.23 volts, which is 4 dB above that 0.775 volts. Should you run into a "dual reference" situation, well, it adds up.

Let's start with that spoken sentence and its level difference between average and peak of 15 dB. Now add 10 more dB (bringing the level difference to 25 dB) when the program material has a dynamic range more akin to popular music.

If that 25 dB is referenced to a 0 dB level that equates to +4 dB or 1.23 volts, and its being fed to an input that is referenced to 0 dB equalling 0.775 volts, that input is going to be seeing a level equal to +29 dB. Ouch!

To add to the confusion, just as a DC volt is not the same as an AC volt, a peak AC volt is not the same as an average AC volt, and both differ from RMS AC volts. And as you've seen, levels referenced to dBu and dBm (and for that matter dB"V") are also *not* the same.

NOTE: *Watch out for dBv and dB"V" because they are often interchanged. I've seen "dBv" used to indicate dBu. Some Japanese manufacturers seem to think everyone knows that a 0 dB"V" reading per IEC specifications is referenced to 1 volt, as opposed to the ASA standard 0 dBu reference of 0.775 volts. Although the difference between the two ends up being only about 2.2 dB, it can drive you crazy if you don't understand what's going on when your mixer reading is 0 dB, but the recorder is showing −2.2 dB, or vice versa.*

Reality vs. Theory

All this academic discussion is fine. But now let's take a typical real-world situation, such as in recording a basic session where you're riding herd on a dozen and a half microphone pre-amp levels. You're unaware of the fact that someone has accidentally bumped into a stand and caused a single drum in the kit to be moved in closer to the microphones. If all your inputs are being monitored by peak-reading meters, several might respond to this situation by suddenly reading "in the red." With average- or slower-reacting VU meters, more than likely the one assigned to the closest mic's input will be the only one with an overdriven response.

Let's try and straighten out this whole meter mess by taking a close look at meters, starting with a couple that are no longer in use. As useless as this may seem, you will find the information amusing, and it will help increase your understanding and appreciation of the meters that are more common in our field today.

Electrodynamometer

If you eliminate the magnets from the meter movement and use the magnetic field set up by the applied AC signal fed through a second coil, the current in both of the coils would reverse direction at the same time while turning the moving coil in the same direction. Now, the movement depends on the force, or current (I), setting up the magnetic field and that same force or current (I) acting upon the coil. Thus, the movement equals the current (I) squared. Therefore, the result is a direct RMS reading without any need of AC-to-DC conversion or any scaling. Great! Now we're talking, right? Well, no. You see, even though their frequency limits, which are below those of radio wave frequencies, have no effect on recording usage, these meters have *very* low input impedances. This means they will load down or require a lot of watts from the input signal just to produce a reading. Therefore, they will cause voltage drops in the input signal, meaning our line-level audio signal could drop from that nice +4 dB of 1.23 V to below half a volt. And this drop wouldn't be just on the meter's readout but everywhere! Our line-level signals simply can't handle this kind of load. If you think that meter's weird, check out *thermal meters*.

Thermal Meters

Nicknamed "hot-wire" meters, old-time thermal meters depended on *the linear mechanical expansion of a heated wire*! Still somewhat common today are thermal meters that utilize a filament in a vacuum (such as in a lightbulb). The filament is heated by the applied current, and a thermocouple is used to convert the filament's temperature to voltage. The problem here is that filaments are sensitive to both ambient temperatures and shock. They are also easily burnt out by slight overloads. In other words, do not use a thermal meter on an uncompressed snare drum.

Okay, enough of the joking around. It's time to get back to business and take a look at the real audio workhorse, the VU meter.

The VU Meter

Many people believe the least accurate meter is the D'Arsonval type, such as found in a common VU meter, because its movement is lazy and slow, and the actual maximum levels of the input signal will at best be an educated guess. Here, due to the VU meter's inherent mechanical nature, the pointer will quite often overshoot the actual level so that when the needle "pins" against the right-hand stop, you have no idea what the actual level is.

Even though, as specified, a VU meter's 0 dB reading equates to an input across 600 ohms of 0.775 volts, this should still be verified, because most professional recording meters use a 3,600-ohms resistor placed in series with the meter's input to provide a 0 dB reading referenced to +4 dBu or 1.23 volts. I actually once saw "0 VU = 1.23 volts = 7,500 ohms at 1 kHz" printed right on a VU meter's faceplate.

VU meters with a 0 dB reading referenced +4 dBu are composed of the normal needle movement indicator (which has an internal resistance of 3,900 ohms) in series with a 3,600-ohms resistor and a full-wave rectifier. The variable attenuator used must constantly apply 3,900 ohms of resistance (totaling 7,500 ohms) to the meter so that any changes in attenuation has no effect on the meter's ballistics. If a meter level reading relied only on a source impedance of 600 ohms, a 1-milliwatt input would render a 0 dB reading, but problems would arise when it was connected to bridging interfaces with ratios of 10 to 1 or higher. To retain consistent readings and still allow the gain to be settable, the 3,600-ohms resistor is placed ahead of the attenuator circuit.

This causes a 4 dB loss. Now that 1-milliwatt input will read *negative* 4 dB and you'll need to input +4 dB or 1.23 volts to achieve 0 dB VU. *This* is what a unity level that's referenced to 0 dB equaling +4 dBu is really all about: the changeover from matched 600-ohms inputs and outputs to bridging, and low-output- and high-input-impedance interfacing. Transistors need high input impedances, which makes the meters need added resistance so as not to load down the signal, but this means that we'll need a higher input for a 0 dB reading. *Now* you can see how important input impedance can be. A simple thing like changing from tubes to transistors and their need of a higher input impedance means that all the meters had to be changed!

Is the meter you're reading referenced to 1 milliwatt (0.775 volts), 0 dBm = 0 VU, or is it referenced to +4 dBu or 1.23 volts = 0 VU? Does it seem unreasonable to expect an audio "professional" to know the answer?

Don't worry too much about this, because running into a VU meter without that 3,600-ohms series resistor today in a professional recording facility is about as likely as running into a fully matched system with all inputs and outputs set at 600 ohmimpedance— still possible 20 years ago but extremely rare today. Besides, there's a way to check every meter "in the house" all at once that will reveal *any* discrepancies.

I've never been shy about readjusting a meter's reading as long as I could feed it a verifiable input level. In fact in some situations, especially before the proliferation of peak-reading meters in studios, it was very helpful in avoiding overloads to set up one or two VU meters to have a 0 dB readout that was 12 to 15 dB below the system's maximum allowable level. When you start getting close to 0 VU, look out. They were nice and simple to use as long as they were clearly marked as having this special calibration.

With complex waveforms, which are what we generally deal with in audio, a VU meter will average the overall level. Now, because of its natural movement ballistics and pointer mass and the fact that it cannot accurately read short time-based signals (less than 0.3 seconds in length), transient material, whether outright percussive in nature or part of a complex waveform, might not "sound" long enough to be able to cause the VU meter to display its *real* level. This inaccuracy could easily cause a gain-setting mistake leading to distortion, such as in pre-amp clipping or, in the old days, tape saturation. This is because volume unit levels can be 10 to 15 dB lower than peak levels. This may seem technically inept, but the VU meter is actually quite accurate, considering it was designed specifically with the visual limitations of humans in mind.

Here's a bit of history. Before the VU meter was developed jointly by Bell Labs, NBC, and CBS at the end of the 1930s, there was no standard metering to monitor audio levels. First called a "volume indicator," it used a set standard reference level of 1 milliwatt into 600 ohms or 0 dBm. In May of 1939 the electronics industry adopted it as its standard and the meter was given the name volume unit meter.

The movement itself or its ballistics are factory set, because pointer movement adjustments are made via mechanical means, which includes varying the weight of the pointer's tail and adjusting its horizontal balance, that is, by soldering on added weight. The front (face) panel set screw is only for setting the no-input, or left-hand-most, needle position.

NOTE: *For you to perform this adjustment, the meter must first be removed from its mounting and lain with its back down.*

Electronic calibration adjustment involves simply changing the internal attenuation or external amplification of the input signal so as to obtain the required zero (0 dB) reading.

The actual standard called for the VU meter to be calibrated in volume units that are referenced to a level of 1 milliwatt. The use of 1 milliwatt as the reference level makes everything easily adaptable to the decimal system because this level is related to watts by a factor of 10, and it also meant most measurements read in the positive.

Every VU meter is so alike in its electronic response characteristics that when many are fed the same signal, such as through a mixing desk to a multitrack recorder, it is expected that *all* the meters will read exactly the same. If not, recalibration, repair, or replacement is in order. Their uniformity in ballistic movement, frequency response, and level readout allows them to be used throughout an audio system to verify correct operating levels. Impressed?

This uniformity is no accident. The VU meter, while average-reading, responds almost like an RMS meter (depending on the waveform and the amount of harmonics involved). It's ballistic specifications call for the pointer to overshoot by at least 1 percent but less than 1.5 percent (or 0.15 dB) and to be able to reach the 0 VU mark in a third of a second. The VU meter's frequency response should not deviate from the 1 kHz level by more than 0.2 dB from 35 Hz to 10 kHz and by 0.5 dB from 25 Hz to 16 kHz.

Anyone who's ever witnessed a VU's pointer slamming about will attest to this meter's durability. But did you know that it was specifically designed to withstand an overload of 10 times the 0 VU level for half a second and levels five times the 0 VU level, *continuously*?

If you think that's impressive, check out PPMs.

The Peak Program Meter (PPM)

The most impressive in terms of exacting meter ballistics requirements has to be the peak program meter, or PPM. Formulated and licensed by the British Broadcasting Corporation, it was designed strictly to display transient audio signals accurately, not only in terms of the level, but also to have defined meter movement with regard to overshoot, rise, and fall time.

Because sounds can easily have unsymmetrical (that is, complex) waveforms, causing levels that vary from average readings by as much as 10 dB, the PPM's standards were drawn up so as to define attack (rise time) and decay (fall time) characteristics with logarithmic scaling and to show both positive and negative peak detection levels (that is, peak to peak). All the specifications are set to international (approved by the IBA, EBU, and BPO) "Peak Program Meter" standards, as licensed by the BBC.

Because of this adherence to standards, all peak program meters, like VU meters, will give *identical* readings and movement, regardless of meter or faceplate sizing. Here, even the fall time is matched so closely that dual meters can be used without worrying about calibrating the two to each other. This is critical with PPMs because they have their own unique "quasi-logarithmic" scaling "law," which was "designated as the optimum" for displaying audio signals with a wide dynamic range.

The specifications call for "quasi-peak" rectification characteristics so as to have correlation between the meter's reading and "perceived" distortion. The faceplate scale markings are completely unadorned (no numbers, just slash marks) to give an "uncluttered appearance" to help us humans make quick and accurate evaluations even with just a glance.

All the specification requirements make calibrating these guys no joke. I know because I've done it, and the proper calibration procedure will run about two hours *per meter* with 20-turn potentiometer presets for "zero level," full-scale reading, and overall input gain, and a single-turn pot for the exact setting of the lowest level, just for starters. The specifications also call for an isolated 10-millisecond burst of a 5 kHz signal with a level set 14 dB *below* the lowest mark to still cause a "perceptible deflection" of the pointer.

The faceplate markings at positions 2, 4, and 6 have to be within (±)0.2 dB of the spec, marks at 3 and 5 within (±)0.3 dB, and marks at 1 and 7 within (±)0.5 dB.

Frequency response seems simple enough at (±)0.3 dB from 30 Hz to 20 kHz, but further reading calls for a much more exacting, 10-Hz point being down no more than 2 dB, and 40 kHz to be down no more than 1 dB. Full-scale decay time is to be between 2.5 and 3 seconds.

Rise time requirements allow for

A tenth of a second or one-hundredth of a second burst of 5 kHz to be off by only (±)0.5 dB.

A two-hundredth of a second burst of 5 kHz tone has to be within (±)0.75 dB of the spec.

A two-thirds of a millisecond burst of 5 kHz signal can deviate from the norm by a whole decibel, and, believe it or not, a one five-thousands of a second burst of 5 kHz is allowed to be off by a full 2 dB.

Want more?
According to my notes, PPM standards ballistics calibration called for a steady-state 5 kHz signal to be fed into the meter and then raising the input level until the pointer reached mark 6. This same level was then fed to the meter in short-duration bursts. With a 100-millisecond

burst the meter was expected to reach that mark 6 level with a maximum variation of (±)0.5 dB. Does that seem rigid? How about a 10-millisecond burst of 5 kHz varying from the steady-state mark 6 level by only −2.5 dB (±)0.5 dB. A 5-millisecond, that's right, a five one thousandths of a second burst of 5 kHz signal is allowed be down from the steady-state level by a maximum of −4 dB (±)0.75 dB. A 1.5-millisecond (0.0015 seconds!) burst of 5 kHz tone should reach a level that's 9 dB (±)1 dB (8 to 10 dB) below that of the steady-state 5 kHz tone that was adjusted to a level to reach mark number 6 on the meter's face! There was also a standard-level check using a DC input of varying time duration. Thankfully, I did not have to get involved with that.

The facing's center position is at mark 4, and a level corresponds to a 0.775-volt RMS sign wave, but internal gain adjustments along with internal component change can increase the meter's sensitivity by as much as 20 to 40 dB. "Although with 40 dB of gain there'll be some loss in high-frequency response." Finally, fall time from a full scale reading of mark 7 is to take at least 2.5 seconds but not more than 3.2 seconds.

You may have detected a bit of sarcasm in the foregoing description, but the meter I had to calibrate was dual concentric, meaning it had two pointers within the same meter, so two exactly matching calibration routines had to be performed!

Everything Has Its Limitations

While quite observable, a single PPM is still a little harder on humans than a VU meter. Don't get me wrong; in that British custom-made six-channel mobile broadcast mixer I was calibrating, they performed outstandingly, but there's no way I would want to try to scrutinize 24 of them at the same time.

While both VU and PPM meter types are more than adequate as far as giving ample warning of overload situations, they were not designed to take any variations in a system's abilities to handle full frequency bandwidth into consideration. Therefore, in situations where higher levels at particular frequencies present a problem, an audio spectrum analyzer with a fast response time is unbeatable for eyeballing the complete audio content along with the levels of individual frequency bands.

NOTE: *Slow-responding spectrum analysis is used for room equalization work.*

Spectrum analyzers don't seem to be used as often in today's CD-dominated digital world as in the days of vinyl cutting, where any excesses in either the level or phase shifting at low frequencies could be disastrous (usually calling for a "do-over" remix). In those days a spectrum analyzer and a phase meter or oscilloscope were standard additions to most of our "recording engineer tools" when mixing.

Thankfully, we do not have to make the choice between accuracy and readability because there are meters that display both average or VU and peak levels. There are standard (slow-response) VU meters that include an LED that gives peak indication. If the threshold of this

LED's turn-on point can be calibrated, it becomes a viable, highly useful, and therefore beneficial tool. However, without the ability to calibrate or set its level reading, any LED type of peak-reading meter can end up being no more useful than a light show. It may impress the clients, but it's not going to be something an audio recordist can bank on.

Your comprehension of the capabilities of the different meters currently available will be aided by going over a few real-life examples that were common in recording studios during the 1980s.

Two Meters in One

Many of TEAC/TASCAM'S products utilize standard VU meters with a single, red, peak-reading LED mounted in the upper-right section of its faceplate. The level at which this LED illuminates is internally settable and thus provides the user with the benefits of both the slow, easy-to-follow movement of a VU meter and some of the easily noticed "Look out!" warning capability of a peak meter.

Harrison Systems, a recording console manufacturer, came out with a very effective "bar graph" meter during the early 1980s made up of 12 LEDs. The lower eight covered the range of -20 to -1 dB VU (-20, -10, -7, -5, -3, -2, and -1) and were yellow in color, while the four top LEDs (0, +1, +2, and +3 dB VU) were dressed in red. The actual bar graph effect was caused by a plastic light diffuser strip mounted in front of the LEDs. This gave the meter readout a more continuous look that was readily visible over a wide angle. However, the real big deal with these meters was that the whole bar would illuminate and the top four red LEDs would go to twice the normal intensity mode for 50 milliseconds when the input exceeded a preset peak level. That level could be set from +4 to +24 dBm, on a per-meter basis. You could not help but notice that warning sign, even while simultaneously monitoring 36 meters. As with most LED meters, a "peak hold" circuitry enhancement caused the highest LED illuminated to remain lit for a short time after the level decreased, to aid our slow human eye/brain response time.

Trident, the British console manufacturer, along with another company called A & R, took this dual-usage metering concept a step further. They also used LEDs (14) in their "Peak Level Meter," but this one could be switched between PPM and VU indication. It used green LEDs for normal levels from -30 to 0 dB and red for +2 to +8 dB (a dynamic range of 38 dB). Furthermore, 16 of these meters fit into a self-powered, three-space-high (5.25-inch) rack-mountable casing. The LED column's brightness was adjustable via an easily accessible rotary pot mounted on the faceplate's right-hand side.

This setup completely eliminated the problem of trying to scan across a console-long array of meters, plus it gave the choice of either easier-to-monitor VU response for general audio signal monitoring or switching an individual channel that required closer scrutiny to PPM response. This was *very* handy.

The next step for viewing multiple meters at the same time came with using a video monitor for the display. NTP, a manufacturer out of Copenhagen, was one of the first to take advantage of this method. This type of metering allows for total electronic control over what's seen. Since the scale is electronically generated, not only is it very accurate, but it can be switched for readings that conform to *any* standard. Using a color monitor gives additional visual assistance. This device could display up to 36 channels of metering at once. These were divided into groups of four, with the individual bars being background illuminated. Upon overload, the color of the full bar changed to red. Each individual channel could also be changed in color to indicate those channels that were being recorded, soloed, or recorded over (overdubbed) or in input mode. All this information could be gained by simply glancing at a 25-inch screen. I've used this type of display in a classroom situation, and even those 20 feet back had no problem gleaning every bit of information this display made available.

MCI made 101-segment plasma (neon gas) displays an option on their 500 series consoles back in the late 1970s. This metering system was capable of displaying levels in PPM or VU standards, although each channel was not individually switchable. But it was the brightness and 101-segment resolution that made them useful for presenting fader levels and movement, such as in (nonmotorized fader) VCA automation.

You see, 101-segment plasma displays offered the best-looking resolution of all the segment-type displays. Unlike the jumpiness or limited dynamic range with some LED displays, these meters offered a dynamic range of 60 dB and smooth movement and had the benefits of brightness and accuracy and gave immediately recognizable level warnings. Such meters also became available from a Danish company called NTP and later from the console manufacturer SSL.

Yet the real tour de force was MCI's, and later SSL's, spectacular light show across the full meter bridge of 31 bands of audio spectrum analysis, in which each channel's meter handled an individual band of the audio frequency spectrum. I'm talking clients in awe, showtime here. But its usefulness during recording and mixdown cannot be overstated.

Dorrough Electronics makes another very popular dual-reading meter especially favored in the field of broadcast. These have LED arrays that are curved much like the travel arch of the common VU meter. Therefore, they more readily give the appearance of VU meter ballistics. This meter puts the VU reading, which they term *persistence range*, more to the left side of the LED scale, with a peak reading more toward the right. If you think about it, this makes perfect sense because at lower levels the VU's easygoing average response has a nice, calming look about it, while the response of a PPM is all about a "Hey, look out!" warning. Dual 40-LED sets of different-colored LEDs and a single peak-hold (lingering) LED help in readily discerning between the two simultaneous readouts so that getting used to judging signals with this meter configuration takes little to no time. It's hard to beat these meters for monitoring dual channels (which probably accounts for their popularity in broadcast).

But although they do make a straight-line (uncurved), narrower version so that it could be possible to view many of these (one for each channel) on a recording console, multiple VU meters would be both easier on the eye and certainly less expensive. But as their popularity in broadcast proves, they are great level-monitoring tools for a stereo bus.

The Bottom Line?

Hey, the answer for me was easy. Most studios are full of VU meters, so I'd purchased a set of peak-reading meters and added patch bay connectors to them so I could easily scrutinize any signal I thought might require it. This was important because while I personally like D'Arsonval VU meter movement, standard PPM meters have storage circuitry that holds the energy level of complex waveforms or transient inputs so that they have the time to react and display a waveform's true peak levels. With most LED meters the PPM movement characteristics are not simply mimicked but enhanced by this "peak hold" circuitry, which causes the highest illumination to remain lit for a short but extended period of time. You see, while oscilloscopes give an instantaneous indication of peak levels due to the rapidness of these signals' appearances and disappearances, they can be difficult to read because the decay time needs to be slowed down so as to give the human eye/brain combination the time to perceive these changes.

Even with a slower decay rate it is still difficult to notice peaks that are approaching distortion levels when scanning across 36 meters' worth of level information, no matter what type of meters they are, unless their display changes color when it reaches "zero" level.

One argument in favor of average or VU metering is that using peak-reading meters can cause an engineer to be overly cautious or "gun-shy" in setting recording levels, to the point of possibly trading a lesser chance of distortion for an increase in the S/N ratio. Yet PPM meters are very helpful for individual channel referencing via patching and are great for mixdown duty. Let's face it; whether you're mixing to a lacquer-cutting lathe, an analog tape recorder, or the latest digital converter, they all only understand peak levels, as far as their input limitations or distortion are concerned. Unfortunately, the typical PPM reading meter, be it of plasma, pointer, or LED indication, not only is fairly expensive, but provides only peak-reading response.

So what meters did I choose, and why? Well, the key factor for me wasn't just technical superiority; like everyone else I had to look for cost effectiveness. With a little research I came up with a state-side company called Dixon whose multisegment (29-LED) bar graph meters have the capability to be switched between PPM and VU and cost about one-third what their plasma or neon counterparts cost. The trade-offs? Well, those were no big deal at all. Prior to any recalibration, the meter's range is only from −50 dB VU to +5 dB VU, with zero VU calibrated to 1.23 volts (or +4 dB above 0.775 volts). Calibration is performed by adjusting two multiturn trimpots that allow the upper and lower limits to be set to almost any "zero" meter reading. The standard configuration is green below 0 VU, yellow at 0 VU, and red above 0 VU.

The meter's frequency response is within 0.5 dB from 20 Hz to 16 kHz and within 1 dB from 20 Hz to 20 kHz. Display resolution is far from the infinite of the VU or cathode ray tube, and way lower than that of a 101-segment plasma display, yet its 29 segments do the job while showing indications at −50, −47, −44, −41, −38, −35, −32, −29, −26, −23, −20, −18, −16, −14, −12, −10, −8, −6, −5, −4, −3, −2, −1, 0, +1, +2, +3, +4, and +5

Because its solid-state electronic design requires no moving elements, this meter is not subjected to any physical ballistics limitations, so overshoot is nil. The response time varies, as it should, between VU and PPM display, taking 300 (±)30 milliseconds of time to reach 99 percent of full scale in VU mode or a 10-millisecond attack with a 1.5-second decay time in the PPM mode.

Those are fine meters, but I still wasn't satisfied. I kept thinking about the dual concentric capability of that British PPM meter I once calibrated, because "clocking" a stereo signal on that single meter had been such a revelation. A few years later, when a technician friend decided to sell his Leader LMV-185A dual-channel concentric-pointer AC RMS-reading millivolt meter, I jumped all over the chance. The meter's 4-inch-wide face has scales that read 0 dB at 0.775 volts (0 dBm) and voltage scales that read 1 or 3 volts. The range or multiplier of these scales is from −60 dB to +50 dB or 1 millivolts to 300 volts. Want to check the relationship between input and output levels instantly? This is the meter to do it with.

When I taught a graduate class in basic audio maintenance, over a dozen students not only could easily see, but they automatically understood any point I was trying to make concerning different levels, types of displays, or the effects different types of processing had on signal levels.

Now, I know that many compressors, for instance, provide both input and output meter indication on their front panels, but scrutinizing those same levels on a dual concentric RMS meter while also having the ability to switch between VU and PPM on a multisegment display renders a whole world of information that's just not possible with two rows of several LEDs. But let's be fair. How many units of *any* effects device could a manufacturer expect to sell with $300 worth of metering added to the price tag? Still, just one look is convincing enough so that anyone would agree with the viability of bringing a meter setup like this to every session.

It should be obvious that having peak and VU combination meters (such as the Dixon example) with their capability to instantaneously switch between peak and VU readouts gives an engineer an undeniable advantage in controlling gain stage setting and achieving optimum levels. A benefit that isn't as obvious is that it provides the engineer with a standard by which all other meter readings can be verified. How so? Well, you know that 0.775 volts is equal to 0 dB when referenced to 1 milliwatt into 600 ohms. Peak = 1.571 × the average, and therefore average is peak ÷ 1.571, which means that peak should read 1.27525 dB. Please do not take this seriously. I'm not suggesting you're trying to set up an input signal for a 1.27525-dB reading. Hey, don't worry, I wouldn't have led you on unless I was able to say that a square wave signal (which can be easily had from almost any synthesizer) will produce equal level readings on both VU (average) and peak meters. With the type of combination metering we've been discussing, all you have to do after deciding on your unity gain or operating level (the 0 dB point) is to connect a square wave signal at that level into the meter and set both meter readings for 0 dB. Now that meter becomes the standard by which you can compare all other meter readings. Not that important? Well, try sending the same square wave signal to several peak and VU meters in your facility and see how many give you the same reading.

I have to admit that I also used my meters for technical repair and calibration work, but I still considered them to be valuable and essential recording tools as well. Hey, would a fine cabinet maker show up at the job site with a rusty hacksaw?

Audio Gain Stages and the Weak Link

Whenever audio equipment is interconnected it forms a signal path, or chain. Just like any other chain, an audio chain is only as good (clean) as its weakest link. That weak link could be one of the pieces of equipment, the cabling, or connections between them or the individual who is setting up or adjusting the gain stages of each piece of equipment. Yeah, *you* could easily be the weakest link!

Gain stage settings are generally performed at each device in a chain by adjusting that unit's input and output level controls. Sometimes it calls for adding balanced-to-unbalanced interfacing devices between them and occasionally incorporating resistor pads (such as in-line adaptors) to provide needed attenuation. Proper gain stage setup is required to achieve maximum S/N ratio, minimum distortion, and an optimum amount of headroom. The S/N ratio is the difference between the noise floor and the average signal level. Headroom equals the difference between the average level and the onset of clipping.

Although the unit that has the worst noise specifications will affect the chain's S/N ratio the most, because it increases the level of noise, any equipment operated at too low a level will also adversely affect the S/N ratio, because these operating levels will bring the average level closer to the noise floor. Meanwhile, the device that has the lowest onset of clipping or is the easiest to overload will have the greatest affect on the chain's headroom. To reduce the amount of accumulated noise, the unit with the lowest noise specifications should be used to achieve the most gain, as long as it has the headroom to handle the peak levels.

No matter what, there will always be some noise. But while recording, conscientious engineers do whatever it takes to keep the level of the program material well above the noise floor. Now, any background noise from air circulation rumble, nearby activities (jackhammers, transportation), light fixtures, and other sources will be lower in level than the program material, to the point where the average level of the program material when reasonably loud will easily "mask" the noise "floor" content of the signal. Obviously, this starts with using the best microphone placement for the particular situation. Here, using a mic with a narrow polar (or pickup) pattern will help, as does increasing its proximity to the sound source. Attaching pickups or a miniature (lavaliere-type) mic directly to a music instrument might be necessary. If there isn't an audience, you might get away with using baffles, or a move to an isolated area may be called for. This all may take time and be a lot of work, but if it results in a cleanly recorded signal it's worth it.

In fact most of the time, background noise is not as great a concern as (clipping) distortion. However, if during quiet passages the background noise level is close to that of the average signal level, there will be no place for you to hide and there's no "fix" that will do the "trick."

Adjusting Levels Is More Than Positioning Faders

It is very important to preserve a signal's best possible S/N ratio as it passes through the system. This takes constant vigilance. A signal's susceptibility to additional noise degradation increases whenever you drop its level below unity gain or line level. This is because the next stage the signal passes through will always add some noise, even if it's 150 dB down from the average level. Remember, we're dealing with a ratio here, so if the level of the signal is set at +4 dB the S/N ratio of that stage will be down 154 dB, but if the signal's level is dropped −10 dB that S/N ratio becomes 140 dB. No big deal, but again once when noise becomes part of the signal, there's little you can do to get it out.

Increasing a signal's level before it passes through a noisy stage lessens the amount of gain required from that stage. Here, if a unit's noise is proportional to the amount of amplification required from it, noise might no longer be a factor. This is what's meant by saying that "increasing the signal level decreases the S/N ratio, while decreasing the signal level increase the S/N ratio."

But how do you know where this point is? You increase the gain control of that stage with no input connected and monitor its output at a higher-than-normal listening level. You'll hear when the noise becomes prohibitive. Manuals and manufacturer's specifications can't possibly indicate the history or condition of every particular piece of equipment, which is why it's a good idea to check out older signal-processing equipment this way.

Most of the equipment a recording engineer uses today is flexible enough in level-setting allowances for the gains of each piece of equipment to be set so that they *all* arrive at their clipping points with the same input level. This provides the maximum amount of headroom and at the same time minimizes the amount of distortion. This is not as simple as it sounds, because we are talking about audio, the land of constant trade-offs.

In order to comprehend this fact you must first understand that since the S/N ratio (in decibels) is the difference between the background noise "floor" and the average level, it will certainly have some bearing on the signal's dynamic range, though not as much as headroom does. The S/N ratio dictates how low the average can be set, while headroom dictates how high the average level can be set. The ideal would be to set the average level for maximum headroom and the greatest S/N ratio. This seldom occurs because there is a trade-off between the two.

The Trade-Off Between Headroom and S/N Ratio

The S/N ratio affects the usable dynamic range below the average level and headroom affects the usable dynamic range above the average level. With low headroom capabilities, the average level must be decreased to avoid the possibility of distortion. This puts the signal closer to the noise floor. With a high noise floor, the average level must be increased to avoid the possibility of hearing the noise. This puts the signal closer to the clipping level.

On the other hand, before passing through a noisy or distortion-prone stage, consider the following: In order to increase the S/N ratio, the average level should be increased. This lowers the amount of available headroom because now the average level is closer to the clipping level. You can increase the safety margin against clipping (more headroom) by lowering the average level. This moves it closer to the noise floor, increasing the S/N ratio.

Any way it's stated, the choice is easy. It is wise to be safe and prepare for those high-level peaks often found in music program material, even if they occur only once in a while. As every audio engineer learns very quickly, average sound levels do not come close to presenting the whole picture as far as signal dynamics are concerned. However, anything can be overdone, even caution, especially if it causes the background noise to become perceivable during a quiet passage.

So how can a recording engineer know what's going on between all the outboard gear, the mixer, and the recorders in a studio? By checking the "calibration" of the room. No, I do not expect you to calibrate the equipment. The calibration of a control room is performed by a technician to ensure that all the elements that make up the recording system have matching reference levels. But, as you'll see, a single square wave signal can be used to check if *any* meter in the control room is reading incorrectly.

Technicians are required to have a pure-tone generator to inject a signal with a specific waveform into the recording system at a defined level. If recordists are really serious about making quality recordings, they should at least be able to set up an in-house (facility-owned) synthesizer for a square wave output at unity gain if they don't own a test CD as part of their arsenal of recording tools.

NOTE: *Due to the 20 kHz bandwidth limitations of compact discs, square waves at frequencies above 1000 Hz will not be shaped symmetrically.*

This means that a tech would have to augment CD use with a bench-style function generator. But for recording engineers just trying to confirm that all meters they're about to rely on are speaking the same language in terms of reference level, a test CD is still about as handy as it gets. I use three such CDs: *Sound Check* by Alan Parsons and Stephen Court, the *National Association of Broadcasters* (NAB) Broadcast and Audio System Test CD, and the *Studio Reference Disc* (SRD) from Prosonus. Their usefulness is remarkable.

After using a calibrated multimeter to confirm the proper line output of the CD player being used, a technician could actually utilize these CDs' tones for calibration purposes. I know this doesn't really concern the recording engineer all that much. But hold on, because we haven't even scratched the surface in terms of all the different types of signals they provided. The NAB disc was in fact designed to be a tool for testing and has typical parameters, like frequency response and distortion checks, but it also includes a dual-phase linearity test tone that can be used to check individual phase connections in crossover/monitor setups. It also provides a set of PPM, VU, and even peak LED "flash" indicator calibration tones. The Prosonus disc adds several control room listening evaluation checks and instrument tuning pitches of not only A-440, but also 442 and 444, along with a full octave of both piano and sawtooth tone pitch reference notes. The Sound Check disc is my favorite. Since it was put together by two recording engineers it has some tracks that are uniquely essential to the audio recordist.

How about separate bands of third-octave pink-noise tones so that you can adjust a 31-band graphic EQ one band at a time? Five minutes each of SMTE time code at EBU (25 fps), nondrop (30 fps), and drop (29 fps) frames are provided, just in case you're in a situation where a time code generator is not available for striping purposes. The feature I use most is the selection of music instrument tracks recorded utilizing no equalization, reverb, or any other effects. These are great for checking out outboard effects. Using a track that already has reverb on it to set up a reverb device makes no sense. What are you going to do instead, pull down a client's tape and start running it without permission? Not I, sir, not I. These three compact discs have a total of over 250 tracks of information among them. You don't really need three, but at least one should be in every recording engineer's arsenal.

Using a CD or an in-house synth to check out a control room's meters still requires your setting up the output level to the proper reference level. This means using a calibrated multimeter or comparing it against an audio level meter you know to be calibrated correctly (preferably your own). The latter is a lot easier to deal with than trying to set up an output level to 0.775 volts (for 0 dBm operation), 1.23 volts (for +4 dBu operation), or even 1 volt (for 0 dB"V" IEC operation).

Instead of trying to set the level to a point in between an analog multimeter's markings or to the constantly fluctuating numbers of a meter with a digital readout, all you'll have to do is raise the level until it reaches your meter's 0 dB mark. The tone can then easily be dropped to −10 dB when starting a signal path checkout with inputs designed to handle the lower line levels associated with music instruments. For starting with microphone input levels I'd suggest using Shure Model No. A15TG pocket tone generator. It not only is small, but has a built-in XLR connection, and when the internal button (watch-type) battery eventually starts to weaken, the *frequency* of the tone is affected long before any level change occurs in its 10 mV (0.01 V), −40 dB output. So, matching your signal level output to this device puts you right at mic output level.

NOTE: *Once you get your level set up, leave it alone until you've adjusted (calibrated) or marked the meters with a grease pencil or at least notated any discrepancies in your journal.*

Congratulations, now you at least know where all the equipment you're using is at, level wise. Feels good to be confident, right?

Okay, you've taken the time to check out the control room and made notes in your journal. You know the system's capabilities and where you need to take extra precautions, and you certainly won't be fooled by any inaccurately calibrated meters. You've done all your homework, and then during the session, all of a sudden you get broadsided by a sneak attack. Remember that 6 dB jump in level obtained by passing an unbalanced signal through a differential stage? That +6 dB of gain will appear in your level and can cause you a major problem by decreasing your available headroom. But before we get into the upcoming quagmire, it's best to reiterate. Along comes this audio signal down an unbalanced two-conductor cable (carrying both "high" and "ground," or low). It's usually around −10 dB in level, but let's make its level 1 volt for the sake of easy mathematics. Remember, it's alternating current, meaning that at any point in time the conductors can be positive or negative in polarity.

In Figure 1-14 the signal is connected to a differential input. In this case the ground (or low) conductor is passed directly to the output of the differential gain stage (also see "non-floating" ground connection in Figure 1-15D). Meanwhile, the high, or hot, signal conductor is shipped to two op-amps. One is set up as an inverting amplifier, while the other is noninverting. As you can see from this figure, it doesn't matter if the signal is positive or negative (at any particular point in time) because it will be going through op-amps set up for inverted *and* noninverted operation. Between the two, they will always put out both the original signal and its mirror image polarity-wise no matter what the input polarity is.

One thing must be cleared up. Although the signal is passing through an op-amp, it is not necessarily being amplified. But again, just to make things easy, we'll say these two op-amps were set up design-wise to make up for any attenuation the signal would have been subjected

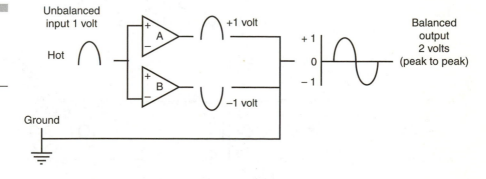

Figure 1-14
A simple dual op-amp differential stage

Figure 1-15A
Questionable
unbalanced-to-
balanced
wiring scheme

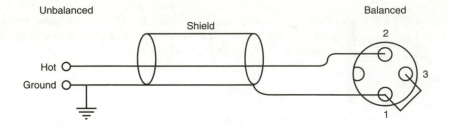

Figure 1-15B
A better
"wiring-only"
method of
balanced-to-
unbalanced
interfacing

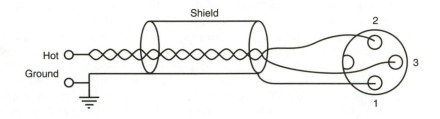

Figure 1-15C
Examples of
"floating
grounds," such
as in MIDI
wiring (top),
isolation
transformers
(middle), and
direct box
"ground lift"
switching
(bottom)

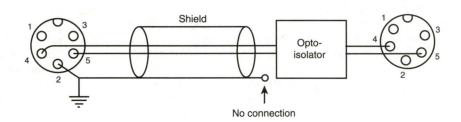

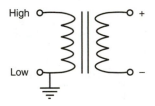

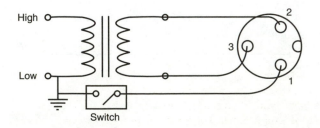

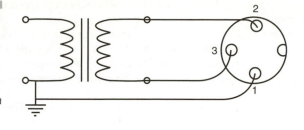

Figure 1-15D
A two-way street-transformer-type balanced-to-unbalanced interface with a ground that is nonfloating, but connected across the interface

to between the device's input connectors and its reaching the input of these op-amps. Thus, the output of each of the two op-amps will be a level that's the same as the input, 1 volt. When the two op-amp outputs, $+1$ volt and -1 volt, are combined they will produce a peak-to-peak voltage of 2 volt because the difference between -1 volt and $+1$ volt is 2 volts.

So a signal with a level of 1 volt that is passed through a differential stage has its level increased to 2 volts. This level is now twice as much, or double the amount of voltage it was, which equates to 6 dB of voltage gain and a 6 dB improvement in the S/N ratio. "How can this be, $-1 + 1 = 0$, doesn't it?" This is true for those noise signals that are identical but not for a constantly changing audio signal, which is why differential stages only have "common"-mode rejection.

This all sounds like a great deal, but if the unbalanced signal at the input was -10 dBu and the differential output stage of the device is set up for $+4$ dBu-operation, you would end up with 14 dB of gain! "Nothing to be alarmed about," you say. Well, -10 dBu (as referenced to 0 dB = 0.775 volts) is 0.316 volts, and $+4$ dBu (with the same reference) will be 1.23 volts.

I know that's only a little over five times the original level. However, if you are not expecting the 14 dB loss in headroom, when the signal increases to a nice refined 0 dB during a fortissimo (very loud passage) it will be $+24$ dB at the differential output stage. Now, when that signal reaches a gain stage further down the audio chain that was set up for an amplification of only 10 dB, the $+34$ dB level might just happen to cause a little distortion. And this is not even the quagmire I was alluding to!

To completely understand the real problem involved and how it all came about will take a bit of historical background. You're in luck because you'll be hearing it from someone who was not only a witness but was in the thick of it. During the early 1980s I was handling the technical duties for a handful of single-room studios, engineering on recording, live music, and theater projects, as well as doing a lot of overdub work, substituting for several engineers who were overbooked. I had been recruited by both the Institute of Audio Research to teach part time (they prefer their instructors to be actually involved in the field) and the chief editor of *International Musician and Recording World* magazine to fill the position of technical editor, doing equipment reviews along with writing tutorial articles. I was also involved in what was called sales support for a major east coast Pro-audio dealer. This translates to giving technological explanations; taking care of the interconnecting, calibration, repair, the procurement of manufacturer's support and parts; and anything else it took to make a bad situation smooth and happy. My nickname was "The Torpedo," and they'd fly me a couple of thousand miles away to places where a recently purchased professional audio equipment package costing a million and a half U.S. dollars was not operational. Yeah, I was very busy, and no, I do not miss it.

At this time musicians were purchasing a steadily increasing amount of equipment that had previously been reserved for studio use. They wanted to have the sounds and effects they heard on records or had gotten in the studios available to them while onstage. This is called "demand," and to the marketplace it meant big money. So manufacturers responded with a slew of products.

The retailers were delighted, but started reporting back through their distributors to the manufacturers that many of their customers did not understand what the three conductor XLR connectors were about. Musicians used ¼-inch connector guitar cords on stage and with their PortaStudios, unless they were hooked up through RCA-type connector patch bays.

You see, home studios, which before this time had been extremely rare, were becoming more prevalent. This was due to the innovations in recording that really began to take off when Tascam's "PortaStudio" cassette-based multitrack/mixer combination made musician's recording "notebooks" a reality. This was a high-impedance, unbalanced world into which these new "cost-effective" recording processors strode. MI (music instrument) was MI, semi-pro was semipro, and professional was professional Everything was great, for about a minute or two.

However, we're talking about a living, breathing competitive marketplace that says that if you, Mr. Manufacturer, incorporate the correct impedance matching (here meaning unbalanced-to-balanced electronic or transformer-isolated differential stages) into the effects processors you are trying to market, you will either not be able to recoup your expenses or not be able to compete with the price point of your competitor who offers the impedance matching as a separate device.

So manufacturing management told the design department to drop the differential input and output stages. This reduces the cost, which translated to a lower price point, making for a more competitive product, which increases sales and profit. So it was off to the races, because now demand totally ruled.

Purchasers don't like it when they have to pay extra for a box containing a set of unbalanced-to-balanced networks that cost about $100, and they don't want to have to change all their connectors to XLR, which don't work with their music instruments onstage, and they want the same piece of gear to work for them in both live and recording situations. The answer? Manufacturers came out with devices that could be used "in every situation." They utilized ¼-inch three-conductor jacks like those used for stereo headphones. These connectors were internally wired so as to be able to handle unbalanced operation via the use of ¼-inch two-conductor connectors (where the low side or ring of the three-conductor connection gets shorted to the ground) as well as balanced operation by using three-conductor ¼-inch "stereo" connectors. Now, plugging in a ¼-inch mono guitar cord automatically provides a −10-dB unbalanced operation, while using a ¼-inch stereo headphone cable automatically results in +4-dB balanced operation. An elegant solution?

How about the products that have both XLR and ¼-inch unbalanced connections on both inputs and outputs to give "everyone the capability to handle any situation." And as if that were not enough, while doing calibration work in a broadcast facility (I also was doing work for a subsidiary of the old RKO Radio Network at this time) I ran across a device that had not only all of the above connectors, but also rear panel switches that set the operational levels at either −10 dB or +4 dB.

When using outboard equipment that has both balanced and unbalanced inputs and outputs, things can get a little bizarre, especially when they utilize ¼-inch stereo jacks for both balanced and unbalanced operation or have unseen rear-panel-mounted switches with the capability to change the unit's 0 dB reference levels.

The Audio Chain from Hell

How's this for an audio chain from hell?

A. The initial device in the chain has balanced inputs and outputs that are referenced to +4 dBu or 1.23 volts, equaling 0 VU level on its front panel metering.

B. The next unit is set up to handle unbalanced inputs of −10 dB with 0 dB referenced to 1 V (dB"V"), and it provides both unbalanced and balanced outputs with the levels of −10 dB (0.316 volts) and 0 dB"V" (1 volt), respectively.

C. The third processor in the chain offers the choice of balanced (+4 dBu) or unbalanced (−10 dBu, 0.245 volt) operation at both the input and output stages, selected by the use of either 2-conductor or 3-conductor ¹⁄₄-inch connectors.

D. The final device in the chain has a choice of balanced or unbalanced operation for both inputs and outputs. It uses ¹⁄₄-inch jacks for unbalanced operation and XLRs for balanced operation. Level-selection switches (−10 dBu or +4 dBu) are provided on the rear panel.

"What's this madness all about?" you ask, "Aren't there supposed to be standards?" Well yeah, of course there are:

The first (A) unit's standard calls for balanced input and output professional operation, with 0 dB referenced to +4 dBu, making it a ProAudio piece of gear. Prior to the early 1980s, when life was fairly uncomplicated, this was the predominant interface in professional audio facilities.

In answer to their buying power, the next manufacturer has aimed its product (B) at the recording musician. Therefore, its inputs had to interface with the standard *music instrument* (MI) level −10 dB"V" unbalanced operation. In order for the output of this device to interface with this same standard (such as when used live) and to have balanced 0 dB operation (when used with professional recording equipment), an interface to balanced operating levels of 0 dB"V" was added at the output.

The third is an example of a manufacturer's trying to make his piece of gear (C) capable of covering all the interface situations a musician might encounter either onstage or in the studio. Therefore, they've decided not to go with any single standard, but to offer the musician every possible combination of balanced and unbalanced operation and MI and professional reference levels. It utilizes a wonderfully simplistic interconnection scheme using cabling familiar to musicians. They have the choice of readily available two-conductor ¹⁄₄-inch guitar-type cables for use with −10 dBu level, unbalanced (MI, live, and "semipro") operation or three-conductor ¹⁄₄-inch headphone-type cabling for +4 dBu balanced, professional-level operation. The insertion of the connector automatically selects the interface used.

The final device in the chain (D) is a piece of gear that is trying to be all things to all people on a slightly more professional level. This manufacturer decided to fulfill every need the user might have by providing balanced and unbalanced operation on all inputs and outputs. To add to the convenience, both ¹⁄₄-inch connectors for unbalanced operation and XLR connections for balanced operation are available at *each* connection. Additionally, to cover any possible situation, four rear-panel switches are furnished so that −10 or +4 dBu operating levels can be individually selected for each input and output!

As you've probably already surmised, there's a chance that there might be a problem with this scenario:

- Your audio levels might not be high enough to drive the next stage.
- Your audio levels might be high enough to distort the next stage.
- Your audio levels could jump about like crazy as the signal passes from one piece of gear to the next.
- Your audio might have its S/N ratio seriously degraded because of the attenuation needed for it to interface with the next piece of gear in the chain.
- Your audio levels could end up with severely diminished headroom because of the amplification required for it to be able to interface with the next stage in the chain.

However, I'm not worried about any of these things, because they're *your* audio levels. And I know you'll be sitting behind that mixing desk patching in and out of each device, bringing those outputs up to a spare meter to find out exactly where any problem might be coming from, as you check the results against a chart much like the one in Table 1-2, which you compiled when you checked out the calibration of the studio long before the start of the session.

Always remember that a level referenced to 0 dB"V" (0 dB = 1 V) is 2.2 dB higher than a level referenced to 0 dBu (0 dB = 0.775 volts). Therefore, to compare voltage levels referenced to 0 dBu and 0 dB"V":

$$dBu = dB \text{ "V"} - 2.2 \, dB \quad \text{and} \quad dB\text{"V"} = dBu + 2.2 \, dB$$

Table 1-2
Input and Output Levels of the "Chain from Hell"

Input	Device	Output
Balanced +4 dBu ref. (1.23 volts)	**A**	Balanced +4 dBu ref. (1.23 volts)
Unbalanced, −10 dB "V" (IEU 0 dB = 1-volt reference) (0.316 volts)	**B**	EITHER Unbalanced −10 dB"V" (0.316 volts) OR balanced 0 dB "V" (1 volt)
EITHER Unbalanced, −10 dBu (0 dB = 0.775 ref.) (0.245 volts) OR Balanced, +4 dBu (1.23 volts)	**C**	EITHER Unbalanced, −10 dBu (0 dB = 0.775 ref.) (0.245 volts) OR Balanced, +4 dBu (1.23 volts)
EITHER Unbalanced, −10 dBu (0 dB = 0.775 ref.) (0.245 volts) OR Balanced, +4 dBu (1.23 volts)	**D**	EITHER Unbalanced, −10 dBu (0 dB = 0.775 ref.) (0.245 volts) OR Balanced, +4 dBu (1.23 volts)

So:

$$-10 \text{ dBu } (0.245 \text{ volts}) = -12.2 \text{ dB“V”}$$

$$-6 \text{ dBu } (0.3875 \text{ volts}) = -8.2 \text{ dB“V”}$$

$$0 \text{ dBu } (0.775 \text{ volts}) = -2.2 \text{ dB“V”}$$

$$+4 \text{ dBu } (1.23 \text{ volts}) = +1.8 \text{ dB“V”}$$

and

$$-10 \text{ dB“V” } (0.316 \text{ volts}) = -7.8 \text{ dBu}$$

$$-6 \text{ dB“V” } (0.5 \text{ volts}) = -3.8 \text{ dBu}$$

$$0 \text{ dB“V” } (1 \text{ volt}) = +2.2 \text{ dBu}$$

$$+6 \text{ dB“V” } (2 \text{ volts}) = +8.2 \text{ dBu}$$

Now before you decide to abandon your pursuit of a career in audio engineering, don't take all this too seriously. I only brought the whole thing up to make you aware of the potential problem of combining differing operating-level references before it blindsided you in front of an important client.

Let's examine a couple of the possible results of passing a signal through the preceding fictional audio chain. Using a console's unbalanced −10 dBu send to feed the first piece of gear in the chain may require a slight, 4 dB increase in level using the main send potentiometer to help drive that device's input. This not only increases the level from −10 dBu (0.245 volts) to 0.3875 volts, but decreases the available headroom by 4 dB. Hopefully, not a tragedy.

Due to the differential input of device A, that 0.3875-volt level is doubled to 0.775 volts or 0 dBu. At the output of device A, which is also balanced and referenced to +4 dBu, the signal's level as set to 0 dB is now 1.23 volts. So far the signals experienced a total gain of 14 dB.

This is then routed to device B's unbalanced input, which is referenced to −10 dB“V” (0 dB = 1 volt) and expects only 0.316 volts (or −7.8 dBu) at its input. In order to avoid distortion the signal is attenuated by 10 dB using device B's input control. This brings the level down to −6 dBu (0.388 volts) or −8.2 dB“V,” which is very close to where it should be. But now in terms of any potential noise added by the next stage you've just decreased the signal part of the S/N ratio by 10 dB.

Since you already know that the next device in the chain, C, can accept either an unbalanced input at −10 dBu (0.775 volts) or a balanced one at +4 dBu (1.23 volts), your choices are simple if you're able to climb around the back of the equipment rack and reconfigure device B's wiring. You could take the unbalanced −10 dB“V” (0.316 V) output, which equals −7.8 dBu, and feed it to device C's −10 dBu unbalanced input so that only 2.2 dB of gain would be required (decreasing your headroom by 2.2 dB).

You could take device B's balanced 0 dB“V” (1 V) output and feed it to device C's unbalanced input using a wiring scheme that connects the low conductor to ground or common. This would decrease the level by 6 dB, making it −6 dB“V” (0.5 volts), which equals −3.8 dBu but would still require a further attenuation of 6 dB. This would, when compared to any potential noise added by the next stage, subtract 6 dB of signal from its S/N ratio and, as you'll see, possibly cause ground noise leakage into the signal.

You could take the 0 dB"V" (1 V) balanced output and feed it to device C's balanced +4 dBu (1.23 V) input. This would require increasing the level by 1.8 dB, thus decreasing the headroom by that amount.

This is a tough choice. You already have a potential 10 dB degradation of the S/N ratio as well as a decrease in headroom by 4 dB. The lesser of the three evils may be the second, because cutting the available headroom to one-half (−6 dB) could cause serious distortion during loud passages.

Take heart. The rest of the road is a downhill slide. Simply route the balanced +4 dBu output of device C to the balanced +4 dBu input of device D and then route the unbalanced −10 dBu output of device D to the −10 dBu aux return of the console. But don't forget those switches, and make sure you notate any wiring changes you make on the unit's faceplate so as not to trip up the next engineer who uses the room.

Not so bad. You just halved the available headroom and added a potential increase in the noise part of the S/N ratio of 10 dB. It could have been much worse. You see, the major problem here is not the varying reference levels or combinations of balanced and unbalanced gear. Those can be dealt with fairly easily. No, it's the surprises that can take a bite out of you that irritate me. Let me explain.

Any device that utilizes both two- and three-conductor ¼-inch jacks to provide unbalanced and balanced operation includes a problem that stems from their nonlocking plugs either being incorrectly seated or having been jarred or pulled from their seated positions. So what? Well, if we're dealing with a two-conductor ¼-inch connector it simply means no signal would be carried because the tip, or high, would not be engaged.

On the other hand, if it's a three-conductor ¼-inch plug that's not correctly seated, the tip, or high side, could easily end up connected to the ring, or low contact of the jack, with the plug's ring connected to the jack's sleeve, or ground. All of a sudden, that balanced +4 dBu line is unbalanced, and the signal is now one-half its original level!

Don't believe that this can't happen. I've run into it more than a few times, and the risk of it increases with every search-and-destroy mission made behind a studio's rack by someone trying to find dropped coins or to see where things are connected and to add or subtract another piece of gear or additional wiring.

I guess it's part of human nature to monkey about with things. But the special attraction humans have to playing with little switches is beyond explanation. Returning to the previous audio chain example, imagine you take a short break and while you're out of the control room there's a "visit" behind the rack. How much could the S/N ratio, headroom, distortion, and level of a signal suffer from a few dislodged connectors and a repositioned switch or two? Mucho, no? You see, it's those unseen switches and ¼-inch plugs not fully inserted that can be the real culprits because interfacing balanced and unbalanced lines just calls for a standard converter, and widely differing reference levels will at the very most require utilizing an additional gain stage.

Standard Unbalanced-to-Balanced Converters Always Work

So the answer is to place an unbalanced-to-balanced interface at every nonprofessional-level audio connection. Look, you saved about a hundred bucks by buying a piece of gear that does not provide balanced inputs and outputs. Now you can spend that on an impedance-matching device, use it whenever you need to, and when the piece of gear you purchased loses its usefulness you'll still have the interface. Nice move.

Single units housing eight interfaces are available. Whether they're passive transformer-based or driven by active electronics it's always important to check the manufacturer's specification sheet for frequency response and power-handling capabilities.

NOTE: *There are three main things the term impedance matching* will refer to: 600-ohm input and output impedances that were utilized to achieve the most efficient transfer of power between tube electronics, the newer "bridging" low-output-to-high-input impedance interfacing, which suits transistor-based electronics just fine, and interfaces that match an unbalanced, high-impedance operation to a balanced, low-impedance operation.

You don't really have to worry much about the first two types of impedance matching because one is rarely utilized today and most op-amp-based inputs and outputs can drive loads well above 600 ohms without any signal degradation. So outside of incorrect signal-level adjustment the main problem will be mismatched unbalanced-to-balanced interfacing. What we are talking about here is the last type of impedance matching.

Remember, in balanced operation the two signal conductors have an equal amount of impedance between themselves and the common or ground conductor. They are only referenced to ground and not actually connected to it. With unbalanced operation the "hot" conductor has a high impedance between itself and ground, and the other, "low" signal conductor has no impedance to ground because it's *attached* to it. Due to this signal-carrying conductor's connection to ground, unbalanced lines are more susceptible to power supply leakage (hum) noises and shield-induced (buzz) noises.

Using an unbalanced-to-balanced wiring scheme that calls for the unbalanced "hot" conductor to be connected to the balanced connector's high (XRL pin 2) while the unbalanced ground conductor is connected to the balanced connector's ground or shield (XLR pin 1) *and* tied to its low (XLR pin 3) allows for the introduction of any ground line noise right into the signal line. This connection method also negates the benefits of using balanced lines and differential stages, such as less frequency response degradation, lower susceptibility to noise, and higher CMRR. (See Figure 1-15A.)

Why Would Anyone Want to Do This?

You might do this to save a little money by avoiding the purchase of an interface. If that's the case, there's another inexpensive method of balanced-to-unbalanced interconnection that uses only balanced-type cable (a twisted pair of conductors surrounded by a shield). (See Figure 1-15B.) Here the unbalanced hot is connected to both of the twisted-pair conductors, with one of them connected to the balanced low and the other to the balanced high (XLR connector pins 3 and 2). The high-impedance, unbalanced ground, shield, or common is connected to the ground, shield, or common pin (1) of the balanced connector. Now the ground remains separate from the signal and you're able to retain the CMRR. On the other hand you now may run into noise problems from ground loops.

Simply put, no single wiring scheme will work in every interface situation. Don't believe it? Try it out for yourself by disconnecting a piece of gear that's operating noiselessly and substitute several others. After listening to the increasing and decreasing noise levels that result you'll become convinced that the only way to deal with this situation is to use unbalanced-to-balanced interfaces at every nonpro-to-professional audio interconnection.

A transformer allows for the balanced low side or even the hot side to be connected with the ground without any detrimental effects as long as the ground is not connected at the secondary. This is referred to as an *isolated floating ground,* and the common "direct box" usually offers a switch for this purpose. (See Figure 1-15C.)

On the other hand, the level difference is 11.8 dB between −10 dB"V" (0.316-volts) consumer unbalanced operation and +4 dBu (1.23-volt) professional audio balanced operation and 14 dB between −10 dBu (0.245-volts) and +4 dBu operation levels. If there is not enough gain available in the equipment itself to make up for this discrepancy, active electronic interfaces or separate amplifier gain stages (like those found throughout a mixing console) must be used. Trying to achieve this kind of jump in level using only small line transformers could easily result in saturation levels, meaning low-frequency loss and distortion. At the very least, it calls for your checking out the frequency response and level-handling specifications of the 1:4 turns ratio transformer you're about to use. This is precisely how transformers got their bad reputation, incorrect usage by individuals disregarding this component's limitations in processing low frequencies at higher-than-specified power levels.

An unbalanced output can be fine without using an impedance-matching interface if it is connected to a balanced differential input properly. A differential input actually improves the unbalanced signal's quality in terms of its S/N ratio. Here the signal is combined with its mirror image (from an inverted op-amp's output) and is referenced to ground, or zero. It's peak-to-peak level will now be twice as great as it originally was before going through the differential stage (see Figure 1-14). This 6 dB of gain translates to a 6 dB improvement in the S/N ratio. Unbalanced signals double in level when passing through differential stages; noise levels do not. In this case the differential stage will output a cleaner signal, that is, *if* the system's available headroom exceeds 6 dB.

Why Use Impedance-Matching Interfaces?

I know I'm harping on the topic of impedance-matching interfaces, but it's because they provide optimum performance by matching up differing impedances and levels. Whether they are transformer or op-amp based, they *will* handle the interconnection difficulties caused when an interface between two pieces of equipment result in mismatched balanced-to-unbalanced operation. Again, a wide variation in reference levels is another story. If you have to use an added gain stage to match up the levels, an active electronic interface may be the answer.

These are often composed of bidirectional buffer amplifiers, meaning they are made up of four separate paths that handle a pair of input and output signals. They not only provide load and impedance matching between unbalanced and balanced equipment but can offer added isolation and level correction as well.

They are designed to match "standard" *Institute of High Fidelity* (IHF) "consumer" (RCA connectors), at −10 dB"V" unbalanced 0.316 V levels or "semipro" (¼-inch connector) −10 dBu unbalanced 0.245-volt levels to "professional" (XLR connector) +4 dBu balanced 1.23-volt levels. Some even have the ability to vary the gain, allowing a boost or cut in the signal level by as much as 6 dB or more. This doubling of level is enough to handle just about any interface dilemma, even when feeding signal over long (greater than 100-foot) runs.

This is precisely why distribution amplifiers are as popular as they are in broadcast facilities. These units generally have a pair of inputs (stereo) that pass through an isolation amplifier stage that has an attenuation/gain adjustment control connected to dual front-panel meters. The signals are then fed through additional op-amps to half a dozen dual-output connections to be "distributed" anywhere they are needed throughout the facility, even to other distribution amps. Transformer-based units get a little heavy, and the power supply must be isolated to reduce hum pickup, but they work well when only a small amount of gain is required.

What Are Distribution Amps Used For?

Say a radio station is receiving an important feed from an out-of-facility source. Let's say it's being carried over a *plain old telephone service* (POTS) line that's routed through an impedance interface to the distribution amp. The engineer can set the level and feed the signal to the console for immediate broadcast, to a recorder to preserve it for later repeat broadcasts, to the "talent's" headphone cue system so that this "personality" can comment on the content of the feed, to a production studio where parts of it will be added to other program material, such as background information for a "special" news segment, and to anywhere else it might be used throughout the seven-story 500,000-square-foot facility, all without any worry about signal degradation, variations in level, or problems with impedance matching.

This kind of interconnection capability is exactly why old-timers went through so much trouble making the inputs and outputs of every piece of equipment in a facility 600 ohms, to the point of even adding 0.5-watt resistors to the rear-panel connectors — because it made *all* the equipment work perfectly when interconnected. It's a dream come true, and I don't understand why it should be a secret.

In my 38 years of experience I have yet to run into an unbalanced-to-balanced interface situation, outside of the occasional need of an additional gain stage, that was not easily handled by the use of an impedance-matching interface. One hundred bucks can bring a lot of peace of mind, and *every* piece of nonprofessional equipment should have one connected to it. I'll die waving this flag, because I know it to be a fact.

Am I saying that no other method of interconnecting balanced and unbalanced equipment will work? Of course not. Hey, the fact is if one of the noninterface wiring reconfiguration methods mentioned works out in a particular situation, great! But if there are any subsequent changes (such as another addition to the rack), I wouldn't bet on your retaining noise-free operation. Unbalanced-to-balanced interfaces utilizing shielded transformer operation are impervious to these types of changes.

As stated, we don't really have to worry much today about impedance matching because most op-amp-based inputs and outputs can drive loads above 600 ohms without any signal degradation. Instead the main problem is mismatched balanced-to-unbalanced operation and improperly adjusted signal levels.

So now that we've taken care of impedance-matching and interface problems, let's concentrate on the job of setting levels.

Gain Stage Setup (Level Setting)

Generally, a signal will end up passing through several gain stages in the mixer plus a few outboard pieces of equipment on its journey from the preamp output to the multitrack recorder and then back again, until it reaches the final mixdown medium. Improper gain stage setting not only can cause distortion, but could easily double the amount of background noise.

There are other ways to give a signal a higher presence in the mix without relying on an inordinate amount of level increase. Using a compressor can add as much as 20 dB of level without increasing the peaks, which helps avoid distorting a subsequent gain stage. Limiting will cut off the peaks, thereby allowing for an increase to the average level, and with brick wall limiting you'll be able to cut back on the amount of compression used so as to retain more of an instrument's natural dynamics (at least what occurs beneath the limiter's threshold setting), making for a more "open" sound. Equalizers will change a signal's tonal balance to offset system deficiencies of extra sensitivity at certain frequencies or of resonance problems. Units that handle multiple processing functions or a combination of individual processors can do all of these things at the same time. One of the best aspects of using some modern processors is their capability of accomplishing their tasks automatically. This type of signal processing will be covered elsewhere, but just to give you an idea of what's possible I'll let you in on one of my favorite tricks.

A fact of life when using analog tape recorders is that the bias levels (the signal that the audio is carried on) affect the signal's frequency response. As bias is increased, the overall output increases until the "peak bias level" is reached. After that, as bias is increased further (called *overbiasing*), the output level decreases.

This decrease is more gradual for low frequencies than it is for high. Additionally, the output decreases proportionally as the frequency increases. To match this, early analog tape recorders used more equalization boost at higher frequencies when there was a need for a higher amount of bias for low-frequency accuracy.

Enter Dolby noise reduction systems, which used a combination of compression- and expansion-level adjustment along with equalization preemphasis and deemphasis. Compression reduces the higher levels and increases the lower levels. Expansion (used during playback) provided the opposite of compression by decreasing the lower levels while increasing the higher levels. Preemphasis increased the amount of high-frequency equalization boost when that signal content is low in level, and during playback deemphasis corrected this tonal imbalance. These systems were constantly improved over the earlier units, which were themselves standard "must have" equipment in the 1970s, to the point where by the 1980s the Dolby HX ("Headroom Extension" system) had the capability to automatically adjust the amount of both the bias and the equalization per the input signal's level and high-frequency content.

When the input levels were low or there was less high-frequency content, it both increased the amount of bias and added a boost in high-frequency equalization. This increased the level of both areas of signal content. On the other hand, when there were higher input levels or high frequencies dominated the signal, there was an instantaneous reduction of bias levels and a lowering of the amount of high-frequency equalization boost. The result is that low frequencies are boosted by a bias increase when the level is low, and high frequencies undergo a boost in equalization only when the high-frequency content is low. This automatic adjustment of level and equalization according to the signal's content translated to an increase in headroom without any added distortion.

I found no rules stating that I had to use *both* the record and playback sections of the HX system. You see, due to their low sampling rates, early digital effects were very sensitive to high-frequency content. Analog bucket brigade and early digital delays and reverbs ran into trouble when it came to processing ("glitch"-inducing) higher frequencies. At the same time, here was this three-head cassette deck that could furnish me with not only a delay, but delivering Dolby HX as well. I ran into a dead end because the HX effect could only be had after the output of the tape playback head. This meant not being able to have only compression and preemphasis but also the inclusion of expansion and deemphasis, plus both the tape's delay and noise. But the unit did come furnished with separate send and return connections for use with a DBX noise reduction system. There wasn't one available in the facility, but a simple shielded wire jumper from a point just before the record head to one of these connectors gave me exactly what was needed.

Granted, listening to the "soloed" signal that had undergone only one-half of the intended processing did leave a bit to be desired. But when used with an older delay or reverb effect and then combined with the original signal, the resulting signal had a sound several years ahead of its time quality-wise. It's easier to use only half of a noise reduction system that's not actually built into a recorder. Try going through just the output section of a DBX noise reduction unit to utilize its expansion and high-frequency roll-off. This was not a universal effect by any means, but it was still interesting. On the other hand, many of today's processors automatically handle dynamics control like no human could ever dream of doing.

The point? When you understand the capabilities of the equipment you're dealing with you have the advantage of being able to manipulate sound to a far greater extent. I'm not talking about diving into a tape recorder and tapping into the signal conductors just before the record head here; this fact holds true for more commonly used equipment, such as mixing consoles.

Mixing-Console Gain Stages

Charting out the signal flow through the gain stages of a full-blown multitrack recording console, as was done with the "audio chain from hell," seems like an impossibly daunting task until you take into consideration that all the input channels are exactly the same. This means that on a console having 96 I/O (combination input and output) channels, charting out a single channel takes care of 90 percent of the work.

Once a signal reaches a mixer, it generally passes through a differential stage that converts it to unbalanced operation. It would be cost prohibitive and wouldn't serve any useful purpose to run a signal throughout a mixer in balanced mode. Therefore, signal routing within a console shouldn't be a problem in terms of mismatched balanced-to-unbalanced operation. Neither will you have to concern yourself with impedance matching because most op-amp-based inputs and outputs will drive the loads encountered without any signal degradation. All the levels *within* the console will have the same reference even if the levels vary at different output interface points, such as sends and multitrack outputs. However, due to the number of gain stages a signal must pass through to get from the microphone input to the final two mixor stereo-bus output, improper level adjustment along any of the console's routing paths might severely degrade signal quality.

I believe the ability to properly set up gain stages is the recording engineer's most important function. That's why I decided to write the technical section of this book. I specifically wanted to provide you with the knowledge you'll need to perform this task. I'll finish it all off by laying out a signal's flow through a console in chart form and add the manufacturer's specifications for each gain stage. This visual representation not only will make it easier to understand gain stage setup throughout that particular mixing board, but it will also serve as a model for you to follow when you chart out a representation of the mixer you work on. That's right, more homework!

But before you can make an attempt at laying out a console gain stage chart, you first have to gather some information. You'll need a signal flow routing chart, information on all input and output connections, including the patch points, as well as a diagram showing the location and function of all controls, switches, attenuators, pots, and indicators. You'll need to know the capabilities of every gain stage as far as its amplification and limitations in terms of distortion and S/N ratio and the extent to which onboard processors, such as equalizers, gates, and compressors, can affect the signal level. Hopefully, all of this information will be found in the patch bay layout, channel flow charts, schematic diagrams, specifications list, and outboard connection or installation sections found in the console's manuals. But I've gotten ahead of myself, because in order for you to gather all the information you'll need to chart out the signal flow through a console, you first have to know how to use its manuals.

Using Audio Equipment Manuals

The last thing anyone seems to want to do today is read an operations manual, let alone go through the service manual. It's gotten to the point where many manufacturers don't even bother including a manual with their products because it just ends up as expensive packing material.

In my case, I couldn't wait to get my hands on the manual. Not only does it guide you through the initial learning curve of a new piece of gear, but it opens the door to many possible uses you may not have considered. Yet the biggest plus is when you discover unique applications to serve your own specific needs, some of which the manufacturer hadn't even considered. Now it becomes *your* piece of equipment, which you use in your own unique way to your own advantage ("Hey, you should have heard what that engineer did with that thing!").

As you go through examples of using various effects devices that I'll be giving, you may wonder how I came up with some of the ideas. I have to be honest and say that for the most part my ability to take full advantage of any piece of equipment usually stemmed from my climbing all through its owner's and service manuals. It will amaze you how you'll be inspired to try things you never thought possible just by going through a unit's manual.

The best way to understand how to take advantage of manuals would be for me to go through one I considered ideal. While there's no such thing as a perfect manual, I've used enough service, repair, and operational publications to know what they should comprise. Hey, maybe someone who reads the following layout on how to prepare a manual will end up as a technical writer (it *is* one facet of our field, you know) and use this outline in his or her work. The designation *manual* applies to both the owner's (operational) manual and service manual, whether incorporated into one publication or printed as two separate documents.

The quality of the equipment manuals provided today range from in-depth, highly accurate, and helpful tools to, well, almost joke books.

The most important aspect of putting together a manual is proper organization. If this is attended to, important information can be more easily accessed. Most subtitle headings are traditional: Introduction, Specifications, Installation, Operation, Calibration, Maintenance,

Troubleshooting, Parts List, and signal Flow Charts or Block Diagrams. When information is located under the proper headings, it helps deter confusion and delays caused by questions concerning usage or service. The practice of, for instance, dividing a calibration routine between Operation and Maintenance sections of the manual invites incomplete setups, hinders rapid calibrations, and could result in hit-or-miss experimentation, which might end up being harmful to the equipment. All documentation should be under the proper headings, if for no other reason than to shorten the length of equipment downtime.

The introduction should provide a brief synopsis of the device's purpose, applications, and interfacing. In other words, a good intro will describe a device's personality. This section should make the parameters covered by controls and external connections understood by providing a good overview of the device as a whole. This will increase user comprehension of and rapport with the equipment. Oddly enough, most of the information a good introduction should include has often been provided by the advertising literature of the manufacturer's marketing and sales department, though it is not often put to use by consumers after the purchase has been made.

Specification lists are often one of the most used sections of a manual during calibration, maintenance, and installation. They should include data on system accuracy, power requirements, interface, connections, and parameters such as clock frequencies. The standards used to derive the specification results should also be listed, as well as the type of measurement equipment used with makes and model numbers included to facilitate end user verification of proper operation.

Installation procedures that are laid out logically via a step-by-step connection routine with nothing left out will help avoid customer frustration and equipment damage. I repeat: Nothing should be left out. Front- and back-panel illustrations should be clearly and accurately labeled to avoid confusion. The operational checkout procedures should list front-panel settings, as well as a "source of error" listing, and both must be clearly laid out for easy referencing.

Operational instructions should be in a three-part format. The first should be a simple, short explanation of control settings and functions, as well as any needed warnings for quick referencing. A good example of this was the small fold-over card included with the Fluke 8060A multimeter, which fits inside its carrying case. For nonportable devices, an attached pullout plastic card, such as found on Hewlett-Packard and Sencore test equipment, is ideal. Should both of these approaches be deemed too costly to implement, the information could be included in some format within the manual. Next, several control setting illustrations should follow, depicting commonly used applications.

A blank illustration should be included to allow for the copying and filling-in of user-defined settings. These illustrations also help users familiarize themselves with the equipment even when it's not actually being used, while plainly illustrated hookup procedures with application/usage examples result in quick out-of-the-box functioning and instantaneous consumer gratification.

Finally, a more detailed documentation of operation and applications should be included. Nothing should be left out. If there is not enough room in the manual for all the information available, the text should at least mention reference sources and a list of catalog numbers where further information can be obtained.

If you think this is asking too much, check out what I had to say about what a service manual should consist of. At the time of this writing I was totally immersed in all aspects of audio engineering, teaching audio technology, and repairing and calibrating equipment. You'll easily pick up on the fact that my background includes a stint of work in military electronics, where lives may depend on the outcome of the work you do. What I had to say was right on target, but I admit to getting a bit too commander-in-chief about it all. This kind of thing can happen when you get totally locked into something.

Service manuals must contain precise circuit descriptions. Many maintenance personnel, let alone end users, are not able to ascertain individual circuit functions via schematic diagram analysis. This process should be aided by block diagrams. Such diagrams, as well as the circuit descriptions, should include all relevant outboard connections. Calibration routines, as stated earlier, should be completely listed in a single section. If certain calibration procedures have already been given in either the Installation or Operation section, they must be repeated here. Illustrate test equipment connections, specific equipment models, any generic counterparts that are recommended, and any of the specifications that allow for these substitutions.

The documentation must explain what's being accomplished electronically; this ensures both proper calibration and repeatability. First, give front-panel control settings. Illustrate circuit board adjustment locations. Since mechanical adjustments are difficult to translate into words, they require illustrations as well. Give access routes to obscured adjustments. List any special accessories needed to accomplish proper calibration. To avoid adjustment interaction mistakes, provide a detailed, step-by-step routine. Give troubleshooting procedures that isolate malfunctions to specific areas that will point out problem sources.

The manufacturer's service staff should be consulted before documentation is completed because their knowledge of a specific device and any common sources of problems will help cut end user service time greatly.

Flowcharts with systematic fault-finding procedures that list all test points and give test responses with clear references to waveforms and voltage levels can reduce equipment damage due to improper or unnecessary servicing. Remember, every time soldering, desoldering, or testing is performed, there is a risk of service damage. Well-laid-out troubleshooting routines help to avoid these search-and-destroy missions.

Dismantling and reassembly of some equipment can be a puzzle-like ordeal. Access to assemblies, switches, meters, and power supplies should be included in the data via exploded illustrations and detailed, step-by-step dismantling instructions.

Parts lists must be complete and accurately numbered. A single list format makes for faster referencing and minimizes replacement order errors. Generic information should be included whenever possible, giving, for instance, wattage, tolerance, type, temperature coefficient, matching data, and other vital selection parameters for components.

Component layout illustrations are essential, especially when printed circuit boards or subassemblies are not labeled with screen printing. All adjustment locations, component values and designations, as well as test points should be clearly labeled and accurate. There should be no duplication of numbers and no discrepancies between the parts list, component layout prints, and schematics.

Schematics are extensively used, along with block diagrams, during servicing. It is therefore most helpful to have them on facing pages. A schematic's value is determined by many factors, such as sections lightly blocked in and captioned per function when partial schematics are not possible. It is extremely helpful to have labeling of important component functions and operating voltages listed at certain IC and transistor pin outs for identifying improper circuit functioning. A good schematic will also provide voltage notations, waveforms (with frequency), and amplitude, in addition to an indication of ganged switches, with their position settings as well as those of any interacting front-panel controls.

With multiple schematics, the interconnection lines should be coded and captioned for easy traceability. Clear and correct component part numbers and values, along with information on encapsulated module pin outs, should be given. When all this information is prepared in advance, it will avoid handwritten notations and corrections which are often unreadable and inaccurate.

Professional-quality printing is essential, because illustrations, photos, and schematics often must be reduced in size. Detail and sharpness are a must, and drawings must be directly proportional with the actual device. Illustrations are important to overall quality judgment by the user and service personnel because they are generally the first pages to be looked at. Paragraphs and illustrations should be numbered and cross-referenced. Manuals should lay flat and not have metal (or conductive) bindings, and the paper used should not reflect light.

Putting together a manual in this way is expensive, and the only readily apparent source of return on this investment is either the sale of duplicate manuals or separate service data sales. Often because of highly competitive pricing, costs for producing a manual cannot be added into equipment purchase pricing. This can make detailed manual preparation seem like a wasteful investment. However, this view is both shortsighted and incorrect. Often a manufacturer's image is substantially enhanced (or marred) by the quality of their manuals, especially in the eyes of service personnel. Documentation that enables fast and easy servicing enhances the service person's reputation and thus their opinion of the manufacturer. Remember, these maintenance and service people are often very influential in most purchasing decisions.

Since the overall quality of equipment is judged by its performance, consider that a properly written manual that helps avoid incorrect calibration and maintenance can thereby help avoid costly downtime, while improper operation data makes the end user's job more difficult and fosters a negative opinion of both the product and the manufacturer.

Quality manuals, whether they have operational information combined with service data or are produced as separate documents, are therefore essential in adding to both the respect of the manufacturer's image and the subsequent sales that are bound to follow due to consumer confidence in the manufacturer's full product line.

With that said, we can now move on to the task at hand. Again, before you can put together a layout of a console's gain stages, you'll need to gather as much of the following information as possible from the manufacturer's operation and/or service manuals.

Flow Charts, or Block Diagrams

These present the console's signal routing, with the channel signal flow made up of blocks depicting each individual stage the signal passes through. If you're lucky this one document will provide a wealth of information concerning signal routing throughout the console, but it will not provide everything you'll need to complete the task of charting out the console's gain stages.

Patch Bay Layout

This will include a listing of all patch points and sometimes a patch bay designation glossary. Here the patch points are named or labeled and a brief description given of their usage or function, with at least a hint at the location point where they "break out" from the console's routing path. A patch bay layout can sometimes be used as a map of the signal flow through the console and beyond.

Operation Guide

This section of the manual is devoted to explaining the various functions and capabilities of every part of the console, both individually and grouped together, so as to achieve practical goals such as setting up a cue send, grouping channels, assigning channel outputs to specific recorder tracks, microphone and line input level setup, remixing, overdubbing, equalizing, mixing down to stereo or mono, "bouncing tracks," as well as describing any automation and "recall" capabilities. Seventy-five percent or more of the information you're after in terms of functioning will be here. For example, it could tell you if an attenuation switch affects both the mic and line levels and to what degree.

Control and Indicator Layout Diagrams

You may have to refer to multiple drawings showing sections of a module with its controls and indicators to get the location and function of every control, or this could be a single illustration depicting one whole section of the console, such as an input module, in which all its switches, rotary pots, faders, and light indicators are shown. Often they'll have arrows point from each of these to very brief descriptions of their functions.

NOTE: *A module refers to all the switches, faders, amplifiers, and light indicators that make up a mixing console channel. The electronics for that channel are assembled onto a large printed circuit board that is attached to a metal faceplate onto which all the necessary controls and indicators are mounted.*

Hopefully, all of this information will be found in the console's manual. Even so, you may still have to utilize the "dreaded" service manual for its Installation section.

Installation Section

This should show all the input and output connections and whether they are balanced or unbalanced, giving some indication as to input, output, and patch point levels.

Specifications List

This will tell you the console's capabilities as far as amplification and its limitations in terms of distortion headroom and the noise floor (S/N ratio) at each gain stage, as well as the extent to which any onboard processing, such as equalizers, gates, and compressors, can affect the operating level.

Schematic Diagrams

Schematic diagrams are the last word as far as how specific control functions affect the signal routing.

NOTE: *Make sure the drawing's reference number matches the console version you are dealing with. A single change can have a significant effect on how a console operates. Manufacturers constantly update their consoles, because as one put it, "Specifications subject to change as innovative advancements in technology are incorporated."*

Reading schematic diagrams is almost an art in itself. There isn't enough room here to cover this subject, but don't worry because there'll be plenty of handbooks available at your local library. While you're there, take the time to examine their catalog of works on music instruments, electronics, acoustics, noise reduction, the physics of sound, and public address systems. Audio engineering requires broad-based knowledge.

Examples of the Charting of Gain Stages

I'll use a couple of consoles to illustrate the charting of the gain stages that a signal passes through when it is routed from, as an example, the mic input to the multitrack recorder. One will be the MCI 600 series console. This desk was chosen simply because I already have all the documentation for this mixer. The other had to be a *solid-state logic* (SSL) board, simply because of its worldwide popularity. It just makes sense for you to be a little familiar with the console you're going to run into most often.

But before we take on these "big boys" in the world of mixers, we'll start with smaller formats that have simpler routing configurations. Even though these mixers are not very complicated in terms of their routing structures, they'll still give you a taste of the variety of busing features that are found in consoles. This is something you really need to understand.

The Console Bus

Once a signal enters the console and its level is brought to unity gain, it is assigned to various bus routes. Some buses are very direct in their routing and offer a kind of door-to-door service, that is, straight from a mic preamp out to a particular recorder track. Others buses might pick up additional signals and form a group trip in which multiple signals are ferried to the same recorder track. Detours such as side trips out of the patch bay into the big world of out-board processing can also be available, often with round-trip (return patch) ticketing provided.

Some only take part of the signal's passenger load and veer off for tours of the studio by way of cue or headphone sends. These, like send buses that route signals directly to outboard amusement park–like destinations such as high-cost reverb chambers, take signals from across the whole board and mix or combine them so as to end up with a well-balanced group feed.

The two most important bus routes are those that bring the signal to the inputs of the multitrack recorder and to the main stereo, four-channel, or quad, and now 5.1 mix outputs.

These routes can take side trips as well, such as through EQ wonderland and straightlaced dynamics with military school, "Hey, you kids in the back settle down" filtering and "you guys in the front step behind the white line" dynamics control.

The Altec 250 T3 Control Console

The Altec 250 T3 control console is the first mixer we'll discuss. While there are many other decks I could have used, I decided to go with the 250 T3, for several reasons. In addition to the fact that it was the first professional console I recorded on, its functioning is simple, its operation is logical, and it was built to professional-quality standards. But the most important reason is that you'll get the whole ball of wax as far as its gain stage layout. No credit is due me for this, because the manufacturer actually furnished everything, *including* a signal gain stage chart in its manual. It is the only example of this that I know of.

The Altec 250 series mixers were designed to fulfill the needs of public address announcement and music mixing for sound system venues such as convention centers. They could also be used as program mixers for radio broadcast stations and as mixers in recording studios.

All application-specific signal-routing interconnections through this console were performed by the end user. For example, all the outputs could be connected from the hardwired (punch-down or soldered) terminal block, either directly to other pieces of equipment, like reverberation processors, monitor amplifiers, and tape recorders, or brought out to a user-assembled patch bay. I know this seems like a lot of work today, but you have to understand that until the early 1970s, most studios had to build their own consoles *from scratch*!

The "all solid-state" Altec model 250 T3 control console dates back to the 1960s, and at the time it was top of the line. Not only was it fully modular in construction, but it used a single type of amplifier (the Altec 9475A) for preamp, for line-level (unity) gain stages, and for driving its output. These solid-state push-pull direct coupled class A amps provided 45 dB of gain and could handle an input of up to +24 dBm "without danger of peak overload." Their maximum output of 27 dBm (at 1 percent THD, 20 to 20,000 Hz) doesn't seem like much by today's standards, but since 0 dB is here referenced to 1 milliwatt into 600 ohms, we're talking about a pretty hot level. The frequency response of this console's "old-time" technology was within ±0.5 a dB from 20 Hz out to 20 kHz. The amp's noise level was −127 dBm, unweighted, 10 to 25 kHz bandwidth, input unterminated (meaning with no input). Altec was rightfully proud of this console and furnished just about every possible bit of information except for a schematic diagram of that amplifier.

It was undoubtedly the best 12-input three-track mixer around. Three track? Yeah, between stereo and quad there were three-track recorders that were actually used quite extensively in live classical recording. Think about it a second: left, center, and right. Makes sense, no? To help understand why the Altec 250 T3 could be utilized for classical recording, here are the manufacturer's specifications.

The Altec 250 T3 9475A Amp Specifications

Microphone Inputs:
Gain at line out, 99 dB

Frequency response, ±0.5 dB, 30 to 20,000 Hz

Distortion, 0.5 percent THD, 30 to 15,000 Hz at +18 dBm output

S/N ratio, 70 dB (+17 dBm output with −55 dBm input)

Line Inputs:
Gain at line out, 40 dB

Frequency response ±1 dB, 20 to 15,000 Hz

Distortion, 0.5 percent THD, 30 to 15,000 Hz at +16 dBm output; ±1 at +18 dBm output

S/N ratio, 72 dB (+17 dBm output with a −4 dBm input)

NOTE: *Comparing the Altec 9475A amplifier's performance in handling both mic and line inputs confirms Altec's achievement as far as eliminating the need for a variety of amplifiers within a single console. This allowed for the plug-in interchangeability, which helped keep these mixers online at radio and other broadcasting facilities 24/7. I still think it's remarkable that the same amp could boost inputs with levels ranging from −55 to −4 dB for gains of 99 dB and 40 dB, respectively, with the two resulting S/N ratios differing by only 2 dB (70 and 72 dB).*

The signal flow and gain stage chart for the Altec 250 T3 console in Figure 1-16 reveals a couple of what seem like surprising facts. Even though these amplifiers have a preset gain structure, unlike most modern mixers the 250 T3's faders and attenuators are placed after the amplifiers, as opposed to before them. You may also notice that all of the circuits outside of the amplifiers cause an (insertion) loss to the signal's level. This is typical of passive electronic circuitry. For instance, true passive equalizers are cut-only devices. You have to add a gain stage or active electronics for them to be capable of boosting a frequency. That's just the way it is. However, this console was designed so that you could wire another amp in line after the equalizers or anywhere else to boost the level if necessary.

The layout of Figure 1-16 is fairly straightforward, but it won't hurt to run through it.

A. The mic output comes in at −60 dB. It is connected to a 9475A amp and used here as a mic preamp for a 51 dB gain (resulting in a level of −9 dB).

B. The stepped attenuator that follows the mic preamp is the channel mix fader (a large rotary dial knob), again providing only cut, and its *minimum* amount of gain reduction, 22 dB, results in a signal level of −31 dB (the maximum gain reduction was infinity or the signal turned off).

C. The signal loss through the bus assign switch is 14 dB, and this drops the level down to −45 dB (only 15 dB above that of the mic output).

D. The booster amp's gain of 51 dB brings the signal level up to +6.

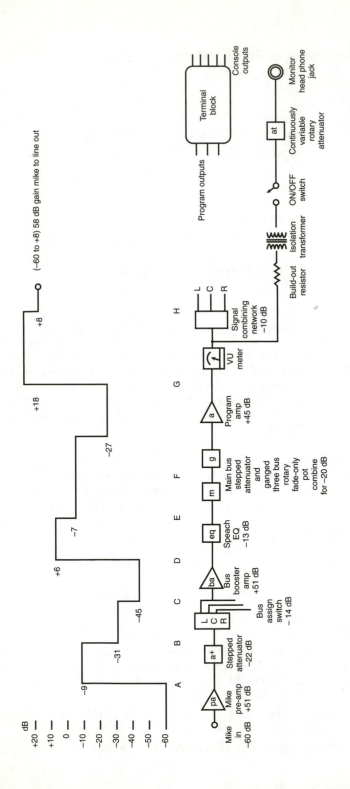

92

Figure 1-16
The signal flow and gain stage chart of the Altec 250 T3 console

E. When switched in, the EQ's insertion loss of 13 dB drops the signal level down to −7 dB.

F. Two more attenuators follow, the main fader for each of the individual buses and a ganged (three-stage) potentiometer that was the console's main cut-only fader, which had control over all three buses. These two faders combined to cause at least a 20 dB loss in level, dropping the signal down to −27.

G. The final program amplifier provides 45 dB of gain and brings the level up to +18 dB.

H. The output combining network's load causes a loss in the signal level of 10 dB. It then routes the resulting +8 dB signal level to the program outputs on the terminal block.

In this situation, the overall gain is +68 dB from the mic input to the bus output.

This diagram also shows you that the post-VU meter, monitor split-off bus has a resistor/isolation transformer combination in line. This allowed for the insertion of headphones into the monitor output jacks while a feed was taking place without its affecting the program outputs. You can also see that the headphone feeds are equipped with on/off switches and a continuously variable, instead of a stepped, attenuator.

Turns out there's an awful lot going on in this simple little old-time mixer after all. Yet for recording it required a huge amount of rewiring work when changing over from recording to bouncing or combining tracks and then to mixdown. During a single session the complete audio chain might have had to be repatched from record mode to accommodate each individual overdub, for the bouncing or the moving and combining of tracks, and finally for mixing down to the final mono or stereo format configuration. I know!

One of the next console advances was the greatly needed addition of switch-activated bus feeds to the multitrack inputs, along with switch-controlled multitrack output routing to the console's monitoring and mixing buses.

The Tascam M-50 Mixer

I've chosen the Tascam M-50 mixer for our second example. This was one of the consoles that helped initiate the proliferation of "home," or "semipro," recording facilities during the early 1980s, and many of them are still in operation. It was designed to feed input signals to its eight-multitrack buses, two stereo auxiliary "send" buses, and the stereo mix bus, all via switch activation as opposed to repatching.

Most important for our purposes is that its manual provided block diagrams of its signal flow. Here we get to deal with a flow chart drawing similar to the ones in Figure 1-17. It shows the console's signal-routing scheme with its switches, gain stages, attenuators, and the EQ section (the only onboard processing) in a nice sequential order. This type of "block diagram" allows you to trace the signal flow and to see where everything occurs, such as exactly where a patch point breaks out from the signal chain and exactly where it returns back into the routing flow.

Figure 1-17 shows you the following:

- Input connections
- Phantom power on/off switch
- Mic attenuation and a phase reversal switche that affects only the mic input

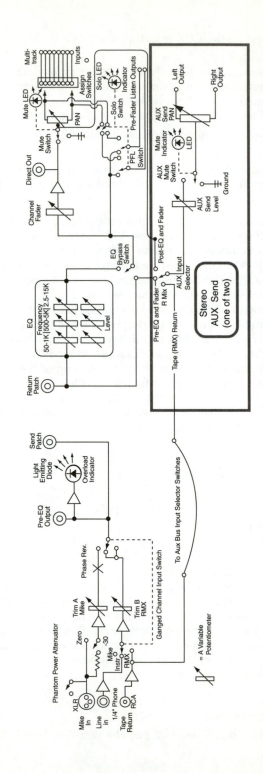

Figure 1-17
The signal flow and gain stage chart for the Tascam M-50 mixer

94

- Instrument and remix input gain stages
- The instrument and remix input switching, which is ganged to the channel input switch (as designated by the dashed lines)
- The dual (as designated by its sections being labeled a and b) instrument or remix, and mic trim pot
- The channel input mic, line, or tape return selector switch
- An LED overload indicator
- A pre-EQ out patch point
- The send patch point
- The remix routing feed to the auxiliary bus input selector switches

We then pick up at the channel's patch return and also show the remix (tape return) feed to the aux buses:

1. The signal from the patch return splits off into the EQ and to the prefader input of the aux input selector switch.
2. The signal taken pre- or post-EQ (as determined by the EQ bypass switch) then moves on to the channel fader gain stage.
3. Then it moves to the direct output patch, which is followed by a mute switch and the multitrack assign switch matrix with odd/even pan capability, which feeds the signal to the multitrack inputs.

Meanwhile:

1. The aux input selector switch picks up either the channel signal pre- or post-EQ and fader signal or
2. The tape remix return signal and feeds it to an aux level control, which is followed by
3. A mute switch and pan control, which feeds the aux send's left and right outputs.

At the same time, solo and prefader listen switches pick up the signal from just before the assign switches (SOLO) or just before the fader (PFL) prefader listen. They route these signals to the headphone and monitor speaker feeds.

You may think that is a lot of information to get from a couple of simple block diagrams or flow charts, but it doesn't end there. A closer look reveals a great deal more.

While there are separate controls for the mic and remix levels, the remix input shares its level control with the line input. This makes sense because both would be coming in at a level that would require less gain than a mic.

The overload LED indicates the initial input gain stage level and appears before the patch return point, so you'll have to pay closer attention, watching out for excessive return levels from outboard processors.

That simple-looking EQ is actually a semiparametric (it has no "Q," or bandwidth, controls) with overlapping frequency selection. It allows for the use of dual controls in the sensitive areas of the audio spectrum between 500 Hz to 1 kHz and 2.5 to 5 kHz.

The aux input selection can be prefader and pre-EQ or postfader and postmute. This means that if you want to mute the main signal outputs and still feed the output aux sends so that you are able to mute a rhythm guitar during the chorus and still have it feed into the reverb, simply select the prefader (and premute) position. This has a haunting effect when used on a flute or woodwind in a ballad.

As you can see there's a lot to be gained by going through this small amount of aggravation. It should be obvious that the addition of the multitrack returns bus and switching provided meant the M-50 circuitry saves the operator a lot of the patch bay rework that had been required with consoles like the 250 T3 when changing from recording to overdub or mixdown duties.

In order to survive in the marketplace, consoles designed specifically for audio recording need to have a great deal of flexibility to be able to handle multiple recording applications. Even though things get much more complex with large-scale professional console formats, their flow charts are still very helpful for understanding their functioning possibilities. Flow charts help us focus our thinking on the logic involved in a console's signal routing per the various functions that it must be adapted to. For instance, the stereo mixdown and monitor outputs can simply be thought of as two of the possible destinations for the mic, line, and multitrack playback (return) inputs, others being the patch bay to outboard processors and the send buses that feed the headphone cue system.

A Side Trip for More Background

If you've read everything up to this point you already understand gain stages, headroom, S/N ratios, decibels, and the parts that make up a manual. However, it would be incorrect for me to assume that you also understand the complex routing schemes provided by modern multitrack recording consoles. While we will be going over the individual parts that make up these consoles, I feel it best now to explain the routing strategies that are commonly used because clarification in this area is required before you can tackle charting out their gain stages.

Format

A mixer's format most often refers to its numerical layout, in terms of inputs and outputs. All mixers have a format, and this simply tells you in plain numbers what the mixer is capable of handling. For instance, a $36 \times 24 \times 2$ mixer would have 36 input channels, 24 group outputs for multitrack feeds, and a stereo output. This is not set in stone because live mixers utilize group outputs for separate stage monitor feeds. For a spilt console with a separate multitrack return/monitor section, this format would relate to a 36-channel input section and a 24-group output/monitor return section and a stereo mix bus out. Today it's common to find formats with numbers more like $64 \times 48 \times 5.1$. Another aspect of the term *format* has to do with the way in which the console's signal routing is laid out.

All of the following information will be easier to digest by looking over the module signal flows as they are depicted in Figures 1-18, 1-19, and 1-20. They show the routing configurations used with split, common I/O, and SSL I/O consoles to increase the number of inputs available during mixdown.

■■■ ■■■ ■■■
Figure 1-18
Configuration used by split consoles to increase the number of inputs during mixdown

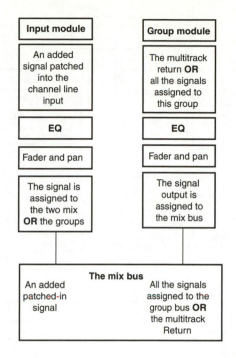

Input module	Group module
An added signal patched into the channel line input	The multitrack return **OR** all the signals assigned to this group
EQ	**EQ**
Fader and pan	Fader and pan
The signal is assigned to the two mix **OR** the groups	The signal output is assigned to the mix bus

The mix bus

An added patched-in signal	All the signals assigned to the group bus **OR** the multitrack Return

■■■ ■■■ ■■■
Figure 1-19
Configuring a common console I/O to increase the number of inputs during mixdown

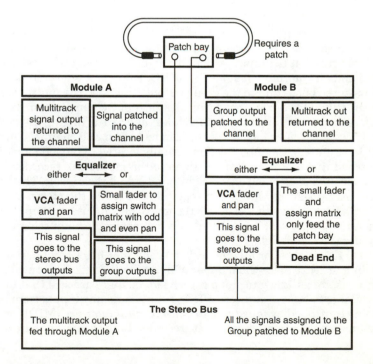

Patch bay — Requires a patch

Module A		Module B	
Multitrack signal output returned to the channel	Signal patched into the channel	Group output patched to the channel	Multitrack out returned to the channel
Equalizer either ←→ or		**Equalizer** either ←→ or	
VCA fader and pan	Small fader to assign switch matrix with odd and even pan	**VCA** fader and pan	The small fader and assign matrix only feed the patch bay
This signal goes to the stereo bus outputs	This signal goes to the group outputs	This signal goes to the stereo bus outputs	**Dead End**

The Stereo Bus

The multitrack output fed through Module A	All the signals assigned to the Group patched to Module B

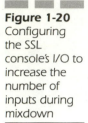

Figure 1-20
Configuring
the SSL
console's I/O to
increase the
number of
inputs during
mixdown

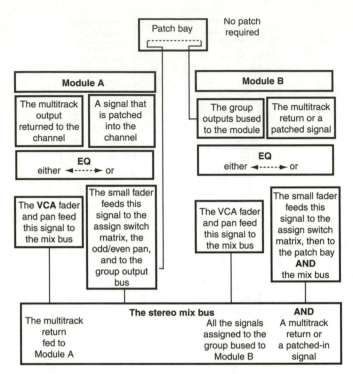

Split consoles divide the signal routing paths between two separate sections of the console. Upon arrival at the input section, the signal will generally first see a gain stage made up of a pad and an amp, often followed by an EQ. The signal is now prepared to be fed to both the two-mix stereo buses and to the group outputs that feed the multitrack inputs.

In this way, during mixdown all the input channels can handle additional inputs and effects returns while the group modules take care of mixing the multitrack returns. Therefore a $36 \times 24 \times 2$ split format console can be used for 60 channels during mixdown.

In summary, the input section (usually located to the left) handles all the microphone and patched line inputs and routes them to the multitrack recorder through the "group" output section, usually located on the right-hand side of the console. This is also where the multitrack returns appear and are mixed together and fed to the stereo and monitor buses. Switching allows for the selection of either the multitrack input or group output signals to be fed to the monitors, often with some EQ and a pan control for adjusting the placement of that signal across the stereo field.

All this makes for a large desk, but one that has the advantage, during mixdown, of providing at least twice the number of inputs to the mix. This means that once all the recorder returns are assigned to group return faders, there are still that many or more input modules available to handle outboard effects returns and any inputs from instruments that are either being controlled by *musical instrument digital interface* (MIDI) or played live during the mix session so as to save tracks on the recorder. Both sections will have equalizer capabilities, but only one set (usually on the input section) will offer extensive functionality such as multiband or parametric control, as opposed to simple high- and low-frequency boost/cut controls.

Split vs. In-Line (I/O) Console Layouts

These designations are pretty much based on where the returns from the multitrack recorder and the monitor-level controls are located on the console. In-line consoles are made up of input/output, or I/O, modules. These contain the main channel inputs, group assigns, the monitor-level controls, and the multitrack tape returns, all on the same module.

The I/O module layout is utilized by both of the two consoles I'll be using next as examples (the MCI 600 series and SSL 4000G consoles), in contrast to the split console system. These recording consoles' I/O modules are designed to handle inputs from two sources, live inputs and those that are prerecorded. These inputs are then routed to two basic places, mix buses and the multitrack recorder. Two-level controls are required to accomplish this. Their EQ's not only can be switched in or out, but can be used to affect just the signal routed to the monitors, those sent to the multitrack recorder, or both. It's all up to the operator, and that's just a part of the flexibility that's offered by modern console configurations.

In a nutshell, I/O consoles have two signal paths running through each module. During initial recording and subsequent overdubbing, one path feeds the input signal to the multitrack recorder while the other routes the recorder's prerecorded returns or outputs to the musician's headphone cue and to the mix monitors.

Mic or line input signals are mainly routed to the multitrack recorder and the monitors when recording. During overdubbing, the tracks already recorded, along with the new mic and line inputs, are routed to the headphone monitors. In order for the musicians to play along with prerecorded tracks, the console must have the ability to be set up to play back prerecorded tracks at the same time, as it is being used to bus newly played signals to the multitrack record.

To accomplish this they must have dual-level controls, such as a rotary pot generally used to feed signal to the monitors and a fader for adjusting the signal level sent to the multitrack recorder and later to a mixdown recorder. Nothing here is written in stone, so the level controls can be switched (*reversed*, or *flipped*) should you prefer to use the "long throw" (4.5-inch) faders to set up monitor mix levels and the rotary pots, or the "short throw" (2.5-inch) faders to feed recorders.

NOTE: *One reason that non-voltage-controlled amplifier (VCA) faders are used for setting signal levels going to tape is to avoid the small amount of noise VCAs can add to the signal.*

When tracks are being "bounced," or moved and added together with or without other newly recorded signals, then the multitrack returns will be fed back into other multitrack inputs. All these inputs and returns can be routed to the mixdown buses as well. All of this occurs within a single I/O module.

Assign Switches

Using a combination of a fader and a pan control, the operator can set up the signal's level and its position between the left and right spread of the two-mix bus. The input section's assign switching controls the feed to the output buses. Here the level of the signal and any pan or split between odd and even assign switching is used to mix and route the signal to one or more of the group outputs.

This section of the mixer is called a *routing matrix* and is made up of a number of switches that route the signal to the group number they are related to. In multitrack recording it's mighty handy to have a group output for every multitrack input. The reason these buses are called groups is that many signals can be routed to them and thereby be "grouped," or summed together, before being sent to the multitrack. In this way the engineer can commit a mix of a four-piece horn section to a single track. I say *commit* because thereafter they'll be stuck with that mix.

During basic recording, the assign switch matrix is used to feed mic or line inputs to specific multitrack inputs. When overdubbing, the assign matrix is used only for the new tracks being recorded. While prerecorded tracks come into the I/O module's tape return input and are fed to the cues, mix buses, and monitors.

The assign matrix is also used to route mic or line inputs along with multitrack tape returns when "bouncing," or combining, two or more prerecorded tracks into a single track and when mixing a new track with a previously recorded one onto a different track in order to free up more track space. This "bounce" function also utilizes switches with names like *dump* and *Ping-Pong*. For the most part these just route the signal present in the channel to the multitrack bus assign switches and therefore to the group outputs that appear at the patch bay. You can use the I/O's second level control to feed additional signals (whether they are being recorded, prerecorded, or patched in) to the group output/multitrack input buses by way of the assign switches. You can also patch this signal out the bay.

Bingo, you win!

During mixdown you *always* need more channels for effects returns and/or additional MIDI instrument inputs. With I/O modules you can create an additional submix within the console by using the assign switches to route additional line inputs to the patch bay group output locations that feed the selected multitrack inputs. This signal, when patched out of the bay, can be diverted to another channel's input.

In fact many signals can be assigned to this same point, and you can use one of the two faders on each channel to mix them. Furthermore, by using two assign points, two channels, and two patch points along with the assign matrix section's odd/even pan control, you can set up a stereo submix. The cost? For every submix, you must give up an input channel in order to utilize its routing to the two mixes. With stereo submixes the cost will be two channels because one will be panned hard left and the other hard right, thus providing a stereo feed to the two mix buses. Thank you very much!

Personally, I never heavily favored one type of console format over the other. But because some folks get adamant about it, I'll brief you on some of their arguments.

Split consoles can be more flexible and easier to operate, they give producers their own section of the console, which allows them to set up the mix. In-line consoles take up less room and can be less costly to build. These arguments are strictly for console purchasers. Engineers work with whatever the studio has. But if given the choice, I prefer an I/O console for tracking (or recording) because everything you need to deal with is right there in front of you. During mixdown, a split console's flexibility and its ability to "double-up" the channels, that is, also use the input channels for effect returns or additional inputs, can be mighty handy. However, as you've seen, it is possible to use the I/O console's channel assign switching and pans to set up multiple submixes of additional inputs during a mixdown.

The SSL I/O

An exception to this definition of console configurations is the method utilized by SSL. Their G series I/O modules split the signal routing more completely than common I/O configurations. This means that they can be operated during mixdown in a way that's closer to the separate input and output routing schemes of split console designs.

Like most I/O modules, the tape returns appear at each module, and, via switching, the EQ and dynamics effects can be applied to either the channel output signal going to the multitrack bus or to the monitor bus routed to the monitor and mix buses. But the SSL console can also bus all the group outputs directly to the corresponding I/O modules. Pressing a switch on each module places it in control of all the signals that have been assigned to that bus. Therefore, each group bus can feed multiple inputs to a single module, which can then be used as a group leader, meaning that it will be capable of controlling those signals' overall level, EQ, dynamics, and further routing.

The result is that all the signals assigned to, say, multitrack input bus 1 will appear at I/O module 1's "group mix" amp. These group amps have attenuators that can be used to trim the level when multiple signals are bused to the same group. This signal can be routed through either fader to that module's assign switches.

With a common I/O scheme this would merely be a convenience, because it only saves you from having to insert one bus-output-to-channel-input patch cord when you want the inputs you've assigned to a bus to appear at a channel. However, the real difference is that the SSL's assign matrix also includes assign switching that routes the signal to the two-mix (or quad-mix) buses as well as the group outs. This one design improvement makes it possible to route many additional inputs through each channel directly to the two-mix outputs.

Now instead of just being able to route the multitrack returns *or* line inputs to the stereo bus, each SSL G series module can be used to handle a multitrack return *or* some other input *and* control the two-mix level and pan position of the multiple inputs assigned to it. This does not mean that the number of inputs the SSL G series consoles can handle during mixdown is infinite. I cannot place a finite number on the limit, but it's high.

This sure is a far cry from the functional capabilities of consoles like the Altec 250 T3, and it all came about in less than 20 years. I'm a witness. Again this information will be more readily digested by using the module signal routing flow charts shown in Figures 1-18, 1-19, and 1-20.

Back to Examples of the Charting of Gain Stages

Now we can get back to charting out gain stages without fear of getting thrown off track due to a lack of knowledge about console routing functions.

The MCI 600 Series Console

Since I have all the MCI 600 documentation, we'll get to deal with an actual patch bay configuration layout, an I/O module faceplate control and indicator layout, the specification list, and the I/O block diagram or signal flow chart. They also provide something they called the "logic" flow diagram, which is almost identical to the "signal" flow diagram but a little more spread out and slightly more detailed. Additionally, specific sections of the I/O module's schematic diagram can be used to point out anything noteworthy about the signal routing paths that run through this console.

NOTE: What you had to work with.

■ *A complete layout of the console's five connecter panels showing each individual contact point. Because with the MCI 600 series console every connection was balanced with a normal operating level of 14 dB "v" (later called dBu), there is no need to go through it all.*

■ *A systemwide, console block diagram, flow chart sized at 11" × 17", which was pretty difficult to use until you divided it up into sections, enlarged each on a photocopier, and reassembled it at 3' × 3'. The rest of the drawings were also 11" × 17" but needed to be enlarged to only 17" × 24".*

■ *A signal flow diagram of the I/O that provided more detailed information than the systemwide flow chart.*

■ *An illustration showing all the controls and indicators on the I/O module face plate with short descriptions of their functions.*

■ *A drawing of the I/O PC board layout, which showed sub-boards such as the mic preamp, equalizer boards, as well as locations of buffer, summing, send and channel output amplifier ICs (integrated circuits).*

■ *A complete schematic diagram of the electronics of the I/O module showing all components and interconnections.*

Plus they repeated all of those for each of the other console modules (communication, monitor, and main control). If this seems like a lot, they also included parts layout and schematic diagrams for both the transformer and transformerless mic preamps as well as all four of the interchangeable equalizers; the three band peak/dip EQ, the fixed or switchable Q parametric EQ, the continuously variable Q parametric EQ, and the 10-band graphic EQ. You got the whole ball of wax here!

Because this is everything you could possibly ask for, you'd think with all this information there wouldn't be any problem putting together a simple gain stage diagram. Guess again. While looking at the "logic" flow chart, I noticed a gain stage that seemed odd. It was labeled the "channel busing amp" and was located between the fader output and the assign matrix pan control. It didn't bother me that it was not noted in the "signal" flow chart or that it wasn't listed in the specification section. But when it did not turn up in the I/O's schematic diagram, it automatically became a "drawing mistake" as far as I was concerned. All this research took time, and the conclusion reached still left a small question in my mind as to how this "mistake" could have occurred.

Many times things won't show up on an I/O's schematic even if they are shown in the signal flow diagram, simply because they are not actually located on the I/O module. Yet in these situations the schematic should have routing lines that extend to the edge of the document, which are labeled as to their function and provide information as far as where it will be "picked up." That is not the case here.

What this shows is that even in the best of situations charting out a console's gain stages is not by any means an easy task. The fact is you'll more than likely never fully complete any console's gain stage chart. I tell you this up front so you won't be disappointed, but it doesn't matter if you don't, because now at least you'll know how to use all of this documentation. I for one admit to appreciating their value and usefulness.

I've found that a good place to start when trying to figure out an unfamiliar piece of equipment is its interconnections. How something interfaces with the rest of the world tells you a lot about its operation. All the hard-wired connections on this console are balanced low impedance, and, as you'll see when we get to the specifications, with the exception of the mic inputs everything operates at a line level of +4 dB as reference to 0.775 dBv or 0 dBu, so there is no need to list every connection.

I had to live through the days when all the dB references (dBm, dBV, dBv, and dBu) were being used at the same time and often interchanged incorrectly. This happened at the same time as the discrepancy over whether an XLR connector's pin 2 or 3 should be connected to the "hot" wire. You could run into either at any time and often both in the same facility. I'm glad that's all over, so there's no sense making you live through it too.

All of the MCI 600 console's decibel ratings were references to 0.775 volts being equal to 0 dB when connected to a load greater than 600 ohms. They and many others at the time called this "dBv." Later this reference became "dBu," so I've changed the MCI decibel specifications to read "dBv(u)" to reflect both the original designation and its replacement. Just for the record, though, dBv and dBu refer to the exact same 0.775 volts into a load greater than 600 ohms. When dealing with specifications it's always a good idea to remember that

$$0.775 \text{ volts} \ = \ -2 \text{ dBV} \ = \ 0 \text{ dBu (or, as here, dB followed by a "v")}$$

$$1 \text{ volt} \ = \ 0 \text{ dBV} \ = \ +2 \text{ dBu (or here, dB "v")}$$

$$1.23 \text{ volts} \ = \ +2 \text{ dBV} \ = \ +4 \text{ dBu (or here, dB "v")}$$

You might have already figured this out yourself, but when a console has an internal operating or "unity gain" level that's −2 dB (re: 0 dB = 0.775 volts) when the signal passes through a differential output stage, it is doubled, or has +6 dB added to it (remember this is volts, not watts), which results in an output level of +4 dB. You see, all this stuff really works out once you understand the basics.

We'll go over this console's patch bay layout first.

The MCI 600 Series Patch Bay

The Channel Section The mic inputs are wired directly to a specific module, but there was an optional bay section that allowed for them to be patchable to any module.

The MCI 600 Series Patch Bay Layout—The Console Input Section

Row	Designation	Use
The Console Input Section		
1	Preamp output	This point allows patching the mic preamp output signal. Note: Since this patch occurs before the channel input switching, it provides access only to the microphone signal, which then normaled to the following:
2	Preamp return	If the mic signal has been selected by the channel input switch and the equalizer has been placed in the channel routing, a signal patched in here would proceed to the equalizer input. Only when all of these conditions are met can this patch be used as an input to that channel's EQ.
3	Equalizer out	This is the post-EQ output, which then normaled to:
4	Fader input	This insert return is fed directly to the main channel fader.
5	Channel line outputs	This is the feed from the channel to the groups, which is normaled to:
6	Multitrack inputs	This patch point could be used to route the group outs to any multitrack input.
7	Multitrack outputs	While these were normally routed to specific channels, this patch could be used to reroute the multitrack outs to any channel return. This is normaled to:
8	Channel line inputs	Feeds the multitrack outputs to the channel line inputs.
The Console Output Section		
9	Prefader stereo patch out	These patch points give access to the main stereo bus by providing an insert point just before the console's main fader. It is normaled to:
10	Stereo fader input patch,	This patch point when combined with the previous one could be used to equalize or compress the entire mix.

Starting with row 9, the bay contains the system's overall console connections, such as aux outputs, the two-mix returns from stereo recorders, reverb or echo "chamber" returns, and the patch points, which are here called "tie-lines," which connect all the outboard processing gear to the console through the patch bay. We'll deal with patch bays some more when we get to Part 2, the "studio tour" section of the book.

As you can see there's nothing really astounding here. However, the fact that the preamp return patch point feeds directly into the EQ is nice to know. Now whenever an EQ isn't being used, you know it can be patched in and out to double up on a signal that might need some extra help with its frequency content.

You may already know about using the group outputs that appear on row 5. Or are they the feeds from the channels to the two mix? Not likely, because as you can see by the way the manufacturer laid out this bay, all the outputs appear right above the points where they normally are returned. While there is *no* two mix or stereo out from the I/O modules, the signal can be assigned to two group outputs, which provide you with another stereo feed from that channel if you need it.

The Manufacturers Specifications for the MCI Series 600 Console

Actually two specification tables were produced for the MCI JH-600 Series Console (see Tables 1-3 and 1-4). There was a single-page list that was part of a salesman's product brochure, which also provided configuration variables including frame sizes (18 to 52 I/O modules), VU or 101 segment plasma display metering, equalizer types, patch bay layouts, automation systems, VCA grouping capabilities, accessories, spare modules, parts, and other options. A second, more comprehensive list of specifications was provided with the operation and service manual.

Because the specifications in both lists are the same, there is no reason to duplicate both of them in their entirety here. However, the mic and line input amplifier gain and other specs as laid out in the salesman's brochure are easier to comprehend. This just shows that even having more than one set of specifications for the same piece of equipment can often end up being helpful.

Table 1-3
MCI JH-600 Specifications from Sales Brochure

Mic Preamp	
Input impedance:	Greater than 1500 ohms differential input
Output impedance:	Less than 10 ohms
Gain range:	
Low range:	12 dB to 35 dB
High range:	30 dB to 65 dB
Equivalent input noise:	(Per preamp) -129 dBv, ref. 0.775v rms
Frequency response:	$+0-\frac{1}{2}$ dB, 15 Hz–20 kHz
I.M. distortion:	Less than 0.05%, 3 dB below clipping
Max. output level at patch:	$+22$ dB, ref. to 0.775v rms into 2 k-ohms
Nominal output level at patch:	-2 dBv, ref. to 0.775v rms
Phantom power:	Balance resistors provided, individual on/off switch on each module

(continued)

Table 1-3 *(continued)*
MCI JH-600 Specifications from Sales Brochure

Line Input

Input impedance:	Greater than 7 k-ohms differential
Gain range:	Switched −6, 0, +6
Nominal input level:	+4 dBv, ref. 0.775v rms
Max input level:	+28 dBv, ref. 0.775v rms

Line Output

Output impedance:	Less than 120 ohms
Nominal out level:	+4 dBv, ref. 0.775v rms
Max output level:	+27 dBv, ref. 0.775v rms

Channel Output Bus

Headroom:	24 dB
Noise floor: One channel assigned	
JH-618:	Less than −86 dBv, ref. to 0.775v rms
JH-636:	Less than −82 dBv, ref. to 0.775v rms
JH-652:	Less than −80 dBv, ref. 0.775v rms

Channel Direct Assigned: All frames, less than −88 dBv, ref. 0.775v rms

Bandwidth:	20 Hz–18 kHz
Cross talk:	20 Hz–greater than 80 dB 16 kHz–Greater than 70 dB

Monitor Mix Busses

Headroom:	24 dB
Nominal levels at:	
Fader patch:	−2 dBv, ref. 0.775v rms
Line output:	+4 dBv, ref. 0.775v rms
Maximum levels at:	
Fader patch:	+23 dBv, ref. 0.775v rms
Line output:	+27 dBv, ref. 0.775v rms
Separation:	16 kHz–20 kHz greater than 65 dB
Noise floor:	
JH-618:	Less than −88 dBv, ref. 0.775v rms
JH-636:	Less than −84 dBv, ref. 0.775v rms
JH-652:	Less than −80 dBv, ref. 0.775v rms

Bandwidth:	20 Hz–18 kHz
Output:	Differential active balanced output
Nominal level:	+4 dBv, ref. 0.775v rms
Max level:	+23 dBv, ref. 0.775v rms
Outputs:	Transformered, impedance less than 150 ohms

Oscillator

Three oscillators are provided.

Slate:	Fixed 20 Hz @ −18 dBv, ref. 0.775v rms
Noise:	(White and pink) variable level from −40 dBv to +10 dBv, ref. 0.775v rms
Wien:	20 Hz–20 kHz, Auto-Leveling Output variable from −40 dBv to +10 dBv ref. 0.775v rms

Communicate Functions

Four provided:

Talkback:	To studio
Comm:	To cue mix
Slate:	To TKS and mix, sending to TKS adjusts mix level
Conductor:	Output to separate conductor headphones

Automation

Automation will be functionally identical to all MCI JH-50 systems.

("Specifications subject to change as innovative advancements in technology are incorporated.")

Table 1-4
MCI JH-600 Console Specifications (As Provided in the Operations/Service Manual)

Channel System

The channel system refers to that portion of the circuitry in the I/O module used for amplifying, mixing, and routing audio to the multitrack recorder. The channel level control has 12 dB of gain available. The two primary functional blocks are the nominal output level at patch and the max output level at patch.

Mic Preamp

The JH-600 mic preamp is a transformerless design

| **Nominal Output Level at Patch:** | −2 dBv, ref. 0.775v rms |
| **Max Output Level at Patch:** | +22 dBv, ref. 0.775v rms with 2k ohms load |

(continued)

Table 1-4 *(continued)*
MCI JH-600 Console Specifications (As Provided in the Operations/Service Manual)

Phantom Power

Phantom power resistors and individual on/off switch provided on each I/O.
Optional phantom supply provides 48V D.C. @ 1 amp with ripple less than 2 mv.

Input Impedance:	Greater than 1,500 ohms measured across differential inputs Constant with frequency over the audio band of 30 to 18 kHz
Output Impedance:	Less than 10 ohms
Gain Range:	
Low:	12 dB to 35 dB
High:	30 dB to 65 dB

Equivalent Input Noise:

  −129 dBv, ref. 0.775v rms

  Measured at full gain with a bandwidth of 20 Hz to 18 kHz

  Input terminated with 150-ohm resistor

Frequency Response:	+0, −1/2 dB from 15 Hz to 20 kHz
I.M. Distortion:	Less than 0.05% SMPTE Std.; Measured 3 dB below clipping
C.M.R.R.:	60 Hz, 65 dB; 10 kHz, 75 dB; 100 kHz, 75 dB

Source impedance 100 ohms at each input, measurements referenced to 775v rms.

Channel Mix Bus System

Summing Bus Headroom:	+24 dB
Output Trim:	On summing amplifier, +2 dB to −15 dB range
Noise Floor:	Referenced to +4 dBv output

# of I/Os	18	36
No channels assigned	−88 dBv	−84 dBv
One channel assigned	−86 dBv	−82 dBv

Measured with channel fader at unity, line-in mode, input terminated with 50 ohms, bandwidth of 20–18 kHz.

Cross Talk Between Buses:	Measured between a bus direct assigned and any other bus.
20 Hz	80 dB separation
16 kHz	70 dB separation

Cross Talk into Channel System from Monitor:

Measured with:

    Channel and monitor faders at unity

    Sends at -6 dB, pan at center

    Mic level at 35 dB gain, 150 ohms input termination

    $+4$ dBv into channel line input

Measured at the channel line output with channel assigned to its own bus.

20 Hz	-80 dB down from input
1 kHz	-80 dB down from input
16 kHz	-75 dB down from input

Output Amp:

Type:	Active differential output (No transformers)
Nominal output level:	$+4$ dBv
Nominal output impedance:	120 ohms balanced
Max. output level:	$+27$ dBv balanced, $+22$ dBv unbalanced
Distortion:	Less than 0.5% SMPTE STD; measured from line in to line out 3 dB below clipping.

Monitor System

The monitor system refers to that portion of the circuitry of the I/O module used for "mixing down" to stereo and sending echo and cue feeds.

Line Input:

Input impedance:	7k ohms differential
Gain range:	Switchable -6, 0, $+6$ dB
Nominal input level:	$+4$ dBv, ref. 0.775v rms
Max. input level:	$+28$ dBv, ref. 0.775v rms

VCA Fader Package:

Available gain:	$+12$ dB
Turn off:	Greater than 80 dB (20–20 kHz)
Distortion:	0.04% IM, SMPTE, measured with $+4$ dBv, ref. 0.775v rms into the channel input

Measured at channel line output.

Noise floor of fader module:	-86 dBv, ref. 0.775v rms, input shorted, 20–20 kHz bandwidth

Monitor Mix Buses:

Summing bus headroom:	$+24$ dB

(continued)

Table 1-4 *(continued)*
MCI JH-600 Console Specifications (As Provided in the Operations/Service Manual)

Pre-Main Fader Patch

Nominal level:	−2 dBv, ref. 0.775v rms
Max. level:	+23 dBv, ref. 0.775v rms

Main or Overall Fader:

Turn off:	Greater than 80 dB
Tracking:	0.5 dB between channels, from +6 to −20 dB
Cross talk:	Between 2 mix buses 65 dB separation (20–16 kHz)
Noise floor:	All channels muted; 618, −88 dBv; 636, −84 dBv; measured within 20–18 kHz bandwidth

Output Amp:

Type:	Active differential output (no transformers)
Nominal output level:	+4 dBv
Nominal output impedance:	120 ohms balanced
Max. output level:	+27 dBv balanced, +22 dBv unbalanced
Distortion:	Less than .05% SMPTE STD, measured from line in to line out 3dB below clipping

Send Buses:

Nominal output level:	+4 dBv
Max. output level:	+23 dBv
Level controls:	+6 to −20 dB range
Outputs:	Transformer coupled, 150-ohms impedance
Noise floor:	Same as mix bus

Equalization

A choice of two equalizers are available in the JH-600. Both are of state variable design. Complete curves can be found in the JH-600 manual.

Standard EQ:

Low pass filter:	18 dB/oct. slope, 3 dB point at 16 kHz
High pass filter:	12 dB/oct. slope, 3 dB point at 45 Hz
Low shelf:	
Frequency range:	30 Hz–250 Hz
Boost/cut range:	±14 dB

Standard EQ *(continued)*

High shelf:

Frequency range:	4.6 kHz–16 kHz
Boost/cut range:	±14 dB

Mid peak:

Frequency range:	180 Hz–10 kHz
Boost/cut range:	±14 dB

Phase reverse provided

Tolerance:	±10% on all controls
Nominal level:	Entire EQ operates at −2 dBv, ref. 0.775v rms
Max. level:	+23 dBv, ref. 0.775v rms
Noise floor:	−100 dBv, ref. 0.775v rms
Bandwidth:	20 Hz–18 kHz

All bst/cut controls on zero,
input terminated with 100 ohms
HP and LP filters not engaged

Optional Parametric EQ:

Low pass filter:	18 dB/oct. slope, 3 dB point at 16 kHz
High pass filter:	12 dB/oct. slope, 3 dB point at 45 Hz

Low shelf/peak:

Frequency range:	30 Hz–250 Hz
Boost/cut range:	±14 dB
Q:	2 steps, 0.4 and 0.8

High shelf/peak:

Frequency range:	4.5 kHz–16 kHz
Boost/cut range:	±14 dB
Q:	2 steps, 0.4 to 0.8

High shelf/peak:

Frequency range:	180 Hz–10 kHz
Boost/cut range:	±14 dB
Q range:	Continuously variable from .3 to 3

Phase reverse provided.

Tolerance:	±10% all controls
Nominal level:	Entire EQ operates at −2 dBv, ref. 0.775v rms
Max. level:	+23 dBv, ref. 0.775v rms
Noise floor:	−100 dBv, ref. 0.775v rms
Bandwidth:	20 Hz–18 kHz. All boost/cut controls on zero. Input terminated with 100 ohms, HP and LP filters not engaged.

(continued)

Table 1-4 *(continued)*
MCI JH-600 Console Specifications (As Provided in the Operations/Service Manual)

Communications Functions

Four communicate functions are provided:

1. **Talkback** to studio

2. **Communicate** to cue mix

3. **Slate** to tracks and mix, sending to tracks automatically adjusts mix output level

4. **Conductor** to separate conductor headphones

Additionally, three oscillators are provided:

1. **Slate** is a fixed 35 Hz oscillator providing a slate tone at −18 dBv, ref. 0.775v rms.

2. **Noise** source provides both white and pink noise variable level from −50 dBv to +10 dBv, ref. 0.775v rms.

3. **Weinbridge** oscillator provides 20–20 kHz auto-leveled balanced output variable from −50 dBv to +12 dBv, ref. 0.775v rms.

Control Room and Studio Monitor Systems

Inputs:	Differential, 5k ohms, balanced; nominal level of +4 dBv, ref. 0.775v rms
Outputs:	Transformer balanced 150 ohms output impedance
Available Gain:	12 dB
Mute/Dim:	
On control monitor:	Dim of −40 dB
On studio monitor:	Mute of greater than 80 dB (30–18 kHz)
Separation L-R:	Better than 65 dB (20–16 kHz)

Power Supplies

Two supply chassis are used:

Audio Supply:

Cables:	Std. 35 feet long; max. 60 feet long
Outputs:	Two ±8v @ 8 amps supplies:
Ripple: less than 1 mv	(1 618 only) (1 636 only)
O.V.P. and current limit at 8 amps	
Fused @ 10 amps	
Optional—48v @ 1 amp supply:	
Ripple less than 2 mv	

Power Supplies *(continued)*

Current foldback protection @ 1 amp

Fused @ 1 amp

Power Inputs:	100, 120, 240 VAC 50–60 Hz
	Max. power in: 1,100 watts

Peripheral Supply

Cables:	Std.: 35 feet long
	Max.: 60 feet long
Outputs:	1. ±8 V.D.C. fused at 4 amps
	2. +8 V.D.C. fused @ 2 amps and @ 8 amps
	Ripple approximately 30 mv (local regulators in console automation system)
	1. +24 V.D.C. @ 4 amps Ripple less than 3 mv Current limit @ 4 amps and fuse protected @ 6 amps
Inputs:	100, 120, 240 V.A.C.
	Max. power in: 800 watts

Automation

(The automation system is the JH-50 subsystem.) Faders, control buttons, and LEDs are built into the fader packages.

Information Storage System:
Any two or more tracks of an audio recording system that meets the following criteria:

1. is able to sync with the master tape

2. is able to record and play back a 14 kHz signal

3. has at least 15 dB isolation between sync tracks at 10 kHz

System Accuracy:
A to D and D to A is adjustable to better than system resolution.

System resolution (0.4 dB becomes system accuracy when this adjustment is properly made.

Dither:	(Exclusive of system 0.4 dB maximum resolution) 0.4 dB maximum
Cumulative System Error:	(Unlimited passes) Total error is 0.4 dB. This error does *not* accumulate.

(READ mode ONLY). Other modes are subject to operator changes.

(continued)

Table 1-4 *(continued)*
MCI JH-600 Console Specifications (As Provided in the Operations/Service Manual)

Automation *(continued)*	
Scan Time:	102 milliseconds. The *scan time* is *not* a delay and does not accumulate.
Bounce Delay:	1.2 ms per pass

The bounce delay is a delay that accumulates with each update pass.

Data Rate:	9,600 Baud (14 kHz bandwidth)
Compatibility:	Tapes made on a standard automated JH-400, JH-500, or JH-600 console may be played back on any other standard automated JH-400, JH-500, or JH-600 console.

Optional Auto Remote:
Provides recording track selectors and second system clear button

(Main system clear button located on Automation Main Control Panel)

Thermal Dissipation

It is often helpful to an air conditioning engineer to know how much heat is generated by the equipment in a particular room.

The following is the heat dissipated by the JH-600 Console:

JH-600 console:	88 BTU/min. max.
w/automation:	22 Kh calories/min. max.

Like I said, I had every bit of information you could want on this console. So in this case we get a chance to use everything including the patch bay layout, both specification lists, the I/O face plate layout diagram, the console signal flow charts, and section of the I/O schematic diagram that cover the mic preamp and line input gain electronic workings, as well as the channel assign routing network to help us put together the gain stage chart for this console.

Before we deal with going through the console specifications and the I/O signal flow, we'll check out the controls and indicators relevant to our task by taking a stroll along the I/O's faceplate layout diagram, shown in Figure 1-21.

The *assign switches* in the section on top are not necessarily the beginning of the signal flow, but the mic and line input controls located there can be.

The *DUMP* switch is used for bouncing or submixing signals. It's effect on the signal flow is purely operational because it is not a gain stage, so it does not apply to our task.

The EQ will not be part of the gain stage chart because it is active and does not load down the signal. However, it's still important to understand the gain limits and all the control functions such as the "IN" switching along with "CHN" (channel) and "monitor" routing assignment switches, if only because this knowledge is necessary for proper use. Yet not understanding how they affect the EQ filter could result in disaster.

The sends: In order to set up proper send levels, it is essential that you know their output level capabilities, impedance, and whether they are set up for balanced or unbalanced operation.

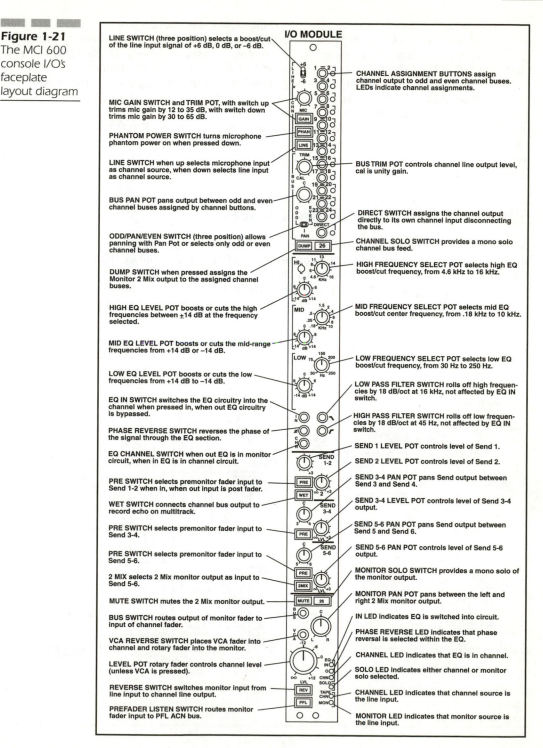

Figure 1-21
The MCI 600 console I/O's faceplate layout diagram

I/O MODULE

LINE SWITCH (three position) selects a boost/cut of the line input signal of +6 dB, 0 dB, or –6 dB.

MIC GAIN SWITCH and TRIM POT, with switch up trims mic gain by 12 to 35 dB, with switch down trims mic gain by 30 to 65 dB.

PHANTOM POWER SWITCH turns microphone phantom power on when pressed down.

LINE SWITCH when up selects microphone input as channel source, when down selects line input as channel source.

BUS PAN POT pans output between odd and even channel buses assigned by channel buttons.

ODD/PAN/EVEN SWITCH (three position) allows panning with Pan Pot or selects only odd or even channel buses.

DUMP SWITCH when pressed assigns the Monitor 2 Mix output to the assigned channel buses.

HIGH EQ LEVEL POT boosts or cuts the high frequencies between ±14 dB at the frequency selected.

MID EQ LEVEL POT boosts or cuts the mid-range frequencies from +14 dB or –14 dB.

LOW EQ LEVEL POT boosts or cuts the low frequencies from +14 dB to –14 dB.

EQ IN SWITCH switches the EQ circuitry into the channel when pressed in, when out EQ circuitry is bypassed.

PHASE REVERSE SWITCH reverses the phase of the signal through the EQ section.

EQ CHANNEL SWITCH when out EQ is in monitor circuit, when in EQ is in channel circuit.

PRE SWITCH selects premonitor fader input to Send 1-2 when in, when out input is post fader.

WET SWITCH connects channel bus output to record echo on multitrack.

PRE SWITCH selects premonitor fader input to Send 3-4.

PRE SWITCH selects premonitor fader input to Send 5-6.

2 MIX selects 2 Mix monitor output as input to Send 5-6.

MUTE SWITCH mutes the 2 Mix monitor output.

BUS SWITCH routes output of monitor fader to input of channel fader.

VCA REVERSE SWITCH places VCA fader into channel and rotary fader into the monitor.

LEVEL POT rotary fader controls channel level (unless VCA is pressed).

REVERSE SWITCH switches monitor input from line input to channel line output.

PREFADER LISTEN SWITCH routes monitor fader input to PFL ACN bus.

CHANNEL ASSIGNMENT BUTTONS assign channel output to odd and even channel buses. LEDs indicate channel assignments.

BUS TRIM POT controls channel line output level, cal is unity gain.

DIRECT SWITCH assigns the channel output directly to its own channel input disconnecting the bus.

CHANNEL SOLO SWITCH provides a mono solo channel bus feed.

HIGH FREQUENCY SELECT POT selects high EQ boost/cut frequency, from 4.6 kHz to 16 kHz.

MID FREQUENCY SELECT POT selects mid EQ boost/cut center frequency, from .18 kHz to 10 kHz.

LOW FREQUENCY SELECT POT selects low EQ boost/cut frequency, from 30 Hz to 250 Hz.

LOW PASS FILTER SWITCH rolls off high frequencies by 18 dB/oct at 16 kHz, not affected by EQ IN switch.

HIGH PASS FILTER SWITCH rolls off low frequencies by 18 dB/oct at 45 Hz, not affected by EQ IN switch.

SEND 1 LEVEL POT controls level of Send 1.

SEND 2 LEVEL POT controls level of Send 2.

SEND 3-4 PAN POT pans Send output between Send 3 and Send 4.

SEND 3-4 LEVEL POT controls level of Send 3-4 output.

SEND 5-6 PAN POT pans Send output between Send 5 and Send 6.

SEND 5-6 PAN POT controls level of Send 5-6 output.

MONITOR SOLO SWITCH provides a mono solo of the monitor output.

MONITOR PAN POT pans between the left and right 2 Mix monitor output.

IN LED indicates EQ is switched into circuit.

PHASE REVERSE LED indicates that phase reversal is selected within the EQ.

CHANNEL LED indicates that EQ is in channel.

SOLO LED indicates either channel or monitor solo selected.

CHANNEL LED indicates that channel source is the line input.

MONITOR LED indicates that monitor source is the line input.

The *"bus"* and *"revenue"* switches are used for signal routing and are thus not part of our gain stage chart. So as far as our present task goes, they are purely operational features.

The *mute* switch and the *monitor pan* have obvious uses, but here it's important to note that they affect only the two mix and monitoring buses, although the mutes can be utilized during a "bounce," such as when combining signals to the group outputs and the multitrack.

The *rotary* and VCA *faders* (the latter being shown on a separate diagram) are important for their control range over their signal's gain level.

While determining the level values of our gain stage chart, we'll refer to both the manufacturer's *signal flow chart*, shown in Figure 1-22, and their specifications, as listed in Tables 1-3 and 1-4 as well as the schematic diagrams for the mic preamp, channel assign, and line input sections of this console.

The MCI 600 Gain Stage Chart

As you'll remember from our brief discussion of I/O functioning, they are able to handle two inputs, a multitrack recorded output or return, and either the mic or the line "live" inputs (see Figure 1-23). They also provide two basic output feeds, one to the groups that normally feed the multitrack and a "mix" output feed. We'll go through the charting of only a single routing possibility here, the most important one, from the mic output to the multitrack input.

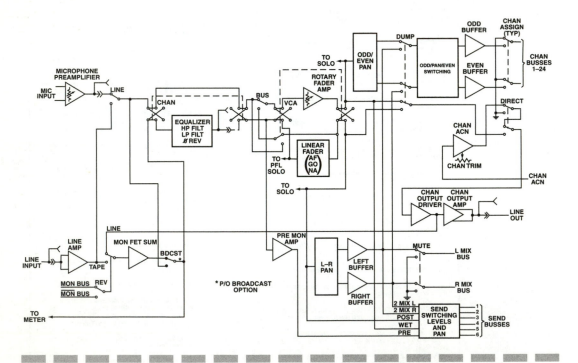

Figure 1-22
The MCI signal flow chart

This will be accomplished using the I/O faceplate layout, the schematic diagram, the specifications, and the signal flow diagram. Together they give us the following information.

Mic Preamp   Unlike what's depicted in the manufacturer's signal flow diagram, the mic preamp gain control is not just a simple continuously variable resistor (pot) but a combination of a dual potentiometer whose series connection with fixed-value resistors is engaged by the "high/low" gain switch (as seen in the schematic diagram of Figure 1-24A). This preamp's gain, according to the specifications, has two switch-selectable ranges, 12 to 35 dB and 30 to 65 dB. This equates to a gain of 65 dB from the preamp stage for our chart.

NOTE All decibel specifications are referenced to 0.775 volts into a load greater than 600 ohms equaling 0 dB. This would equal −2 dB referenced to 0 dBV, where 0 dB = 1 volt, and −4 dB referenced to 0 dB = +4 dBu or 0 dB = 1.23 volts.

The *mic preamp out and return patch point level* is rated as −2 dB "nominal" with a maximum output level of +22 dB referenced to 0.775 volts RMS (or +18 referenced to +4 dBu nominal operating level).

Since the distortion rating was measured at a level 3 dB below clipping, it is safe to assume that clipping would occur at +25 dB. This means you might actually get away with cranking the mic preamp output up for a +22 dB level increase at this patch point should you need a lot of gain.

However, the patch should be marked at the nominal level of −2 dB on the chart, with a notation of the possibility of as much as a +22 dB output.

As we've already discovered, while the mic preamp return patch enters into the circuitry after the mic preamp, it can deliver the patched signal to the EQ only if that channel's source selector switch is in the microphone input position and the EQ input is switched into the channel bus routing.

According to the specifications, the EQ has a nominal operating level of −2 dBv(u), referenced to 0.775 volts, with a maximum level of +23 dB. In addition to the −2 dBv(u) rating, just to be safe, the return patch is noted on our chart to have a maximum level of +23 dB, whether it ends up going through the EQ or not. This fact should be marked down because being able to slam a signal into a console's EQ at 18 dB above the normal professional audio level of 0 dB referenced to +4 dBu without the risk of incurring "unfortunate" consequences is definitely noteworthy.

Backing up a bit, the *line input* has a normal level that is rated in the specifications at +4 dBv(u), referenced to 0.775 volts RMS.

NOTE: *See how using the lowercase "v" in a decibel reference when it is also used to indicate or stand for the term volts can be confusing? Try adding in a bunch of "dBV" referenced to the mess.*

■■■ ■■■ ■■■

Figure 1-23
The MCI 600
I/O gain stage
chart, in four
sections: the
mic input, the
multitrack or
line input, and
the rest of
the I/O

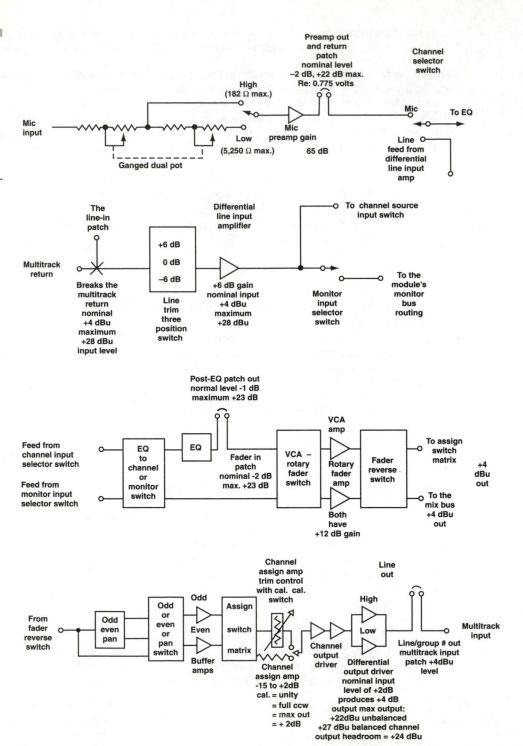

Figure 1-24
Sections of the MCI 600 Series I/O schematic diagram showing: A. the mic preamp; B. the line input gain controls; C. the channel assign gain control. You may find the mic gain control as drawn in the console gain stage chart.

A

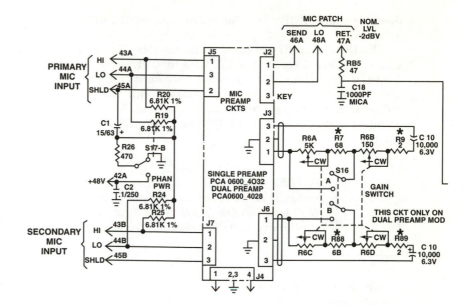

B

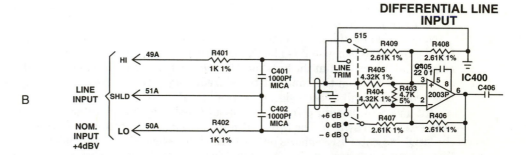

C

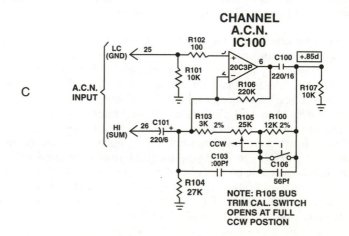

Just remember: 0 dBV = 1 volt, 0.775 volts = −2.2 dBV, 0.775 volts = 0 dB(u) and 1.23 volts = +4 dBu.

The *input* circuitry is rated to be capable of handling a maximum input level of +28 dBv(u), again referenced to 0.775 volts (or 0 dBu). The differential line input amp (see Figure 1-23B) is more than likely provided more for its common-mode noise rejection than any gain because only +6 dB of boost is available. This gain stage will be marked in our chart as +6 dB, with a notation stating that this level switching provides for a normal (at +4 dBu operating level) input range of −2 dB (sound familiar?) to +10 dB and a maximum input of +28 dBv(u).

NOTE: *Looking at the line input circuitry we find that while the multitrack return is normaled to this input, the line input patch breaks this normal, so the insertion of the patch is the determining factor as to whether the signal continuing on will be the multitrack return or a patched-in signal.*

We see from the I/O faceplate descriptions and the schematic diagram in Figure 1-24 that the line trim is not a continuously variable pot as shown on the manufacturer's signal flow chart, but in fact is a three-position (−6, 0, or +6 dB) level *switch* that feeds the differential line amp. After the line amp, the signal goes to the channel source input selector switch and to the monitor selector switch.

Next in line for the signal is the EQ selector switch. We will not be charting out this equalizer's gain stage because it is electronically active and therefore causes no insertion loss. We can tell by the specs that the boost and gain of each of the three bands is 14 dB, but it is impossible to speculate on the bandwidth and frequency settings. We therefore cannot predict how this processing will affect a signal's overall level.

But a very important point to note about this EQ is that its filter switching is independent of the EQ on/off switch. The filters also have no light indication as to their on and off status, the "EQ" indicator LED shows only whether the whole EQ section is on or off. Because these filters (as we see from the I/O faceplate layout switch descriptions) are not affected by the EQ In/Out switch, they require constant vigilance as far as their status, or as my Jamaican friends would put it, "They need checkin', mon."

The EQ Out and Fader Input Patch Points   There are patch points between the EQ output and the fader input on the channel routing. But because the fader select switch follows this patch, it is strictly a "channel" routing patch. We know from the specifications that the EQ's normal operating level is −2 dBv(u) and that it can handle levels up to +23 dBv(u) "ref .775v rms." Therefore, the rating of the "normal" output at this patch is −2 dBv(u), with a notation on the chart of the +23 dBv(u) maximum. The *channel fader input* patch level is a little tricky because no specifications are given for its input level range.

But so far this console has held true to form. It's patch points are −2 dBv(u) with a maximum of +23 dBv(u), while its inputs and outputs are generally +4 dBv(u). However, the fader input patch may not be considered an "operating level" or a "line" level patch. Luckily, there are specifications for the console's main fader patch inputs, and they are −2 dBv(u) nominal with a maximum rating of +23 dBv(u). So it's safe to chart the fader input patch level at −2 dBv(u) and note a maximum of +23 dBv(u).

The *fader amp's* gain specifications are only listed for the VCA (here "+12 dB"), but it would not be out of line to assume that both the VCA and the rotary faders would be designed to provide equal gain structures. So the two faders are rated for +12 dB of gain on our chart.

After the fader reverse switch, the signal reaches the assign section (at the top of the module). Here, the "odd-even-pan" switch selects how the source will be fed from the bank of selector switches that feed the output buses.

This output signal then passes through the *channel assign amp*. It's trim control is handy for adjusting or mixing levels when multiple signals are being fed to one location, such as when bouncing multiple tracks down to a single track. The schematic diagram points out that this continuously variable pot has a switch that is activated at its full CCW position and sets the signal level to "unity gain."

The specifications do not list the channel assign amp's gain stage, but they do list a summing bus output amp with trim. The only "summing" amplifier on the I/O is part of the input circuitry that feeds the monitor bus input, and it does not have a level trim control. The fact is that the only other trim control on the I/O module is the mic preamp "trim." However, this trim does not have its full *counterclockwise* (CCW) position designated on the faceplate to be "cal" (or calibrated so as to provide unity gain) level. Therefore, it's safe to assume that the *channel assign amp* trim located in the same (bus assign) section as the mic trim (see Figure 1-24C) is what the specifications are referring to as the summing bus output trim, which has a range of −15 to not just 0 dB but +2 dB. Without all this investigation, there would be no way to tell that this trim is not just cut-only control when out of its full CCW unity gain or "cal" position.

On our chart this channel assign amp and trim combination will be labeled per the specification rating of −15 dB to +2 dB, with a notation that at its full CCW, or switched-off, position, the resulting output is "unity gain." But what is the "unity gain" level?

The channel assign amp is followed by two more amplifier stages, an *output driver* and a dual op-amp *differential output amplifier*, which is used to balance the output. Neither of these stages has gain controls. However, the specifications show the active "No Transformers" differential output amp to have a nominal output level of +4 dBv(u).

The line or group output and multitrack input patch after the differential amp is not shown on the I/O schematic. Instead the two differential output amplifier conductors are brought out to the edge of the drawing and are labeled simply as "line out," with a note that the line output nominal level is +4 dBv(u). Therefore, it is reasonable to assume that the group output that feeds signal to the multitrack inputs at a level of +4 dBv(u) is here considered to be unity gain.

NOTE: *At other route locations within the console, −2 dBv(u) referenced to 0.775 volts RMS can be considered "unity" gain.*

According to the specifications, the output amp's maximum ratings are +27 dBv(u) for balanced and +22 dBv(u) for unbalanced operation, thus the channel output bus has 24 dB of headroom.

While not part of our charting exercise, it is still important to note that the send buses have a nominal output level of +4 dBv(u), a maximum output level of +23 dBv(u), and a range of −20 dB to +6 dB.

This fact reminds me of several owners of TRIDENT 65 consoles who wanted me to cure noise problems due to that console's unbalanced sends. Here attempting to feed outboard processors that have balanced inputs and operating levels of +4 dBu with unbalanced low-level (−10 dBu) signals by cranking up the send output amps could result in overdriving those amps to the point where their output noise becomes audible.

Yet these studio owners were bewildered when I explained to them that they needed to purchase a balancing device (active or passive) for each of the sends. I guess they just didn't understand a couple of the basics, like gain stages, impedance matching, balanced and unbalanced operation, amplifier maximum output specifications, and the realities of cost-effective equipment having lower prices because they do not provide every bell and whistle. So when you end up needing something like balanced outputs, you must purchase the needed interface. At least this will never happen to you, now that you have a grasp of the basics.

This exercise showed that even when you have all the available information at your command, putting together a complete Altec 250 T3 type of gain chart is next to impossible. On the other hand, knowing how to use all of the manufacturer's operational and service documentation results in gaining a huge amount of information, some of which will lead to previously unimagined operational usage.

The Solid-State Logic (SSL) 4000 G Series Console

Putting together a gain stage chart is just about impossible when all you have access to is an operational manual, a patch bay layout, and a system-wide signal flow chart, as I had of the SSL 4000 G series console. At the very least a list of specifications also is needed. But this exercise was still valid because most of the time this is all you will have at your disposal.

In this case, the information provided by the manufacturer, SSL, in their operations manual is quite extensive. Within this text, which describes the operation of this console, resides a segmented but complete I/O faceplate control and indicator layout, plain English explanations, and examples of complex signal flow routing options with descriptions of the various control and routing functions. See Figures 1-22 through 1-25. This section also covers the patch bay layout completely.

The SSL 4000 G Series Patch Bay Layout   The first two horizontal rows of patch points on this bay route the microphone inputs to the I/O modules.

   ROW A: This provides all the mic outputs as they are hard-wired to the console and they are normaled to ROW B.

   ROW B: This row gives access to each microphone and allows relocating their outputs to different modules.

These first two rows are usually optional (added at additional cost) on many consoles, but they are so useful it is often worth the price. For example, when one instrument has to be physically isolated (a loud trumpet moved away from the woodwinds) for separation, even if the microphone connector on the studio input panel routes the trumpet mic to a console location far from the rest of the horns, you can use these patch points to keep the location of the whole horn section all together on adjacent channels.

ROW C: Multitrack returns

ROW D: Channel line inputs

The multitrack outputs are normaled to the channel inputs of their respective module (i.e., track 1 to module 1). While this is by no means unique, the fact that the multitrack returns appear again on ROW J means that you can (as with most I/O consoles) not only route a multitrack return's to either the channel or the monitor bus, but here you're given a patch point at *both* input points. The access this provides to each module's inputs is significant.

ROW E: The channel "insert" sends (channel patch out)

ROW F: The channel "insert" returns (channel patch return)

While this is the main patch point for adding effects to a signal, it is by no means the only one. But SSL has still provided this patch point with an amazing amount of variability, in terms of locating this patch point within the module's routing in relation to the position that's been chosen for both the EQ and dynamics sections (see the description of the EQ).

ROW G: The group or bus outputs normaled to corresponding number on ROW H

ROW H: The multitrack inputs

There is no row I. I don't know the reason for this, but there must be one. Otherwise they could have skipped H and made it G for the group outputs and row I for the multitrack inputs. But that's simply my take on it.

ROW J: Multitrack returns normaled to ROW K

ROW K: The channel's tape monitor inputs

ROW L: Provides the quad-mix output buses and the four pre- and postmix bus fader patch points.

All the quad mix feeds are normaled to ROW M.

ROW M: The inputs to all the mixdown recorders and input to the monitor speaker amplifiers.

What is most noteworthy here are the pre- and postfader insert patch points. These four faders are also used by the main quad output compressor/limiter dynamics circuitry. This means that these pre- and postfader insert points give you access to two stereo compressor/limiters located right there at the main output section of the console. While you may want to have a stereo compressor riding levels on your full stereo mix, there will be times when you're not mixing to quad or doing a 5.1 mix. Now, this dynamics processor can be used on another audio chain such as when delivering sends to the delicate inputs of some older spring and plate reverberation units. During recording you can utilize the second set of dynamics controls on all kinds of things. For instance, I like placing it on the cue mix to reduce the chance of sudden abusive levels in the musician's headphones.

The main limitation in all this is the fact that all four of these dynamics processors share the same control setting. In fact, since the ratio only goes to 10 to 1, this processing actually provides no limiting, just compression.

That's about it for the inputs because the rest of the patch bay, ROWS N to U (there's no row "O"?), handles the various auxiliary and other outputs, echo returns, automation, SMPTE, and all the outboard interconnections.

Again, adding a service manual with complete schematics and specifications listing would have made charting this console's gain stages fairly easy. But you work with what you have.

As already stated, SSL does not provide an I/O faceplate control and indicator diagram; instead, the faceplate segments are presented in the sections of the manual where their functions are described. This is logical, but I find looking at a complete layout helpful in understanding the signal flow through the module, so I took the time to copy, cut, and assemble all the sections into a single I/O faceplate control and indicator layout (see Figure 1-25).

No big deal, I've done it many times before.

Back when I was an assistant engineer, I'd used a copy machine to shrink and multiply an I/O layout after whiting out all the control knob pointer positions. I'd then assemble a mock-up of the complete console on an 11 by 16-inch sheet of paper. At the end of a session, I would mark off all the control settings and place the sheet in the box with the multitrack tape. When the engineer and producer showed up for the next session I'd have everything reset to where it was when they left off. This was before "total recall" was available.

Engineers and especially producers have so much to deal with that anything you can do to lighten their load will be appreciated (This usually equates to more work opportunities for you).

Starting at the top of the module in Figure 1-25, you'll notice that the beginning numbers of the assign switching is chopped off. This seemed like a superior alternative to reducing the size of everything as I felt it was best to have the complete module on a single page.

The *assign matrix* makes it possible to route the channel signal to any one of the 32 group output buses *and* the quad-mix outputs. A pan on/off switch and a rotary control are used for

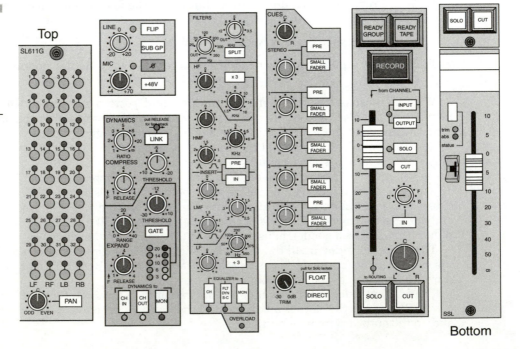

Figure 1-25
The SSL 4000
G series
faceplate,
starting at the
top of the
module

odd/even positioning between the selected group outputs and to place the signal within the stereo left/right field and the front/back of the quad-mix buses.

Channel input selection is provided by the console's main controls and the channel "FLIP" switch. Access to the line in is via row D on the bay, and the normalized mic inputs can be changed by using the row B patches. Green or red LEDs indicate the choice of line or mic input.

Phase reversal and phantom power on/off switching is also provided, but you'll have to check the signal flow diagram to find their location point in the signal routing. Here the phantom power is, as you'd expect, switched in before the mic trim control. Of more interest is the fact that the phase reversal occurs after the line/mic input switching. This means that it affects the input signal whether it's a mic, a patched line, or the multitrack return, this offers the possibility of use for some interesting effects.

The next question about this section concerns the actual amount of gain provided. The line pot is labeled −20 to +20 and the mic +4 to +70. Yet another illustration shows the mic gain pot designations as +4, +10, +16, +22, +28, +34, +36, +40, +46, +52, +58, +64, and +70, which indicates that it's a stepped control and not continuously variable.

The operational manual describes this option as a 6 dB incremental increase per step on the G series consoles. Further investigation reveals that this stepped attenuator or gain control is for a special broadcast mic preamp input. Transformer isolation is favored in broadcast applications because these folks have to deal with some pretty questionable inputs such as POTS lines! This shows the amount of variability found within just the SSL 4000 G series line of consoles and that it's very important to take the time to know the equipment you're working with.

The *subgroup* switch is the final control in this section. It interrupts all other signals routed to the channel input, kills both the green and red LEDs, and feeds the group output bus signal into the channel through the group output section.

This seems like it's a good time to deal with the group output section.

We've seen that the input to this group output section originates from just before the group output patch at row G on the patch bay, which is normaled to and feeds the multitrack. By pressing the *subgroup* switch in the assign section, the group bus out is fed to the channel input through the group output section. It has a trim control that is cut-only and can attenuate the input by up to 30 dB. This is useful when many signals have been assigned to the same group. The *float* switch is used when bouncing, or combining, signals to one track because it routes the signal through the mix fader and therefore allows it to be used to set the level. Pressing the *direct* button causes the channel's prefader signal to be sent directly to the multitrack input; this bypasses the assign matrix and all subsequent trims and is used for a cleaner signal feed. It also can be used to free up the channel so that it can be used to handle a subgroup input.

The 4000 G *dynamics section* is composed of a compressor/limiter and an expander/gate. The compressor *ratio* range control sets the processing amount from compression (less than 20:1) to limiting (here, up to infinity to 1).

The *threshold* sets the effect's activation point to occur anywhere between −20 and +10 dB. The operational manual reveals that the threshold control also affects the output level, which explains the absence of a separate output control.

The limiter/compressor's *release* range is continuously variable from 0.1 to 4 full seconds. All I can say is that a 4-second release time at an infinity-to-1 ratio is serious (guitar sustain) dynamics squashing. There's a notation on the faceplate that states that the release control when pulled up yields a fast attack.

The operational manual confirms this and goes on to provide further details. This fast attack equates to 20 dB of gain reduction in 3 msec. With the release knob in the down position the attack time ranges from 3 to 30 msec and is controlled automatically, dependent on the dynamics of the signal. The right hand of the two sets of five LEDs indicates the amount of gain reduction to be 3, 6, 10, 14, or 20 dB.

The *gate* switch selects either the expansion or gate function (gate when depressed). We find the common range, threshold, and release controls. Yet there's a discrepancy in the operation manual concerning the range or degree of the expansion or the depth of the gate. The expander is rated as having a 2:1 range, and the gate is stated to be capable of delivering an infinity-to-1 reduction in level.

However, the *range* control, which also functions as the on/off switch (off in the full CCW position), is described as providing only 50 dB of depth, which is obviously not infinity. I bring this up to show the importance of checking specification lists, because it would quickly clear this up.

The threshold is noteworthy in that it incorporates some automatic compensation circuitry, which adjusts the turn on/off point of this effect in accordance with the signal's dynamics. Release time is variable from 100 milliseconds to a full 4 seconds.

The *release* control knob also serves a dual function because it incorporates an up/down position switch that changes the *attack* time (or speed of the onset of the affect). In the down setting it takes 1.5 msec for 40 dB of gain change, but when pulled up, the 40 dB of gain change takes only 100 microseconds (that's one one-hundredth of a second), fast enough to "catch" any signal you'll ever come across.

The amount of gain change is indicated by the left-hand column of the two sets of LEDs.

NOTE: *According to the faceplate markings, the level range of the two columns of five LEDs is from 3 to 20 dB. So they can at most only indicate the amount of gain reduction (or expansion) taking place in a very general manner. Yet for critical situations, patching the signal to dual peak-reading meters would be the norm anyway. These two sets of LEDs, the left, expander/gate, and the right, compressor/limiter, are easily differentiated with a quick glance because the left column utilizes all green LEDs and the right uses three yellow LEDs for the lower levels and two red for the upper levels.*

This dynamic section also has a *link* switch that connects its side chain to the side chain of the dynamic section located on the module to the right of it. Thus, when two channels' dynamic sections are linked together, they can be used to process a stereo signal in which both the left and right sides will be affected equally.

The SSL 4000 G series *Equalization* section really handles three functions: filtering, equalizing, and signal routing. Since the routing function is so important and a bit out of the ordinary, we'll take that on first.

Some of the EQ assign switch functions are fairly obvious. The monitor (*mon*) switch places the EQ in the monitor routing path. The channel (*ch*) switch puts the EQ into the channel routing, but this switch also positions the filters after the EQ. The *flt dyn sc* (filter to the dynamic's side chain) switch places the filters in the dynamic's side chain. The *split* switch puts the filters between the channel input and the dynamic sections, but the EQ can still be placed in any of the other positions. Pretty versatile, no?

The EQ section also contains switching that affects the channel insert patch point return's routing. According to the operational manual, this insert's send level is 0 dB "relative to a normal operating level of +4 dB." It's safe to assume that this means that the send level is +4 dBu, but again, it's nice to have the specifications available to confirm this.

The insert patch sends appear at row E on the patch bay and are normaled back into the module via the insert returns on row F. This patch point is prefader but can be switched at the EQ section to return either pre- or post-EQ. But that wasn't enough variability for the SSL G Series designers because this console's motherboard has jumpers that also have an affect on the insert return's routing. These jumpers are used to set the insert to either always return to the channel bus or for them to return to wherever the EQ has been routed, be it the monitor or the channel bus.

The filtering section should be easily understood since it comprises only two simple bandpass networks. The high-frequency filter reduces the level by 18 dB per octave and the low by 12dB. The two frequency roll-off points are controlled by rotary knobs, but there's no indication in the operational manual as to whether the frequency selection is stepped or continuously variable. Again a set of schematics diagrams or a specifications list would clear this up in no time, as would having the console right there in front of you with some signal routed through it.

The operational manual does indicate that the rotary control's full CCW position turns the filter off.

The I/O's faceplate labeling shows the range of the low filter to be from 20 to 350 Hz, increasing with clockwise rotation, while the high-frequency ranges from 12 kHz *down* to 3 kHz also with clockwise rotation. I do not know the reason for this.

The EQ itself has four bands: the high-frequency band has variable cutoff frequency and a continuously variable boost/cut control but no bandwidth variation. It is therefore semiparametric. The high-midband is full parametric because it has a variable boost, or cut, center frequency, and bandwidth, or "Q," control. The low-midband is also full parametric, while the low-frequency band is semiparametric. It would be a waste of energy for me to list every frequency, level, and bandwidth range of this equalizer.

However, two additional switches on this EQ must be mentioned because they make a big difference in its capabilities. The high-midfrequency band has a "multiply by three" switch and the low-midfrequency band has a "divide by three" switch. Both of these affect the center frequency that's been selected. Hitting the "times three" button makes this EQ capable of "doubling up" in its high-frequency content coverage. This is mighty handy when dealing with particularly annoying sibilances, resonances, buzzes, or any nastiness that's part of the signal. The "divide by three" switch does the same for the low midband frequency coverage.

This may not seem like a big deal, but I've personally tested the phase response of many equalizers and while there have been both bad and good results, I know that routing a signal through two separate equalizers usually ends up in unacceptable audible time smear distortion.

Being able to attack a problem from two fronts within the same EQ is one of the main benefits of using a well-designed parametric EQ with fully overlapping bands. With this console you have this capability right there on every module.

At the very bottom of the EQ section is an LED marked on the faceplate to be an "overload" indicator. The operational manual tells us that it "monitors" the signal path at three different points and indicates when the signal is within 4 dB of clipping. The three points are at the channel "front end pre any signal processing," after the insert point and after the channel fader.

The cue, or *send*, section is composed of one stereo (with pan control) and four mono send level controls. Each one has switching that allows that send bus to take its input signal from either the small or VCA fader routing as well as pre or post that fader's level change. Another difference with the SSL sends is its odd mute switching. Instead of a separate mute switch for each send, the send level pot also provides the (push-on/push-off) mute functioning. Personally, I'm a little thick, so I like separate mute switches *with* some kind of indication as to mute's functional position or state.

I also had a question about the on/off switching capabilities of the main send section located at the center of the console. These level attenuation controls have range designations on the faceplate that start at "infinity" in the full CCW position. These controls are also marked as ranging up to 0 in the full clockwise (CW) position.

To my way of thinking, zero equates to no attenuation. The achievement of infinity, or complete, with no signal present, requires a switch, and in fact a switch is shown on the console flow chart.

I've used this function, for example, to mute all the rhythm instrument sends to a very chunky phased and flanged effect during the chorus, because it proved to be just too much during that part of the song.

The question about the main send section's on/off switching (or mute) capabilities as well as its EQ and output was solved by my getting the opportunity to look at this section's schematic diagram. The schematic shows the main send level control to be connected at its full CCW position to ground but with no switch to absolutely ensure "infinite" attenuation. The specifications would be needed to find this control's maximum attenuation. But while I'm sure it's not infinite, it is very doubtful that SSL's customers would tolerate *any* leakage here.

These level controls are also shown feeding the signal to an op-amp stage. Therefore, I'd also be surprised to find that this circuitry is not capable of producing some amount of gain, as opposed to the designated maximum output of no attenuation.

This schematic diagram for the main cue mix bus section also shows the high- and low-equalization circuitry and this section's final gain stage. But the operational manual provides more information concerning the EQ, giving the level variation amount and center frequencies, "15 dB of HF (10 kHz) and LF (100 Hz)" of both bands. The schematic diagram explicitly shows a dual op-amp differential output stage with its balanced ("0 degree and 180 degree") outputs. To me this indicates that the sends have an output level of +4 dBu. (See the simplified drawing in Figure 1-26.)

Delving into the versatility of the SSL 4000 G's small and VCA *fader* sections, as far as feeding signal to the multitrack recorder, monitor, mix, and other level-setting usage, let alone automation control of level and mute, is beyond the scope of this book. But I must point out that the SSL VCA fader grouping functioning is about as simple to use and as variable as it gets.

Each VCA fader can function on its own, meaning it will be independent of the overall VCA trim as well as any group leader control. It can be assigned to follow any other module's fader or one of the main "group" control faders, located at the center of the console. As complicated as this selection routine could have been, SSL has made it as simple as turning a thumbwheel switch.

In summary, I've personally witnessed the advancement of audio recording consoles from the Altec 250 T3 to the SSL 4000 G series. I hope I've been able to give you a glimmer of the insight living through this progression has given me.

The value of having and using all the available documentation, including the operation manual, patch bay layout, schematic diagrams, signal flow chart, and specifications, should be

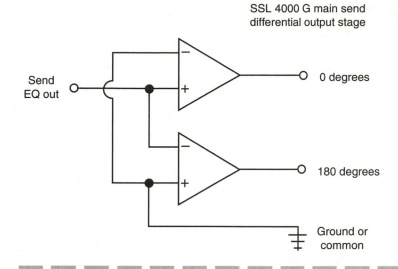

SSL 4000 G main send
differential output stage

Send
EQ out

0 degrees

180 degrees

Ground or
common

Figure 1-26
A simplified schematic diagram of the SSL 4000 G series aux send's differential output stage. The 0- and 180-degree output line connotations are from the original manufacturer's drawing and, along with the ground conductor, indicate that a balanced output is provided.

obvious by now. If not, just look at some of the things we were able to discover via quickly running through this information.

Take, for example, the MCI 600's channel assign amp's trim control. The schematic diagram points out that this continuously variable pot has a switch that is activated at its full CCW position and sets the signal level to "unity gain." But the specifications do not list this channel assign amp's gain capabilities. A summing bus output amp that has a trim control is listed in the specifications. However, the only "summing" amplifier on the I/O is part of the input circuitry that feeds the monitor bus input, and it does not have a level trim control. The schematics also show that the only other trim control on the I/O module was the mic preamp "trim," but this trim does not have a switching function. Therefore, it's safe to assume that the channel assign amp trim is what the specifications are referring to as the output trim on the summing amplifier. Its range is from −15 dB to not just 0 dB but +2 dB. Before going through all the available information this trim appeared to be a cut-only control.

The MCI 600's group output/multitrack input patch after the differential amp is not shown on the I/O schematic. The differential amplifier output conductor traces are brought out to the edge of the drawing. Here they are labeled as "line out," with an added note that the line nominal level is +4 dBv(u). So this informed us that the group outputs feed signal to the multitrack inputs at a nominal level of +4 dBv(u).

There seems to be a discrepancy in the SSL operation manual concerning the depth of the gate. The gate is stated to be capable of delivering an infinity-to-1 reduction in level, yet the range control is described as providing only 50 dB of depth, which is not infinity. This shows the importance of having a specification list, because it would have quickly cleared this up.

Figure 1-27
A signal flow
chart from the
SSL 4000
G series
operator's
manual, which
provides an
overall picture
of the signal
path routes
within the I/O
module

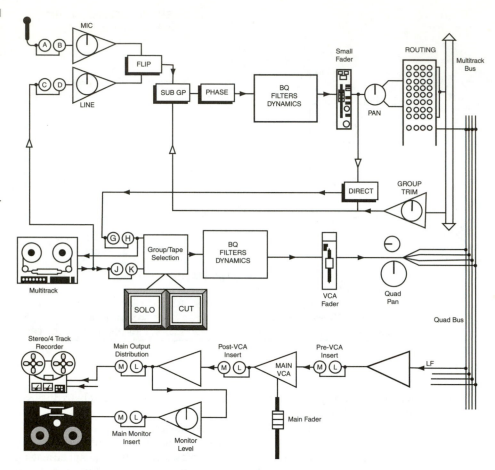

The SSL operation manual informs you that phase reversal and phantom power on/off switching are provided, but you have to check the signal flow diagram to find their location point in the signal routing (see Figures 1-27 through 1-29). The phantom power is exactly where you'd expect it to be, just before the mic trim control. Of more interest, however, is the fact that the phase reversal occurs after the line/mic input switching, meaning that it affects all input signals, whether mic, multitrack return, or a patched-in line level signal.

According to the SSL operational manual, the channel insert's send level is 0 dB "relative to a normal operating level of +4 dB." It's safe to assume that the send level is actually +4 dBu, but it sure is nice when you have the specifications available to confirm this.

The SSL's schematic diagram for the main cue mix bus section explicitly shows the send output circuitry's final gain stage to be a dual op-amp differential output stage with balanced ("0 degree and 180 degree") outputs. This indicates that the sends have an output level of +4 dBu.

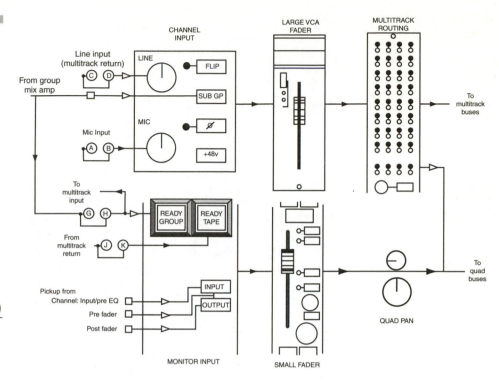

Figure 1-28
A signal flow chart from the SSL 4000 G series operator's manual showing the I/O setup so that the small fader adjusts the mic input level sent to the multitrack recorder (top) while the VCA fader adjusts the monitor level of the multitrack's return (bottom)

The MCI 600 series schematics showed us that the input section's +6, 0, −6 level switch has no bearing on the mic input because it only affects the line input signal. However, we found out that it affects the line input whether it's a multitrack return or patched-in line level signal.

The final example will be the SSL module's overload indicator. According to the operational manual this LED kicks in 4 dB below signal clipping at any one of three places: before the processing, after the insert point, and after the channel fader. The signal flow diagram gives more detail about these locations. The "before the processing" point is actually right after both the channel input selector and the phase reversal switch. The insert send or return can be located before or after the EQ and dynamics processing, depending on the *pre* switch on the EQ section and the motherboard links (see description of the EQ section). The signal flow chart also confirms that the final point along the signal path feeding the overload LED is after the channel fader. But it goes on to show the exact position to be after the assignment section feeding the group outputs, which are normaled to the multitrack inputs. This simple-looking LED overload indicator proves to be a potent warning system once you know what its indication refers to.

The signal flow diagram also has a "+20 dBu" notation under this LED's depiction. Because we now understand how a console's gain structure is set up, we know this is more than likely a printing error. The +20 dBu referred to on the flow diagram should have been

Figure 1-29
A signal flow chart from the SSL 4000 G series operator's manual showing how to configure the I/O so that the small fader can be used as "an extra effects send"

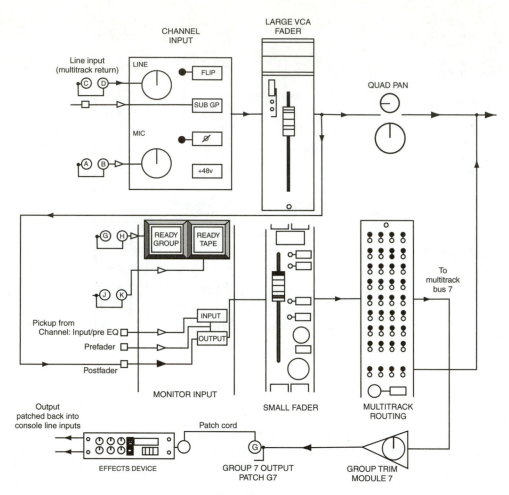

+20 dB with no "u," meaning 0 dB is referenced to a level of +4 dBu, which is written as +20 dB (referenced to 0 dB = +4 dBu).

Okay, I'm flying without visibility on this one, making my way solely by instinct and 35 years of experience. This is due mainly to the fact that the partial information I had to work with furnished only four references to decibel levels within the SSL 4000 G series console:

■ The mic gain ranges from +4 to + 70 dB.

■ The line gain ranges from −20 to +20 dB.

■ The group trim range is from −30 dB to 0 dB (or no attenuation).

■ The overload LED indicates a level within 4 dB of clipping, which, according to the notation on the signal flow diagram, occurs at "+20 dBu."

We already know that

$$+4 \text{ dBu } = +2 \text{ dBV } = 1.23 \text{ volts}$$

$$+2 \text{ dBu } = 0 \text{ dBV } = 1 \text{ volt}$$

$$0 \text{ dBu } = -2 \text{ dBV } = 0.775 \text{ volts}$$

Sharpen your axes, I'm about to stick my neck out! The most important of the three points "monitored" by the overload LED is the one right after the fader and the assign switches because this is just before the output feeds to the multitrack inputs and the mix buses. We have already concluded that the sends and multitrack outputs from this console are balanced and have an operating level of +4 dBu. I'd find it hard to believe that these output levels would, in this day and age, be anything other than 0 dB referenced to +4 dBu (or 1.23 volts into a load greater than 600 ohms). I also find it hard believe that SSL's circuitry is not able to handle levels of +28 dBu before clipping.

Therefore (stick with me now!), I believe the level being monitored is normally +4 dBu (1.23 volts into a load greater than 600 ohms). And it has a clipping point of +28 dBu. This adds up to 24 dB of headroom. Since the LED indicates a signal level 4 dB below clipping, that point would equate to +24 dBu or + 20 dB (no "u"), that is above the unit gain level of +4 dBu, which is 4 dB below the maximum amount of headroom.

It just amazes me how those lowercase letters v, u, and m can throw you off when they are incorrect. But you know all this because you read the first part of this book. And in case you didn't, you can cheat by going back and reading it anytime you want.

"Enough said?

You do the math,

you read the manuals,

you end up knowing what you're doing."

Then you move on to the recording studio, learn about the tools, and start doing the job.

I may sound like a preacher, but I had to prepare you, because our next stop is the recording studio. And not just any studio but a mid-1980s world-class recording facility, which is still considered by some to be the epitome of state of the art in technical and artistic capabilities.

I know, I know. Why deal with 20-year-old technology instead of the latest equipment available? Today a single piece of gear or workstation can handle the complete recording process. It has the mixer, the multitrack recorder, all the effects processing and often a stereo mixdown recorder all built in.

Bridging the Generation Gap

Today's virtual add-ons can be fine tools, and a single box of software can yield just about any amp sound you can imagine as well as quite a few that are yet to be imagined.

But there's something about calling in a particular session's rhythm guitarist, making sure he or she brings his or her pre-CBS telecaster, and then making a mental note to pull out the Fender Super with the four 10-inch JBL speakers. It's the same as using analog-like limiting, other effects processors, and plug-in solutions for format processing, as opposed to renting a pair of LA-2As and the exact Pultec EQ you need. It's all about what you're used to. You have to go with what you know.

I bring this up just to clue you in on my old-timer, geezer preferences.

To me, setting up virtual and modeling effects is the same as presession, swapping out console EQs. As we've seen, the MCI 600 series had four different ones; a simple three-band peak/dip, a semiparametric with switched-center frequency selection, a full parametric with sweepable counter-frequency selection, and a great little 10-band graphic. I also used to love reloading those mothership-like effects racks so as to have all the processing I needed, all set up and ready to go because, as you know, having the engineering speed to keep the musical flow going is very important when recording.

Today's high-speed, DSP-based, harmony presets can instantly change a solo vocalist into a choir if that's what's called for, but I prefer working with real, live backup singers.

On the other hand, there's a huge difference between today's light pipes, which are capable of carrying thousands of signals at the same time at high transfer rates, and the stone-age bundles of wire harnessed together that I had to deal with. It's become a no-choice situation, as the newer method provides higher quality audio while the older method made facilitywide troubleshooting of interconnections and fault finding close to impossible.

Preset controllers may still have a good deal of versatility, but you're talking to someone who actually made triangular holes in a speaker cone using a screw driver while trying to match Norman Greenbaum's "Spirit in the Sky" guitar sound. Accomplishing this required the depletion of electrons from the plates of two tubes (by inputting an overly high level from a signal generator and watching as the tubes changed color). Stop at red/purple, which occurs just before the overly depleted lack-of-detail sound of a tube when it's glowing blue/purple. Oh yeah, and I found the best speaker to use was one of those old jukebox extension speakers in a cabinet designed to be mounted up in the corner of the ceiling. Then you slowly increased the tear in the speaker cone until the triangle cut-out's movement of air (a "woof" sound) could be readily heard. Nuts? Maybe, but you better believe this teenager was lovin' every minute of it.

When you grow up learning how to build mixing consoles, or when you need a couple more inputs or maybe an added submixer, you just build it by adding a few pots or faders with either pairs of resistors or, if you're lucky, a few spare isolation transformers (they were once readily available) instead of looking through a catalog for a rack-mounted "multiple level translator" as is done today.

Fact is, it's six of one, half-dozen of the other, or it's all the same thing: Making audio happen.

I did have one advantage over today's budding assistant engineer. Back when these large-format consoles were first introduced, it was not about surfing the web and picking up bits of information—catch as catch can. So instead of having to download questionable bits and pieces of information, we had *all* of the manufacturer's printed material right in our hands.

As we've seen with the MCI 600 Series Console, this could take up a lot of room. But if you showed up an hour or so early when you pulled duty for Sunday morning control room cleanup and calibration, you could use the extra time to turn the whole control room into your own personal self-teaching/discovery tool.

The luxury of having time in a $1.5 million control room, plastered with console diagrams taped to the window and walls, checking handfuls of patch points was an absolute and rare delight. Yeah, I *was* lucky, and that's why I felt I had an obligation to write this book.

Don't get me wrong. I have nothing against new technology. It's just that in the mid-1980s all the equipment was separate, meaning nice and spread out, making all the individual parts a lot easier to describe and explain. Now that you've got all the basics under your belt, it's time for some fun. So we're off to a world-class multitrack recording studio circa 1985.

Hold it; don't open that door!

The Tour

The Studio Tour

Every recording studio used to have an illuminated sign marked "RECORDING" above its entrance. Its on/off state was triggered by a relay that was connected to the tape deck's record light. When the engineer pressed the record button, the light above the door would automatically turn on. So when the tape was stopped or brought out of the record mode, the light would go off. Although still used in broadcasting, for some reason they are rarely found in recording studios. So it's best to take a moment to listen and to press your palm against a door leading into the studio or control room to feel for any rhythmic vibrations.

The doors that lead into recording areas are often part of a "sound lock" system, and a corridor entrance is constructed so that any outside sound must travel through an additional wall, or two doors in series, before getting into the control room or studio. This virtually eliminates any intrusion from direct sounds. In addition to the physical mass of walls and solid core doors with their sealed perimeters and double-pane glass windows, internal absorption materials also aid the sound lock's function of sound isolation by absorbing any outside noise that arrives from a random incident (a nondirect, reflected sound wave) before it can get through to the studio or control room. Add a couple of cables for mic and headphone lines and you've got a ready-made isolation booth, that is, until somebody opens the doors during a take.

I've found the acoustics of these spaces, for reasons of both size and internal absorption, lend themselves to the recording of percussion instruments like congas and, believe it or not, piccolos. On the other hand, they're often a little tight for horns and a little dead for strings. But due to the deadness of the sound as well as the extreme physical isolation they can provide, vocalists tend to really open up when singing in them. Due to this (and the always crowded nature of basic sessions), sound locks are often the ideal place for recording scratch vocals. I've recorded some great "keeper" takes out of sound locks.

NOTE: Even when you're sure there's no recording going on, it's a good idea to open *all* doors slowly just in case there's someone carrying a heavy or delicate piece of equipment on the other side. I've witnessed more than a few disasters. In professional recording studios, rowdiness is not a virtue. Do you want to be the person who opened a door into the diva's face, broke her nose and possibly ruined her singing career, forever robbing the world of her amazing vocal talent?

The Tour Begins

In this, the "Tour" section of this book, you'll get an overview and some detailed explanations of recording studio equipment.

Tracing an audio signal through the facility's whole audio system would logically begin at the initial transducer used to convert sound into an electrical signal (most commonly a microphone) and proceed through all the ensuing electronic equipment until it reaches the final transducer, the control room monitor, or speaker system. But I've decided to start this little tour in the middle of that chain with the recording studio control room (and it's a 1985 control room to boot).

This is because unless you've spent a lot of time on the flight deck of a star ship, there are not many sights as technologically impressive as a mid-1980s recording studio control room. Your attention is immediately drawn to the most dominant item in the room, the mixer. In fact, in order to design a control room with proper ergonomics and correct acoustic response, the choice of console must be known beforehand. With all its lights, mysterious knobs, and meters, the mixing console is the centerpiece of every control room. It sits under a huge pair of monitors capable of both dance hall levels and wide frequency response. The control room has the ability to mesmerize visitors, clients, and even engineers and producers.

Format

Every mixer has a format and this simply tells you in plain numbers what the mixer is capable of handling. For instance, a $32 \times 24 \times 2$ mixer would have 32 input channels, 24 group outputs for multitrack feeds, and a stereo output. This is not set in stone as live mixers utilize group outputs for separate stage monitor feeds, while a "spilt" console with a separate multitrack return monitor section would correlate to 32 channel inputs and 24 group outputs (as well as 24 monitor returns) and a stereo out. Today, it's more common to find $64 \times 48 \times 5.1$.

Input/Output (I/O) and SPLIT Console Layouts

These designations refer to where the multitrack returns or the "monitor"-level controls are located on the console. With SPLIT designs, the monitor controls, which adjust each of the multitracks' output signal levels going to the monitor mix, are located together in one area called the "monitor section." This area contains the group output controls, which adjust the levels of the signals going to the multitrack and switches that allow for the selection of either the group or multitrack output to be fed to the monitors during overdubbing.

I/O modules contain both the main channel inputs and monitor output controls on each channel module. Personally, I do not heavily favor one type over the other, but because some folks get adamant about it, I'll brief you on some of their arguments. "Split consoles are more flexible, are easier to operate, and give a producer his own section of the console, thus allowing him to set up the mix. In addition they are often more readily and easily repaired." Or "In-line consoles take up less room and are less costly to build and thus purchase." These kinds of arguments are strictly for console buyers as engineers must work with whatever the studio has. But, if given the choice, I prefer an I/O console for tracking (or recording), as everything you need to deal with is right there in front of you. During mixdown, a split console's ability to "double up" the channels, that is, use the input channels as effect returns (or for additional inputs), can be mighty handy. Yet the *Solid State Logic* (SSL) I/O provides much of the same flexibility and you can use other I/O console's channel assign switches and pans to set up multiple submixes to handle additional inputs during mixdown as well.

Every audio signal (often including sync code) passes through the mixer at least once. In fact, with only a few exceptions such as the use of the direct-to-tape method of recording, most signals pass through the console twice. It's utilized in the recording chain both before multitrack recording to control what is being recorded (or overdubbed), and later to control the outputs of the multitrack recorder during mixdown.

The modern recording studio console actually generates several different mixes at the same time. It can combine multiple inputs to a single track of the multitrack recorder at the time of the initial recording or later when tracks are "bounced" together to free up space for additional instruments. It also provides a cue mix specifically for the talent. Its monitor mix allows the operator to hear and adjust the overall sound in the monitors with or without these changes affecting what is being recorded.

The console can also combine signals from all channels to make a feed for reverberation or echo effects devices (often called "echo sends"). Many channel signals can also be "grouped" together to have their levels controlled by a single fader. During mixdown the console is used to create the stereo and mono mixes fed to the program recorders. Things may look very complicated, but upon closer inspection you begin to notice that certain features are repeated. Once you understand a single input channel, you've got all the others in your pocket as well.

I used to tell students to think of the mixing board as sort of a locomotive central switching station for audio signals. Some signals will have their levels adjusted, just as trains are made to slow down or speed up. Some may require a change in their tone or have an effect added, just as trains may need to pick up and unload carloads of goods or be hooked up to faster or stronger engines. Yet the primary function of these switching stations is the rerouting of trains or, in our case, signals. Splitting off and sending the signal somewhere else may be done to get the signal to a place such as a musician's headphone submix or into another piece of equipment such as a tape recorder.

All of these submixes are essential for modern multitrack recording sessions where all the musicians, arrangers, and producers are seldom gathered together, and all parts are not recorded at the same time. This can make the kind of patching that would be required to finish a complete project with the use of something like the Altec 250 T3 economically unacceptable.

The Mixer

Consoles, boards, mixers, desks, audio control surfaces, whatever you call them, their function is the same. They control the level, tone, mix, and pan of all signals coming into their inputs via mics, direct input, and recorder returns. These signals are routed to the monitor speakers, the cue or "foldback" (headphone and/or studio monitors), and the multitrack tape deck. The console is the heart of any audio system; whether it is location film recording, live *public address* (PA), or multitrack recording. You can have the best mics, tape decks, and speakers in the world, but if your console is funky as far as noise or flexibility is concerned, you will end up being handcuffed.

What constitutes a good board? Well, as far as specifications go, we'll get into that a bit later. For now, we'll deal with abilities only in terms of flexibility. Let's look at two extremely different mixers: the Shure M67 Professional Microphone Mixer and Solid State Logic's 4000E Series Master Studio System. If I'm going to do an on-location recording for film, the last thing I want to think about is how I'm going to transport equipment. All I want is a clean mixer, a small tape deck, and a pair of headphones, all of which should fit into a medium-sized suitcase. I don't want to think about a console that's 12 feet long.

The M67 is a simple yet professional four-input, one-output audio mixer that's very clean, small, and light. Because of this, it's one of the most widely used audio mixers for field work. On the other hand, if I'm running two 24-track recorders interfaced with video machines, a digital two-track recorder, and a rack full of processing equipment, the M67 will be almost worthless. In some applications what's needed is a board that will almost operate itself in terms of re-setup of mixing memory, time code-locked manipulation of 56 channels and 32 groups, automated control of tape decks, and a lockup interface with multiple external effects. The SSL is one of the most advanced computer-controlled boards available today and does all of this very easily. As you can see, choosing the right board depends on the functions it must perform. With this in mind, let's follow a signal through a console and see just what we are able to do.

Once into the console, the signal should be assignable to any channel. With PA boards, this is simply done by plugging the cord into the proper input, but if you're holding an XLR plug and all jacks are $\frac{1}{4}$ -inch phone, you may be in trouble. With multitrack recording consoles, all input connections are hard-wired, and via "busing" an input can be assigned to any channel with the push of a button. The signal is now brought to a level that corresponds to that of unity gain so as to match all other signals, thus achieving some kind of normalized setting of input faders. This feat is accomplished by using the console's high/low gain input switching, the input attenuator/gain control, and the pad switching. A phase reversal switch is often provided to correct improper wiring. After the signal is set to the proper line level we can move on to the equalizer.

The equalizer section of the console can be made up of anything from a simple dual band bass/treble to a full parametric or even a graphic EQ (see "Equalizers"). You often have the option of just monitoring the signal as affected by the EQ's adjustment or committing to the change in the signal as sent to tape. An overload indicator is helpful here, as a boost in EQ may add level to the point of requiring a lowering of the overall gain.

There are all kinds of meters, from common VU to 101-segment *light-emitting diodes* (LEDs), with some having the ability to show peak or true *root mean square* (RMS) values. To check on an individual signal along its routing path, several solo or *pre-fader listen* (PFL) buttons allow the operator to zero in on individual stages to verify proper signal content.

Next, the signal needs to be sent out into the world. It may go to a cue (or "foldback") bus for musicians to hear either through headphones (recording) or to (stage) monitor speakers (live). This should be a separate mix, or these sends must have switching that enables the signal to be sent pre or post the EQ and fader adjustments so that they will be unaffected by any changes made to the PA or control room monitor mix. Other sends provide access routes to reverberation units and other commonly used effects. Via in/out patching, the signal can be put into its own individual processing devices such as a limiter or a gate.

Pans are provided for setting the signal's placement across the stereo field and each channel has a fader to control its level in the mix. Often group assign capabilities allow for the control of many signals via a single fader. Additionally, some form of mute button should be provided for switching the signal in and out of the mix at any given time.

At this point, our signal appears in many places—tape deck, effects inputs, control room monitors, headphones, and so on. At the output section of the board, it should be possible to check on how it sounds at any of these locations. Via switching, the tape deck returns can be placed into monitors to see if the signal is getting to the recorder properly. The cue mix can also be brought up to the monitors to see if the musicians are being fed a mix that gives them a good idea as to how their instruments fit in with the whole. The main effects returns are also located here. These can be soloed, just like the original signal, to see if there is any unwanted coloration being added. Overall level controls such as for the cue and monitor mixes are found in the output section along with mono and secondary speaker switching.

All this, and our recording console has only done half its job. Once all the tracks have been "put to tape" (or recorded to disc or computer memory), the console must now function purely as a mixer with many of the same functions used again. Now, input will be mainly from the multitrack returns and while all input levels should already be set at unity gain, the assigns, EQs, groups, pans, effects patching, send buses, and output level controls may be used again.

Several hours can be spent on mixing a single tune only to have the producer return (after listening to the finished product for a day or two), wanting to redo the mix to make a single correction. Resetting all of the various parameters of the console can be tedious. For this reason, many consoles are provided with automation. While many people originally feared computer-controlled mixdown, it has proved to be extremely helpful in these types of situations. They originally used a single track of the tape deck for the recording of the positions of the faders and the status of all the mute switches. In this way automation can handle a good deal of the mixdown. More importantly, it can allow for the easy correction of a single mistake.

By automation, we're talking about the simplest form, which is the multiplex type. At regular intervals or "frames" (but not *Society of Motion Picture and Television Engineers* [SMPTE] or film frames), the position of each fader was sampled, multiplexed into a digital format, and recorded on a single track of the tape recorder. In "write" mode, the faders were operated in the normal way by the engineer. In "read" mode, the information was returned and controlled the individual channel levels exactly as the engineer did, allowing everyone to sit back and judge the mix. If a fault was detected, the "update" mode allowed for new data to be recorded and bounced together with the previously recorded correct information to a second data track. In this way, corrections could be made on any number of faders without affecting the other fader levels. It does, however, require the availability of two tracks for data. Fader-level matchup must be visually aided, and this was originally accomplished by using the board's VU displays (here showing DC level). There was a little inherent time lag every time an update was performed with the multiplex system, but for the most part this was not a major problem in normal operation. Some multiplex systems also offer control over channel mutes as well as group mutes and level functions. All this makes for easy resetability as far as mix levels and mutes are concerned.

Some systems allow the data to be stored on magnetic or floppy disks. This method provides more control flexibility than recording it to the multitrack and only uses a single track of the tape deck to store synchronous information. In this way the tape deck's information track is invariant and therefore has no cumulative delay in the system's responses no matter how many updates are made. There is also a vast array of features and options.

Neve consoles, for example, can be obtained with servo-motor control of the faders, which causes them to move up and down by themselves during the "read" process and eliminates the need for special position-indicating meters but at the same time opens the door to the problem of mechanical failure.

Some systems allow control over various other parameters, and some even allow the storage of multiple mixes of the same tune, letting you combine these mixes in various ways. The SSL board has total recall, meaning a memory of the setup of EQ, pans, sends, returns, outboard, and tape deck. Just about every function of the console is recorded, but this will cost a bit more than an M67. Again, this may not be what is absolutely needed for your home recording facility and is certainly not important to live or field work.

Here again, we've reached the extreme in terms of consoles or mixing boards. So let's go back to the Shure M67. This is about as simple as you'll ever want to get. It has four low-impedance mic inputs (one convertible to line), one mono output, a single meter, and no EQ. But if you're doing live PA work, you may need control over many input levels and EQ would be handy to help with both feedback problems and enhancement of the signals.

If you're not running a splitter to an additional stage monitor board, some form of cue or foldback control would be needed so that the musicians could hear themselves, group assigns would be handy for controlling, say, all the drums or backup vocal levels with one fader, and solo buttons along with a headphone output would allow you to check each signal to ensure the correctness of each sound. Additionally, the board may need a send or two for overall effects, input/output patching for individual channel effects, stereo output, and certainly the input connections should match up with what you're presently using. All these additional features (all the way up to total recall) mean money. So the choice may not be what you want, but rather what you need and can afford.

Obviously, the construction of the console is important. Live boards should easily fit into road cases and all channels should be modular, if not in terms of actual front panel access, at least internally plugable PC cards. Here, capabilities like being able to swap out cards or repair individual channels while still using the board can help avoid disastrous situations. The warranty should be checked out, along with the availability of qualified repair personnel, as well as the accuracy of the schematics supplied. Imagine being in the middle of nowhere, experiencing board failure of some nature, and finding out that the schematics supplied are inaccurate! Believe me, it's happened!

The quality of the board is very important. A noisy unit is worthless in any situation; therefore, the specifications must be checked before purchase. Things like frequency response, input and output impedances, and gain do not require explanation since the mixer either matches your system (provides the proper amplification and has the desired response), or it doesn't. However, things like distortion, *signal-to-noise* (S/N) ratio, hum, and noise should be closely examined. Any audio control system is limited in its dynamic range by noise at low levels and distortion at high levels. Noise should be limited to that which appears at the input. In other words, your board's gain stages should provide adequate gain but not degrade the signal.

Console Specifications

Looking over the MCI 600 Series Console specifications in Tables 1–3 and 1–4 in the last section reveals the fact that many of the console parameters pertain to most all audio equipment. This particular listing is about as extensive as you'll find, yet all consoles should have reference levels noted as part of their input and output gains, noise and headroom specifications, along with the frequency of signals used for the tests or the bandwidth of all ratings. For those of you who have read the "Basics" section, most of these specifications will be obvious. Those who did not can always go back and catch up.

The S/N ratio is the ratio of the output level at normal unity gain (IE +4 dBu) of the signal at a set bandwidth as compared with the output at the same level setting with no input signal, so that just noise appears at the output. Every one of the console's outputs, tape, monitors and sends, will have an S/N spec and, while they commonly will all be different, they are *all* related to gain. Therefore, knowing that the gain setting is important.

Equivalent input noise (EIN), like most specifications, is also dependent on the console gain stage settings. This, in effect, is the signal level required to produce a specific noise output. An easier way to conceptualize EIN is that the S/N ratio minus the gain just about equals the EIN.

EIN is expressed as a power level relative to 1 milliwatt (dBm), and most preamplifiers should be rated around −125 dBm EIN. Some manufacturers don't specify at what bandwidth their measurements were taken. This, along with the use of weighted curves and filters, can result in noise values that are lower than those produced using normal EIN calculations.

Distortion is directly related to amplifier clipping. The normal output may be set to +4 dBm, but some transients will cause peaks that are easily 10 to 15 dB higher than this average level. While some peaks may be fast enough so as not to cause pronounced clipping, others will have disastrous results on the sound quality of your mix. Therefore, many console gain stage amps are rated for levels up to +24 dBm before they reach the onset of clipping, thus allowing for plenty of headroom.

Maximum output-level ratings, or the clipping point, as discussed in the "Basics" section, will affect the available amount of headroom as well as the S/N ratio. This spec should be given in decibels for a specific frequency range or bandwidth.

Total harmonic distortion (THD) testing uses a single frequency input called the fundamental. The output gain setting should be specified and is commonly unity, or the normal operating level, but it will certainly be below the clipping level. The original fundamental frequency is then filtered from the output signal and what is left will be all the added harmonic content. Since it is difficult to also filter out any noise, this specification is generally rated as "THD plus Noise."

Intermodulation (IM) distortion uses two steady state signals fed to an input. These two fundamentals are filtered from the output signal, leaving only the added harmonic content and noise to be measured. (Again this is the reason many IM specs include "noise.") There are several different IM distortion test procedures such as those governed by SMPTE, which uses 60 Hz and 7 kHz in a 4-to-1 ratio, and *International Telephone Consultative Committee* (CCIR), which uses two frequencies 1 kHz apart (5 and 6 kHz, through 13 and 14 kHz) mixed together in a 1:1 ratio. You'll need to know which test was used if you want to compare specifications.

Cross Talk or Channel Separation

These are much like noise in that every mixer has some. Here, a signal assigned to one bus (at unity gain?) will cause some amount of leakage to appear on another bus. In addition to the rated level expressed in decibels, the console's control level setting and the frequency or bandwidth parameters should be listed. For recording purposes, cross talk or channel separation is very important. This rating should be −70 dB full frequency range at the very least and −80 is now considered commonplace.

It may seem like I'm chiding manufacturers with all the above "shoulds" and "musts," but I actually sympathize with their position completely. Specification checks are tough on manufacturers. They are expensive to perform as they require specialized test instrumentation and a highly skilled staff. Plus the end results, even when good, can prove detrimental to sales. For example, take the following real-life mixer specifications for hum and noise. The manufacturer begins by stating that they were "Measured with a 6 dB/octave filter @ 12.47 kHz; equivalent to a 20 kHz filter with infinite dB/octave attenuation." Because this filtering is uncommon, that statement alone leaves them open to question and possible ridicule from a competitor itching for a sale. This manufacturer courageously continues to give what would seem to be an overly extensive listing of specification results. Here "PGM" stands for Program and "FB" means FoldBack.

HUM AND NOISE (20 Hz to 20 kHz, 150-ohm source, Input selector at "−60")

−128 dBm EIN

−95 dB residual output noise with all faders down

−73 dB PROGRAM OUT (77 dB S/N); main fader at nominal level and all input faders down

−64 dB PROGRAM OUT (68 dB S/N); main fader and one input fader at nominal level

−73 dB MATRIX OUT; matrix mix and main controls at maximum, one PGM main fader at nominal level and all input faders down

−64 dB MATRIX OUT (68 dB S/N); matrix mix and main-level controls at maximum, one PGM main fader and one input fader at nominal level

−70 dB FB or ECHO OUT; main-level control at nominal level and all FB or ECHO mix controls at nominal level

−64 dB FB or ECHO OUT (68 dB S/N); main-level control and one FB or ECHO mix control at nominal level (Pre/Post Sw. @ PRE.)

Any salesman looking to sell a competitor's mixer to an unknowledgeable customer could have a field day with these specs. Picture the customer's reaction to being asked, "Do you really think there's any such thing as a 12.47 kHz filter that's equivalent to a 20 kHz filter with *infinite* attenuation?" Or when it is pointed out to them that the S/N ratio is only down 68 dB with just *one* input fader and the main fader set at nominal level. Actually, the above specifications are pretty good. Don't believe me? Then try this:

1. Set up a console for a normal listening level with a CD, a recorder or anything with a repeatable output level fed to one of the inputs.
2. Remove the signal and raise the monitor (speaker gain) level until you hear some noise (hum or hiss).
3. Now start raising the faders and echo returns.

The result will no doubt surprise you, but for a real shocker, eyeball the output meters. If they have moved even a slight amount it means that your noise (*not* S/N) level is down less than 20 dB! Like I said, the above specifications are very good, *they usually are* these days, but even good specification results can end up being a pain in the butt for manufacturers.

Console Functions and Terms

Console inputs can be categorized by their configuration and level. These are generally labeled mic and line, but the differences are a touch more complex. This is not the place to get involved with impedance matching, but most consoles have input electronics so tolerant of impedance mismatches that as long as the input is balanced there will not be a real problem. But even though the first thing the console will do to a signal is put it through transformer or electronic circuitry to unbalance it, bringing an unbalanced signal such as a guitar output to the console in its normal (unbalanced) state will result in too low an input level and cause some signal degradation particularly to its high-frequency content. This will depend on both the cable length (capacitance) and the impedance mismatch (resistance). (See "Impedance Matching" and "Differential Inputs.")

Input level, on the other hand, is easy to understand. The input selection options will be microphone level, instrument level, and line level. Everything is handled by the same three controls: a mic/line switch, a pad switch, and a variable-level (rotary pot) control. Setting levels may be aided by an "overload" indicator light. With a microphone input the mic/line switch positioned on "mic" sets up a gain stage range of, say, 25 to 60 dB. Should the output of the mic by very hot, or if the signal is an instrument output that's been put through balancing circuitry (a direct box), a 20 dB pad switch will cause 20 dB of attenuation and reduce the gain range 5 to 40 dB. This is used to avoid overloading the input and any resulting distortion it may cause. For line-level signals that have already passed through a preamp stage, the "line" setting may yield only 10 to 30 dB of gain. That's okay because the whole idea is to get all signals to the same "unity" line level be it 0.775 volts, 1 volt, or whatever happens to be the normal operating level within the console while obtaining the best S/N ratio and retaining a safe amount of headroom. (See "Setting Gain Stages.")

Monitor-Level Controls   The signal on every channel must be listened to and checked by the recording engineer at various level settings without those level changes affecting the signal that is being recorded on the multitrack. So each channel has its own continuously variable monitor-level control that feeds that channel's signal to the monitor speakers but has no effect on the signal being recorded. This feature is necessary when you're attempting to record signals at an optimum level.

Optimum Recording Level   The recording level that achieves the greatest S/N ratio or the loudest level before the onset of distortion. Without this feed being isolated, changing the level of a signal in the monitors could cause it to be fed to the multitrack at a level two to three times normal. This would make for a nerve-racking recording experience.

Input Selector Switch

It all starts here as positioning this switch will route the input to the channel as being a mic (when recording), a line (when recording or playback during overdubbing or mixing), or the multitrack return (again when overdubbing or for mixing).

Microphone Attenuator   This switch inserts a pad that attenuates the signal before it reaches the input transformer and preamp. This in effect lowers the input sensitivity and helps avoid distortion caused by overdriving the input stage.

Microphone or Line "Trim"   A continuously variable pot (potentiometer) used to reduce the gain of the microphone, line, or tape return. This can be used to improve or retain the signal's headroom and S/N ratio. It is also used to set an operating level that gives your input fader's nominal position the greatest amount of variability (meaning that the normal listening level position is not too close to the minimum or maximum extent of its travel).

The solo switch does away with the need to stop the recording in order to investigate a mysterious sound or problem because it gives the operator the ability to zoom in on each individual sound at various locations throughout the mixer. "Soloing" has no effect on the feed to the recorders. The solo switches allow listening to a signal without any other signal in the monitors. Often several solo buttons are located along the signal path through a channel. This makes it easier to troubleshoot problems or verify the EQ settings, effects returns, and that you have an undistorted, noise-free gain setting.

The direct output sends the signal to the multitrack (or anywhere else) from the point in the channel right after the EQ and fader. This avoids passing the signal through all the added gain stages located in the console's output section; thus, it will be a cleaner feed.

Equalization   This processing circuitry is generally located in the channel signal flow chain right after the preamp gain stage because too low a signal could leave no room for a cut and if too high it may not allow any room for boosting. Console EQs come in just about every flavor. You'll commonly find the simplest bass and treble controls, or those with a "mid" section, and possibly two switches that contour the low-mid and high-mid frequency bands. The width of the bands, or the amount of frequency content that these controls affect are usually set, as are the center points of the bands. More sophisticated semiparametric EQs have center frequency selectors that can switch between different center frequencies.

Parametric EQs not only have this feature, but also selectable bandwidths (also referred to as "Q"). Full parametric equalizers have "sweepable" or continuously variable center frequency and Q filter settings. Some consoles even offer the choice of having 10-band graphic-type equalizers built into the channel. This is all real (hardware-driven) circuitry. Software-driven equalizers allow one to literally invent his or her own equalization control parameters should the need call for it.

Inserts or Insert Patch Points   Located at key positions along the channel signal flow routing (such as pre- and post-equalization or fader) these patch points give access to the signal so as to feed it to and return it from "outboard" (outside of the console) effects devices. They can also be used as another way to route signals around the console and for that matter throughout the facility.

Sends

Separate submixing functions are enabled by the additional level controls on each channel. Using these controls allows the operator to add that channel's signal at any level to a mix bus feed. More important is what they can accomplish. Take, for instance, a basic tracking session. You have a drum kit all miced up in the main studio area. To prevent bleed or leakage, you've placed an acoustic guitar that has a microphone on it in an isolation booth and you're recording the bass guitarist through a direct in box to avoid having to use an amplifier. You have done your job very well. Now you can raise or lower individual tracks without it affecting any other track. There's only one problem: The musicians won't be able to hear each other.

Here cue or foldback send buses come to the rescue. Each player is given a set of headphones into which you send a nice blend of all the instruments by adjusting each channel's cue level and pan controls. Later when the musicians are overdubbing a part, they will need to hear what was previously recorded. Now you'll use the same cue-level sends (or the monitor cues on a "split" mixer having separate input and output sections) to get the music already recorded on tape into the cue system and mixed with the instruments being overdubbed. So "cue" is just a shorter name for a separate mix fed to clue the talent in on the overall program content via headphones.

Auxiliary send is just another name for a send. In any case it is a means of creating a submix that can be routed to somewhere else in the facility such as studio monitors, a tape dupe room, another control room, or to a piece of equipment. "Aux" sends are perfect for shipping multiple signals to the same reverb device. These sends are sometimes referred to as "echo sends."

Pre- and Post-Fader Sends   These are very important as they give you an additional feature in that they have a switch added to their circuitry, which allows the signal routing to start from before, or after, the fader-level adjustment. An example of both uses should provide sufficient explanation.

Okay, back to that basic tracking session. Let's say you hear a funny or off-color squeaking sound and suspect that it's coming from somewhere on the drum kit mounting hardware. You raise and lower the level of each mic on the kit in hopes of narrowing it down. All the musicians stop playing because you've also just caused all those level jumps to happen in the headphone mix as well. Hitting the pre/post switch to have the signal sent to the cue mix before the fader solves the problem. You can now do whatever you want with the monitor mix in the control room and not worry about disrupting the session by throwing off the musicians. This is the last thing you want when you have a room full of union musicians not only getting double time, but all watching the clock because they have three other session dates lined up that day.

It's now mixdown time and you are adjusting levels and sends for a chorus section that leads into the instrumental tag, which fades out at the end of the song. The snare was given big-time reverb for that final chorus, but if the send you've used to ship it to the reverb is pre-fader, when you fade out during the tag, that snare reverb will still come through like gangbusters. All you have to do is switch the snare feed to the reverb device to be post-fader by hitting the pre/post fader switch at the start of the tag.

NOTE: First check to confirm that there will be no obvious level variation in the snare's reverb or switching noise when you reposition the switch.

Output Assigns or the Group Output Section   Here we find each channel's signal fed to something like a dual row of switches, one odd number and one even, with a pan control. These are used to assign the signal to any particular multitrack input. Group one will be connected to multitrack input one, group two to input two, and so on. Using this type of switching configuration allows an individual signal to not only be sent to a single track, but also any number of multitrack inputs.

This section also works like a submixer wherein a mix of any number of signals (all the drums, for instance) can be sent as a stereo pair to two tracks on the multitrack recorder. As we will see, this added feature makes "ping-ponging" or track "bouncing" a snap, and when utilized during mixdown some extremely elaborate submixes can be set up to make additional channel inputs and returns available.

Pan or Panning Control   This control facilitates the continuously variable movement of a signal from one side of the "stereo image" to the other. There could be two or more such controls on each channel. The first will be the multitrack or group output routing pan control that is used to place the signal on particular tracks. There will also be a monitor pan control that places the signal in the stereo mix field, which not only appears in the monitors, but also in the stereo mixdown being recorded. Some cue sends also use a pan control along with a level control to position the signal in the headphone or studio monitor mix. Some engineers find this method more convenient than using two separate (left and right) level controls per each signal.

The Mute Switch   The mute switch turns the channel on and off. This switch can affect only the monitor mix or also the feed to the send buses, the two track mixdown recorder, and the multitrack recorder. It all depends on the position of the *pre-fader listen* (PFL) and send switches.

PFL, Solo, and Solo in Place Switching   PFL switches place the signal in the monitors just as it appears *before* the channel's mute and fader. Depressing the PFL button on any channel will cause the signals on all other channels to mute or turn off in level while that channel's signal is taken at its pre-fader level setting and featured as a mono signal in the monitors. The overall level of signal feed to the monitors, as selected by the PFL switches, can often be adjusted at the console's main or output module.

The solo switches on the channels are very similar except that here the soloed signal remains at its original position across the stereo field at post-fader level. Again there will often be an overall solo-level adjustment located on the main or output module. Occasionally, a console's solo function will yield a mono signal in the monitors. When this is the case there will normally be another soloing switch termed "solo in place" that will present the selected signal as it appears across the stereo field.

The fader on each channel can be used to set the level of the signal fed to the multitrack recorder, the stereo mix, or the monitors, depending on routing switch settings.

Fader Grouping   This places a channel fader, a separate "group" fader, or a "wild" (unassigned) fader in control of multiple channel faders. Now a group such as a complete drum kit, a horn section, all the background vocals, or the whole rhythm section can have all its individual levels adjusted by the repositioning of a single fader.

The Output Section

The Main Module   Unlike console channels, there is only one main or output module. This is where the overall output-level controls are found. These can include the final gain stage settings for the stereo mixdown buses, all the send outputs, the solo, PFL and talk-back levels, and even the internal test or slate oscillator's frequency and level controls.

Monitor Select Switching   Usually located on the main or output module, this switching network is definitely a big-time asset. This allows the operator to choose which of the many mix buses will be fed into the monitors. Comparing the "live" or input signal to what is recorded; listening to the cue or the studio speaker mix while adjusting levels; and checking auxiliary returns, two-track recorder outputs, or any of the sends can all be done with the touch of a button. Things were *not* always this easy!

Talk Back   Most consoles have a built-in microphone with level adjustment and momentary on/off switching controls. This makes it easy for the operator or the producer to communicate with the talent through the headphone and studio monitor speaker cue systems.

When the mic is activated (usually by pressing a momentary push button), the control room monitors are automatically muted to avoid any possibility of feedback, as shipping this down the cue lines to headphones can be painful.

Slate controls route this "comm" signal to recorders to audibly mark song positions and for other cue information. In the days of razor blade editing, slate tone generation was also used. A short burst of low-frequency signal (usually 50 or 60 Hz) was recorded on the tape to mark various in and out points or good and bad take positions. Later during rough editing when whole sections of the tape were being "dumped," these locations could be easily found in fast forward and rewind modes as the speed of the tape travel caused the low-frequency tone to be reproduced as a mid-frequency (1 kHz) tone that readily stuck out from the rest of the high-frequency "gibberish" sounds caused by the high-speed cueing.

Auxiliary Returns

These are also often found in the console's output section. Having sent your signal to another piece of equipment, such as a reverb or echo effects device, that unit's output is then returned to the mixer for routing to its intended destination. While these signal "returns" can be connected to an input channel, these specialized "auxiliary returns" are often a better choice. Sometimes referred to as echo returns, these often provide for stereo inputs or have panning facilities in addition to a set of routing switches that are used to readily ship the signal to your monitors for checking the sound, the cue mix to let the musicians hear and play along with the effect, the multitrack, or to the final two-mix during mixdown, all without requiring patch cords!

Phase Reversal Switches

When two identical signals are wired out of phase with each other they will cancel each other out. But when two identical microphones, placed in close to the same position, are out of phase they will not completely cancel each other out. This is because there is no such thing as two identical microphones and two of anything cannot occupy the same identical physical location. Yet some phase cancellation does result and it can cause some serious degradation to the incoming signal.

An out-of-phase condition can occur when one of the XLR connectors or the wiring used between the inputs and the console has its high and low connections wired in reverse. The phase reversal switch, which is usually found at the microphone input section of the channel, makes it possible to quickly check any suspect phase reversal problems without having to swap out every wire or disassemble the studio set up.

History time. This situation should rarely happen today, but for a short time during the 1980s it was actually a common occurrence. At that time, *most* state-side recording studios were wiring their balanced lines with pin #3 as the hot connection, while U.S. broadcasters and Europe recording studios and broadcasters were wiring pin #2 as hot.

Broadcast cables generally stayed in broadcast facilities and transatlantic cable exchange was not common so outside of wiring mistakes, a phase-reversed wiring problem was rare. Then around the middle of that decade, the *Audio Engineering Society* (AES), under pressure from its Los Angeles Chapter, opted for pin #2 to universally be the hot connection. This had become a necessity due to the increase in mixed media (film/video/recording and broadcast) projects. Some recording studios readily went along with this change; others did not. Some technicians continued with their old methods while others accepted the change. Newer employees started repairing the ends of broken pin 3-hot cables with pin 2-hot wiring. When one side of the connections was replaced with different ones to accommodate new equipment purchases, only the new connectors could be wired with pin 2 hot. It was a little bizarre there for a few years, during which those phase-reversal switches were exercised quite often.

Intentional phase reversal should also be mentioned. Aside from it being a necessary ingredient in accomplishing the stereo mid-side microphone combining technique, it is also useful for several other interesting effects. (See "Recording Studio Procedures.") Any cable intentionally wired out of phase to facilitate one of these uses should obviously have a clearly "flagged" marker. A circle with a diagonal line through it (Ø) represents 180 degrees out of phase.

Phantom Powering

Most condenser (capacitor) microphones need an external power source to polarize their diaphragms and/or to power their internal electronic amplifier circuitry (see "Microphones"). This power can originate from batteries internal to the microphone, a separate power supply, or the mixer that the mic is connected to. Consoles ship this DC power back out to the microphone through the same mic cable and XLR connections that carry the audio signal in a way that keeps the DC voltage invisible to the audio passing through the same wires, hence the name phantom. This was discussed in the section on differential inputs. To keep it brief, I'll only use the passive approach as an example here (see the section on differential inputs for more on active CMRR). In Figure 2-1, the DC voltage from the mixer is fed into the microphone cabling.

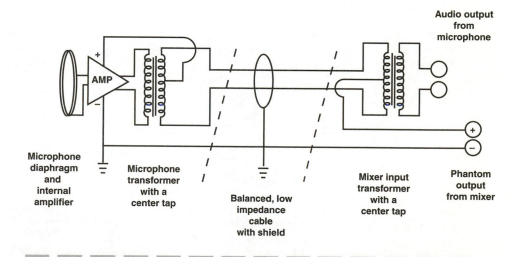

Figure 2-1
The positive or "hot" output from the mixer is fed through a center tap of a transformer, which is connected to both the high and the low sides of the balanced cable, while the negative output is connected to the ground, shield, or common conductor.

The positive or "hot" output is connected to both the high and the low sides of the balanced cable while the negative output is connected to the ground, shield, or common conductor. Notice that the positive is connected to the signal conductors through a center tap in the transformers. Here, the positive DC power enters on the left side of the right transformer, which is the console input transformer. The DC current will *not* be transferred to the right side of that transformer or be fed into the consoles preamp. (Any questions about this will be answered in the section on transformers.) The left transformer represents a microphone output transformer.

Notice how the DC voltage is "collected" from the right side of that transformer and fed directly to the electronic circuitry (amplifier and/or diaphragm for polarization). The DC will have *no* effect on the left-hand side of that transformer, which handles the mic's audio output. Again, because of differential common mode nulling and the nature of transformers, the DC voltage will be canceled out at the points between the two sides of the transformers. This means that the DC power that "jumped across" the microphone output transformer will not be "seen" by the microphone audio signal, nor by the console input. Thus, the DC voltage conducted through the audio cable has no effect on the audio.

Separate (stand-alone) phantom power supplies often incorporate output resistors to help hold the current to 100 milliamps and utilize DC-blocking capacitors, which are helpful should the destination not be a differential input. (See Figure 2-2. Also, review the section on capacitors for an explanation of their DC-blocking capabilities.)

You see, all that crummy basic electronics and math stuff really is useful as it does come into play in everyday audio work. Why else would anyone bother to learn it? To me, phantom powering is almost as amazing as hum-bucking pick-ups. *But there's more!* By now you should understand that it's possible to send a low-current DC voltage down a microphone cable while it's being used to conduct an audio output from a microphone. What if that DC power was fed to a light that could then be switched on and off to cue the person with the mic when to speak? This use of phantom power is called "simplexing" and its uses can get *way* out there.

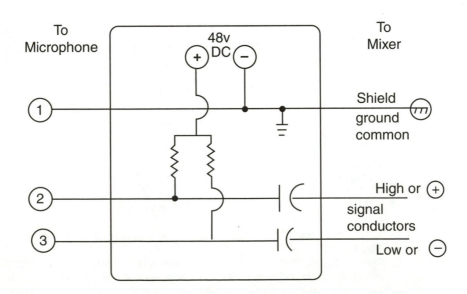

Figure 2-2
A remote phantom power supply with matched output resistors and DC-blocking capacitors

If the DC power and switching are on the talent, or microphone end, the "speaker" can use the switch as an interrupt. Here when they hit the switch the power travels down the mic cable and activates a relay that kills, say, the background music and switches the input to the system over to the microphone input. Chairmen of the board love this, I know. *But it goes way beyond that!*

The credit for this next application of simplexing goes to the pro-audio manufacturer named "Audio-Technica." At least they have the trademark on the name "modu-comm" and I've never run into anyone else using it. Fasten your seat belts, folks!

This methodology allows the AT 4462 Stereo Field Production Mixer's output program to be shipped back out through an incoming microphone line to a belt clip-on decoder, which delivers the signal to the talent's headphones. The program output is not affected by the mixer's microphone gain setting and the process does not affect the incoming mic signal. I'm not privy to all of the actual functioning, but I imagine that the modu-comm circuitry uses the program information to modulate the phantom DC. It will still cancel out at a differential input, but when demodulated (decoded) the program material remains.

There are limitations. As with normal phantom power, there can be *no* use of additional pads, filters, or transformers on the line.

Furthermore, the AT 4522 has a "comm" button that allows for the producer/operator to communicate with the talent to provide verbal cues over the modu-comm system. The program signal sent along the mic cable can also be interrupted by patching into a line input, allowing any signal to be fed to the talent.

Finally, if the talent's microphone input is switched to "cue," a two-way conversation can be held between the operator or producer and the talent without any of it showing up in the program or on-air outputs. *Ain't audio grand when it's done well?*

The Producer's Desk and Side Cars

Many I/O consoles can be ordered with an additional desk-like section added on as an option. Usually, this is added to the right-hand side of the patch bay, but left-side versions are offered as well. This gives the producer his or her own work area and often keeps the producer and the engineer from getting in each other's way. The lower, flat, desk-like space is large enough to spread out charts and the upper "meter bridge" section often contains a reading light. Other accommodations include a talk-back mic with "comm" switch and either headphone monitoring jacks or a small set of stereo "automotive" speakers with amp and level controls built in. An "alternate center" speaker-positioning switch is part of the main console's monitor section. This, along with two trims (adjusted with a small screwdriver through holes in the monitor module's face plate), varies the level of the left and right bus outputs as fed to the monitor speaker amps. This setup is used to reposition or focus the main monitor speaker's center point to the producer's desk location. Because of the "Haas effect" this cannot be accomplished via level adjustment alone. (This can of worms will be discussed in both the monitoring and delay sections).

Due to the fact that today producers seldom work with orchestral sections and charts the way they used to (and when they do, they prefer to be located in the "sweet spot" between the control room's monitors), producer's desks have become a thing of the past. However, that added section (when available) makes for a perfect "side car."

This area can be used to house an additional patch bay. Its meter bridge section is often perfect for effects such as a stereo compressor that's used on the overall mix and specialized mic preamps, which are more handy when located there. The desk section is also ideal for the inclusion of a small submixer. Here a small board can be dropped right in and used during mixdown to combine multiple effects returns, anything from a stereo to eight separate bus output feeds that can then be patched to the main board's channel inputs. Besides increasing the number of inputs during mixdown, the choice of submixer also affords the possible addition of equalizers or automation. With nonautomated consoles, the advantage of having your effects returns, mutes, and levels memorized is considerable.

The Patch Bay   Patch bays are, more often than not, built into the recording console. This was not always the case and many feel that locating the bay in the rack with the outboard equipment is a superior approach. The way they see it, since you'll have to go to the outboard rack to adjust the effect's parameters anyway, why not make all your patches there instead of going back and forth between the console and the outboard rack when setting up a multiple effect chain? Hey, as far as I'm concerned, it's six of one, half-dozen of the other, but vertical rack positioning does reduce the amount of debris that accumulates on the jack's internal contacts. This means you'll have to clean the jacks in a bay that's mounted on a slant more often.

Clean the bay? I know of no bay with gold contacts so oxidation occurs, which requires occasional brushing or burnishing and, as always, dirt finds its way into everything. They make special tools to handle this, but the plugs can be cleaned with just a rag and metal polish. If you don't clean these contacts, you'll have to live with the attenuation and intermittencies dirt and oxidation cause.

Aside from the need for occasional cleaning, the only other thing you need to understand about patch bays are the various types of normaling that are made possible by modifying how the jacks are wired. These include normaled, half-normaled, not (or unnormaled), and multed.

In order to communicate you need to speak the language of the audio professional, so some definitions are needed. First, connectors like those found on the end of cords that are plugged into equipment are called *plugs*. The receptacles they are connected to are called *jacks*. That whole other gender thing is very un-pro. Next you need to know the names or numbers of the individual contacts within the connector and how they are used. Obviously, the best way to visualize a connector would be to look at one, but the diagrams in Figure 2-3 should also accomplish this task.

Part "A" depicts a patch bay jack's frame or base. Its left-hand side connection point goes to the common (also called shield or ground) conductor of the signal. "B" shows the movable spring contact that will come into contact with the tip, which is the positive or hot conductor of the inserted plug. "C" is the part of the jack that provides the "normal" connection with the tip's spring contact when no plug is inserted. "D" is the part of the jack that provides the "normal" connection with the ring's spring contact when no plug is inserted. "E" shows the movable spring contact that will come into contact with the ring, which is the negative or low conductor of the inserted plug.

Figure 2-4 shows a patch bay jack fully assembled with no plug inserted. Notice how a hot signal fed to point "B" would (due to the tip spring connection's contacting the pad of the tip's normal connection) appears at point "C" and that same relationship occurs between the ring's spring contact and normal connections.

Figure 2-5 depicts the result of inserting a plug into a patch bay jack. The plug not only connects to both of spring contacts, but moves them away from the normal contact pads, thus "breaking" those normaled connections. With this understanding, we can now talk patch bays.

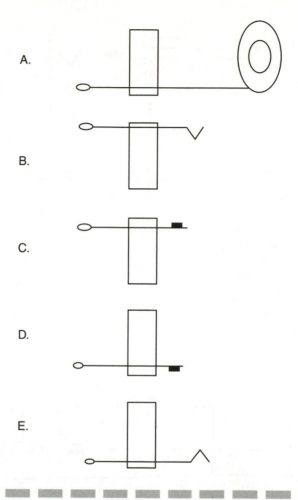

Figure 2-3
The individual contacts that make up a patch bay jack. (A) The patch bay jack's frame or base, which is connected to the common conductor of the signal. (B) The moveable spring connection that contacts the tip, positive, or hot part of an inserted plug. (C) The "normal" connection, which contacts the tip's spring when no plug is inserted. (D) The "normal" connection, which contacts the ring's spring when no plug is inserted. (E) The moveable spring connection that contacts the ring, negative, or low part of an inserted plug.

Go back to the MCI 600 or SSL 4000 G patch bay descriptions and check out the many examples of normals. A group output, for example, will be normaled to an input of the multi-track recorder. Here the console group output's hot feed is connected to the output jacks tip spring contact while the low is connected to the ring spring contact. With no plug inserted, the hot signal is fed to the tip's normal connection and the low to the ring's normal connection.

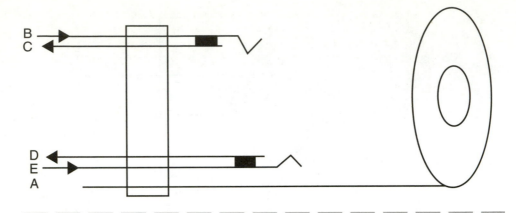

Figure 2-4
A fully assembled patch bay jack with no plug inserted. A signal fed to point "B," the tip spring, makes contact with the pad of the tip's normal connection, causing the signal to appear at point "C." This same relationship occurs between the ring's spring contact and normal connections.

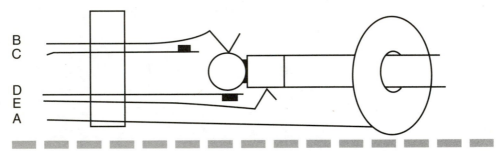

Figure 2-5
The result of inserting a plug into a patch bay jack. The plug not only connects both spring contacts, but moves them away from the normal contact pads, thus "breaking" those normaled connections.

If these two group output jack normals are then wired (or "jumped") down to the two normals of the multitrack input jack directly below it and that jack's two spring contacts are wired to the multitrack input (with no plugs inserted in either jack), the group output signal will flow unimpeded to the multitrack input without the need of any patch cords. *This is called full normaling* (and I'll leave it up to you to mark this routing on the drawing).

As with channel send and return patch points, if the send's high and low conductors are "Y'd" or split and fed to the spring contact of the send's output jack *and* the normal contacts of the return jack, you will have what is called *half normaling*. Now, when you insert a patch cord plug into the send's jack, that signal's feed via the normals to the return jack is not "broken." You can now patch this signal into an effect, bring it back to an open channel, and mix it with the original signal whose normal was not broken so it continued on through the original channel's routing.

An outboard effect's input and output conductors would more than likely be connected only to the bay input and output jack's spring contacts. Here patching into the input jack gets the signal into the effect while the output jack provides access to the signal from the effect's output. Apart from the occasional side chain, two *un-normaled patch points* like these are all that is required.

Let's take that "Y'd" connection we mentioned and run it across the patch bay row instead of to the jack below. Now repeat it twice and you'll have four jacks connected together. Patch a signal into one of these jacks and you can ship it out of the other three jacks to three different locations. This is called a multiple jack connection or a *mult* for short. There's more!

If you continue "Y"ing so that you have two four-way mults sitting side by side with a single jack in between and if you connect the spring contacts of the fourth jack to the spring contacts of the fifth, and the normal contacts of the fifth to the spring contacts of the sixth, you'll have an eight-way mult without a dummy (nonwired) plug inserted into the fifth jack, or two four-way mults with a dummy plug inserted.

Since we mainly discussed the section of the patch bay that handles the input channel's patch points while charting gain stages, a little elaboration is now needed.

The MCI 600 series patch bay's row #9 includes a stereo output that allows pre-fader patching to the main stereo bus outputs. Without any plugs inserted, these jacks are normaled to the stereo input patch just below on row #10. Row #9 also provides "Aux" out patch points and the automation data to tape patch points. Row #10 has a mono output feed and three two-track (L + R) output returns to mixer patch points. Row #10 has a pair of "Aux In" patches that can be used to feed a stereo signal right to the monitor system.

Both rows 10 and 11 have dual four-way mults and rows 13 and 14 combine for another four-way mult. They also have permanent dummy plugs that act like a switch. As stated in the manual, "Five sets of Mult jacks are provided. Each set has four jacks connected in parallel. The tie switch between each set connects the two sets together when pulled into the UP position. With all four tie switches in the UP position, the 20 jacks would be in parallel." No suggestion of an application for a 20-way mult is given. Some send outputs are normaled at this bay to cue amp inputs and others to echo inputs.

Echo device outputs or returns are normaled to "chamber" returns (which is the name given to the separate reverberation return sections on this console). There is also an aux input that routes a patched signal to the pre-fader listen amp.

Two patch points bring a pair of signals up on the console's phase meter. Several patch points are used to feed and return signals through a set of optional "wild" faders. I think they were called wild because being unconnected to any function, your imagination can run wild using them for anything you want, yet they *were* connected to the console's automation system.

The MCI 600 series bay also had a unique pair of patch points that provided "access to two isolation transformers. These transformers are normally used for isolating the console circuits from some external function." They had a one-to-one turn ratio at 600-ohms impedance.

It's not hard to understand that at $150 for just a dual 48-point bay strip, a full-blown custom patch bay with all cabling and outboard connectors can run several thousand dollars. But I think the functional versatility makes it worth it.

At the bottom of the bay, the tie line patch points are located, normally numbering 72, but with the option of having up to 250.

"Tie Lines Jacks are provided so that any external function may be brought directly to the console." Tie lines are used as connection lines between different control rooms, to a tape machine room, the studio, or anywhere else that may need to receive an output from or be used to provide an input to the console. To the recordist, however, the most important use for the patch bay tie lines may be to connect the console to that rack of outboard effects processors.

Equalizers

Hey, this place has a ton of outboard gear! Enough to fill a half-dozen 3-foot high racks. Together their tops make up a 12-foot long work area that doubles up as a keyboard stand for overdubbing in the control room. It's a pretty slick layout in that all the outboard gear is grouped together in the racks as per function: equalization, then dynamics processing, followed by the delays, after which comes the reverberation devices and finishes up with some "special effects" toys at the end. They have a rack-mounted patch bay that can be used to re-route outputs to, you guessed it, a submixer sunk into the console's side car.

In this case it's a little rack-mountable Hill Audio "multimix" 16:4:2:1 mixer. This is a clean-sounding little board that provides 15 dB of headroom on top of its +4 dBu operating level. It's SMPTE IM distortion is rated at less than 0.02 percent. It also has 12-segment (0 = +4 dBu) LED metering that is peak reading, but its decay response is easy to follow because it's been slowed down to VU speed.

However, what's really special about this mixer is the EQ on its channels. These appear to be nothing more than an elementary three-band (100 Hz, 1 kHz, and 10 kHz) "tone control" type of EQ, having fixed center frequencies and bandwidths with the only variable being the amount of boost or cut.

Back in the 1980s while field testing this mixer for an audio publication, I thought I discovered something special about this EQ. Normally when boosting, say, a kick drum at 100 Hz, cranking a guitar's sound up at 1 kHz for more "bite" or using the high frequency to modify the output of a reverb, you hear some added coloration. At the bottom end this is usually described as muddiness or with highs as a harsh whooshing sound. This is caused by the delay and the consequent phase cancellation (comb filtering) often inherent to equalizers. However, with the Hill multimix, the frequencies that were being modified sounded clean and punchy and thus seemed unaffected by phase shift distortion.

When I put it on my test bench this proved to be true. As can be seen by the phase response graphs in Figure 2-6, this EQ is nearly linear with frequency.

In these graphs (printed on a Neutrik strip chart recorder) the center horizontal line represents no phase shift and the top and bottom horizontal lines represent the (+ and −) 180 degrees out of phase points. As you can see from these plots the response was nearly identical with all three bands set at either full boost or cut. You have to get to 20 kHz before the signal is close to 90 degrees out of phase and well past 40 kHz before it's near the +180-degree point. This may not seem like a big deal, but normally when plotting equalizer phase response, the pen making the line is slammed up and down ranging, from the full +180-degree point to −180 degrees repeatedly across the whole graph.

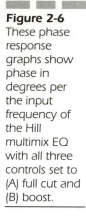

Figure 2-6
These phase response graphs show phase in degrees per the input frequency of the Hill multimix EQ with all three controls set to (A) full cut and (B) boost.

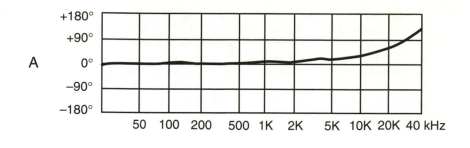

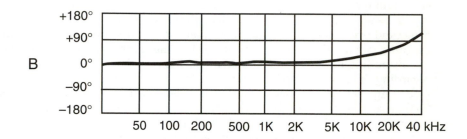

After confirming my initial conjecture about this EQ, I called the manufacturer to find out if this EQ had been designed with a computer and if the EQ on their larger eight-band mixers also had this kind of response. The answers were no and they do not. But I did find out that the capacitors used in this EQ are not enclosed in "cans" but open. They appear to be made up of a polyester dielectric between the two metalized "plates" that have leads attached to them. I was told by the designer that because of this open design, these caps could be ground down (shortened), while a meter was attached to the leads and the machining stopped when the cap reached a desired capacitance.

Because of this, the designer had access to capacitors of *any* value, which is what made this three-band phase coherent equalizer a possibility. The end result is a bass drum that's a straight-out karate kick to the chest and a shimmering hi hat without any added high end harshness. No magic, just hard design work, taking advantage of the latest innovations and a lot of compromises in control parameter variability.

You need to have a basic understanding of the various equalizer types and the extent of their individual imperfections in order to use them properly and to avoid causing the "ugly" sound. As you know from the section on capacitors and inductors, any AC signal passing through them has its time "domain" (its place in time) affected (shifted). I did not want to get too involved with capacitive or inductive reactance in this section and I certainly do not want to delve into this horribly boring topic now. But just to help you to disregard "pseudo-scientific" facts about equalization, I'll reiterate.

Capacitors have the ability to stop DC from passing through them. AC and pulsating DC, because of their varying charge, can induce current to flow through them. How much current flows is *directly* proportional to the frequency or the rate of change. Here, as the frequency increases, the amount of current moving across the capacitor's plates increases. This relationship is termed "capacitive reactance." It is measured in ohms and as the rate of varying increases, the capacitive reactance in ohms decreases. Stated another way, as the frequency of the voltage increases, the capacitor's opposition to current flow decreases.

Since the current flow is dependent on the rate of change in the voltage, current and voltage cannot occur at the same point in time. With capacitors, current is said to "follow" voltage and thus the two are not in phase with each other. With inductors, the rate of *current* change causes variations in the opposition to the flow of voltage, so voltage follows current. These two component's varying impedance with frequency makes them close to ideal for use in filters and crossovers ("close" because of the inherent time delay).

History Time   Little in the way of EQ development transpired during the early days of radio and telephone. It wasn't until the introduction of film "talkies" when demand ($) influenced research that the first real professional audio equalization tools were generated. At that time (and even today), the sound recordist's most important job on the film set was to keep the microphone and all other "sound-gathering apparatus" out of the camera's view. This meant pulling back on the microphone positioning. This in turn caused two problems when recording the actor's speech. With distant mic placement more of the room reverberation was picked up. Not a big problem indoors, just deaden the acoustics of the sound stage. But the farther sound has to travel through the medium of air, the more its high-frequency content is lost. This is a straight-out fact of physics, without any hocus pocus or half-truths.

This was a tough hit on those original film sound recordists, but they were in for some big-time luck. The technicians involved in correcting this problem did not have to answer to a sales department and were yet unaware of any form of techno-hype. They approached the alleviation of this problem from a strictly scientific position. Since a high-frequency reduction was the cause of the imbalance, if an electronic circuit was introduced to decrease the low-frequency content matching this reduction, it would "equalize" the two sounds. That's actually how EQs came about and how they got their name, honest. But the luck continued! The scientists theorized that these equalizers needed to be variable in order to match the different mic distances. Thus, variable, cut-only, low-frequency EQs were born. But there's more. The speed of sound through air is not uniform with frequency. High frequencies are delayed in their passage through the medium of air more than low frequencies. It just happened that the phase shift caused to the low end by the use of these early EQs (though very slight) was just enough to match the time lag of high-frequency sound through air so that at the output everything "equalized" phase-wise too. This was *not* planned; it was actually just a delightful accident!

The rest of this section is really going to point out that the more complex the EQ is, the greater the potential for misuse, or overuse, which can result in bad sound. Don't believe me? Try sweeping a continuously variable frequency control on a parametric EQ. That nasty whooshing sound is caused by the change in phase response as per the changing center frequencies. That's why older fixed-frequency and switched-frequency EQs sound better. Not only did the designers design out the possibility of excessive phase shifting by choosing to not use continuously variable or swept center frequency selection, but they opted to use center frequencies that had the least amount of phase shift. This gave them those legendary "musical sounding" EQs.

An even simpler example of equalization is the dual-band "tone controls" as found on the average stereo system or guitar amp. Here, should the sound seem to have too much bottom end, you can turn down the bass, crank up the treble, or do some combination of the two in order to "flatten out" the response.

Figure 2-7

Figure 2-7
This graphic
representation
of a simple
dual-band
"tone" control
shows band-
width, center
frequency, the
overlapping of
the bands, and
the amount of
boost or cut.

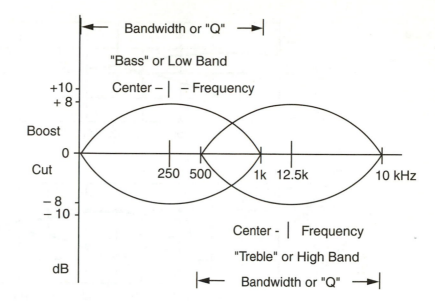

The bass and treble controls have overlapping frequency ranges (see Figure 2-7). The center frequency of each range is fixed and is represented by the peak in the hills (or bottom of the valleys) of the curves. Each range is called a band and the band's width, more commonly called the "Q," is the range of frequencies affected by that control and in this case, it's also fixed. The result is that when the knob is turned up or down, the level change affects all the frequencies within that bandwidth to some degree.

Looking at the graphs of the Hill EQ's frequency response in Figure 2-8, you can see how a change in level affects the whole group of frequencies that make up a band.

These graphs show all the boost and cut curves and thus the bandwidths and center frequencies of the Hill EQ's three bands. They are actually made up of six separate plots, one each of the low, mid, and high bands set at both full cut and boost. The result is a pictorial representation of each band's width, shape, and center frequency.

Notice that the three center frequencies are not exactly at 100, 1 k, and 10 kHz as labeled on the mixer's face plate. The low extends to 50 Hz but still affects 100 Hz equally because it's shape is *shelving*. The mid, which is termed a *peak/dip* curve, has a center frequency that's just slightly higher than 1 kHz. The high band is arguably closer to 16 kHz, but the level difference between the 10 and 16 kHz points is so slight that it's almost nonexistent. Take notice of the way all the plotted lines drop off in level above 20 kHz. This was a function of the way I had the plotter set, not the EQ. Again no magic, just a great sounding EQ due to hard work by the designer. Taking advantage of the latest in technology and accepting necessary compromises to achieve one goal: a sound that needs no hype. They never even mentioned the superb sound of this EQ until I questioned them about it.

The common two-band (bass and treble) "tone control" equalizer can be further subdivided to the extent of having 27 bands as found with some graphic EQs. Here again the bands controlled by each fader are fixed as per both the bandwidth ("Q") and the center frequency. But now, because there are more bands, this graphic EQ will obviously render a more precise

Figure 2-8
These frequency response graphs show the shapes of the Hill multimix EQ's (A) low, (B) mid, and (C) high bands.

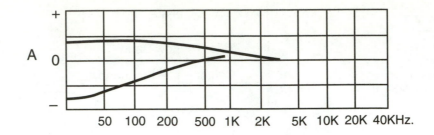

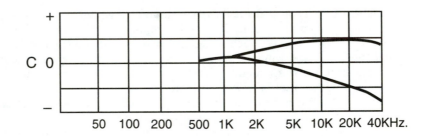

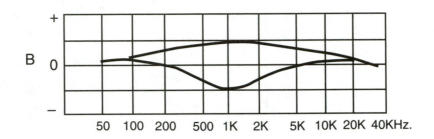

control of the overall sound. This type of device can be a great help when trying to flatten out, for instance, a speaker system where you will have various peaks and dips across the whole response curve. If the bands of your graphic match up perfectly with these peaks and dips in both Q and center frequency, as well as the amount of level to be added or subtracted (boost or cut), you're all set.

But what if the center frequencies don't match up or the Qs of your graphic are wider than that of the problem you want to alleviate? A perfect example of this often occurs in live stage monitoring. When moving from stage to stage, different acoustic situations arise, causing the center frequency of feedback problems to vary. The bandwidth of your graphic, even with 27 of them, is still going to cover a full one-third of an octave.

If the feedback's frequency width is narrower than that of your 27-band graphic, say only one-fifth of an octave, by pulling down a fader you'll be unnecessarily decreasing the level of some frequencies that you may need to help (a specific sound) cut through the onstage din to reach the musician's ears. In this case you may be better off with a parametric.

Parametric Equalizers

Parametrics come in various configurations that can be subdivided into two main groups: full, having control over both Q and center frequency, and semi, having control over only one of these parameters. They can be subdivided even further into those that have sweepable or continuously variable control over these functions, and those that have switchable settings for a particular number of center frequencies and several different bandwidth or Q ranges. Obviously, full parametrics with sweepable controls offer the most flexibility. With this in mind we now go back to our feedback problem. With a full parametrics set to, say, a full octave, you can just cut the level and sweep your center frequency until the feedback ceases. Now, via the Q control you cause the band's width to narrow until the feedback just starts to reoccur. Simply adjust your center frequency again, along with your cut level or "roll-off," until you've tuned out only the problematic area, leaving everything else unaltered. In recording situations, a parametric will allow you to eliminate the ringing of a tom tom so that you don't have to overly dampen it, thereby preserving more of its full and natural sound.

Parametrics are also very good ear training tools. Since you can sweep across the different frequencies in the band as well as either bring them out to the forefront or eliminate them, you can easily pick up on various aspects of the sounds you are dealing with. Take an electric guitar, for example. You can narrow down your Q and sweep across the frequencies until you hear an increase in the sound's bite. Once you read off that center frequency, any time you want to add or subtract this bite sound you'll be able to tune right to it.

One of the drawbacks, as far as parametrics are concerned, is the fact that they have generally only four bands. So just as a graphic may not be the end of all your live monitoring problems, the same is true of a parametric. In some cases you may need both.

Important parameters to check out when you're looking at parametrics, aside from things like controllability and frequency response, are S/N ratio (which should be at least −85 dB), and distortion.

Also, look at its interfacability, in terms of handling what it will be connected to, from an instrument output's low level to a PA's preamp line level. The amount that you can cut or boost the signal is also important. If the device you are using only provides 6 dB of cut and the problematic frequency is peaking at 12 dB, you may come up a little short. While they can be fabulous tools, no equalizer is a cure-all and it's a waste of time to pretend that it is.

The Loft 401 Parametric EQ   With all this in mind, let's now take a look at the Loft model 401 parametric equalizer. It's clean (THD + N is 0.01 percent) and its internal circuitry is well laid out. Servicing should be no problem as all *integrated circuits* (ICs) are socketed. The chassis construction was designed to be rugged enough to take roadwork abuse. Input is via ¼-inch phone jacks, which can be balanced or unbalanced. Its preamp stage provides 20 dB of gain, allowing the 401 to accept almost any level from an instrument output to line level. Maximum input level is +24 dB, referenced to 0.775v. This equates to 20 dB of headroom above the operating level of +4 dBu. The outputs (+18 dB max) also appear on ¼-inch jacks (these are unbalanced) with one providing a line level capable of feeding power amp inputs and one padded down 20 dB for feeding preamp inputs. Complementing this ease of interface is a level control that has a three LED headroom indicator to aid correct gain structure setup.

The 401 is a fully parametric four-band EQ. All of its controls are not only continuously variable, but each has 21 detents. This allows for fast resetability and can aid in determining exactly what frequency, level, and bandwidth you are dealing with. Here the degree to which these four bands overlap (low band, 30–600 Hz; low-mid band, 100–2 kHz; mid-band,

400–8 kHz; high band, 1 kHz −20 kHz) is such that as many as three can be used in one specific frequency area. Add to this the fact that the level can be varied for up to 18 dB of boost or cut and that Q is variable from three octaves down to one-sixth (0.15) of an octave in width and you begin to see how powerful this device is. It can literally increase one frequency while cutting two others with only a couple of notes separating the three. It can actually be used to help "tune out" speaker cabinet resonances. I've done it.

Because of the phase anomalies inherent to equalization, some people are anti-EQ. They usually do not work in the real world of recording. I remember while teaching audio technology hearing someone say that you should never use EQ, but choose your microphones so as to achieve the desired affect in the frequency content. Yeah, sit this person down behind a console with a dozen and a half union musicians on the other side of the glass and when the producer says something sounds "a little tinny," they'll immediately reach to turn down the high-frequency EQ control. Everyone does. It's just part of our "audio instinct" ingrained from thousands of experiences.

Everyone has an opinion; everyone is right, yet not that many really know the facts. I certainly don't know all the facts, but just so you'll know what you're talking about (and when someone else is talking voodoo electronics), a factual run down on the different types of equalizers follows.

Tone Control EQs

Tone control EQs have already been discussed. These have fixed bandwidths and center frequencies. At one time, the only operator variable parameter was the amount of cut. Now it's more common to find both cut and boost.

Shelving EQ or Filter

If the shape of the frequency band's curves is less slope-like than the Hill EQ's mid-band and the affected area rises up and flattens out across the band (as the Hill EQ's low- and high-bands do), they end up looking like a shelf, hence the name *shelving filters*. They are generally not continuously variable, but often have switch control bandwidth, cut, boost, on/off status, and frequency range.

Peak and Dip EQs

Some EQs utilize *only* bell-shaped curves for the slopes of all their bands. The Hill's center band is peak/dip. These EQs are referred to as peak and dip EQs because the center frequency is the point of the highest amount of cut and boost. Your adjustments will, in effect, cause a larger amount of "peak" or a greater "dip" at that center frequency. This obviously will be more significant with narrower bandwidths and steeper slopes.

Switchable Center Frequencies

These allow for a greater degree of variation in a band's frequency setting. The simplest of these has a mid-band center frequency switch that provides the choice of three different center frequencies (i.e., 1 , 2.5, and 5 kHz). These also come in different configurations. They can provide all-cut, all-boost or some of either; have two bands, three bands, or more; and on and on. These added features increase an EQ's ability to vary the signal's frequency content.

Swept Center Frequency EQs

Here the full audio bandwidth is divided up into two or more bands (I've seen as many as four and eight) with each band having two controls, one to boost (+) or cut (−) the frequency content of the band and another that "sweeps" through the center frequencies, allowing you to select the exact area within that bandwidth to be affected. You can tune these while listening to the output almost like zeroing in on a radio station. When doing so, remember to note where the amount of whooshing and/or any harshness is added to the sound and avoid processing in these areas.

Semi-Parametric EQs

This is my own name for *parametric* EQs with *switchable* "Qs." These have everything previously mentioned plus a control that allows switching between different bandwidths (or "Q" settings). This determines the width of the frequencies above and below the center frequency. To recap, you now have any number of bands (two to eight are common) and each one can have its center frequency setting changed by either switching or swept via a continuously variable knob (potentiometer). In addition, the Q or bandwidth around that center frequency can be widened or narrowed via switching. This is a lot of control.

Full Parametric

Again, this is my name for *parametric* EQs that have everything mentioned above plus a "Q" setting that is continuously variable.

Graphic EQs

These are comprised of 10 or more bands and, like tone controls, range from simple, fixed bandwidth and center frequency devices, allowing only a change in level to full blown *paragraphics* (I swear that's what they're called) that provide 10 or more bands of full parametric control. Here the boost or cut is usually controlled by a fader.

NOTE: For the most part, graphic EQs are those that utilize vertical faders for boost and cut controls. Often their center position is detented (has a noticeable stop), which equates to no level change. Once all the faders are set, they provide a graphic view of the equalization that's been set. Other than this there is no real difference between using pots as opposed to faders except for cost considerations (faders cost more than pots). However from a tech's point of view, sealed rotary pots are less prone to the collection of dirt.

Graphic EQs normally have pre-set center frequencies and bandwidths that correspond to the *International Standards Organization* (ISO) specifications. This is an advantage as these EQs directly match up with measurements obtained with analyzers that conform to the same ISO standards. Both are often used together for the adjustment of the frequency response of monitoring systems. Here equalization processing is only used for a slight amount of adjustment (under 6 dB) to avoid causing phase response degradation. They are meant to be used for "touching up" any electronic discrepancies, not as a cure for any faults in the room's acoustics design.

I am repeatedly playing down the use of equalization for correcting acoustic irregularities in rooms because these problems should be handled by making appropriate physical changes to the acoustics. Electronic compensation will not achieve the desired results and can often end up causing far worse phase (timing) and smear (imaging) errors (see "Control Room Acoustics and Monitoring").

As with *all* equalizers while *active* graphic EQs provide both boost and cut, cut-only models are often favored as their specifications are often superior. Either type will be referred to according to the number of bands provided or the octave intervals between each band: octave/10 or 11 bands, one-half octave/21 bands, one-third octave/27 or 31 bands, two-thirds octave/16 bands, and one-sixth octave/64 bands (they do exist).

Some Generalities

The parametric's advantage of adjustable Q means you can adjust only the frequency content you wish to, without affecting other areas of the audio spectrum as well. This is very helpful in decreasing and often completely eliminating PA ringing where you also do not want to defeat the stage monitor's ability to cut through the din. While a parametric is better for the precise adjustment at a specific area of a sound's tone, graphic EQs are better for the adjustment of a sound's overall tonality. Both tools work fine, but each is designed for different purposes and each is better than the other when used for the purpose for which it was designed. Nothing unusual here.

Specific EQs

When it comes to the adjustment via equalization of monitor speaker/amplifier/crossover anomalies, White is the name seen most often in professional recording studios. Their model 4301 not only has thirteen ⅔-octave filters covering the frequency bands from 1 kHz to 16 kHz but below this, where fine tuning is often acutely needed, this device provides 28 ⅙-octave filters covering the 40 Hz to 900 Hz center frequency bands. Newer White models incorporate full parametric and programmable functions.

During the early 1980s, the folks at 360 Systems picked up on the fact that sound shaping via equalization often required changes at different points during a program. For example, a guitar having some of its mids (vocal range frequencies) cut back during the verse sections may require a boost in that frequency area to be able to take the rhythmic lead position during the chorus. This kind of variation, while manually possible, can be cumbersome during a mixdown and close to impossible live. Their model 2800 offered four bands of full parametric equalization. The bandwidth range is from as narrow as ⅙-octave up to 5 full octaves. Each of the bands is capable of anything from 60 dB of cut to 12 dB of boost. Due to this device's frequency response of 20 Hz to 20 kHz, if used correctly it would fit into just about any pro audio application.

However, its ability to silently switch between any of its 28 nonvolatile (lithium battery protected) stored memories of gain, bandwidth, and Q settings, allows this device to be used as a "special effect."

This droning on and on about EQ parameters could easily go on for several more pages. But I do not want to make all the EQ variations, complexities, theories, names, methods, and techniques into some kind of a creed or dogma. I prefer to deal in real-life day-to-day usage. Therefore, this space would be put to use describing several real-life equalizers along with their capabilities. So, on with the tour!

Pultecs Hey, this studio's owner must be a collector as they have a bunch of solid state (non-tube) versions of EQs from Pulse Techniques, Inc. (better known as Pulstecs), of Teaneck, New Jersey. Their Model EQP-1A3 is not your common three-band tone control. The equalization circuitry is all passive, but all losses are restored due to the addition of a solid state operational amplifier. Due to its "no loss passive" design, the "overall result is no loss and no gain." Clipping occurs at +21 dBm. So a "+4 dBm input allows generously for signal peaks without clipping." The fact is since a +18 dB boost is possible, +3 dBm would be the safest maximum input level. This would still provide a S/N ratio that's greater than −80 dBm. (I hope you took note of all those lowercase m's.)

The low-frequency band is of the shelving type and provides boost (0 to +13.5 dB) or cut (0 to −17.5 dB) at 20, 30, 60, or 100 Hz. The mid-band is a peaking filter with switchable center frequencies of 3, 4, 5, 8, 10, 12, and 20 kHz. This band provides only boost (0 to +18 dB). It's "Q" or bandwidth is switchable (five positions) between "broad" or wide-band, and "sharp" or narrow-band. Yet this control has more of an effect on the amount of level change than it does the bandwidth. Here it adds, at the narrowest bands, as much as 8 dB to the peak boost.

Just below, in the same rack sits a Pultec EQH-2. This equalizer looks like it's just a two-band version of the EQP-3, but it's a whole different animal. The low-frequency band is of the shelving type and has a set bandwidth or "Q" but with switchable frequencies of 20, 30, 60, and 100 Hz. This band's boost range is 0 to 13.5 dB while the cut or attenuation range is 0 to −17.5 dB. The high-frequency band has a peak/dip band shape, a fixed bandwidth, and switchable center frequencies of 3, 5, 8, 10, and 12 kHz. The range of boost control is 0 to 16 dB. Surprisingly, this band's attenuation control *only* affects the 10 kHz setting. Furthermore while the boost at that frequency setting is of a peak/dip shape, the cut is shelving. How this was accomplished I do not know, but it was designed to roll off the highs above 10 kHz. This high-cut filtering can be quite radical. At the maximum, the "−16 dB" attenuation setting, the cut is 2 dB at 1 kHz, 5 dB at 2 kHz, 12 dB at 4 kHz, the rated 16 dB at 10 kHz, and 18.5 at 20 kHz. So long, radio airwave and tape hiss distortion. This could be a good trade-off as the resulting phase shift is within the acceptable standards (distortion being 0.15 percent).

The Pultec MEQ-5 provides two peak/dip bands and a cut-only shelving filter. If this were all there was to it, it would be unremarkable. However, the ability to combine these three sections to operate over the same frequency area makes this device a very precise and versatile equalization tool. The first band (like all the rest) has a fixed bandwidth. It's a peak, boost-only filter with the center frequency setting selectable at 200, 300, 500, 700, or 1 kHz. The range of boost is 0 to 10 dB. The next band is a "dip" or cut-only (0 to −10 dB) filter with its extended range of selectable frequencies being switchable to 200, 300, 500, 700, 100 Hz and 1.5, 2, 3, 4, 5, and 7 kHz. The right-hand-most band is another peak boost-only filter, which provides a range of boost from 0 to 8 dB. Its center frequency can be set at 1.5, 2, 3, 4, and 5 kHz.

Due to this overlapping of bands it is possible to boost 1 kHz by 10 dB, cut 1.5 kHz by 10 dB and boost 2 kHz by 8 dB, all at the same time. I didn't go to the trouble of pointing this out because of any possible application but instead to show the amount of flexibility provided. In fact, in some situations this kind of setting could sound terrible.

The Pultec HLF-26 is a dual (stereo) two-band cut-only shelving filter. The low-frequency bands provide a 40 dB cut or roll-off at 250, 50, 100, 80, and 50 Hz. The high cut-only filters yield

−4 dB at 6 kHz and −32 dB at 20 kHz, at the "6 kHz" setting

−4 dB at 8 kHz and −25 dB at 20 kHz, at the "8 kHz" setting

−4 dB at 10 kHz and −20 dB at 20 kHz, at the "10 kHz" setting

−4 dB at 12 kHz and −16 dB at 20 kHz at the "12 kHz" setting

−4 dB at 15 kHz and −12 dB at 20 kHz at the "15 kHz" setting

The Pultec HLF-23C (stereo) is also a dual band and has high and low-cut-only filters. But it has a very wide range of selectable frequencies. The low bands offer a cut of −4 dB at 50, 80, 100, 150, 250, 500, and 750 Hz, as well as 1 kHz, 1.5 kHz, and 2 kHz. The slopes of these cutoff settings end up

40 dB down at 450 Hz at the 2 kHz, setting

40 dB down at 350 Hz at the 1.5 kHz, setting

40 dB down at 225 Hz at the 1 kHz, setting

40 dB down at 175 Hz at the 700 Hz, setting

40 dB down at 125 Hz at the 500 Hz, setting

40 dB down at 50 Hz at the 250 Hz, setting

40 dB down at 30 Hz at the 150 Hz, setting

40 dB down at 20 Hz at the 100 Hz, setting

40 dB down at 10 Hz at the 80 Hz, setting

38 dB down at 5 Hz at the 50 Hz, setting

The high-frequency cut-only shelving filters have selectable frequencies of 1.5, 2, 3, 4, 5, 6, 8, 10, 12, and 15 kHz. Again all of these points represent a 4 dB cut at the stated frequency, but the real story is what happens with the rest of the filter's slope:

The 1.5 kHz setting ends up down 40 dB at 7 kHz.

The 2 kHz setting ends up down 40 dB at 10 kHz.

The 3 kHz setting ends up down 40 dB at 15 kHz.

The 4 kHz setting ends up down 40 dB at 20 kHz.

The 5 kHz setting ends up down 36 dB at 20 kHz.

The 6 kHz setting ends up down 32 dB at 20 kHz.

The 8 kHz setting ends up down 24 dB at 20 kHz.

The 10 kHz setting ends up down 20 dB at 20 kHz.

The 12 kHz setting ends up down 16 dB at 20 kHz.

The 15 kHz setting ends up down 8 dB at 20 kHz.

All of the above information did not come from the manuals. No, you have to pour over all the graphs that are provided with the specifications using a clear ruler and a magnifying glass to dig it all out. I did not do it for this book. I worked for a studio that had most of these Pultecs. I did it at that time (early 1970s) just so I'd know the *facts*. Believe me, it paid off. Because I knew the capabilities of the HLF-23C, I was once able to use it as a *notch filter* on a guitar. I set the low- and high-cut frequencies to 1.5 kHz. This meant that at 2 kHz (the center point) there was no attenuation. The signal was only down 4 dB at 1.5, but because I had carved out a band pass filter that was 40 dB down at both 450 Hz and 5 kHz it was exactly what you'd want to get the biting sound that will cause a lead guitar to jump to the forefront of a mix. This signal was then boosted through one of the console's gain stages and mixed in with the original. The result was that the guitarist made sure I was at the next session he did. Nothing like getting repeat business when you're still in your teens and just learning your chops!

Most folks would eventually figure this out, but since it's not written out anywhere that I know of, and as I was lucky enough to have an old-timer (who knew) explain it to me, I'm duty bound to pass it on.

The "P" in EQP designates it as an equalizer designed to cover the spectrum of the full program.

The "H" in EQH designates it as an equalizer designed to cover the high-end frequencies.

The "M" in MEQ designates it as an equalizer designed to cover the mid-range frequencies.

The "HLF" designation refers to its encompassing high- and low-frequency filters.

You may have noticed that the boost, cut, and center frequency selections do not match up mirror image-wise. For example, the Pultec Model EQP-1A3's low-frequency band provides a range of level boost of only 0 to +13.5 dB, while the cut range is 0 to −17.5 dB. This is no accident as the difference in the range of cut and boost is due to the *sound*. These people were not just manufacturers; they designed and built the first models to use themselves and adjusted their design, making adjustments and compromises until it sounded good to *their* ears.

The point is that these folks designed this EQ so that the operator *could not* adjust the boost level for more than +13.5 dB of gain. You do not go through this much trouble over 4 dB unless it makes a difference. That difference is the main reason many of these older effects devices have reputations for delivering "legendary" sound. Back off on the amount of boost and cut you use and other equipment may achieve "legendary" sound as well. But, if you are a modern manufacturer of audio recording equipment, trying to stay solvent, you must give the customers what they want and that's generally the most processing variability possible. Even though it was meant to be beneficial, in the wrong hands this power can be turned into a hindrance.

A case in point. Tascam used to manufacture a full parametric EQ called the PE-40. It was designed for use in a musician's private studio. It was divided into four sections, making it perfect for four-track recording. This was a powerhouse as each of those sections included four separate bands, all with sweepable frequency and Q plus a continuously variable boost and cut control. Tascam gave each section a high-pass filter switchable to 80 or 100 Hz, a switchable low pass filter, and furnished each with a bypass switch. I know all this because a keyboardist friend had one in his studio. The first time I heard it, it was all fingernails on a chalkboard! I couldn't believe it, so I took a look "'round back." This nut had daisy-chained the wiring, feeding the output of the first band into the second, and so on. He was very proud of his cleverness in turning it into a 16-band parametric EQ until I started hitting the bypass buttons and the sound of his keyboard got better and better. Hey, you can't blame manufacturers for giving people what they demand nor for the way people use their products.

It just seems like only old-time audio manufacturers understood the fact that many folks tend to overuse effects. But more recent makers of "legendary" EQs are aware of the same thing. Rupert Neve's most popular EQ is probably the 1073. This is a simple three-band tone control with a built-in high-pass filter. The low-frequency band offers the selection of 35, 60, 110, and 220 Hz. The mid-band yields a frequency selection of 360 and 700 Hz, along with 1.6, 3.2, 4.8, and 7.2 kHz. The high band is shelving with a set frequency of 12 kHz.

You may have noticed a bunch of frequency settings here that seem to have odd values, but you already know why that is. Things *sounded* better at those settings. Four low-frequency and six mid-frequency possibilities do not provide enough rope here to hang yourself. No wonder everyone likes the sound. You *can't* foul it up. Another console EQ that must be mentioned is API's 550a. Again this EQ provides limited choices and clean sound. I felt a little sad when I heard they were going to increase its capabilities by making it a four-band EQ, but they're still making the original.

Music Concept Inc., the same MCI as the console, made a 10-band graphic EQ as an upgrade for their 600 series console and Sony carried it on with their 3000 series console. API has made their 560 10-band graphic for many years. Angus made one for awhile too. But it really all started in the late 1960s with a Canadian console manufacturer named Olive. Don't believe me; flip through a bunch of "dB" magazines from that era and you'll see their ads. I've used all four of them and they're all basically the same. You *can* make something sound bad with them, but if you keep the boost or cut setting of their thumb-wheel switches to no more than 4 to 6 dB, you won't have anything to feel embarrassed about later on.

Mention must be made of the manufacturer Klark-Teknik as they've also been making a fine line of equalizers for many years. While their graphic equalizers are a staple in many recording studio monitor chains, they are even more widely used in live and contractor applications.

The DN 301 is a cut-only 30-band graphic equalizer with its center frequencies set to ISO $\frac{1}{3}$-octave standards.

Each band provides up to 15 dB of cut and it also incorporates both high- and low-cut shelving filters. The fact that it is cut-only makes it a rarity these days. The DN 300 is a boost and cut version of the DN 301 and it provides for 12 dB of boost or cut. For permanent installation, the DN 330 is a "contractor" rotary pot version of the DN 300. Their DN 360 is a dual (stereo) version of the DN 300. They also make contractor (rotary pot) and fader versions of a dual 16-band, $\frac{2}{3}$-octave graphic equalizer, which are of no great significance in recording situations.

Klark-Teknik made the transition into the digital age adding controllability, memory, and then full digital audio processing. The DN 3600 is used in monitoring system "alignment" just like any other $\frac{1}{3}$-octave stereo graphic equalizer, except this one can memorize 60 different EQ settings. It also has MIDI for bulk dumping and reprogramming the memories and a little function called "crossfade" between the memory settings. This can be highly dangerous if not approached with caution. But when given the respect it is due (like stepping well off a trail when a grizzly passes by) it can produce wonderfully musical sounding EQ changes from one section of a piece to another. For PA and contractor's permanent installation applications the MIDI makes wireless remote control command of up to 49 (that's right 49!) DN 3600s possible.

By the end of the 1990s things got a little scary when they introduced the 4000 series Parametric EQ. This all-digital (20-bit AD/DA conversion) unit features a 20 to 20 kHz bandwidth, 32 memories, 114 dB dynamic range, 5 full parametric bands, high and low shelving high- and low-pass filters, AES/EBU interface, and, as if that were not enough, a delay of up to 340 milliseconds on each channel. In a permanent PA situation this is probably a dream come true, but in the wrong hands, you could have the audio version of nuclear fission.

There seems to be a dichotomy here.

It's essential in some corrective situations for an equalizer to have the precise fine-tuning capabilities required to administer the appropriate remedy. But since excessive manipulation can diminish sound quality, just a few choice settings might be preferable.

Everyone has their favorites. For this reason you'll find that most professional studios have a wide selection of equalizers to choose from.

I only have one rule concerning the use of EQ: As with any effect, if I can hear it, it's too much, so I raise the level until I just hear it and then cut back on the amount I actually use. It's simple makes common sense, but it's very unexciting as far as "client-wowing" is concerned.

What's 12 dB per octave mean? Look at the plot in Figure 2-9.

The drawings depict both high and low pass shelving filters. Everything to the left of the high pass slope is cut off to some degree, while everything to the right of a low pass is affected. If the specification for the amount of cut is 12 dB per octave and the high "pass"

Figure 2-9
Examples of high and low pass shelving filters. The HPF has a cut or roll-off of 12dB/octave.

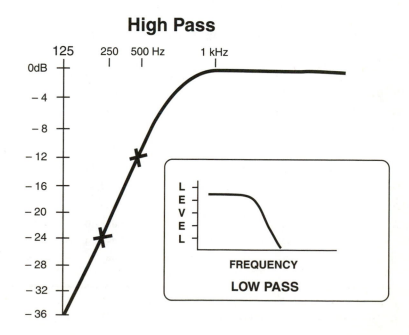

point is set at 1 kHz, then the level will be 12 dB lower at 500 Hz, 24 dB lower at 250 Hz, and 36 dB at 125 Hz. The same device's low pass filter would have an equal effect on the frequencies above the setting of its pass point.

Now is a good time for a different approach to understanding EQ settings. Let's look at it from a musical perspective. An octave is composed of eight consecutive notes on a musical scale. All octaves *almost* correspond to the interval between two of the same notes, resulting in a ratio of two-to-one! (Wouldn't that have made things easy?) But it's not all that bad, as you'll never be required to adjust a single note with a fixed EQ and any adjustable EQ is pretty easy to tune by ear.

Technicians have an advantage in that they have the opportunity to check out EQs "on the bench." Here, by feeding single frequencies to the input and visually observing the output on an oscilloscope, you get a display of everything you'd ever want to know about the way an EQ will process particular frequencies in terms of available boost or cut, phase response, and distortion.

But everyone can glean the same information by carefully listening to an EQ's output in "solo" mode. Comparing the processed signal to the original is also helpful. Try placing one in the left and the other in the right speaker. At this point, simply hitting the monitor section's "mono" switch will clue you in on the phase response very quickly.

Back to the musical side of things. What follows (for your use as a reference) is a list of frequency numbers as they correspond to whole notes on a piano.

Note	Freq	Note	Freq	Note	Freq
A (0)	27.5	E	164.81	B	987.77
B (0)	30.868	F	174.61	C (6)	1,046.5
C (1)	32.703	G	196	D	1,174.7
D	36.708	A	220	E	1,318.5
E	41.203	B	246.94	F	1,396.9
F	43.654	Middle C (4)	261.63	G	1,568
G	49	D	293.66	A	1,760
A	55	E	329.63	B	1,975.5
B	61.735	F	349.23	C (7)	2,093
C (2)	65.406	G	392	D	2,349.3
D	73.416	A	440	E	2,637
E	82.407	B	493.88	F	2,793.8
F	87.307	C (5)	523.25	G	3,136
G	98	D	587.33	A	3,520
A	110	E	659.26	B	3,951.1
B	123.47	F	698.46	C (8)	4,186
C (3)	130.81	G	783.99		
D	146.83	A	880		

I've used low-frequency bandpass filters as follows.

Set the roll-off point to the second octave A or 110 Hz by ear using the studio's piano or your test CD. An octave below that is the first octave's A at 55 Hz. An octave below that is the 0 octave's A at 27.5 Hz. The curve looks like the following:

$$110 \text{ Hz} = 0 \text{ dB}$$

$$55 \text{ Hz} = -12 \text{ dB}$$

$$27.5 \text{ Hz} = -24 \text{ dB}$$

Feed the kick drum signal in, set the level for *no* gain and patch the output to a spare channel. Now raise the fader on that channel and mix this boosted curve in with the original signal. It's a good bet that you'll be able to get 12 dB, or more, of gain before things start to sound "funny." But even at +12 dB that's going to be a serious bass drum sound as it will be not only boosted by 12 dB at that 110 Hz fundamental, but will also have a gentle rolled-off curve of boost to end up with its natural harmonics set in an evenly spaced gain structure.

Here's another trick. Let's say you brought a pre-mixed song with an overemphasized guitar note sounding during every chorus. To match that single note you can sweep across the center frequencies with a narrow bandwidth set for a fair amount of cut until its level is reduced. If you have the charts (and therefore know that it's the second F# above middle C) in a key of G chorus section you can use a keyboard instead. Synths and organs are better than grands here as they provide continuous tones. Play the F# or hold the key down with a matchbook between the keys. Again sweep the center frequency with the gain control to cut. When the keyboard sound lowers, you've hit the correct center frequency. This sweeping is referred to as "nulling" and it can also be done by using the F# from your reference CD.

If the EQ you're dealing with uses digital control and has a numerical readout of the center frequencies, you can play the numbers game. The G above the F has a frequency of 783.99. Subtracting the F's frequency of 698.46 from this leaves 85.53 Hz. Half of that plus 783.99 is 826.755 for the F#.

Once you've found the center frequency, you can adjust the Q for a narrow enough band to lower the excessive level to *just* the F#. After all this work you'll often find that there's something missing as far as an emotional feel. There's a reason the guitarist emphasized that note! Well, the F#an octave up will be just about 1654 Hz. Don't be surprised to find that this content appears and fills that emotional void when you slightly boost this higher frequency.

Like any piece of processing gear, knowing its weaknesses as well as its capabilities goes a long way toward getting a good sound, and reading the manual is key to this knowledge. By the way, just in case you were wondering what "the ugly sound" refers to, well, with low frequencies it's the sound of a gorilla's roar with a sock stuck in its mouth; at the mid-range, it sounds like a guitar with only two strings in tune, the high end has additions that sound like fingernails on a chalkboard, and phase distortion that can peel wallpaper off a wall. This doesn't add to listener fatigue; it causes it instantaneously.

One last point. As stated, the design of equalizer circuitry using resistors, capacitors, and inductors causes phase cancellations due to time differences. In fact, a designer will often start out with just this time delay information and from this determine the value of the components used by time $T = R \times C$, $C = T$ divided by R and $R = T$ divided by C. This time "T" is called the "time constant."

Broadcast transmissions also have a better S/N ratio when the high frequencies are boosted or "emphasized."

The *Federal Communications Commission* (FCC) places (finable) requirements on TV and FM radio broadcasters. FCC specifications call for preemphasis that is applied to a transmitted signal to have an impedance/frequency characteristic that is the same as a circuit consisting of a resistor in series with an inductor having a time constant of 75 microseconds.

So what's this got to do with recording? Well, that 75 microseconds equates to a frequency boost of +17 dB at 17 kHz (or 75 uS = +17 dB @ 17 kHz). You will find the ability to conceptualize this small point handy in the next section.

Dynamics Processors

Dynamics processing is used to make a signal's dynamic range more consistent and manageable. The controlling of dynamic range began with the use of automatic leveling amplifiers. These were set for a desired output and they controlled both the gain and attenuation of the program material (via feedback) to match a required level. Some also included compression and limiting circuitry. They still are used for their primary function, which is to preserve the optimum level of, say, background music, regardless of input variations.

Compressors and Limiters

As natural as we want everything to sound, the fact is that there is no recording system available that can match the dynamic range of human hearing. Furthermore, music is never played at any set level. Who'd want it to be? And expecting an operator to "ride" faders to control the levels of 12 or more channels simultaneously during a recording session is just not practical. So it generally falls upon those tried and true compressor/limiters to help keep things at an orderly, more consistent level.

If the dynamic range (level change) of a signal is too broad, the lowest levels will barely be heard and the loudest may cause distortion, tape saturation, and even damage to subsequent equipment. Dynamics processors can increase the overall output volume by raising the low-level signals while setting a maximum output ceiling so that the loud signals will, in effect, be captured and held to within a safe volume.

By reducing the loud passages, the signal is restricted to a certain range. This overall range can then be reset to a more optimum level or position. By increasing this new overall range, the low levels will be elevated. The range can be adjusted to any level required. This is not easily accomplished when manually riding a single fader and it's certainly less than practical to attempt it during a multitrack recording session.

Dynamics processors come in all kinds of configurations. You'll find limiters protecting quite a few PA systems, and it's not unusual to walk into a professional recording studio and find more than a dozen different kinds of units available. Why so many? Processing the dynamic range of audio is handled in many ways. The nature of different sounds requires it. Peak, RMS, and average levels differ and can change drastically from instrument to instrument. A peak limiter put on a synthesizer to hold down the outbursts in level when changing programs does a great job of preventing overload, but that same setting on drums may cause them to sound overly squashed and unnatural. This is because a drum's attack (the beginning of its sound) is the loudest part of its content and that higher level is needed for the listener to perceive the sound correctly.

In this case, an average responding limiter should be used so as not to overly suppress this important part of the sound. Picture a wind section made up of just a piccolo and a tenor sax. The peak level reading from the sax, due to the complexity of its waveform, could be two to three times higher than the piccolo. Yet the frequency range of the piccolo is such that it may cut through in a mix more than the sax will. If the limiters used on each detect volume as per peak voltage, they may pass the piccolo as is and reduce the already apparent lower level of the sax. The actual amount of gain reduction depends on two factors: the setting of the ratio control and the level of the input.

Ratio   The general consensus is that any gain control processing that is set up with an input to output ratio that requires 20 dB of additional input above the threshold to achieve an increase of a single dB at the output (20:1 ratio) is considered a limiter. An input-to-output ratio that requires less than 10 dB of added input for there to be an increase of a single decibel at the output (10:1) is considered compression. This is not a hard and fast rule, as many feel the dividing line (if there is one) should be above the 10:1 ratio.

The gain reduction begins at the point when the input level reaches the threshold level. The desired amount of reduction is determined by the ratio selected. A 2:1 setting means that for every increase of 2 dB above the threshold level only a single dB of gain increase will appear at the processor's output. An infinity:1 setting means that no matter how much the input signal rises above the threshold setting, there will be no increase in the output level. For comparison, a 6:1 ratio is a good all-around general use setting that will not cause the output signal to sound strangled.

With negative ratio settings (such as −1:1), a decrease in level below the threshold level (by 1 dB) will cause the signal to be boosted (by 1 dB). Thus, a −1:1 setting can effectively reverse a signal's dynamics.

Threshold   This control sets the level, above which the input signal will undergo dynamics processing. Strictly speaking, threshold is defined as the point at which 0.5 dB of gain reduction occurs. But more often it simply refers to the point at which the input signals level begins to trigger the start of gain reduction. This control is usually marked in a range of (+ and −) decibel settings, such as −10 dB. It operates in conjunction with the input level control and the ratio setting to determine the amount of gain reduction.

Attack   This control sets the timing as far as how quickly the gain reduction begins.

Release   This control sets the speed at which the dynamics processing stops once the input signals level drops below the threshold.

Both of these controls can sometimes be superseded by an *auto* switch on units that have the ability to predict the dynamics of the incoming signal. Automatic attack and release settings follow the dynamics of the input, constantly updating the settings much like someone riding faders. For automatic attack and release to function, RMS or feed-forward level detection is generally used.

Feed Forward   While an ideal limiter would allow absolutely no gain beyond a certain set threshold level, only a few (feed-forward) limiters can actually achieve ratios of infinity to one successfully.

Okay, time to get feed forward out of the way. Older compressors sensed the output level and adjusted the gain via a feedback loop. By moving the sensors to the input and having it track the envelope (wave shape) of the incoming signal, changes in the dynamics can cause a gradual onset of gain reduction before the signal actually reaches the threshold. This gentle method avoids the pumping-like switching on and off of the gain reduction caused by incorrect attack and release settings, especially at lower frequencies.

An example of feed-forward compression would be dbx's OverEasy processing. This circuitry is used to anticipate upcoming dynamic changes by following the input signal waveform. This enables the device to apply a gradual or transitional "soft knee" increase in the amount of compression as required for a more transparent and natural sound.

Noise Gates   Gates are used to separate an instrument's sound from the background noise. Any signal level below the threshold setting causes the gate to close and shut off the background noise. By using a compressor you can increase low levels and decrease those that are high. By using a gate you can cut off low levels and pass only the higher levels. By using an expander you can decrease the lower levels (like gating) but increase the higher levels.

Dynamic Range

Dynamic range is something that is often mentioned, but not always understood. Take your ears, for instance. The difference in sounds that you can hear, from the lowest (a pin dropping) to the loudest, or the threshold of pain (an explosion) is extremely vast. In order to discuss this range in a simpler way, different values are described in logarithmic form using decibels (dB). In this way, it's very simple to discuss the loudness of sounds: office talk = 50 dB, street noise = 70 dB, semiloud rock music = 90 dB, loud rock music or a jet taking off or the threshold of pain = 120 dB. Now let's apply dynamic range to a guitar.

The minute you turn on the amp, there's a slight buzz (low dB level); you lightly pick a jazz phrase (medium dB level) and then thrash out a rock chord (high dB level). Wouldn't it be nice if you could stop the buzz yet keep everything else? Well, that's what a gate does. The gate views inputs only in terms of their dynamic range or level. It has a threshold, which is settable, and only signals that are louder than the threshold setting will cause the gate to open.

Back to the buzz. Its level was less than that of your light jazz playing, and certainly less than the rock chords. If you set the threshold to a point that is above the volume of the buzz and below that of the light jazz playing, when the musician is not playing, the gate will stop the buzz from passing through (it's below the gate opening limit as set by the threshold) so the guitar input will be silent. When the guitar is played, the level will be above the threshold setting and will therefore open the gate and allow the signal to pass through. Of importance here is the fact that the buzz will also pass through when the gate opens, but it will be covered up or masked out by the guitar sound. Therefore, even when gating you should still try to eliminate all buzzes and other noises.

In live situations, if gates are put on the vocalists' mics, their outputs will be shut off (gate closed) when they're not singing. Therefore, those mics will not pick up any stage sound leakage during those periods. A gate can trigger this switching off a lot quicker than can be done manually (in thousandths of a second) and will just as quickly turn on again when the vocalists start to sing (no worry about a missed cue). In the studio, gates really shine. I don't use a gate when recording drums because if a tom is hit too lightly, it may not be loud enough to open the gate and thus not be recorded. During mixdown, if a tom was hit too lightly I can boost it. Since the mics on the toms are only a foot apart, when the tom on the left is hit it will also be heard through the right tom's mic. It may take double gating for only one tom to trigger the gate to open, but once accomplished it will give close to ideal separation and the ability to pan the toms hard left and right with little bleed.

When the gate is open the signal should appear at the output exactly as it is at the input. When closed absolutely no sound should emanate from the device. The preset level or trigger point is occasionally called sensitivity, but it's identical to threshold in functioning.

Attenuation operates much like the ratio controls of compressors, but their setting (in number of decibels) indicates the actual amount of gain reduction caused by the closing of the gate.

Attack and release controls are often provided to give the operator control over how fast the gate is opened and closed. These controls can be used to hold off the onset and turn off timing of the processing. Here longer time settings take place before gate activation. (Up to and occasionally even more than 10 seconds are available.) They are helpful when dealing with program material such as speech, with all its natural pauses for breathing, or while sussing out word formulation. If there is any background noise, as is the case with most live outdoor interviews, the listener may hear that noise being continually turned off and on if a gate's attack and release settings are too fast.

The gate can be set to remain open for an extended amount of time. This little fact is very important when gates are used for some dynamics effects. Gates are generally used to help isolate signals from each other during multitrack recording and when signals are being processed differently from each other. But they can also be used in some pretty bizarre situations.

Ever had to deal with a band with two lead guitarists? How about two using the exact same type of guitar and amp setup? They were into trading off their lead lines and each one did back off in volume when the other was playing but never quite enough. I was able set up gates on both guitars so that the slightly lowered playing level was further decreased by another 6 dB.

This same type of noncutoff gating can make a ride cymbal ring die off just as if it had been grabbed by the player's hand. Want to make an organ sound de-swell? You may be familiar with RT-60, but ever think of what the massive amount of early reflections inside of a stone cathedral would sound like if they could somehow be squeezed into a smaller room? I did, and I labeled the track that held the gated cathedral reverberation with a slow attack and short release, RT-20.

Keying   As shown with gating, a device can be cued to turn its processing on and off by the incoming signal level. What if you could cue it from anther source? This is called keying. Here, a second input is used to cue or key the processor to gate the main input signal.

Equalized, short bursts of white noise (the sound that comes from your TV when it's between stations) sound a lot like hand claps. Set a synthesizer to white noise, tape down the key and put it into the input, use the snare track on tape for the key input, set the threshold to that volume just under the volume of the snare, and set the release for the length of time you want each burst to last. Now every time a snare beat occurs, you'll get pseudo-hand claps. The uses for keying are as unlimited as your imagination.

Companding   This was once referred to as the "comprex (compression/expansion) recording process." It was used to expand a signal's usable dynamic range. The original signal is compressed to fit into an optimum dynamic range for the medium to which it's being printed. Later, during playback it is passed through the compander's expansion section, which restores the signal's original dynamics. Setting up these two separate stage's ratios to be equal was no simple affair, so companding never gained great popularity. The expansion was used after the compression to increase the signal's level during louder sections. These processors did not decrease lower levels, and the input signal level was used to determine the amount of gain or expansion. It is possible to use the expansion processing separately (without compression) to provide interesting effects, but be very carefully as expansion can easily cause distortion.

Expanders

Expanders today are generally incorporated into gates. They function in the opposite way of a compressor in that when the incoming signal exceeds the threshold it triggers the start of an increase in gain according the ratio selected.

Expanders, besides increasing levels when signals are above the threshold point, can also lower the level of signals that are below the threshold point. If the ratio is set at 1:2, every decibel the signal drops will cause a decrease of 2 dB at the output. In this way background noise is made lower via a soft curve that does not risk cutting off any of the signal. This is important on instruments such as violins, flutes, and organs, whose sounds tend to swell up and down and do not have defined cutoff points. Playing around with expanders can be big fun.

Side Chains   Some units include a side chain that allows the insertion of additional processing such as an equalizer. This can be used to make the device more sensitive at particular areas of the frequency spectrum by it simply boosting that area of the spectrum on the equalizer connected to the side chain. Here's an example of how this feature can be used.

A radio program was utilizing a live feed. A *plain old telephone* (or POTs) line was feeding an interviewee's mic directly into the mixing board. Every time the guy said anything with the letters "C" or "S" during level check he whistled. Panic set in, but I had run into this before, with a background vocalist no less. It has to do with the way a person's teeth are shaped and the way the lips form while producing those "S"-like sounds. A quick patch through a parametric EQ with not too broad a Q setting, a 6 kHz center frequency, and a boost of 6 dB did the trick (no translation needed for those who read the previous section).

It would have been faster and easier had a de-esser been available, because that's essentially what that compressor with side chain capability had been turned into.

Placing an EQ or other processing at the input of a dynamics processor will have an affect on the overall amount of compression. At the output it will equalize or affect the entire output signal. Placed in the side chain it varies the amount of the compression according to a modified version of the signal. Side chains also allow another signal's dynamics to control the amount of processing.

TIP: Whenever you're lucky enough to get a chance to use a dynamics processor (studio time is not cheap) you should take advantage of it. But now is the time to start preparing for that even. If you don't already have a test CD such as the Parsons and Court "Sound Check" (which has individual instruments recorded with full dynamics), at lease start compiling your own.

Then when you get your hands on a processor you're unfamiliar with, you can use your CD to check out its functioning on a bass, a kick, a snare, vocals, acoustic guitars, and everything else you can gather.

Examples of Dynamics Processors   This will all be easier to understand by going over the parameters and functioning of several specific dynamic controllers. Having an understanding of the operation of different types of dynamics processors is a necessity for those involved in audio recording. Please refer to the compressor/limiter gain reduction graphs in Figures 2-10 through 2-12 while reading this section.

I plotted gain reduction "curves" for both 10:1 and 20:1. As you can see there's only a single decibel of difference between the two even when the input signal exceeds the threshold by a full 20 dB. Yet audio folks (who just love to argue) get all adamant about defining limiting as being either "at least 20:1" or "over 10:1." Go ahead, try it some time. Ask some audio "pro" (especially the older ones) about the "onset of limiting as opposed to compression" and watch them bug out over it.

NOTE: The following is not meant to be an all-encompassing "Buyer's Guide" listing, nor is it a rating system. The presentation isn't even in chronological order. The only importance I attach to the choice of any of these processors over others was their being better examples for illustrating certain points.

Figure 2-10
Hard knee gain reduction at several ratio settings

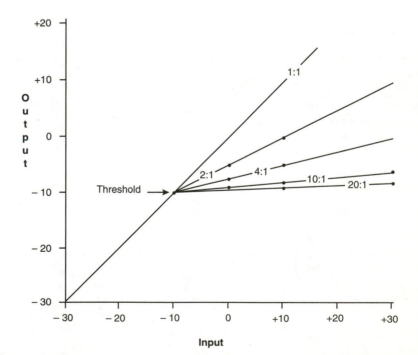

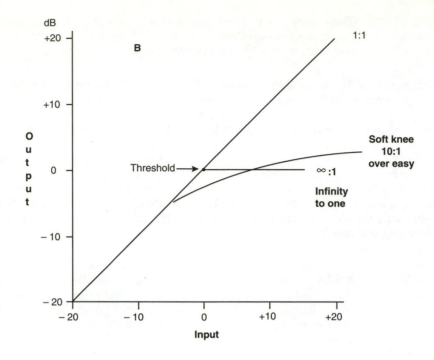

Figure 2-11
Soft knee,
"overeasy" gain
reduction at a
10-to-1 ratio,
and hard knee
infinity-to-one
limiting

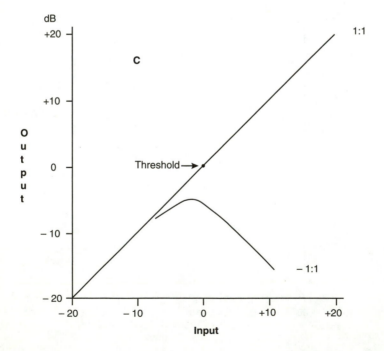

Figure 2-12
Negative gain
reduction at a
negative 1-to-1
ratio setting

Tube-Based Dynamics Control   Older tube leveling and compression amplifiers adjusted the input signal to keep it within a specific dynamic range, according to the output level that the operator had set. Because of the amplifying or "leveling" part of the processing involved, there's a big difference between this and simple compression or limiting. The amount of processing (gain variation) was dependent on the input signal's level and dynamic range, and with dynamics that had widely varying levels, greater amounts of amplification or leveling would be required. This meant the possibility of adding amplifier noise to the signal. Because of their slow-acting circuitry, these devices were also not quick enough to catch rapid transients or level changes. This could not only cause distortion, but these slow-acting processors often overreacted by stopping and starting the processing repeatedly. The resulting thumping sound is commonly referred to as pumping.

This was fine for handling program material like background music that had no sudden or sharp transients, but forget about any special effects dynamics processing. These devices were designed to be set for a specific average output level and to control input signals whose dynamic ranges were not very excessive.

They operate by varying their tube amplifier's bias with a DC voltage that was rectified from the input signal. This causes the gain to follow the input level. Tubes are not as picky as transistors about the amount the load you attach them to, but they do not like their bias levels played with.

When the gain of the signal is controlled by changing a tube's bias or grid voltage, it causes that voltage to deviate from the optimum distortion setting. In other words, changing the bias level from what is considered optimum increases distortion. This increase is also proportional to the amount of limiting and leveling required by the input signal's dynamics. Simply put, more gain change equaled more distortion.

I personally found these devices to be a little slow when used on instruments with rapid transients (snare drums). This slow action also makes them susceptible to those on/off thumping or pumping sounds caused when the processor reacts to the half-wave peaks of low-frequency signals such as slowly played bass lines. Furthermore their frequency response only extending out to 10 kHz, and THD specs that are greater than 1 percent at higher compression settings, may be okay for the purpose for which they were designed, or even for broadcast and film work, but not for modern professional audio recording.

Hey, maybe it's just that I'm not savvy enough to handle their complex intricacies. Yeah, and I'm also not trying to sell you one either. No one can deny that these devices are generally heavy (certainly not portable), take up a lot of (up to eight) rack spaces, run pretty hot, and have parts that cannot be replaced. Yet I know someone who recently purchased a Westrex 1593 for over $2,000.

The only thing I can hope for him is that he makes his money back on client "wowing" and has a lot of fun discussing his new conversation piece, because the sound quality and versatility of a more modern compressor/limiter as well as a tube-based preamplifier (if that's the sound he was after) would have cost him a lot less, worked better, and been more serviceable.

Don't pay me too much mind, one of my job descriptions once included the title of "Feasibility Assessment Engineer." But in pro audio as in anything it's best not to take everything anyone tells you at face value.

It's up to you to decide what is real and what's fluff. Do your homework and you won't be cheated. For example, I once heard a salesman tell a customer, "It's a must-have and a no-brainer." The *only* must-have I know of in audio is some kind of preamp stage to increase a microphone's output and anyone who believes that there's such a thing as a no-brainer better hope so, because they're lacking some. Now, back to tube-leveling amplifiers and the Teletronix LA-2A.

Teletronix LA-2A   On the other hand, the LA-2A program-leveling amplifier is widely popular for recording studio use. This was originally produced by Teletronix Engineering Co. from the early 1960s until 1967, then by Studio Electronic Corp (who produced it until 1969), when it was made by *United Recording Electronics Industries* (UREI), who manufactured LA-2As well into the 1980s. This kind of longevity speaks for itself. There are still versions of the original LA-2A being manufactured today, 40 years after it was first offered!

The LA-2A's popularity was initially due to the fact that it could provide FM and television broadcasters with 100 percent modulation without the risk of overmodulating the transmitters. It even had a multi-(8)-pin jack for plugging in different pre-assembled or "canned" emphasis networks.

FM stereo limiting *must* be matched on both channels or the balance goes out the window. The LA-2A provides for this as well as equalization compensation to prevent high frequencies from causing overmodulation. For stereo lock up the control voltage that dictates the amount of limiting is brought out to the rear panel. Connecting two devices together at these points causes them both to follow the strongest of the two limiting control voltages. Hint: An external signal generator, or music instrument outputs from guitars, drums, a bass, keyboards, and organs, can be connected to this point as well.

The LA-2A has a fast attack and provides 40 dB of gain reduction. It's photoconductive cell's resistance lowers with an increase in light quickly enough to provide a 60-millisecond attack time. The release is slow, often taking over 10 seconds to completely return to the no-gain reduction state. The release range specification is 1 to 15 seconds depending on the dynamics of the input signal. But its main advantage over the early tube compression/leveling amplifiers is the fact that its voltage amp has a fixed gain setting. While most compressors usually have switchable metering to indicate the output level as well as gain reduction, the LA-2A's VU meter only shows the amount of gain reduction.

One important note, however, is that this meter is driven by a second electroluminescence network to make the meter match the processing exactly. This is a good thing to know if the main electroluminescence component should burn out as they can be very difficult to obtain.

The LA-2A's Electroluminescence Component   Photo-optic conductors react to light intensity. The intensity of a light source can be controlled either directly or indirectly by the level of an audio signal and an electro-optic component placed in proximity to that light source will vary proportionately with the input level.

The LA-2A's electroluminescent attenuator takes advantage of these properties. Here the incoming AC signal is used to vary the illumination intensity of a phosphor-coated plate. Phosphor is what is used on the inside of oscilloscope tubes to produce their glowing affect. This light source was chosen over incandescent bulbs because of the time lag in the filament's reaction. The correct term for this delay, thermal inertia, aptly describes this reaction. Illumination via a neon source is also possible, but because of its nature, the changes in intensity would not be proportional to the input voltage. LEDs are just as fast-acting as the phosphor-coated electroluminescent circuits, but they, like neon bulbs, cannot respond proportionally to audio signals. However, as you'll see, they do make good light sources for triggering gates.

The glowing phosphor-coated plate causes the photoconductor's properties to change value. There are different kinds of photoconductors, in the case of the LA-2A, the more intense the illumination, the lower the photoconductor's resistance. This control stage ends up being a direct cause-and-effect control system. So direct, in fact, that the change in resistance takes as little as 10 milliseconds.

The most important difference introduced with the LA-2A was the variable attenuation circuitry being positioned in the signal routing before the amplifier stage. Because of this the amplifier's gain setting could remain fixed. This was an advantage over the previously discussed designs as far as distortion is concerned as the amplifier tube's biasing voltage could now remain at the optimum distortion level. Check out the simplified version of signal routing (see Figure 2-13). The input signal, after passing through an impedance matching transformer is fed to two locations: directly to the input of the optical attenuator and to the peak reduction control.

The reduction control varies the level of the input signal fed to the amplifier that drives the electro-luminescent light. More level equals greater light intensity, lower resistance, and more signal level reduction, thus the name reduction control. Here's where the description of this circuitry usually leaves off. Seemingly, it doesn't make any sense. If you lower the resistance there will be *more* signal flow, so the higher the signal level, the greater the output. That's expansion, not compression.

Having the ability to read schematic diagrams fills in the missing information. Yes, photo-conductors *do* lower resistance and therefore conduct *more* of the signal when its level increases. But they are conducting that signal to ground or common, where it is canceled out. The higher the input level, the more cancellation. Schematic diagrams offer a treasure trove of information to the audio recordist.

The control voltage that is fed to the voltage amplifier (after it is set by the peak reduction control) is also sent to the rear panel terminal connection for locking dual devices together for stereo processing.

The stereo balance trim should only be adjusted as per the manufacturer's calibration routine to ensure equal gain processing to both channels. The complete calibration procedure not only effects the stereo balance between two LA-2As, but also makes it possible to use either gain reduction control for an equal effect on both units. After the optical attenuation circuitry,

Figure 2-13
Signal flow
chart of the
LA-2A limiter/
compressor

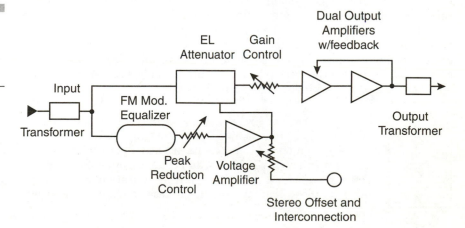

the gain control sets the level fed to the output stage, which ends with an impedance matching transformer.

No attack, release, or threshold controls are provided, only gain and peak reduction controls. But the gain control sets the output level and the peak reduction control effects the ratio, threshold, attack, and release.

The attack is as fast as possible with the level detection circuitry used, and release is fairly fast at 50 to 60 milliseconds for the first half of the return to no gain reduction. The transition then slows down and can take up to 15 seconds to fully return to no level reduction. Again, not the quickest reacting dynamics processor around, but that's not the whole story.

Okay, take note of this! I believe the key to the LA-2A's "legendary sound" is due to its photo electric conductor. Increase the light and there's less resistance; decrease the light and you get more resistance, right? Almost. You see those photoconductors do not operate in a linear fashion. Nor are they logarithmic. In fact, they do not even vary the same way when the light decreases as they do when it is increased. They react more like rubber bands!

The electrons that make up the photoconductor are loosened up pretty quickly by the bright lights of stardom so they almost immediately commence with conducting (causing a lower resistance). Once the bright lights/big city intensity decreases, they begrudgingly return to their previous state at a *slower* pace as compared to their initial change. Furthermore, the taste of stardom does not leave their little electron mouths so quickly. Therefore, when the light intensity increases, again they drop their resistance even more rapidly.

This elastic-like reaction is further extended when high input levels cause greater light intensity. This is what makes the release time dependent on the input level. Should the input level require 10 dB of gain reduction, the release time will only take a couple of seconds to return to the no gain reduction state, but the release after 30 dB of gain reduction could take as long as 15 seconds! This kind of processing transition rarely causes any audible thumping or pumping (on/off) sounds.

This cushion-like reaction extends to the amount of gain reduction introduced. Lower-level increases are only compressed and the changeover to limiting at higher levels is gradual. No wonder the sound of this device is legendary! An operator would have trouble overprocessing the signal, even if they tried!

Yet for the brave-hearted with the technical know-how, an original Teletronix LA-2A can be internally modified to strictly compress or limit; in fact some later versions made this choice internally switchable. One modification allows for continuously variable compression/limiter selection. The high-frequency modulation control adjusts only the amount of low-frequency cut. This effectively increases the high-frequency content of the signal, which makes the processing more sensitive to higher frequencies by as much as 10 dB at 15 kHz. While this pre-emphasis can be varied, I know of one unit that was modified to produce several different EQ curves. As mentioned the stereo (dual unit) input can be used as a control voltage input, thus opening the door to all kinds of triggering possibilities.

NOTE: The value of an original working LA-2A is now thousands of dollars. Some of its internal components are no longer replaceable, internal DC voltages run above 200 volts, and changing any of the internal components or settings calls for a rather in-depth recalibration. So monkeying around inside of one of these is not a good idea unless you know exactly what you're doing.

The UREI 1176 Compressor/Limiter   Another popular old-timer you'll often run into is model 1176, also manufactured by UREI. There are several versions of the original device as well as a few copies and even a stereo unit dubbed the 1178. From my service notes, I have drawings dated 7/9/70, 3/15/73, 7/2/75, 3/18/76, and 4/4/76, which shows how this manufacturer kept updating and improving their products.

While the schematic revisions are labeled A through D, my notes also show these drawings being related to the letter suffixes added to the serial numbers that ran A through H. (I had no G?) There were at least three different output transformers used and between F and H and the differential input stage changed from transformer to electronic (dual op amp) circuitry.

The 1176 differed from the LA-2A primarily in the use of an FET transistor for voltage control over the variable resistance. This model also gave the operator control over the attack time (ranging from less than 20 to 800 microseconds), release time (50 to 11,000 milliseconds) and ratio settings (4:1, 8:1, 12:1, or 20:1). Also, its circuitry was transistor not tube-based. The specifications are

Frequency response was within ± 1 dB from 20 Hz to 20 kHz.

S/N ratio came in at better than 81 dB (30 Hz to 18 kHz).

Distortion was rated at 0.5 percent THD with a +24 dBm output.

NOTE: This distortion rating is good considering that even though second-order harmonics are not greatly affected by compression ratio settings, time constants, or frequency content, third-order harmonics can greatly increase with faster time constants, greater amounts of compression, and higher amounts of low frequencies.

The stereo (dual 1176) linkage required a special circuit that provided a low-level (less than 1.5 volt DC) voltage to both units and it included an offset control for matching the sensitivity of the two. I used to build a lot of these for folks. Instead of mounting it on the rear panel (as was the factory unit) I built them onto a single space rack panel.

This provided ready access to the offset pot, stereo in/out switching, and on the rare occasion when battery replacement was needed, easy access to a power source that was used to provide the stereo control voltage. But the most important detail for stereo hookup was having matched FETs. When the two unit's serial numbers are close (less than a thousand apart) they should be fairly matched, but I found that keeping a half dozen sets on hand made things go very smoothly. While the 1178 is a fine stereo compressor, there's nothing like having two 1176s and a front panel stereo controller. This means there's two for recording individual tracks and a matched pair that can be used later on the stereo mix bus. I heard of someone paying close to $6,000 for one of these rigs. No wonder several copies of the 1176 are still being remanufactured.

Okay, enough of all the old stuff. While they do have their own particular sound as opposed to the transparent dynamics processing of more recent offerings (a blessing or a handicap, depending on your subjective point of view), they did offer gain reduction parameter setup that was fast, easy, and often foolproof.

That covers simple compressors and limiters. But what about expansion, gating, ducking, de-essing, and other dynamics processing? For that, it would be best to fast forward 10 to 20 years to the mid-1980s and take a look at the Valley People Dyna-Mite limiter. But first it's necessary to understand a little history.

I don't remember people thinking it was a serious problem when 8-track recording was (for what seemed like only a minute) state of the art, but when the formats got to 16 tracks it became more than just noticeable. I'm talking about S/N ratio, background, tape, and leakage, bleed, or cross talk *noise*.

As more and more faders were raised during a mixdown session, the control room monitors made all of those noises very obvious. Racks of Dolby noise reduction processors were brought in to help eliminate tape hiss. But it was still tough going and the problems were felt most by the pop music recordist.

I, like many others at this point in time, was the first to deal with a whole new situation. Take two electric guitarists, each running through mega-power stacks comprised of Hi-Watt amps on top of dual 4 × 12-inch speaker cabinets, a bass player using one Acoustic folded horn and one Ampeg front-mount speaker cabinet, both with 18-inch speakers driven by a 250-watt Plush amp. Add in a drummer with a giant (28-inch kick) mega-sized drum kit played with sticks held backwards so as to have the thick end out, and a keyboardist playing a Hammond organ attached to two Leslie speakers (with variable motor speed drives), plus a synthesizer capable of producing subaudible tones plugged directly into the console. Picture all of this in a 20 × 25-foot room, hit record, and you'll have an idea of what I'm referring to.

Due to the excessive levels it seemed like every mic had an omnidirectional pickup pattern and was pulled back in its positioning. Raise the hi-hat mic and you'd hear most of the kit, the organ, and the bass. When the guitarist wasn't playing, the mics on his speakers were also doing a great job of augmenting the low frequencies from the bass, organ, and kit.

Deaden the studio, erect panels, go-bos, and isolation booths, and when you raised the fader you still heard what sounds like the first syllable of the word *first*. The problem was universal enough so that even in those days (1970), when there were only a couple of hundred professional recording studios nationwide, it was still profitable to offer a solution (i.e., Dolby-A multitrack noise reduction).

Another pioneering outfit, Allison Research, at this time offered expanders and limiters. Pioneering because they were one of the initial developers of the *voltage controlled amplifier* (VCA), which changed the way audio was processed (dynamics) and controlled (automation).

During the early 1970s it was common to see more than one rack cage (an eight-module unit) fully loaded with Allison Kepex and Gain Brain devices. Kepex stood for keyable-program-expander and the Gain Brain was a limiter with a continuously variable ratio control. Both made recording pop music more tolerable.

Expanders, while normally used to increase levels above the threshold point (as when replacing levels reduced by compression), can also be used to lower the level of signals that are below the threshold point. In this case if the ratio is set at 1:4, every decibel the signal drops will trigger a decrease of 4 dB at the output.

In this way the background or bleed noise is made lower in such a way as to not risk cutting off any of the music. This is important on instruments such as organs that do not have a defined on or off signal while drums on the other hand do, and their tracks benefited greatly from the use of gates.

It was not unusual during basic sessions of that era to have as many as 10 mics on the drum kit, direct and miced bass speaker inputs, a direct input and a mic on the guitar, several keyboard inputs, and a scratch vocal or two. During these sessions the engineer had to simultaneously ride herd over a dozen inputs. While every studio already had at least one or two compressor/limiters, seeing a rack full of Allison's face (their original logo) sure took a load off one's mind. This company manufactured many products over the years, including a decent equalizer, but their main vocation was dynamics processing, which they developed to a state of high art with their Dyna-Mite limiter.

The Valley People Dyna-Mite Limiter This device handles a slew of limiting functions and other effects. In fact, each of its two channels has the ability to perform 18 different types of signal processing. Its most basic function is as a limiter. It can limit the average level so that it doesn't squash or excessively reduce the level of the drum's attack (the initial hit). It can be set to peak limiting for arresting signals strong enough in voltage to cause damage to the subsequent equipment in the chain. When used to control the dynamic range of vocals or music, it can be set to average detection mode. Here the attack time is relatively slow but the processing is not wave-form dependant, which can cause pumping at low frequencies. However if more rigid limiting control is needed, peak-detection can be utilized.

FM preemphasis makes the Dyna-Mite more sensitive to the high-frequency emphasis common to FM transmission. The result is that the threshold of limiting is lower in the presence of those high frequencies. The detection can provide either rigid (peak) or moderate (average) control. This mode can be used to effectively de-ess a vocal track. In this case, average detection is often preferred, as the "S" sounds are basically sine wave in form. The input signal can be "Y" corded and then patched to the main input and through an equalizer to the external input. By increasing the level of the same bands where problems occur, you have more precise control over de-essing or any other frequency-dependent processing.

The Dyna-Mite can also key. Via the external input, a secondary signal can cause the main signal to be turned on and off. If you have a kick drum that lacks a defined low-frequency fundamental, an oscillator set to 100 Hz plugged into the Dyna-Mite's input can be turned on and off by the kick track itself. Later, the two signals can be mixed together to effectively beef up the kick's sound. A much less radical usage, but still very handy, is to key several background vocal tracks from one singer, and have all come in at the same time. As opposed to just turning on and off, this device can perform envelope following. Here the main signal will actually follow some other signal's level with either fast attack (peak) or slow (average) triggering. Thus, the dynamics of any two signals (such as those in a duet performance) can match exactly.

Ducking is the opposite of keying. It causes the main signal to be reduced in level in the presence of another signal. The Dyna-Mite does this in three different ways. It will follow either the peak or average level of the external source, giving a 1 dB decrease for every dB of increase, or set up differently, it can cause the input to drop out completely (gate). The amount of this reduction is continuously variable between 0 and 60 dB. The release or recovery period to normal level can be set anywhere within a range of 5 milliseconds to 5 seconds per 20 dB of gain change. This feature makes it possible for the engineer to automatically mix speech and background music during commercials.

Inverse envelope following, or soft ducking, renders automatic cross-fading, as opposed to switching. In this mode you have the choice of fast or slow attack time. Here the main signal follows the external signal's level inversely.

As an *expander* this unit is very flexible. Expansion can be handled three different ways. It can gently attenuate background noise with a 1:2 ratio, as opposed to the hard gating of a 1:20 ratio setting. Since it lacks defined on/off switching, this processing is less discernible. In the average mode, the attack time is fairly slow, while for peak it is very rapid. The amount of gain reduction is controlled by the range pot, and this, along with the release, allows the user to process and reproduce all prerecorded signals reliably. Here, any signal above the threshold will be output at its normal level while the background noise during nonplaying sections will be reduced.

Negative Limiting   This effect is rarely used, but still very useful. It causes a signal to be reduced in gain between 0 and 60 dB once it exceeds the threshold. Here, a signal that has a strong attack has its beginning chopped off and, depending on the ratio and release settings, the rest of the note can drift upward in level much like an organ's sound does.

Try this on a rhythm guitar part played with a 12-string "Rick" that's been run through a chorusing effect. The Byrds would have loved it!

This processing can be further modified via the use of the unit's internal filter, making it more sensitive to higher frequencies. Too much cymbal sound in the overheads? De-essing with negative limiting causes S sounds to be cut off with the voice fading back in rapidly instead of jumping on and off. The Dyna-Mite is also a *gate*. When a signal has a defined attack, as with most percussive sounds, and where radical turn-on and attenuation is required (as when trying to prevent leakage), gating becomes very useful. Here the signal, when present, will pass through unaffected while the gate will shut down noise and leakage in its absence. The processing turn-off time is controlled via the release setting, which allows for a smooth transition yet is variable enough to be capable of tuning out the after-ring of a tom. The gating mode can also utilize FM/DS filtering.

This can be helpful when processing percussion that is mostly high-frequency in content, such as a hi-hat. Here the extra sensitivity helps insure the gate will open when the hat sounds. Again, for even more precise control, an external EQ can be inserted to increase the unit's sensitivity to any frequency and bandwidth. In all, the Dyna-Mite performs gating and keying in no less than seven different ways.

As if this were not enough, the two channels can be coupled together by hitting a single button, letting you use both channels on one signal.

However, the thing I liked most about the Dyna-Mite is its portability. It weighs a little over 3 pounds and measures a mere 8 × 3 × 9 feet. What this translates to is actually being able to walk around the control room with the unit in your hands (provided you've used long enough cables). It's always bothered me that I have to turn around and bend down to adjust an effect, not so much because of the strain, but because most of the time when zeroing in on a sound or level the signal would disappear.

Invariably, the performer decides that's the time for a break, and because of the position I'm in I don't see them quit playing. With the Dyna-Mite you can sit with the unit right on the mixing desk, looking out the window, seeing all, having all the solo and PFL switches at your fingertips, and make qualitative decisions sitting in the sweet spot between the speakers instead of facing away from them into a rack. I think this is important, but if you don't, the Dyna-Mite also comes in a rack-mounted version. It may seem that this description of the Dyna-Mite is a bit overdone. But since it performs just about every dynamic processing function, this device ended up being perfect for use as an example.

As far as control settings, there are four knobs for tuning (threshold, release, range, and output), as well as nine LEDs, eight for displaying the level of gain reduction and one to designate clipping to help the operator with level setup. This device also utilizes three 3-position toggle switches whose various configurations add up to its 18 effects. Figure 2-14 shows a simplified version of the signal flow through one of the Dyna-Mite's channels. It was included specifically to illustrate the inter-relationship between the toggle switches, but it won't hurt to follow along with this more elaborate method anyway.

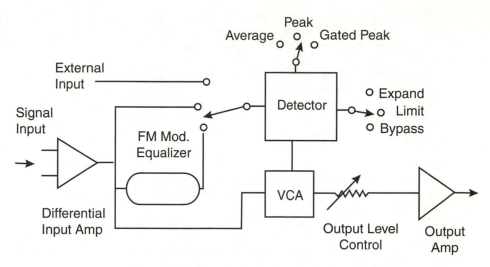

Figure 2-14
The Valley People's "Dyna-Mite" limiter flow chart, showing its three 3-way toggle switch controls

Basically, the Dyna-Mite is a very simple device made up of only an input-balancing circuit (differential amplifier), a VCA, and a line driver (output amp). The input amp gives this device the ability to interface with almost any piece of audio gear (−59 dB to +24 dBu input, referenced to 0 dBu = 0.775v) and still maintain a very good S/N ratio (−90 dB @ +4 dBv input). The heart of this unit, the VCA (Valley People EGG series), allows for very accurate control of gain. Finally, the output stage is capable of delivering up to +21 dBu (into loads of 600 ohms or more).

The VCA is controlled by a detector that can receive its gain-control signals from any of three places: an external signal, the input signal itself, or the input signal fed through the FM/DS high-end booster. Only the signal fed to the detector is boosted, not the signal as it appears at the output. The detector can read the input signal as per average, peak, or gated peak levels.

The choice of this setting also determines the ratio. The processing can be bypassed altogether, or it can cause the VCA to act as a limiter or an expander. The processing is also affected by the threshold, the release time, the gain reduction range, and the output-gain control. Therefore, the Dyna-Mite conditions the signal fed to it and transforms it into the control signal that is fed to the VCA.

The VCA in turn controls the level of the input signal as fed to the line-driver amplifier and the output. The processor also feeds an LED array that shows the amount of gain reduction, and via a rear-panel output this signal can feed either an external meter or another VCA in addition to the one onboard.

Another rear panel connection allows the user to bypass all internal processing and patch an outboard controller directly to the VCA. This outboard signal will then control the gain of the input signal.

Since the range of the Dyna-Mite's VCA is 160 dB (−100 to +60 dB) and is controlled via a DC voltage (−1 volt = +20 dB and +1 volt = −20 dB), a sine wave from an oscillator with a 1-volt peak-to-peak output connected at this point will make the input signal swing (rise and fall) in level over a range of 40 dB. The rate at which this swing occurs depends on the frequency at which the oscillator is set.

Finally, a coupling output from the unit's processor is connected to a switch that allows for the stereo operation of the unit's two processing channels. This can be used to preserve the center image. The processing will follow whichever of the two channels is set to the higher degree of gain change.

As you can see, the user's control over the signal processing is in some instances quite extensive, while in others it's a bit restricted (such as with the ratio and attack adjustments).

Valley People chose not to give the user complete control over the ratio. Instead, it is preset via the function switch settings: low (1:2) for expansion, high (infinity:1) for limiting and also high (1:20) for noise gating and keying, and (1:−20) for negative limiting or ducking. This eliminated the unit's ability to compress, which may not a problem, but the fact that the attack time is predetermined can be. I found the settings (20 dB per 5 seconds for release in peak modes), to be somewhat slow and the 1 to 15 milliseconds (signal dependent) in average modes to be so quick that depending on the setting of the threshold, it can end up being a little too sensitive.

On the other hand, the rest of the parameter controls provided have extensive ranges. Though not effective during normal limiting and de-essing modes (where it's preset at 60 dB), the range control provides contiguous variability of the maximum gain reduction in all other modes and can be set anywhere from 0 to 60 dB. The release is continually variable between 5 milliseconds and 5 seconds per 20 dB of level return.

In the limiting and de-essing modes where the range control is ineffective, the threshold and VCA output are coupled together. Normally, when you change the threshold setting on a limiter, you'll affect the output level. Here the output is automatically readjusted so that when you change the threshold to achieve more limiting, the output will remain constant. Another helpful feature is that (as with the dbx 160X), the output control is calibrated. When de-essing or limiting with average detection, the output control's calibrated markings will correspond directly to VU. In other words, if you want your signal output to be +4 dBu, just set the output control to that marking and forget about it. As always, it seems that simplicity and controllability are mutually exclusive. Valley People's choices seem to be ideal in most cases. If they had provided complete variability of all parameters it would have resulted in an overly time-consuming setup. The Dynamite is a very powerful dynamics processor. While their name changed from Allison to Valley and then Valley People, they are now no more, but many of their products are still being offered by Galaxy Audio.

dbx   Another early pioneer in the field of dynamics processing was dbx. Their model 160 made setting up a compression child's play. The operator simply set the desired amount of compression ratio (from 1:1 to infinity to 1), the threshold (which had signal-level detection "above" and "below" LEDs as guides), and the output level (which ranged from + to −20 dB). The front panel VU meter was switchable to display the input, output, or the gain reduction.

Easy, yes, but no matter how much people rave about it, there's nothing special about these features. The greatest contribution dbx made to the field of dynamics processing was not ease of setup but the Soft-Knee compression they termed the *OverEasy* mode of threshold setting. This helped dbx achieve a truly elegant solution to the problem of audible compression processing noise whenever the state of the processing changes between on and off. But the complete solution to compression processing being audible during its switching on and off required both the overeasy threshold setup and another dbx first, feed-forward circuitry.

The dbx 160X is a worthy successor to its now classic predecessor. It incorporates many of the same features as the 160, such as both OverEasy and hard-knee compression. By hard knee, it's meant that when the input level reaches the threshold point, the full amount of compression is introduced. Switching to the OverEasy mode causes the compression to slowly activate as the signal level approaches the threshold, thus providing a smoother, and less noticeable, transition. The dbx 160X utilizes true RMS level detection, has wide bandwidth and low-distortion circuitry, and is capable of more than 60 dB of compression.

The compression ratio range is from 1:1 (no change above the threshold) to infinity:1 (no further gain or a brick wall above the threshold) and beyond. Infinity + ratio settings of up to −1:1 are possible with the 160x. With these, every decibel of increase above the threshold results in a decrease in the output level and with every decrease in level below the threshold there is an increase in output. With low-threshold settings, it is possible to reverse a signal's dynamics.

The threshold control is calibrated and it ranges from −40 dBm (7.8 millivolts) to +20 dBm (7.8 volts). The output control is also calibrated with a range of −20 dBm to +20 dBm. Attack and release are internally set and are program dependent. This process seems more remarkable once you understand the facts. The 160X controls these parameters by a process they call feed-forward detection. Alright, by now you should know what RMS detection is, but simply put, it means the input signal is continually sampled. The sampled levels are squared, added together, and from this total the square root (the root mean square) is taken. Obviously, this sampling gives a more accurate picture of the input signal's level changes than an average response will. Since this picture is of a larger portion of the waveform, variations as far as upcoming gain reduction requirements can be "predicted" more easily. This processing constantly updates the attack and release settings. All I can say is that it would be difficult to set these parameters more accurately by hand.

The dual or stereo mode switch (also with LED indication) allows two 160x units to be wired and controlled together via the control settings of one. This provides stereo processing and it's also ideal for A/B-ing (either/ or-ing) two settings. The 160X's bypass mode connects the input directly to the output.

An external "detector" (key input) connection allows the input's dynamics to be processed according to the dynamics of another signal. This patch point can also be used for frequency-dependent dynamic-range control. Operationally, the 160X is as simple to use as the 160. The same three rotary pots set the threshold, compression ratio, and output level. Thus, the operation of this sonically improved unit was familiar to almost everyone.

As far as the dual LEDs located above the threshold control, which show overeasy functioning, dbx had also improved the metering. Here three (green, amber, and red) LEDs are used to show when the input signal is below, in the OverEasy range, or above the threshold setting. I consider this an improvement as it makes it easier to confirm the state of the processor's functioning with just a glance. However, the improvement as far as the metering, which displays input level, output level, or gain reduction, is debatable. The 160X has the ability to indicate both gain reduction *and* either the input or output levels at the same time.

For metering, it uses two rows of calibrated LEDs. The top row of 19 LEDs display a range from −40 dBm to + 20 dBm and can be switched to show either input or output level. The lower row of 12 LEDs gives the first 40 dB of gain reduction. Due to this wealth of visual information and the fact that the controls are calibrated, it is possible to position oneself at a side rack and set up dynamic processing without having to continually turn around and check console VU readings.

There's even a meter calibration trim on the rear panel that allows for the setting of the 160X level display's 0 dB point to anywhere between −15 dBm (1.38 millivolts) and +10 dBm (2.45 volts). So why is it debatable?

Well, there's something about seeing a VU meter's needle slammed against the upper (excessive level) or lower (constant maximum gain reduction) extremities that's just straight out hard to ignore. This is not the case with LED displays. Maybe some operators do not fully understand this type of display, or it could be because the units were positioned out of the operator's line of sight (as in two of the situations I've witnessed).

One instance was a live venue. A half-dozen 160X limiters were being used between the stereo three-way crossovers and the multiple amplifiers powering this PA system. I was called in to check out the system because the sound was, well, horrible! Totally distorted and this was a brand-new, very expensive setup! One look at the rack of dbx 160Xs told the story. *Every* level and gain reduction LED was illuminated. No blinking anywhere, just solid *on*.

The first place I looked was the crossover output, which was +26 dBm. Someone had, for some unknown reason, jacked up its output control. I asked the operator if he had noticed the limiter's meter readings. There was a shrug of affirmation, but it didn't hide the glaze in the eyes.

The second instance I mention only because it is even more remarkable. It was a syndicated radio production facility. Four dbx 160X units were the final stage before the house feed to the bird (satellite up-link), for nationwide distribution no less. Granted they were situated at floor level in a rack in what was called the terminal room, which was seldom used for any audio outside of my repair bench, but those meters could be seen by anyone walking down the main production area corridor. It ended up that the problem was a mispatch. One of the engineers had decided to place an equalizer on the mix bus. Instead of using a console patch point that was before the output bus meters and the control room monitor feeds, the EQ's return was inserted into a post VU and monitor patch point. The end result was that no one heard what was going on. Luckily, it was only a repeat (backup) feed, so no harm was done. But in both cases those LEDs were putting on a light show and a half! *The point?* I believe the sonic difference between the dbx 160 and 160X is due to their metering.

What's this nonsense? People expect LED metering to look radical since they usually indicate peak response. But VU meters are supposed to have a nice, slow, relaxed-like movement. I've watched people set up levels using both. This is my impression of the way people approach dynamics processing setup while viewing LED and VU metering:

With LED metering: CRANK. Yeah, that's a bunch of gain reduction, and CRANK, okay, we've got a seriously good output level now.

While with VU metering: OOPS! That's a lot of gain reduction, I better back the ratio off a bit and, oh no, the needle is occasionally as high as the red zone; better reduce the level.

One look at the specifications will explain why the dbx 160X is not audible. It's pretty much noise-free. This device has a very flat frequency response from unity gain to 40 dB of gain reduction. The 160X's phase-per-frequency response is exceptional. On all counts, the 160X is sonically superior to the 160. Why would anyone think that 10 years of design and manufacturing experience would produce anything less? It must be tough as a professional audio equipment manufacturer. Give the people what they want: more control, better sound with more in-depth display, and all your efforts may be disregarded.

Internally, the 160X is nicely laid out with the transformer well-shielded from the main circuit board. A second PC board for the unit's metering is piggybacked to the first. All part locations are well labeled, as are the test points, some even with voltage notations. The removal of a mere four screws allows access to the component and test-point side of the PC boards; however, replacement of components will be a little more complicated, as ICs are not socketed and getting to the solder side of the boards requires a whole host of disassembly steps. On the other hand, since this unit is not really user-serviceable, this problem will usually be left to the manufacturer's service department.

Input is via a rear panel ¼-inch phone jack (tip, ring, sleeve balanced, or tip and sleeve unbalanced) or barrier strip. Output connection is the same except having balanced output is optional. The detector input is via barrier strip only, while the strapping (stereo) connection is TRS ¼-inch phone.

This unit's sound quality, along with its ease of setup, the fact that its controls are calibrated, the wide range of its processing, and the smooth OverEasy compression, make it a joy to use. Put a VU meter in it and I think dbx would have had another classic. But as with *any* individual stage in the audio chain, if the maximum input (headroom) level is exceeded, the result *will* be distortion. dbx continues to develop its line of compressors and limiters, and its range is now quite extensive.

Audio and Design Recording's F769X "Vocal Stresser"   As you can see, more than a single compressor or limiter may be needed when dealing with difficult program dynamics. Often a combination of units, such as two different limiters or compressors, will be required to achieve the desired dynamic range. Furthermore, an equalizer or two may be hooked up before or between these units, or in a side chain if available.

Once you have the proper dynamics setting, another EQ may be patched in, followed by a gate or an expander for further enhancement. This can be used to give drums added punch, to reduce background noise, or cut down on the hissing or breathing noise, which is often a result of compression. Obviously, this is an exaggerated example, but for our purposes it demonstrates the extent of possible equipment patching used for dynamic-range control.

There are two problems with this approach to signal processing. Even if the engineer is pretty fast at getting things all patched and set up, the client may be getting a bit impatient, especially if it is a basic session that has already eaten up time due to equipment alignment and instrumentation and mic setup. No budget-conscious producer wants to spend time on anything at $150 per hour, especially if the date is a jingle and there are half a dozen musicians standing around while getting paid scale. Secondly, all these ganged-up effects have amplifiers and this kind of series connection will always degrade a signal to some extent. Additionally, many studios do not have enough outboard equipment to allow for this type of setup on more than one or two signals and still have enough remaining to allow normal dynamics processing for the rest of the tracks.

Audio and Design Recording came out with a device that to some degree solved both of these problems. The Vocal Stresser is actually a package deal, being a combination of their F760X compressor/limiter and an E900 equalizer mounted in one unit. In all, the F769X is composed of a peak limiter, a separate compressor/limiter, and an expander/gate, and these three effects can be used alone or in combination.

Along with the equalizer section they comprise a complete dynamic-range processing package. Because all these units were combined in one device, it saved setup time, and since only one VCA was used throughout, signal degradation due to series connection of multiple amplifiers was eliminated.

The variability of this device's processing options is pretty extensive. So much so that the best way to explain it all is by describing its controls. The initial control function is a common bypass switch (here called "system in/out"). This is followed by what seems to be a simple input level control. This input attenuator's range is from unity (no attenuation) at its full clockwise position down to infinity (no signal) at its full counterclockwise position. Its normal level setting for unity gain with flat response is the midway point, which is −20 dB; any less than that yields only peak limiting while more causes increased compression. This attenuator therefore allows for optimum S/N ratio. Furthermore when used with side-chain-mode operation it acts as a threshold control.

Next is the equalizer section. This is a four-band semi-parametric having fixed Qs (set bandwidths) with continuously variable center frequency selection:

Band #1 has a width of three octaves and a center frequency that ranges from 40 Hz to 1400 Hz.

Band #2 has a width of one and a half octaves and a center frequency that ranges from 80 Hz to 1600 Hz.

Band #3 has a width of one and a half octaves and a center frequency that ranges from 400 Hz to 14 kHz.

Band #4 had a width of two and a half octaves and a center frequency that range from 800 Hz to 16 kHz.

Thus, the range or sweep of each was about half of the total of this unit's (35 Hz to 18 kHz) bandwidth. This allowed more than one band to be used in the same area of the spectrum for more precise control. The cut, or boost, for each band was 20 dB. Here each band had a level control with a range of 0 to 20 dB. Via the very intelligent use of three-way ("dip-off-peak") toggle switches, the level control's full arch was used to set the amounts of either boost or cut. The "off" position can be used to defeat any band but was also a big help for verifying proper settings as it encouraged comparative evaluation. The equalizer section can also be used as a separate unit by patching in and out through the rear panel "auxiliary" connections. It can also be routed via a front panel switch into the side chain, or it can be positioned before or after the dynamics processing section. Versatile enough for you?

The dynamics section also had a bypass ("system in/out") switch as well as input and output level controls. Four toggle switches helped define the parameter setup of the dynamic section. The peak limiter can be switched to off/on, or on with preemphasis, which was set to a 70-microsecond time constant, but Audio and Design made time constants of 50 and 100 microseconds available as an option, to meet specific broadcast requirements. Another three-way toggle switch ("exp-off-gate") was used to activate expander or gate processing.

The expander's attack time was switch selectable for 20 microseconds, 2 milliseconds, or 40 milliseconds. The compressor's attack time was switch settable to 250 microseconds, 2.5 milliseconds, or 25 milliseconds. The peak limiter's attack was pre-set at less than 250 microseconds.

Both the limiter/expander and the compressor sections had variable release, threshold, and ratio controls. The threshold ranges were as follows:

Peak limiter, −20 to +14 dBm

Compressor, −20 to 0 dBm in 2 dB steps

Expander/gate, −40 to +14 dBm

The release range for the expander/gate was continuously variable from 25 milliseconds to 5 seconds. The compressor release time was switchable between 25, 50, 100, 200, 400, and 800 milliseconds plus 1.6 seconds, 3.2 seconds, or automatic.

The peak limiter's release was set at 250 milliseconds. The peak limiter's ratio was set at 2:1.

The compression ratio was switchable to six positions: 1:1 (no compression or off), 2:1, 3:1, 5:1, 10:1, or 20:1. The ratio of the expander/gate depended on the setting of the expander-off-gate switch. The expander had an infinitely variable range control that allowed for up to 20 dB of attenuation. The expander's ratio was fixed at 1:2 (2 dB output drop for every dB of input drop). The gate had a maximum of 20:1 attenuation ratio (20 dB of output attenuation for every dB of input level drop). Thus, this device was capable of ducking.

As you can see, Audio and Design gave this device a lot of variability, but not enough to overburden the operator. Try looking at it that way: The peak limiter's attack and release were both set at 250 milliseconds, which Audio and Design Recording considered to be optimum for avoiding repeated program ducking and keeping low-frequency modulation (pumping) distortion to a minimum. The threshold was 20 dB below the input's unity gain. Meanwhile the compressor's attack time was switched to either 250 microseconds, 2.5 milliseconds or 25 milliseconds, and the threshold ranged from 0 to −20 dB in 2 dB steps. Release had nine settings: 50 milliseconds, 100 milliseconds, 200 milliseconds, 400 milliseconds, 800 milliseconds, 1.6 seconds, 3.2 seconds, and an automatic ("tracking") setting. Thus, the combination of these two sections allowed for slow compression while enabling fast peak limiting to maintain distortion-free levels.

The addition of an onboard equalizer made for a well-rounded package. All sections were color coded and parameter setup was further eased by the multiple in/out switches, which again aided comparative judgments. Metering was via a reverse-direction VU that showed gain reduction (−20 to 0 dB). The unit also had a red LED peak indicator and a green expander/gate-function-on light. Signal path setup was a logical step-by-step process and the interrelationships between control settings were not overly extensive. The result was that it was hard to mess up the sound using an F769X, but like *any* audio processor it is possible.

NOTE: There's a lot of modeling going on theses days due to the new software-based approach to processing control. I don't mean to sound like an old fogy, but the difference between tube electronics and tube modeling is simple. Modeling affects input signals. Input signals affect tube electronics.

I believe that being familiar with a processor's parameter control layout and even the effect of its distortion on the sound has a lot to do with operator confidence and therefore usage as well as purchasing decisions. This would certainly account for why Drawmer had to reissue its legendary 1960 series tube preamp/compressor, a device unto itself and a legitimate classic.

Input can be at line or mic levels and provisions were made for an added auxiliary input. The preamp has 60 dB of gain and switchable phantom powering. This unit also incorporates a high pass shelving filter, switchable to 50 or 100 Hz and a dual-band equalizer with in/out and brightness boost switching. The compressor has a soft-knee response and variable threshold (infinity to −24 dB), attack (fast, medium, or slow), six release settings, and an output level control with a range of −20 to +20 dB. Metering is via an LED clip indicator and a VU meter that is switchable to show either the output level or the amount of gain reduction (reverse VU).

A feature unique to this device is the output monitor S/C Listen-Output-Bypass switch that allows the operator to fine-tune the Side Chain by ear.

The Drawmer model #1960 is about as comforting as it gets. It's right up there with the LA-2A as far as it's a warm, natural sound. It's as fast and easy to use as the 1176. And it has the quick VU meter verification of the processing parameters of the dbx 160.

While the 1960 is an outstanding dynamics processor, it's familiar layout has a great deal to do with its success.

On the other hand unfamiliar parameter control, functioning, and, yes, even metering can result in operators having an unjustifiably negative opinion of a processor. A case in point is the Aphex Compellor.

I've been raving about this device since I reviewed it for *Recording World* magazine over 20 years ago. To this day I still have to lead people by the hand showing them what an incredible job this unit is doing. Why? Because you can't hear it.

Furthermore, its metering, while extensive as far as the information it provides, is very much like that of the dbx 160 X. As stated I feel the 160X would be as much of legend as the 160 if it had VU metering. If the Compellor had two VU meters, Aphex wouldn't have been able to keep up with the demand.

I believe this because I've set it up that way countless times in both class and control rooms. The effect of seeing two VU meters exhibiting the visual evidence of substantial compression while you're hearing none is staggering.

The Aphex Compellor

In terms of strictly controlling dynamic range, without any apparent change to the signal, the Aphex Compellor is the best I've run across. Little by little, dynamic-range-processor manufacturers not only improved the transparency of their circuitry but also added intelligence (read microprocessors) to their effects devices, which increase the ease of parameter setup. With the introduction of the Compellor, Aphex took both of these approaches a quantum leap forward.

Not only is this device's sound transparent, but its basic operation can be accomplished by adjusting a single dial, but don't be misled by the fact that there is only one main level adjustment control. The Compellor functions as a compressor, a peak limiter, a leveling amplifier, and a processing gate, and the first three functions all work *at the same time*.

The leveler automatically increases the input signal's level when it drops. This improves the S/N ratio. The process gating stops the leveler from increasing background noise when there is no input signal. This ends the common breathing problem heard with many *automatic gain controls* (AGCs), wherein the leveler does not increase the level of a signal but actually jacks up the background noise level and adds additional self- (amplifier) noise.

The gate is not overtaxed in performing this function, as the leveler operates at a fairly slow average detection rate. So the leveling increases do not climb up and down out of control, and the gate is not repeatedly opened and closed. Instead you have a slow increase that is natural sounding and most impressive to watch when displayed on an oscilloscope.

The job of the compression circuitry, which handles the bulk of the dynamic range processing, therefore has to deal with less erratic signal dynamics. So dynamic range control can be accomplished using a gentler average response. Again this type of gain reduction makes the compression appear more natural to the ear. But the real evidence of this natural sound is detected by the knowledgeable eye, whether registered on a scope or a pair of input and output meters.

In order for this device to handle sudden transient or even sustained high increases in input level, a peak limiter is incorporated as a brick-wall-type protection device to prevent internal distortion and any overloading of subsequent equipment in the audio chain. This is exactly what you want a dynamics processor to do. Because all three of these functions are controlled at the same time by intelligent electronic circuitry, the operator need only choose the amount of leveling (level increase) as opposed to compression (gain reduction). This is again accomplished by a single rotary balance control.

No longer is there any need to sit at the mixer and continually adjust the mix bus levels. The Compellor handles this function perfectly and automatically, freeing the operator from fader riding duties.

The Compellor is so clean and transparent sounding that the first couple of times I brought it into a facility to use, test, and get opinions on, people felt that it wasn't doing anything. Thereafter, a dual trace scope, as well as a pair of input and output (both Peak and VU indicating) meters were hooked up so that everyone could see, if not hear, what this device was doing to the input signal's dynamic range. I plotted frequency and phase-response graphs of this device, and they were just about perfect. Ruler flat-frequency response from 20 Hz to a single decibel of increase at 20 kHz. Square-wave testing indicated no ringing or overshoot, a very fast rise time, and only the slightest rounding off of the leading corners of a 10 kHz signal.

In fact, the only manufacturer's specification that I do not consider to be an understatement is maximum level; I personally would not push this unit a full 3 dB beyond +24 dBm.

However the real cause of the audible transparency is not just the high quality of this device's signal path electronic circuitry, but the ingenious way in which it handles dynamics processing by combining old-fashioned leveling and compression with a new way of gating and brick-wall peak limiting. The Compellor automatically controls all of these functions simultaneously according to the signal's dynamics and how the operator sets the leveling or compression balance control.

Many people found it difficult to understand the integrated workings of the Compellor's dynamic processing. This was due in part to the new terminology used to describe it. *Dynamic recovery computer* (DRC), silence gate, *dynamic verification gate* (DVG), processor balance control, drive control, and stereo enhance switching seemed like a lot of hooey until you take the time to understand what is meant by these odd parameter names.

Instead of the familiar input-level control, you're faced with a drive control. Well, this rotary pot does not just attenuate or boost the input signal, it determines the output level of the amplifier that feeds all of the detected information to the VCA. Lower this and the VCA does less processing, crank it up and the processing increases. Therefore this control does literally "drive" the processing.

The silence gate, or threshold, does not function like a normal gate, which cuts the signal level off by some predetermined amount when it falls below the threshold level. Instead, it gates the processing when the signal drops below the threshold. Here the processing remains set so it's ready when the signal level returns to its normal level. Therefore, radical processing changes are avoided.

The processor balance control ships the audio to the compression (gain reduction) and the leveling (gain increasing) circuitry. Like a pan control, dead center position results in an equal split. A counterclockwise position adds more signal to the leveling (gain) circuitry, while clockwise positioning increases the amount fed to the compression or gain-reduction circuitry. So this control actually does "balance" the two processing functions.

Dual processing models such as the 320, which incorporate a stereo enhance switch, are capable of feeding some of the compressed and leveled signals out of phase to each other. This has the affect of causing the stereo image to be slightly widened, augmenting, and *yes* enhancing the stereo spread.

The dynamic recovery computer is another story. This is a proprietary module. It is made up of the level detection and other circuits that control the Compellor's processing. Therefore, detailed information about its inner workings was not published. But what *is* stated fits in with the overall functional approach of this device.

The purpose of the DRC is to follow the signal's waveform. It increases the compressor's release rate (it's connected to the compressor's side chain) when the input signal has a waveform that's complex or has multiple peaks. Due to this the compressor will not over-process low-power signals because its release rate changes too slowly for it to fully recover after processing the initial peaks. The signal fed to the dynamic verification gate comes from the output of the VCA and is fed back to the leveling and compression controllers. Part of this signal is also rectified and fed to the gating controller. Sound familiar?

Aphex has used every trick in the book to achieve a single goal. A dynamics processor made up of interactive leveling, compression, and peak limiting that utilizes program-controlled side chains to achieve easy parameter set up and processing transparency. The end result of all this is that you do not hear any processing. Forget about pumping and breathing—no squashed sounds here. In fact, try as hard as you like, you cannot mess up a signal with this device. The worst thing you can do is set your input too low or your output too high. Even this should not occur easily because of the extensive metering provided.

Again blame it on the metering not being VU.

All information concerning input, output, and processing levels can be obtained by pushing two buttons. Processing is shown by the 10-segment metering, not only in terms of gain reduction, but also in terms of the amount of leveling and compression, all at the same time.

Like the DBX 160X, LED bar graphs are used, but here Aphex decided to give the operator complete visual indication of the processing by a method that is unique. Many operators found this metering (like that of the 160X) to be unfamiliar and thus lacking the comfort of a common VU meter.

The Compellor's LEDs show VU (average) and peak levels at the same time. An additional bar graph display mode shows the gain reduction. The front panel labels the 10 LEDs as showing a range of -15 to $+12$ dB in 3 dB steps for level, and 2 to 20 dB of gain reduction in 2 dB steps.

Nothing difficult there, but these LEDs are bicolored—meaning they can glow in red *or* green.

When displaying levels, the last LED of those glowing green indicates the peak level, and the last LED of those glowing red indicates the VU or average level. The gain reduction display is all green except for a single LED that glows red to indicate the amount of leveling. The green LEDs to the left of the single red LED show the amount of gain reduction; and the green LEDs to its right indicate compression levels.

To provide all-inclusiveness of the displayed parameters, a single red LED flash indicator located to the right of the bar graph display designates peak limiting, while to the left is a yellow LED, which ignites whenever the signal dips below the DVG's threshold setting.

It does take a little getting used to, but this display makes it possible to grasp everything the Compellor is doing with a single glance. A lot of people just shut off when they have to deal with something that's unfamiliar. And when the processing is audibly undetectable, light-show metering seems more incomprehensible. You should try doing exactly what I do, when you get your hands on a console or a pair of VU or peak reading meters and a Compellor. Feed a signal to the Compellor's input and one meter. Feed the output of the Compellor to the second meter. Without listening to the signal, adjust the Compellor for an output meter movement that's ideal. Then switch between listening to the input and the output. Now mute the signal and adjust the Compellor for a radical change in the signal dynamics. Unmute the signals, and switch between the input and output again. You can repeat this process as often as you want and I bet you'll hear no difference in the Compellor's signal quality at *any* setting with inputs below +27 dBm. If Aphex had included two VU meters on this device, it would have done away with any scepticism, but it could have added too much to the price tag. But anyone can add the two VU meters themselves.

The Compellor's internal workmanship is superior, all ICs are socketed, and all parts clearly labeled. The manufacturer not only provides 24 test points but has wisely opted for large hoop-style contacts, which are perfect for attaching "easy-hook" test leads. The electronics are divided into five PC boards, with one for the power supply and a separate I/O and control board for each channel. The meter boards are mounted piggyback to the I/O boards, but instead of simply using standoffs, Aphex used hinges. This is exceptional. Interconnection of these boards is via both DIP-socketed ribbon cable and Molex. Should a failure occur, it is possible to disassemble this unit, grab a common multimeter, call the factory's service department, and be guided through a test routine. The malfunctioning board could then be shipped back to the factory, or a replacement part could be ordered. Very likely this would leave one channel operable in the interim, avoiding complete unit downtime.

In my opinion, due to this unit's functioning, ease of operation, and lack of sound, it comes close to being the ideal dynamics processor. Close? Well it does not act as a true limiter, and the compression ratio extends only to 3:1, so this is not what you'd turn to for that squashed big drum or unlimited sustaining guitar sound. But for controlling difficult vocals and managing the dynamics of instruments like basses and violins, it simply cannot be beat. Yet it's true strength is in processing a mix bus. Outside of radically excessive levels, it fits everything into a nice neat dynamics pocket. When you combine the Compellor with Aphex's Dominator (multiband limiter), nothing escapes the dynamic range you've set up and the sound is still unaffected. Which brings us to multiband limiters.

Multiband Limiters

Imagine feeding a signal into a three-way crossover and then feeding those outputs into three separate dynamics processors. This is fairly common when setting up live PA systems, but here the three post-dynamics processor signals are recombined.

With normal limiting, while using an EQ in the side chain, if you've heavily compressed the high frequencies, moderately controlled the mids, and hardly touched the bottom end, the sound will be contoured to increase the gain reduction when high-frequency peaks occur. If the gain reduction is set for limiting, the result may make it possible to increase the overall level without overloading subsequent equipment that may be sensitive to high frequencies.

Multiband dynamics controllers make it possible to limit only a single band. Here you can limit just the high-frequency content, and then boost the recombined signal. In both instances, the bass has been increased without using any equalization, but with multiband processing the program's dynamic range is under three-way control. This has become a fairly common setup in broadcast systems.

Additionally, these processors come in various configurations, with as many as eight bands. They can be used on individual tracks or mix buses and provide a lot of processing power. With all of this coverage of dynamics processing, I haven't even scratched the surface.

Compressors built into consoles are a whole category unto themselves.

Neve compressors like the 32264 are so popular that two units mounted along with a power supply in a single rack space housing can fetch thousands of dollars. The same goes for the venerable API 225 compressor limiter or a pair of SSL limiters.

I hope I have given you some insight into the vast field that dynamics processing has become. The real deal is to just dive in. Read the manual and suss out how the parameter controls affect the processing. Make something sound *bad*. Then correct your settings so that you'll understand what you did wrong and definitely make a note of it. That "bad" sound may be just the thing that's needed to add sparkle to something dull. You never know.

Limiters, compressors, levelers, gates, expanders, multiband processors, attack times, release times, ratios, thresholds, and metering are all things recording engineers love to argue about. But the most debated subject is (as usual) their subjective likes and dislikes as far as the sounds various processors render. As usual, put some audio folks together and there will some arguing about sound.

It's really all about demand, whether it's caused by the flavor of the month or the need of a specific feature that keeps you out of a jam. The way I see it, the more the merrier. Luckily, arguments over subjective opinions about sound quality do not apply to the next rack of outboard gear. The goal of time domain processing can range from completely changing a sound to leaving it exactly as is, just positioned a little later in time.

Stop the Presses!

The Human Hearing System

In writing this book I purposely chose not to presume any prior knowledge on the part of the reader. Even so, I was about to start the topic of delay processing by discussing the "Haas effect" (also referred to as precedence). This was originally going to be covered in the section on control room monitoring. Then I realized that I also had to cover the delay before the onset of reverberation. That explanation was going to be handled in the sections covering reverberation as well as in control room and studio acoustics. It then occurred to me that tackling the subject of human hearing, or psycho-acoustics, in dribs and drabs as we went along would be a disservice to the reader and would also drive me nuts. Therefore, I decided it would be best to deal with this subject before getting into delay, reverberation, control room acoustics/monitoring systems, and all of the other areas that apply to hearing. Are there any that it doesn't? The fact is that every audio recordist needs to add the knowledge of how the human hearing system works to their engineering toolbox.

Our brain/ear combination is trained to process audio information right from birth. Ticking clocks are now placed in cribs to avoid the chance of infant death and is there anything more important than the sound of a mom's voice? As we're still discovering, our hearing system has learned to handle a lot of information and it is continuously on duty evaluating our acoustical environment.

How Human Hearing Detects Sounds   The classic method of beginning this discussion is to divide the hearing process into its three main components. The source of the sound (such as a music instrument or voice), the medium it travels through (in most cases, air), and the receiver (our hearing system). In a nutshell, sound travels from a source through a medium to the listener. The medium sound travels through is a little more complex than simply air. If that were the case (as in an anechoic chamber and to some degree free space), the only factors we'd have to consider would be frequency content and humidity.

Air absorbs high frequencies more than low and this attenuation increases with humidity. However, in addition to a direct path from the source to the receiver, our brains have all the reflective paths to process. Each reflection will display a differing amount of frequency absorption depending on the material that makes up the surface it was reflected from. Our hearing system uses dual pickups that interface with our brain. We also incorporate vibrational sensory perception into our composite make-up of a sound because we feel low frequencies.

Yet, even a large number of microphones supplemented with a vibrational pickup cannot duplicate our hearing capabilities, simply because they do not have a brain to interact with. Our brain ascertains a vast amount of information as to both the location of the source and the environment into which the sound is being dispersed from the aural clues given to it by our two ears. Ask anyone who's visually impaired how they know details of a room's dimensions before they've had a chance to scout it out physically.

Clues about the location of the sound source come from the difference in a sound's intensity and arrival times (phase) at our two ears. The ear closest to the source will receive a louder signal, earlier than the ear that is more distant, is turned away from, or has the sound partially blocked from it by our face. These time or phase differences further aid us in locating a sound source. Simply put, if the sound arrives at our right ear first we know it's to the right. These clues are used by the brain to narrow down the position of the source. Only those sound sources that are located at exactly the same distance from our two ears directly in front of us (or on the "median plane") will have identical arrival times and equal intensity at both of our ears. While the pickup pattern of human ears is almost omni-directional, our brain is able to process the sounds we hear so that we can discern one sound from another in far-off locations as if we had a pair of shotgun mics for ears.

Our audio perception also includes the ability to learn about the acoustical environments and the sound source's location, and adjust our perception so we end up being able to pick out exactly what we want to hear. Think of speech emanating from a transistor radio in a restaurant full of people talking, glasses clinking, babies crying, and dogs barking. Add in a constant drone of the building's *heating, ventilation, and air conditioning* (HVAC) system and you'd still be able to differentiate between all these sounds, lock in on the sound of interest, and filter out all others so that you can perceive and understand what's being broadcast by that radio's little speaker. All accomplished in a heartbeat. Great tools those ears!

Our two ears' focusing ability operates much like the two base points of a triangle with the sound source being the third point. Think of how you turn your head to focus in on and hear one particular thing more clearly. Our two base points are set since we generally do not have the ability to individually move our two ears, as dogs and other animals do, which gives them a big hearing advantage. Yet we all naturally move our head when we hear faint sounds coming from behind us, just so we can determine if, whatever it is that's coming up on us, is doing so from the left or right. Preservation is a major factor. I do not believe that it's a coincidence that our hearing is most sensitive to the frequency range that includes the bandwidth of human cries of warning.

We learn to distinguish between sounds, ascertain their direction, discern their origin, and thereby establishing its location. We not only comprehend the source as it relates to our own location, but we perceive and begin to understand the space around us, all in a matter of moments.

The outside or visible part of our ear (the pinna) is subject to the laws of physics. Because of its size, its irregular shape, and the size of the canal opening, it ends up working like a high pass filter. This means our hearing can focus in on higher frequency sounds more easily than low. The classic way of explaining this is to determine the distance between our two ears (about 6 inches), equaling a wavelength of roughly 3.5 kHz. Below this frequency, our ability to locate sound sources begins to diminish. Before I knew about this, I figured it had to do with the size of our ears. My calculations for a 3-inch ear ended up at 4.5 kHz. I thought of it this way: Say our ears are about 3 inches (or 0.25 feet) in length. That size wavelength corresponds to a frequency of about 4.5 kHz. Therefore, anything below this frequency will be somewhat less pronounced. I was wrong, and actually human hearing does better than that. Yet our localization capabilities are still greater at higher frequencies (above 3 kHz) than lower. This explains our less than directional low-frequency pickup characteristics.

This is also why the use of a single subwoofer or low-frequency driver is a viable possibility in stereo reproduction systems. Just one example of ways we can take advantage of the peculiarities of our human aural perception. Don't believe it? Disconnect a speaker's high-frequency driver, turn out the lights, and have someone try to point to its location. Do the same with the low-frequency driver disconnected and you'll find the high-frequency driver will always be the hands-down winner in this localization event.

High-frequency localization is also aided by the irregular shape of our ear, which diffracts these frequencies and by the natural resonances of our conch canal, which causes phase differences that the brain uses to help calculate source localization and environmental information details.

Frequency-dependent attenuation caused by air and reflective surfaces helps clue us to the type of area we're in. There's no question that you can tell the sound difference between a room with wood as opposed to drape-covered walls. There's a "cozy" feel when we hear a sound deadened by a fur, carpet, and heavily draped room.

Our brain begins to learn all this from birth. We get to the point where a sound can be attenuated due to the distance of the source, the absorptiveness (absorption coefficient) of the material covering the walls in the room (or even a wall between us and the source), and yet we can tell which is the case with just a moment's listening.

Furthermore, in the case of a wall between us and the source, we can instantly tell when a door in that wall is opened. "Way" good tools! And pretty powerful sensory reception! The brain-ear interface is so complex that we still do not fully understand how it operates.

Okay, let's get down to brass tacks. We first hear the sound via the direct path from its source to our ears. Next come the early reflections. When speaking about these later arrivals, each successive one is given a higher "order" number, the direct being the first order, the next, the second, next the third order, and so on. As the time delay increases they become harder to discern as individual sounds. The early reflections are most important to our perception of the source within its immediate environment (such as the stage area). The longest delayed reflections end up as part of the reverberation information, which tells us more about the room's overall size, shape, and acoustic treatment.

Timing also plays another role in our perception of sound whereby a sound that's heard corresponds with our other sensory information pickup. If we hear the clap before we see the two hands come together we instantly know something is wrong. We can find this confusing and it may even cause alarm! How long does it take you to notice when the audio and video portions of a dialog are not in (lip) sync? Pretty quick, right? Yet single events (such as a hammer hitting a nail) when seen before heard, can sometimes trick us into assuming that the event took place at a distance farther away than we had originally judged it to be.

A sound received directly from its source will usually arrive at our two ears at slightly different times. This is called "inter-aural time difference" and it's due to the physical difference in the path length to each of our ears. The resulting delay is short, sometimes not even 10 milliseconds at the ear that's opposite from the sound source.

This time difference lessens at higher frequencies especially with signals that are more steady-state (sign-wave-like) in nature. But transient signals like those normally found in music, which have defined attacks, increase our hearing system's ability to process interaural time differences.

As mentioned, localizing is also aided by level differences. These are termed interaural amplitude differences. Our perception of level differences increases at higher frequencies. So our hearing system uses low and high frequencies in different ways for information on a source's location.

I don't pretend to be an expert on this subject, but it's not that difficult to understand how the human hearing system uses perceptions of amplitude and time variations (augmented by years of experience) to pinpoint a sound source's location. It's just plain logical that interaural time differences provide angle information and interaural amplitude differences provide information about distance. Yet I still can't fully conceptualize our ability to detect elevation, but I know it to be a fact.

As stated, we can localize or tell the position of a sound source in the real world, even when there's all kinds of other sounds going on. In the pseudo-real world of sounds emanating from speakers, we localize the sound as being at the speaker's position. Delaying a signal and sending it to the right will reinforce the signal's apparent positioning to the left. What?

The Haas effect is named after Helmut Haas, who was part of a group of researchers in the 1930s who determined that the first sound heard takes *precedence* as far as our hearing system is concerned. Therefore, additional sounds delayed by as much as 50 milliseconds will appear to us to be *part* of the original sound and to also be emanating from the first location even when panned to an opposite speaker. This is our ear/brain system trying to eliminate any confusion that may occur due to time smear and thus preserve the intelligibility of what's heard. That's *some* helpful tool!

Okay, the location trick. Send a mono signal to a pair of speakers at equal levels and the sound will seem to emanate from in between them. If you move closer to the right the sound appears to be coming from the right speaker. Move back to the center position and send a delayed version of the signal to the right speaker. The signal's source will then appear to be located at the left speaker's position. To defeat this "precedence" effect, the delayed signal would have to be 10 dB louder than the original. Remember, steady state sounds are not affected by the Haas effect as much as transient signals.

The automatic blending together of sounds that have close arrival times is called "temporal fusion." With it we have another positioning tool we can add to our pan pots and level controls. Not only does combining the delayed signal with the original increase its level and broadness across the stereo field, but the original signal's intelligibility is enhanced as well. So delay can be used to cause an increase in level, a change in the perception of a sound's location, and to broaden the width of a sound across the stereo field. How could I discuss delay processing before explaining that?

Temporal fusion also accounts for our not being distracted by all the initial delays we hear due to the sound's reverberation within a given environment. The explanation of arrival time is also divided into three sections. First there's the direct sound coming from the source, then *early* reflections that bounce off of the close (stage) boundaries around the source, and finally the full room's reverberation.

Early reflections give us a lot of information about the sound source. *Every* music instrument from a hammer strike on a steel I-beam to the human voice on up to a huge cathedral-sized church pipe organ has its own *unique* sound radiation patterns. For instance, a trumpet will have a higher direct level than a upright bass simply because it is more directional in its output. So each instrument projects its sound differently, and this part of the information is conveyed to our brain by the early reflections. They also clue us in on the acoustical makeup (liveness) as well as size of the stage area.

Since the early reflections blend with the original sound they have a lot to do with its propagation, timbre, sustain, and frequency content. This is why good halls become legendary among musicians. They enjoy playing there because the early reflections not only improve the sound, but make it possible for the musicians to hear themselves and each other more easily.

These early and later reflections quickly build up to a point where they become part of the overall room reverberation, which arrives at our ears from all directions and provides us with information about the space we're in.

By this point, the reflections have had to travel a bit of a distance through air and have rebounded off many soft boundaries. This causes a greater amount of attenuation to the high-frequency content. That's why natural reverberation seems to add warmth along with level and sustain to the sound. As is often the case, the lessening of a sound's high-frequency content is equated to the addition of warmth.

The amount of attention paid to reverberation decay time (RT-60) is not inordinate considering the fact that unless we are very close to the source, most of what we hear in live situations will be from reverberation. Overly short reverb times tend to make music sound dead while longer reverb times can destroy the intelligibility of speech. A lot of time is spent on determining the *critical distance*. This is the distance from the source where the direct and reverb levels are equal. To the audio recordist this is merely an aside as far as mic placement is concerned.

But for the sound contractor faced with installing a PA system in a cathedral where speech recognition requirements and architectural (acoustic) surface reflectiveness are both very high, knowing this distance *is* in fact critical.

Easier said than done. Equations for calculating *critical distance* (Dc) take the following into consideration:

The total reflective surface area, as per each individual surface area's absorption coefficient

The loudspeaker or source's "directivity" (find *that* in a specifications listing)

The internal volume of the room

The minimum and maximum reverberation times

The *sound pressure level* (SPL) within the room

This is not something that the studio-only recordist needs to worry about. I only brought it up so you could see how deep this subject area can go. Although that example is a bit specialized, it comes into play when doing live recording; therefore, I highly recommend you do further reading on the subject of acoustics.

Audio engineering is a combination of art and science. The science part explains the limitations of the technology. By understanding this, we can better deal with the limitations we're faced with and thus do a better job of performing the artistic aspects of sound engineering. I know, I know, "Pee-You." Yet you at least have to acknowledge the value of understanding the way the human hearing system operates *and* the need for you (if you plan on spending any time in this field) to protect *your money makers*!

Hearing Damage   But you're not a heavy metal musician, you say? Well even classical musicians are often exposed to sound pressure levels of 90 dB or more and can suffer from hearing loses that are unusually higher than normal for people of their age group. Just as with musicians, any loss of aural perception will only detract from the quality of the audio engineer's performance. In some cases it has ended careers.

Audio Perception   It's pretty obvious when a single incident of exposure, such as an explosive level of sound pressure, causes severe damage such as eardrum rupture. But accumulated damage caused by repeated and/or prolonged exposure to high sound pressure levels is less obvious. This is because short-term exposure to loud sound merely affects your audio "perception."

Many audio engineers are unaware of this problem. I remember when I once used a 1/3-octave spectrum analyzer to perform a quick check on the house PA system's EQ a live sound mixer asked me, "What's the matter? Don't you trust your ears?" The fact is, you can never be too sure.

For instance, while subway sound levels normally reach only 90 dB, the 100 dB sound level of a 4 kHz screeching (metal on metal) brake squeal can throw your perception off. This is called "sensory cell fatigue" and it causes what scientists call a "temporary threshold shift." Here, your threshold of hearing is changed or shifted upward in the offended frequency range to make your ears less sensitive to those offensive sounds. It's all part of our hearing system's natural protection, which is constantly adjusting to the environment. However, now you may need to hear twice as much (+3 dB sound *pressure*) level at 4 kHz in order for it to match the levels of the rest of the audio spectrum. If that's the case, *please* don't touch the sound system's EQ without the help of an analyzer!

Furthermore, it may take days for you to completely recover from this exposure. Next you'll start to have trouble picking out a single voice in a crowd of people talking. Then you won't be able to focus your listening on a single member of a brass section. In cases where exposure is repeated, that temporary threshold shift will last longer and longer and it *can* become permanent.

Warnings   Hey, it's not like you don't get warnings. If you hear buzzing, ringing, or hissing in your ears, watch out. Note that high-frequency losses occur from both high listening levels and old age, so youthful damage *will* worsen with time. Yet convincing audio engineers to wear flat response in-ear hearing protection is a tough sell, even though you'll find construction workers commonly wearing ear plugs when they're exposing themselves to loud sounds. I had one buddy who ridiculed me for wearing heavy duty electronic hearing protection with sound-activated limiting while at the shooting range, and this guy professed to be a "classical purist" engineer. Two to three hundred rounds of 45 ACP through a compensated race gun (which redirects the explosive sound right back to the shooter's ears) just before engineering a session does not add up to purist recording. Shocked at this guy? Well, if someone else can hear your headphones while you listen to music, you are playing your music much too loudly!

Ear Training   Don't get me wrong. I don't have anything against a good set of enclosed (non-open air) headphones that can be used at lower listening levels. In fact, I think they're perfect for ear training. Ear training?

F. Alton Everest put out a set of cassettes that were designed to help the audio engineer to distinguish between the sound of different frequencies, levels, and even phase shifts. To this I would add recordings that were arranged by people such as Nelson Riddle, Quincy Jones, and George Martin. Now you'll be listening to full string, brass, and reed sections made up of *real* players and instruments. Also get your hands on the charts to popular pieces conducted by Leonard Bernstein and movie scores such as "West Side Story" and Broadway hits like Pipin, The Music Man, and so on, which are all fairly easy to come by. This is big-time ear candy.

Soon you'll be able to respond properly when a producer asks you to raise the level of the viola (not the violin) over that of the cello during a certain section and you'll know if it's the coronet or the French horn that's beginning to sound a bit "spitty," that is, unless you've already blown out your money makers.

Time Delay Processing

While critical distance equations are not terribly important to the audio recordist, time delay calculations are, and luckily the equations used are about as simple as it gets. We know that sound travels at roughly 1130 feet per second, so dividing the distance in feet by 1,130 will give you the delay time in seconds. But most delay times are measured in milliseconds, so we move the decimal point over three places and the distance a sound travels in feet divided by 1.13 equals the delay time in milliseconds (mS). The inverse is also true so that when the time delay in milliseconds is multiplied by 1.13 the result will be the distance the sound traveled in feet.

$$DISTANCE = \text{Delay time in milliseconds and}$$
$$\text{Delay time in milliseconds} \times 1.13 = \text{Distance in feet}$$

Outside of dual or ambient mic placement, calculations such as these do not readily lend themselves to recording studio applications. In recording situations we are more often attempting to match tempo, change a pitch, or increase the apparent "room size" of our artificial reverberation with time delay. In all of these cases setting the delay time by ear will usually suffice.

Setting Delay Time

There is one method, however, of determining the correct delay time that's simple to set up and provides accurate results:

Choose a track with a hard transient, rhythmic sound such as a snare, kick, tambourine, or even rhythm guitar chords. Gate it, if needed, to end up with a short percussive sound. Let's take as an example a snare that sounds on the "2" in a $\frac{4}{4}$ measure. Patch that signal into a variable delay.

Patch an out of phase (make the patch with an out-of-phase cable or use the phase reversal switch on the delay processor) and place the delayed version of the snare and the original snare to two channels that are feeding a mono bus. Adjust the two faders for equal levels on both channels. Solo those two channels or mute all the other tracks. Adjust the delay time on the processor until the signal just about disappears from the monitors. The delay setting will now be equal to the timing of a full measure. In other words, the delay will be set to the length of time it takes for four beats to occur. Following along to the end of this example, which is laid out in beats, should clear up any questions.

(Measure #1) beat 1, beat two (snare hit and input to the delay), beat 3, beat 4

(Measure #2) beat 1, beat 2 (snare hit and the out-of-phase delayed snare hit), beat 3, beat 4

The rest of the measures will be the same as Measure #2.

Additional Drum Beats   Okay, let's say the time between each beat is 0.5 seconds (120 beats per minute). The delay time you'd have set would equal 2 seconds. But no matter what that delay time is, when you delay the snare by half of that value and mix it (in phase) with the original, you will double the number of snare beats, making them occur on both the 2 and 4. These delays or echoes will return to the mix not only as additional beats, but also in perfect tempo.

Now, if the snare is delayed by one-quarter of that full measure setting, and both the original and the delayed signal are then put through a second delay processor that is set up for a delay of one-half of the full measure's delay setting, the two delays will cause snare hits to occur on *every* quarter note. Extrapolate this and imagine this effect on something like a pair of congas.

We've discussed the use of delay to add intelligibility, level, and localization cues to program material. Let's take a look at how delay processors are put to work in live PA systems. "Hey, I thought this book was supposed to be about *recording*!" Well, some of the earliest delays as well as some of the most popular multiple delay devices were designed for use in live venues. Here, processors with multiple (individually adjustable) delays were manufactured for aligning the time/distance variations of different speaker locations in live PA setups. These multidelay units make wonderful recording tools too, as they can produce amazing stunt-like performance results particularly with percussion tracks. Understanding how they are put to use in live situations clues you in on ways that their specialized functions can be put to use in the studio.

Using Delays for Intelligibility

Acoustical consultants try to conform their designs to a set of intelligibility standards called "the articulation index." In brief, this states that dual speech signals arriving *within* 10 milliseconds of each other will increase the sound's level and intelligibility while any delay longer than 100 milliseconds could result in some cancellation, frequency degradation (from comb filtering), and "garbled" intelligibility. These guidelines are often further restricted as to keeping delay times within 30 milliseconds.

Auditorium (a definition in itself) loudspeaker systems require a delay to preserve intelligibility in seating areas that are exposed to both the direct sound and loudspeaker coverage. Again, sound travels through air at about 1,130 feet per second (0.884 milliseconds per foot), but sound travels through electrical wiring almost instantaneously (at 186,000 miles per second).

If the stage from which the direct sound emanates is at a distance of 70 feet and the PA speakers are set at a distance of 30 feet from the listener, the two sounds would arrive within 33.5 milliseconds of each other [70 feet divided by 1.13 = 60 milliseconds, 30 feet divided by

1.13 = 26.5 milliseconds]. Depending on the program, the signal may not be garbled, but the sound will appear to be coming from the speakers instead of the stage. To have the two signals arrive at the same time requires that a delay of 33.5 milliseconds be added to the PA system.

Systems that utilize "distributed" or multiple spaced loudspeakers may require several variable delays if the unamplified "stage" signal can be heard everywhere in the room. This will "synchronize" the speakers so that they all sounded at the same point in time throughout the room as does the unamplified signal.

Another live usage that also applies to recording and broadcast audio is the "apparent" increase in level derived from the addition of a short delay added to the onstage monitors mix. The same thing works on an individual track or to broadcasted program material.

Because temporal fusion causes a "perceived" increase in level it will, in the case of a broadcast, increase the station's area of coverage but not incite the FCC to levee fines for actual broadcast level violations. In live applications, the monitors will be able to cut through the onstage din more effectively without causing any additional feedback. Wow, an extremely rare case of something for nothing, times two!

What's it got to do with recording? How about increasing the apparent level of the vocal track without it causing any added gain reduction to the overall mix either by the dynamic processing controlling the mix bus, or later when played over the airwaves? Times three. That's right, live PA usage my friend, live PA usage!

Short-Time Delay Effects   Many times processing effects are based on using a short delay time to only slightly change the time domain of a signal (its place or length in time). The changed signal is then recombined with the original version and the interaction that occurs between these two signals causes the effect. The most obvious examples of this would be comb filtering and metallic tunnel and ringing effects. Phasing and flanging time shift effects are also possible. The addition of feedback results in pseudo-reverberation effects. Adding negative feedback to a longer delay will produce fading repeats. Just citing examples of delay usage would make for dull reading and tedious writing, and wouldn't provide an overall perspective of the development of delay processing.

Delay Processors   The original delay processors were acoustic spaces. These "chambers" were configured to any size or shape that was practical, but their inside wall surfaces were generally hard and reflective. These spaces produced such short delays that they were inseparable from the ensuing reverb. Chambers will therefore be covered in the section on reverberation.

Acoustic Tubes   The first and only one of these that I've had experience with was the Model 920-16 "Cooper Time Cube" from UREI. Get ready to smile, shake your head, and wonder how things could have ever been *this* archaic. Yet I'll never forget my excitement nor how proud the geezer was who explained its operation to me. It was almost as if *he* invented it, but he was just so tickled by its cleverness. *You,* on the other hand, will not believe it.

Okay, we've all been in airplanes and experienced sound through those plastic "headphones." Ever wonder how they operate? They don't have any wiring or speakers so don't bother trying to use them with your MP3. Back where the sound "originates," be it a video deck or a distribution amp that also handles flight crew announcements, there's a transducer. It operates much like a speaker in that its main function is to move, excite, or vibrate air. This transducer is physically different than a speaker in that it has a small diaphragm and short throw because it has to move air inside of a $3/16$-inch tube. That's right; this "speaker" is moving the air that's directly coupled to your ear drum via the tubes connected to those plastic headphones.

What would happen if I ran an electric feed to one side of those "cans" and hooked up a real headphone speaker? There'd be all kinds of problems. Sound travels through air at 1,130 feet per second while electrical signals scream along at 186,000 *miles* per second. If it's a 100-foot plane and the source and receiver are at its two extremes the electric signal will arrive (for all intents and purposes) instantaneously while the "air-born" sound would take 89 milliseconds (100 feet divided by 1.13 feet per second) to get there, depending on the cabin pressure!

NOTE: *The UREI 920-16's time delays would decrease slightly at high altitudes, but not significantly until 5000 feet! So by feeding an audio signal through 100 feet of tubing we get 89 milliseconds of delay. Who's to stop us from winding the tube up and putting it in a box? Nobody stopped UREI. I swear, that's how it worked. You have to give the inventors credit though. All kinds of effort was put into developing the transducers and the pre-equalization. Remember, this isn't coupled directly to the ear drum. There had to be a microphone-like transducer to convert the sound pressure back to an electrical signal as well.*

There were two lengths of tubing in the Cooper time cube, one 16 feet long for 0.014 seconds (14 milliseconds) of delay and a second 18 feet long that provided 0.0159 seconds (16 milliseconds) of delay. No, not a lot of delay time, and it wasn't variable though you could cascade the signal through both of them, for a total of 30 milliseconds of delay. The specifications also weren't any great shakes by today's standards:

Frequency response plus or minus 2 dB from 40 Hz to 10 kHz

Channel separation a little over 40 dB

The S/N ratio more than 70 dB

THD less than 0.5 percent

Input sensitivity range −20 to +20 dBm (for +4 dBm output)

This device was not rack-mountable as its size was both deeper and wider than 19 inches. (It was also about 10 inches high.) But it wasn't all that susceptible to outside noise interference and it was a lot easier to keep out of the way in a control room than a second two-track deck. One caution though: I've run into a couple that had been opened and "adjusted" and the result wasn't pretty.

I thought the Cooper time cube was heaven-sent since it didn't have any background noise, which built up every few minutes requiring reels of tape to be flipped and bulk erased. Its delay times were shorter than a tape deck's and were ideal for the intended use of broadening a signal's spread in the stereo field, delaying a send to a reverb to extend the effect's "dimensions," and beefing up a signal's apparent loudness without actually increasing its level. You see the people who designed the Cooper time cube knew about the Haas effect and its limitations of 35 milliseconds, which is roughly the point where two sounds become distinct and the delay is no longer temporally fused with the original but begins to sound more like an echo. Hey, there had to be a good reason because it wouldn't have been that much trouble for them to use longer tube lengths.

Electronic Delays   The reliability of electronic delay as compared with older mechanical tape loops may not have been enough of an impetus for studios to convert, but the added control (over the processing parameters) and superior sound quality they provided created quite a demand.

Analog Delays   Some *analog delay lines* (ADLs) were limited to short delay times as their noise and signal degradation increased prohibitively with longer delay times. Some *bucket brigade delays* (BBDs) became unusable at any settings over 100 milliseconds. Think of these "buckets" as a row of capacitors with the first being filled with the input signal. Upon a clocked (timed) control, the first cap dumps its load into the second and then begins refilling with more input signal. The delay ends when the last capacitor unloads the signal to the output. This was not very quiet or fast circuitry and with the first of these devices, all the processing artifacts that were added at longer delay times could cause the signal to sound pretty bad. Yet these devices worked fine for shorter time-domain processing like *automatic double tracking* (ADT), phase shifting, flanging, and tunnel effects. In fact they often sounded better than some of the earliest digital devices that were prone to producing a lot of glitches.

Flanging was named due to the slowing of tape travel by applying hand pressure to a tape reel's flange. With digital flanging the pitch variation is not only controlled by turning a knob (easier on the hands), but it could also be automatically varied using an oscillator's signal. Here the shape of the waveform controls the sweep or change. This will cause a jump (with a pulse) or gliding (with ramps) to the partial (square) or full (sine) extent of the delay range. The frequency of the control signal's waveform determines the rate of the change.

Digital Delay Sample Rate and Noise   The manufacturer Delta Lab used what they termed "delta modulation," which quantisized only the difference between the current sample and the previous sample. This meant smaller amounts of information to convert and thus better resolution at affordable pricing. Maximum delay time was limited to about one second, but at 15 kHz bandwidth these early digital effects devices handled the shorter time domain processing nicely. Since the information is fed to RAM memory it can also be manipulated to a far greater extent. The memory can be output in a looped fashion for continual repeats.

Digital delays could yield cleaner sounding delays at longer time settings. It all depends on the device's "sampling rate" or speed at which samples are taken. To avoiding the metallic ringing caused by aliasing distortion or quantizing, which produces that familiar gritty sound, a sampling rate that is twice that of the highest frequency being processed is required. By the mid-1980s companies like AMS were concurrently achieving 18 kHz bandwidth and 32 seconds of clean delay.

Using Those Longer Delay Times   When digital delays provide many seconds of delay without any added signal degradation they can be used for effects that require longer delay times such as repeating whole phrases. Here the signal is loaded into *random access memory* (RAM) and fed out again at a later point in time. Since the information is fed to RAM memory it can also be manipulated to a far greater extent. The memory can be output in a looped fashion or cyclically so that a number of repeats occur.

Pitch Changing   If the signal fed into the RAM is output at a faster speed than it was input, the pitch will be shifted upward; if slower, downward. When the upward amount of shift reaches close to an octave, the speed of the output will require a repeat of some of the memory so as not to run out of information as this would leave gaps in the output. Every time the complete memory cycles and jumps from the end to the beginning it could add a click or "glitch" to the sound.

Another popular effect when reading out of the RAM is to output the information starting at the end of the memory and proceed toward the beginning for a "backwards" sound. This is obviously easier to accomplish with sounds that are short in nature, and it works wonderfully when used to get backwards drum sounds.

Delay Processor Specifications

Frequency Response   Having a bandwidth range that extends much beyond 12 kHz is not all that important in live PA work as much of the signal content above this will end up being attenuated by the air it's moving through. On the other hand, since the function of studio monitors is to aid subjective judgment and as their location is close to the listener, attenuation by air is not a factor. In recording applications the bandwidth of every piece of equipment should extend out to 20 kHz.

Noise   The earliest electronic delay units were not much better than tape delays or acoustic tubes in terms of noise. An operator using a delay processor may be able to get away with a dynamic range of 60 dB on a heavy metal guitar, drums, and even vocals, but this amount of background noise occurring during quiet passages of program material with a wider level variance (dynamic range) is not acceptable. This problem was further exasperated by the fact that digital aliasing noise (which is caused by the use of sampling rates that are to slow to accurately process high frequencies) is harsher sounding and thus more noticeable than common amplifier distortion noise. The use of companding and limiting helped increase headroom, thus widening the dynamic range and once digital delays reached the point where they had a dynamic range of 90 dB, these problems diminished greatly.

Due to the fact that early digital electronics had a more difficult time handling overloads than their analog counterparts (instant and massive distortion as opposed to a gradual increase) their position within the audio chain can be important. In studios it was best to place them after equipment that reduces level (compressors) and before those that could increase it (equalizers). This is not always possible in PA work. For example, in situations where the same type of speaker is utilized throughout a hall, retaining intelligibility and precedence requires that a single equalizer be placed before the delay so that all the speakers will sound the same.

Meters   The need for adequate level indication is increased by a unit's susceptibility to overload distortion. Overloads are more readily indicated by peak reading meters than by VU. Therefore it was, and often still is, important to use peak meters to monitor the inputs of digital devices.

Delay Effects

Ambience Enhancement   The space around an instrument can be enhanced by panning or repositioning it in the stereo field, adding reverberation, and by adding a short delay. The delay will make it seem louder and broader, and it often adds warmth to the timbre. If the delay is short enough it will not cause phasing, flanging, or comb filtering problems when summed to mono. "Mono compatibility" is obviously a major concern to broadcasters.

Delay Before the Onset of Reverberation   When we hear a delayed signal our brain automatically combines it with the original and uses the two to place both ourselves and the sound source within a space. Since the acoustics and size of the space are perceived to a great extent by early reflections, adding a single delay to the signal becomes an important tool for augmenting the overall spaciousness of a sound. Delays are normally perceived after hearing the sound directly from the source and before the onset of the room's reverberation. Natural reverberation does not have the instant attack time of springs and plates; therefore, by adding a short delay to the console's "echo" or reverb sends causes these effects to sound more real and increases their apparent spaciousness.

Automatic Double Tracking (ADT)   It's obvious that two guitarists playing the same part will sound much more powerful than one. Since they will not play *exactly* in time the two guitars will have continually varying amounts of delay between them. This increases their ambiance. A single guitar recorded twice (double tracking) by overdubbing or adding delay to a single guitar can simulate two guitars played together, hence the term *automatic double tracking* (ADT).

The delay should be kept within 25 milliseconds unless more than one delay is used. In this case the delay (depending on the attack of the source) can be longer if another delay is placed in time, between the source and the longer delay, as this will increase temporal fusion.

Using delays that are set to at least 10 milliseconds will avoid comb filtering and if slow (under 1 Hz), *low-frequency oscillator* (LFO) modulation is added, the modulated signal will have slight pitch variations that further differentiates it from the original. This has the added benefit of decreasing flutter echoes, comb filtering, and phasing.

Chorusing and Movement   Multiple delays and varied pan placement in the stereo field, along with different amounts of modulation and feedback, can augment the perception of stereo, quad, and 5.1 or turn a duo into a chorus (see Lexicon 224X). The generation of additional signals via delay processing are not only used to surround the listener, but additional LFO modulation of delay times can cause the source to rotate from one speaker to the next, a favorite effect in the days of quad mixdown, and 5.1 mixes are ripe for this kind of effect.

Flanging   It's very hard to accurately describe the sound of flanging. There are so many types that "a jet plane taking off" or "swishing sound" descriptions are inadequate. It may help to understand a little bit about how this sound is produced. First, flanging and comb filtering are close to the same. When you combine a signal with a version of itself that has been delayed by a short amount (under 10 milliseconds), the two signals will at times conflict with each other.

Picture the original signal as a nice, slow-rolling sine wave. At some points in time the two added signals will be out of phase with each other. At these points the resulting cancellation will cause deep cuts or valleys in those rolling hills. At other points in time the two signals will be exactly in phase. At these points their levels will add together, causing sharp peaks to rise above the sine wave's rolling hills. A graphic representation of these two sine waves combined together would have an appearance somewhat like a comb (hence the name).

Delay time also affects frequency. When the delayed signal is added back out of phase with the original, those cuts occur at the fundamental, or base frequency, and at every odd and even harmonic of that fundamental ($f \times 2$, $f \times 3$, $f \times 4$, $f \times 5$, etc.).

When the delayed signal is added back with the original in phase, the cuts occur at the fundamental and only the odd harmonics (f × 3, f × 5, f × 7, etc.). Not only can in-phase feedback be used to warm up a sound, but you can use the delay setting to tune right into the fundamental and thus the harmonics you want to affect (as in "tuned filtering").

The equation for figuring out the delay time (t) for a specific fundamental frequency is

time (in milliseconds) = 1/2 × (the frequencies time base divided by 1,000)

where t = the time difference between the fixed and variable delays. For example, for a filter that nulls a 3 kHz fundamental and all its odd harmonics:

$t = 1/2 \times (3,000 \text{ divided by } 1,000)$

$t = 1/2 \times (3)$

$t = 1.5 \text{ milliseconds}$

Delaying a signal by 1.5 milliseconds and adding it back in phase with the original will cause nulling to occur at 3 kHz, 9 kHz, and 15 kHz).

For a filtering that nulls a 1 kHz fundamental and all its odd harmonics

$t = 1/2 (1 \text{ kHz}) \, t = 1/2 (1 \text{ milliseconds}) \, t = 0.5 \text{ milliseconds}$

Time-Based Modulation Vibrato can be simulated with pitch modulation set to roughly 5 Hz with a depth (level variation) of about 5 percent of the signal's peak-to-peak level. A simpler method is to set up a *voltage control oscillator* (VCO) to control the modulation of a 10-millisecond delay at a rate of 5 Hz. Changing the depth and frequency changes the effect.

Different LFO waveforms also produce different effects. Sine renders a sweep, and triangle gives an evenly increasing (upward) pitch shift followed by an evenly decreasing (downward) one. Instead of the expected pitch shift on/pitch shift off sequence, square waves actually result in an increase-in-pitch, no-change-in-pitch, decrease-in-pitch cycle that continually repeats. As always, the repeats or rates are determined by the frequency of the modulator's waveform.

Outboard, Input Signal, and Foot Pedal Control Synthesizer control voltages utilize LFOs and VCOs. "Envelope followers" follow the changes in the input signal's waveform to cause a change in the parameter settings. Changes in frequency can be used to adjust the "speed" of a variation and level changes caused by the waveform's shape can be used to adjust the "depth," "range," or "amount" of the variation. Performer's foot pedals (such as common level controls) can vary parameters in real time or "on the fly."

Pitch Shifters Remember how the signal fed into the RAM could be output at a faster speed than it was input, causing the pitch to be shifted upward and, if slower, downward? Well, we do not perceive pitch in a linear fashion. Notice as you move up in pitch on a keyboard the difference between successive notes becomes wider and wider in terms of frequency. That's how our ear/brain combination likes it. In order to keep things "harmonically correct," pitch transposers must also affect their changes exponentially with greater and greater amounts of "shift" as input frequencies increase.

Harmonics are multiples of the "fundamental" frequency. If A 440 is the fundamental, A 880 will be the second harmonic (one octave higher), 1,320 Hz would be the third (odd) harmonic (one and a half octaves up in pitch), and 1,760 Hz would equal the A two octaves up from A 440.

Taking a look at some individual delay processors gives a more rounded picture of capabilities, parameter variability, and usage. Plus when you run into one of them you'll know what to expect.

Digital Delays

The Lexicon Model 102-S Delta-T Digital Delay System   The Delta-T line of DDLs were designed by a group of computer electronics engineers in the very early 1970s. The 102-S was the second generation of the Delta-T delays and was the granddaddy that spawned just about every digital audio device that followed. It was personally responsible for the replacement of many mechanical tape and acoustic tube delays and truly did take a "system" approach to audio delay. It consisted of a rack-mountable mainframe that holds up to four delay modules, two output modules, and a single VCO module.

The input modules incorporated limiters with a "triggering" (threshold setting) that could be varied from +4 dBm to +22 dBm. Each output module's level could also be varied over a range of +4 dBm to +22 dBm. Setting these was aided by the system's "headroom" indicator. This display was made up of five LEDs that ranged from "0," the point at which limiting began, to −40 dB in 10 dB steps.

The delay time was accumulatively set by two rotary switches on each output module and was limited to 48 milliseconds per each delay module. Therefore, the maximum delay time was increased by adding delay modules. A four-delay module mono system had a maximum delay time of 192 milliseconds, while the delay time maxed out at 96 milliseconds with stereo operation. However, you could (although I never witnessed it) cascade two fully loaded (192 milliseconds) 102-S systems for a total mono delay of 384 milliseconds.

The system also had options for remote metering, delay time setting, and an external VCO input. But its own VCO module was a whole new kick for most operators. Besides the set (crystal oscillator) control of timing, it allowed for manually changing the system's clock frequency that triggered the pitch shifts. This pitch shifting could also be continually varied by an internal function generator for repeated (clocked) frequency, or pitch modulation, according to the sine, square, or triangle waveform chosen. This may not seem exceptional by today's standards, but we're going back to the early 1970s here! This was wild stuff.

Specifications:

Dynamic range: 95 dB

THD: Less than 0.2 percent

Frequency response: Plus or minus 2 dB from 29 Hz to 15 kHz

Channel separation: Better than 75 dB

By the mid-1980s delay processors were everywhere. This was due to the fact that musicians working in the studio had the opportunity to record using all kinds of mega-buck processing. Their problem was matching these sounds when playing live. The recording engineer could have used less processing, but then the client may not have received the best possible sound. Many of the effects devices used in studios were simply not suitable (roadworthy) for live applications. Many effects units that were designed specifically for live use did not match the sound of the best studio processors. But players wanted the sound they heard on records, so manufacturers responded to this demand.

Many guitarists and certainly bass players may not see the significance of a delay unit that has a frequency response of up to 20 kHz. But my opinion is, why not purchase a unit that you wouldn't be ashamed to bring into a top recording studio?

Pitch Changing, Harmonizing, and Frequency Shifters

Eventide Harmonizers   You can't discuss pitch-shifting devices without mentioning Eventide. Their model H 910 was the grandfather of all "harmonizers;" in fact they coined the phrase. This unit was fairly limited by today's standards. It had one input with two outputs. The pitch could be shifted up or down by as much as a full octave and a delay in the neighborhood of 80 milliseconds could be added to this output. The unharmonized output could also be delayed up to 112.5 milliseconds. If this wasn't a complete enough pitch change effects package, Eventide also made a keyboard controller available, the Model HK 940. Instead of dialing in the amount of pitch variation you could simply press the appropriate key for a third up, a major seventh down, or whatever. This combination's reaction time was quick enough to correct single misplayed notes during mixdown. (I've done it.) Its main function was to set up musically correct intervals and it allowed this setting to be changed right at the point of a key change with a simple keystroke. They actually made a second polyphonic (three-note) keyboard that when connected to three harmonizers could produce chords, but I never had the opportunity to try it.

I can remember sitting in a 24-track control room (back around 1975), playing with the H 910. Picture this: I was sitting at an MCI 500 series console playing sax into a Sony C-37 tube microphone with the signal patched through the harmonizer, which was set up right there within easy reach. I had the speakers muted and was listening through cans. My plan was big: I was going to use three harmonizers and hire myself out as a one-man horn section. The first problem was the cost, then mobility (dragging three 910s from studio to studio would prove difficult), but the third strike was the dreaded "saxophone glitch." Instead of being all disappointed and weepy, I got right into the study of this phenomenon. The clicks seemed to follow the up and down extremities of the reed's movement. They also produced two tones.

Okay, record the dry sax and the effect on separate channels. Mix some of the dry sax (wired out of phase) with the effect's output with no time delay added. Adjust the levels until you're left with *just* the glitch sounds. Run this signal through a gate, an equalizer, a compressor, and then a reverb. I swear I made those glitches sound like bongos. Once this was done, no matter what I played, those bongos were right there with me as if I were playing with the most intuitive percussionist of all time. When the rest of the "crew" at that studio found out what I was up to, they just shook their heads as if to give up on this poor, misguided youth.

Eventide also made straight delay lines and flangers, in fact their 1980s model H949 was a combination effects device that harmonized, delayed, repeated, flanged, chorused, and more. When the 1990s rolled around Eventide turned to software-based processing with their

SP2016. Descriptions like "powerful," "extremely flexible," and "unlimited capabilities" were pretty accurate as Eventide hadn't turned their back on any of the old processing capabilities while they continued to improve sound quality and user interface.

Along with other improvements added to the H3000 "Ultra-Harmonizer," including reverberation, flanging, phasing, echo, delay, filters, and selective frequency delaying, all the rest of Eventide's growing list of effects-processing parameters can be loaded in from a software library. It seemed that they had reached the maximum of time-processing development. Fat chance, because the later 2004 version, the Eventide "Eclipse," can perform all the same effects, but now we're talking about a dual unit whose routing can be configured for stereo, dual mono, *and* series or parallel usage. Other improvements include the ability to use a MIDI foot switch to "tap in" a tempo and use it for setting delay times that are more musical and precise at the same time! The reverb sounds are richer and fuller than ever. Program change is now just about instantaneous. User interface? How about an onboard software "search engine" designed to guide the user through Eventide's vast library of effects programs by either general categories or individual applications and then download these "patches" (or programs) from their web site for free?

Spearheaded by Eventide, the use of time correction with pitch consistency became a boon to advertising profits. *Time compensation* changes the speed of the output without causing a change in the pitch. It increases the speed of the output, which makes its length in time shorter and its pitch higher. It is then passed through pitch-shifting circuitry to drop it back down to its normal tone.

A 30-second spot that has an over-run in time of 6 seconds would be output one-sixth of the time faster and then dropped one-sixth lower in tone for a perfect 30-second spot. It didn't take long for advertisers to realize that this could be used to increase an ad's word message amount without any additional air time costs. Lots were sold.

I felt I had to go through this short history of Eventide since they are unique in their longevity. I can't think of too many other manufacturers of recording studio processors that have survived intact from the early 1970s. This longevity is due to quality, not marketing strategy.

The MXR Pitch Transposer and Optional Display   All connections to this unit, with the exception of the display, are ¼-inch phone plugs. The unit provides two sets of inputs and outputs, these being used for different levels of input. The front panel set is intended for easy patching of music instruments, while the rear panel connections are utilized for hotter line-level outputs more common to a mixer's sends. Additionally, the unit provides an auxiliary in and out patch for use with an external signal processor, such as an EQ or a delay line. To aid this interface, the front panel provides an aux-level switch for choosing between line or instrument levels. There's a remote select jack for use with an optional foot switch as well as a jack-designated "external-preset one," which I'll discuss later.

MXR designed this unit to be very easy to operate. Yet it gives you more control over pitch change selection than many other units. They utilize the common rotating knob form of upward or downward pitch shift control, as opposed to keyboard or fader control. Not only did they allow for easy setup, but they also provided four front panel controls. Let's check this out for a minute, as this amount of variable pitch selection along with switching is impressive. The pitch shifting range is anywhere from one octave below to one octave above the incoming signal.

The amount of pitch shift is controlled by DC voltage, with 0 volts equaling an octave down, 5 volts rendering no pitch shift, and 10 volts providing a signal that is a full octave above the original. The "external-preset one" jack at the rear panel gives a DC output that, via this stereo ¼-inch jack connection to a common "volume" pedal, yields external foot control over the amount of "preset one's" pitch shift. Another "external voltage control" jack connection was designed for use with an outboard controller such as a synthesizer. With this, you can use a keyboard to designate the amount of pitch shift, rendering a more logical and understandable setup. The four front panel preset pots, as well as the bypass switch, provide capacitive touch sensitivity. This means that you can set up each for different amounts of pitch transposition and change from one setting to another by simply touching the corresponding knob, which is all it takes to cause the capacitive switching. Each control has an LED above it that indicates which setting is being utilized. How fast is the switching? I once nailed a 32nd note flaw with an MXR Pitch Transposer.

If two or more knobs are touched simultaneously, the one with the highest priority (starting with preset one) and descending down through the other three (with bypass having the least priority?) will be activated. I guess a choice had to be made and that one's as logical as any. The front panel also incorporates the aforementioned level switch, a peak-reading LED, a power on/off switch, a mix control for determining the blend of the effect, and the original signal output, and a regeneration or foldback control, which was designed so that even when set at maximum there would be no signal runaway or feedback.

I don't want to dwell on the theory of this unit's operation, but some description of the internal workings is needed in order to understand its capabilities and limitations. The incoming signal first goes through an amplifier stage. This acts as both a buffer between the source and the rest of the circuitry and as an aid in setting proper levels (the input level can range from −10 to +12 dBm). A low pass filter is utilized to avoid aliasing distortion, which can occur when the incoming signal is comprised of frequencies that are twice the sampling rate. This obviously limits the high-end response of the effect.

Next the signal is converted to digital data, stored in RAM at a set rate, and then retrieved at a rate and amount corresponding to the desired pitch change. By "amount" it is meant that when processing at a faster rate (upward pitch shifts) the same amount of time cannot be filled. In order to retain the correct time signature, or amount, this device adds in or duplicates some of the information stored in memory.

This is a common procedure, but MXR's circuitry also provides a smooth fade in and out between these additions or subtractions. This yields a more natural sound. The signal is low pass filtered before the output to reduce digital noise.

As far as the actual production of harmonies is concerned, the setup is straightforward. For a harmony one major third up (which corresponds to two whole steps or four half steps) you play a "D," for instance, and in this case adjust the pitch change to give you an F#. Here, the bypass switch can be utilized to compare the F# played in bypass to the harmonizer's F# signal when not in bypass. Simply adjust the setting until the two are identical.

The optional display makes this process even simpler as it gives a numerical readout of the ratio between the input and output (0.50 to 2.00). By hitting the "mode" switch, a more music-oriented display in the form of steps (−12 to +12) is given. To accomplish this, the unit incorporates a microprocessor that continuously calculates the required logarithms. The display also indicates any malfunctioning and needs only to be plugged into the main unit (no AC plug) so the two can be permanently attached together.

So, what can this device do? Well, the actual use of pitch transposition is pretty much up to your own imagination. However, I'll discuss a couple of standard effects to illustrate its capabilities. Obviously, these units are used for producing harmonies, and because there is an inherent short delay in the output, the result is like a second instrument or singer following the original perfectly. These devices can also be used to correct the pitch of an instrument, which for some reason or other is not tunable.

Another common use is for commercial spots where the program material must be squeezed into a shorter time period. Here, the recorded spot is played back at a faster speed via vary-speed and the resulting higher pitch corrected by transposing it downward in pitch.

By regenerating the signal (feeding back the output to the input), a type of "barber pole" flanging (the pitch cycling upward or downward) can be had. By using external devices such as equalizers, flangers, phase shifters, delay lines, and so on, the effects become almost limitless. I found this device to be not only easy to use, but a lot of fun, too.

The use of four touch-sensitive pitch selectors was a great idea, and to my surprise, the changing from one setting to another was not only noiseless but yielded no transition sounds. I expected at least a bit of a sweeping pitch sound, but that wasn't the case at all! In fact, because of this and the instantaneous speed of the switching, you can easily change the pitch of a single note in a relatively swiftly played passage.

This fact alone makes this unit very useful. As far as specifications go, the unit was very much noise free and yielded little or no distortion when input was set up properly. The frequency response is limited to under 11 kHz, due to the previously mentioned filtering, but I found this to be no problem as I never really needed that much high end. You see, pitch transposition for any of the above-mentioned effects does not suffer from a limited frequency range to a great degree. All in all, the MXR Pitch Transposer and Optional Display are top shelf, easily suitable for the most demanding studio and road use. The unit is sturdily constructed and carries a one-year warranty.

It worked as well on vocals as it did on instruments, and the ability to use a foot pedal and keyboard in addition to the presets and bypass controls, increases its versatility.

My own personal acid test of pitch shifters is their use with reed instruments. Here, the unit performed surprisingly well. There were none of the glitches normally caused when the reed reaches its extent of movement! The sound I set up was an alto, and when listened to separately, it could easily be judged as imperfect sounding, but once mixed in with other horns (as it normally would be), the sound was acceptable to most listeners, which is all you can ask from a pitch shifter, transposer, and harmonizer.

MXR Pitch Transposer Manufacturer's Specifications

Maximum signal level: Dry: +22 dB (referenced to 0.775 volts, with no load)

Effect: + 16 dB (referenced to 0.775 volts, with no load)

Dynamic range: More than 80 dB

THD less than 0.25 percent @ 1 kHz

Frequency response dry: +0, −3 dB, 30 Hz to 40 kHz

Effect: +0, −3 dB, 35 Hz to 11 kHz

Publison   I have to mention Publison simply because their DHM 89 B2 stereo digital delay pitch shifter was so much fun to use. This was a dual-delay, dual harmonizer that produced smooth, glitch-free sounds. The two processors could be used separately or on a single input. Pitch change ranged from two octaves down to one octave up. Frequency response per delay length was fair; 20 kHz at a quarter second, 10 kHz at a half, and only 5 kHz at 1 second. But it had *all* the bells and whistles, including dual feedback arpeggiation and echoes that could be played back in reverse from its RAM memory. When combined with their Model KB2000 keyboard this processing became very musical.

Not only did the keys cover three octaves and control the amount of pitch shift, but it had dual envelope followers, could generate vibrato, and "play" via keystroke any sound in its RAM at any pitch within the three-octave range. At trade shows you'd be handed a mic so that you could sing along with yourselves while using the keyboard to change intervals. This often ended in your thanking the demonstrator only to hear yourself, as you walked away, repeating "thank yous" that ranged from a lower than possible bass voicing up to a Minnie Mouse pitch to the delight of everyone around. But let's get back to delays.

The Lexicon PCM-42 Digital Delay   Beginning a review with an application was generally not my style. I'd usually go over the control functions and rear panel interface patches, discuss a few common applications, and finish with the more technical end in dealing with the device's specifications. However, the PCM-42 wasn't your usual device. I've seen units designed with the musician in mind, for ease of use onstage, ones that are fun and exciting to use, and others that are truly tool-like, providing many functional uses. Lexicon's PCM-42 is a combination of all of these and more.

Before I began my review of this device I skimmed through the manual and then placed the unit in a studio. The following two weeks I received reports like, "As good as the PCM-41 (its predecessor), but better" or "The best modulation device I've yet to use" and the customary "It's much better than . . . ."

I wasn't booked for a session in that studio until three weeks after I had dropped off the PCM-42. The date was a simple rough mix for the client's promoter, grandmother, or who knows. Having not been the recording engineer I brought up the tracks that had been laid down (here basics, a couple of overdubs and a scratch vocal) and heard a god-awful ring on one of the toms. I knew equalization wouldn't help much, but started to run the poor thing through a parametric anyway. Then suddenly it clicked! I remembered something that I saw in the PCM-42's manual about varying the pitch of resonant rings (filter tuning). With short delay times and a lot of feedback, delay units are able to duplicate the kind of metallic ringing I was hearing from that tom. By varying the length of delay time, you can change the pitch of the ring.

My idea was this: run the tom through the PCM-42, and adjust the time delay and feedback to get the same ringing at double the volume. Then flip the tape over and run it backwards through the unit to an open track. Rewind, and run it through again with no feedback, just straight delay. Flip the tape and bring these two new tom tracks up to the board out of phase with each other, thereby canceling out everything but the ring. Run the ring through the unit at the same delay setting, reverse its phase, and mix it with the original tom. This, in theory, should eliminate the ring. The process took about 15 to 20 minutes to do, but it worked. Not perfectly, but enough so that I bounced the product to another track. With a little more time spent resetting the feedback level, I think I could have had the ring eliminated completely.

This is the kind of crazy stuff that the PCM-42 tickled your imagination into trying. The number of new doors that it opened up was amazing. Don't get me wrong. It could produce all the standard effects expected of a time delay/modulation processor very easily. You'll get ADT, flanging, pseudo-reverb, vibrato, slap-back delay, echo, very long echo (up to a half-mile away), multiple echoes, and even standing waves.

Now consider that you're dealing with delay settings up to 4800 msec (almost 5 seconds). It accepts any input from −12 dBV to +20 dBV, balanced or unbalanced. It has metering for setting up the proper level, but messing up this setting is difficult as this unit has built-in limiting that prevents overload.

I reviewed a unit with the extended memory option so I was able to switch via the "X 2" button from 2400 milliseconds at 16 kHz bandwidth to 4800 milliseconds at 6 kHz, helpful in checking the effect of doubling the delay. The feedback can be in or out of phase as well as low pass filtered. The delay itself can also have its phase inverted and be mixed with the direct signal to any proportion. There's infinite repeat that utilizes the full memory but can be shortened by the × 2 button and the manual VCO sweep setting. The modulation section's VCO sweep has depth (up to 3 to 1), waveform (sine to envelope and envelope to square), and rate controls.

These interact well, giving a great deal of variation and control. Outside of the × 2 button, delay is set by the manual VCO sweep with a range of × 0.5 to × 1.5 (the amount of delay being shown on a four-digit numerical LED readout) and set by two (up and down) buttons that step through setting increments when pushed or held. By hitting the second button while the other is held, the unit zips through the delay times, spanning the full range within a few seconds.

With this device, output jacks are plentiful. These include a direct, a delay, and a main (mix) output that has a nearby pot for adjusting its level from 0 to 9 volts RMS. That's not all. You get connections for five, that's right, five remote controls. Infinite repeat can be foot-switch controlled, as can bypass. Modulation can be controlled by using the VCO control jack with either a potentiometer foot pedal or a control voltage from a synthesizer or any other audio generator. *Any* control voltage waveform patched into this point can replace the sine, square, or envelope modulation waveforms controlled at the front panel. The mix control jack has three functions according to the schematics I received with the unit. It'll fade the delayed signal in and out of the mix, and it can also be used to inject another signal into the mix output as well be used as a patch point to allow the delayed signal to enter and return through another signal processor. The feedback control jack sets up a remote control of the amount of feedback. It also provides an additional insertion point for another signal into the mix and provides patch points for running only the feedback through an external signal processor. This amount of versatility can become mind boggling!

There's also a clock output jack for synching up external devices to the delay period. Devices such as rhythm machines, sequencers, or synthesizers will receive an audible click. It can be used as a metronome for synching up humans, too. What's a time delay/modulation processor doing sending synch signals out to the world?

Remember, we're dealing with a device that has a very extensive memory, close to 5 seconds. This means that at the maximum delay setting you could lay down over 4 seconds of rhythm and have it repeated as soon as you're done. Not impressed? Then turn up the feedback control and your rhythm will take over 3 additional minutes to die out. Still not enough? Hit infinite repeat and come back at some future date, and barring a power failure, it will still be running! What you have is a tape loop whose sound doesn't become degraded! You can add onto this bed of sound as it fades away (in feedback mode) or play over it during infinite repeat. You can even come out of infinite repeat, add another part, go back to infinite repeat, and both parts will be played!

It's never been easy to play along with a loop as its length and tempo can be difficult to match with. With long loops, this can become almost impossible. That's where the metronome comes in. It divides the memory up into exact time periods. The number of time periods in musical or nonmusical notation is selectable at the front panel and shows on the readout. Now you not only have a metronome *and* a flashing LED to cue you, but a clock output to other devices, keeping everything in time. I have tried this type of overdubbing with the PCM-42. All I can say about playing along with myselves is that we all sounded just fine.

As you can see from the specifications, the Lexicon PCM-42 is easily suitable for any studio use. This is one of my all-time favorite delay processors and I'm not alone. Whenever you *can* find a PCM-42 for sale, the price tag is always well over a thousand dollars. It's a huge favorite in video production facilities where its accuracy, setability, and readout are well appreciated. I just can't understand why some manufacturer hasn't stepped in and offered an identical knock-off replacement for this long-abandoned, consummate time processor. "Go figure!"

Lexicon PCM-42 Specifications

1. THD = NOISE; 0.1 percent max @ 1 kHz, = 9 dBu input (0.06 percent typical)
2. Frequency response: × 1 setting; +1 dB, −3 dB, 20 Hz to 18 kHz (referenced to 1 kHz) × 2 setting; +1 dB, −2 dB, 20 Hz to 18 kHz (referenced to 1 kHz)
3. Dynamic range: More than 90 dB, 20 Hz to 20 kHz bandwidth
4. Delay time with extended memory: × 0.5 = 800 milliseconds, × 1.5 = 2400 milliseconds (16 kHz) × 0.5 = 1600 milliseconds, × 1.5 = 4800 milliseconds (6 kHz)
5. Display accuracy: at 100 milliseconds = plus or minus 1 millisecond (above 100 milliseconds = 1 percent)
6. Input level capacity: −12 to + 20 dBu (peak)

The Klark-Teknik DN700   This delay followed in the success of the (three rack space-sized) DN70, which fed one input into three separately set delays. Each delay time was variable from bypass to 652 milliseconds. It had a single regeneration or feedback control for all three inputs and a switch fed any one of the three delays to the unit's digital readout. The noise was low and the bandwidth ran out to 15 kHz. It's perfect for improving intelligibility and sound localization in public address (live) systems.

The DN700 had a delay range of only 434 milliseconds but takes up only a single rack space and was priced lower than the DN70, meaning that a lot of DN700s were sold. So what's this PA delay got to do with recording? One, they are easy to come by in the used equipment market; two, they are quality delay lines; and three, they are not yet popular with the masses of buyers now involved in "pro" audio so they can still be purchased at a decent price point. Want four reasons? How about it's just about the best thing you can have for multiplying beats per measure, filling out percussive tracks, and beefing up background vocals by broadening their spread across the stereo field.

Today Klark-Teknik's DN7103 provides each of the three outputs with up to 1.4 seconds of delay. It's cleaner sounding, has 18-bit AD/DA conversion, and provides 20 Hz to 20 kHz bandwidth and an increase in dynamic range of 10 dB. This device includes a peak limiter (handy when one of the delays is feeding a subsequent processor that's very sensitive to overloading), and both dual parametric and a single-shelf EQ for *each* output. It has MIDI capability and there's 32 user-definable memory banks. The metering shows headroom level. If 1.4 seconds is not a long enough delay, the DN7204 has a dual input, four output version with 2.7 seconds of delay for each output.

The Lexicon Model 97 Super Prime Time   The Super Prime Time must be one of the most versatile time processors I've ever used. A +20 dB gain switch makes it possible to plug a music instrument directly into the Model 97 as well as to use the more common method of feeding the input with line-level signals from a preamp output. Furthermore, two different instruments can be mixed together using the Model 97's two separate inputs and volume controls. The Super Prime Time easily connects to PA mixer sends and returns to allow for the processing of multiple signals.

Interface was no problem as both XLR and ¼-inch phone connections were provided for the inputs and mix output. In addition the Model 97 provided access via a ¼-inch phone jack to a mix of only the inputs and it had separate outputs for both the A and B delays. All this, and the sounds were studio quality.

Basically, this device was a digital-delay time processor consisting of two individually controlled delays. As a result, this unit renders a very wide range of effects processing. When it comes to flanging, or variable phase cancellations and reinforcements (comb filtering), it ranks among the best. By adding in a little foldback, you'll get deep, resonant flanges. Phase inversion changes the harmonics, and adding modulation results in what's termed "talking flange." On most delay lines, you can double (ADT), but due to its having dual delays, the Model 97 also tripled. Medium delays allow for perfect syncopation when adding rhythmic effects: vibrato, pitch twisting, and even those out-to-lunch metallic ringing sounds are within the 97's power. The long delays possible with the Super Prime Time allow you (using foldback) to have your parts repeat so that you can play along with yourself, adding harmonies or counterparts. Its delay range can render anything from a short slap-back delay to very long echoes. Increase the foldback, in order to create pseudo-reverberation, was augmented by the built-in filtering, which also helps control the oscillation when modulation is used (to add realism).

Its delay time is long enough so that, when using the infinite repeat function, you could play along with yourself, adding parts, almost as if overdubbing with a studio multitrack, but in real time. The Model 97's delay tuning adjustments are precise enough to achieve chromatically tuned resonances with varying pitch from a single note, a drumbeat, hand clap, or a vocal. Because of the dual processing and all the added controls and features of the Super Prime Time, this outline of applications could go on forever.

You may be thinking, "It may be easy for an engineer accustomed to dealing with complex equipment, but I'll never get those sounds." But Lexicon designed the Super Prime Time to be user friendly. Even first-time users have little trouble instructing the computer memory to gather, store, and recall programs. It's easy since this unit comes with eight permanent factory presets and 32 blank memories just waiting to be filled. The fixed programs contain common effects like flanging, resonant flanging, doubling, tripling, chorusing, slap-back delay, and short and long echoes. These provide foundations for building more elaborate programs.

The front panel is divided into two sections—audio processor and computer controller. It's two operating modes are manual and memory. In manual you program your sounds, and then store all the settings, which allows you to later recall this effects "patch" at any time.

The microprocessor resets all the control settings. Although the knobs and faders don't physically move, the parameters they control are internally adjusted. But here's the real kicker: If you want to tailor one of the factory presets to a specific need. No problem. After recalling a memory, all the controls remain adjustable. Their positions may not correspond to the internal settings, so they will have no effect until they are moved to the preset position. At this point the display blinks. You can now adjust that parameter as desired.

The best way to learn about this device is through experimentation. Should you go overboard in exploring the various settings and their corresponding sounds, simply recall the program and start again. An effect (when tailored to your satisfaction) can then be stored in another memory register. This "editing" process is one of the most useful and flexible aspects of computer-controlled processing. The internal battery power supply maintains the memories when you power down. Once all 32 blanks are filled, you can load these to tape in groups of eight or all at once. The 97 even tells you whether or not the unload process has been completed properly. You can also reload programs, rearranging them into sets to comprise a line-up of effects that will be used sequentially during a single song or for over a complete live set.

To truly understand this unit's ease of operation it is necessary to run down some of the controls. The main input control is the only front panel control not memorized by the computer. It's not practical to commit to a specific input level as gain must be set on a per-input basis. This input control is easily set to the correct (undistorted) level with the help of seven peak-reading LEDs. Next to this is the second input (aux) level control, followed by two fold-back level controls (one for each delay). Wrapping up the input section is a filter for adjusting the frequency range of the foldback loops. Output is equally straightforward, with dry (unaffected) output level controls for both main and auxiliary, individual A and B delay output controls, and a main overall output control that has an LED overload indicator above it. Simple, yet flexible. Speaking of flexibility, the two delay inputs, two delay outputs, and the main out can each be individually switched to be out of phase to achieve harsher or warmer sounds.

The modulation section is easy to understand and operate even though its parameters are highly variable. The manual sweep control sets the two delay times. It is variable between ×.5 and ×1.5 times and works in conjunction with the depth control to adjust the amount of automatic sweep (via low-frequency oscillation). How this sweep modulation changes the delay setting depends on the volume changes of the LFO's waveform. If the "shape" pot is tuned to envelope, the change will follow the volume changes of the source signals. When tuned to LFO, it will follow either the built-in sine or square wave generator (as chosen by the "sine/square" switch). The LFO waveform's shape can be infinitely varied between envelope and either sine or square waves. A rate control adjusts the waveform's frequency and thus the speed of the modulation. It's infinitely variable between 500 times per second (500 Hz) and once every 20 seconds (or 0.05 Hz). Now that's *low* frequency oscillation!

An extremely stable crystal oscillator can set the delay time, defeating all modulation. The *voltage gain attenuator* (VGA) is a very important built-in control as it adjusts the levels of the two foldback loops, providing this device with automatic "dynamic recirculation." When you're using repeats while playing complex passages, things can get a little cluttered and confusing.

The VGA automatically reduced the foldback as long as there is new information at the input. Once playing stops, the unit is automatically switched back to the longer foldback setting. The result is a lingering tail-end decay. This is one of my favorite effects to lay on a slide guitar.

There's more! By holding the VGA while hitting the register #1 button, you cause the programs selected to cross-fade rather than instantaneously jump to the new setting. The infinite-repeat feature lets you capture all the information available in the delay's memory and continually repeat it (without sound degradation). Here it's possible to play over previous lines. The bypass switch overrides all processing at any time.

Setting delay time is simple. Each delay has a rotary pot that runs through the available amount of delay in steps. At shorter settings the steps are closer, and at longer, they're wider. Each delay has its own accurate floating decimal point LED readout showing its time setting in milliseconds or seconds. Outside of the longest and shortest delay settings, the two delay settings are internally given slightly different step points in order to increase the fullness of the sound, but a manual sweep control allows you to set the delay time in between steps. What makes this studio device flexible enough for live usage is its ability to be outboard controlled in real time. By this I don't just mean the common bypass and infinite repeat foot switch jacks, the Super Prime Time has these, but the ability to control the delay settings, the amount of LFO modulation, and the rate of modulation. All of which can be set with foot pedals controllers. The rate and sweep are adjusted via voltage scaling. This also enables sequencers and synthesizers to control their settings. As mentioned the programmed memories can be changed sequentially via a foot switch.

All this control applies to studio use as well, not only in terms of foot control, but also automation during mixdown. By recording the tape-out cue tones on a spare track of the multitrack, you can patch this track back into the unit and cause any changes you've made in programs to be repeated automatically. You can practice, check, and punch in your various effects changes until they're perfect and then store those moves forever. The 97 will link up with another 97 for four delay at once time-processing or with any other Lexicon processor for added flexibility and range of effects. Everything I've pointed out so far would be meaningless if the Super Prime Time were not a quality product.

The warranty lasts for one year. However, due to its built-in diagnostic software, ease of access to, and in replacing parts (due to socketed ribbon cable connections between PC boards as well as just about every IC being socketed, a hinged upper PC board and very clean workmanship), any required servicing should be relatively easy.

Now down to the brass tacks of the Lexicon Model 97's sound quality. It's full bandwidth and low-distortion specs are what make the difference. As the frequency response graphs in Figure 2-15 show, this unit's bandwidth is impressive.

I do not expect everyone to get involved in this kind of bench testing, but a little time reading the manual will enrich your knowledge of a device's processing capabilities. The Lexicon's Model 97 Super Prime Time was an extremely versatile time-processing tool.

Reverberation Explained

Critical distance is the distance from the sound source where the direct sound energy is equal to the reverberant sound energy. Unless we are very close to the sound source, much of the level we hear can be due to reverberation. As important as the study of reverberation is, this field is fairly new. It only goes back to the beginning of the twentieth century to the "father" of modern architectural acoustics Wallace Clement Sabin, who came up with a formula to determine reverb time. He found that reverberation was dependent on the room's size and the absorption of its surfaces. He proved his point by having students haul cushions in and out of a lecture hall until its horrible acoustics were deemed acceptable. His formula was reverb time equals the room's total volume in cubic feet first divided by 20, and then divided by the product of the total surface area of all the walls, floor, and ceiling multiplied by the "average absorption coefficient of their surfaces."

$$0.5 \times VT = \text{reverberation time}$$

$$T = \frac{0.5xx}{S \times a} \quad V = \text{the volume of the room in cubic feet}$$

$$S \times aS = \text{the surface area of the room in square feet}$$

$$a = \text{the average absorption coefficient of the surfaces}$$

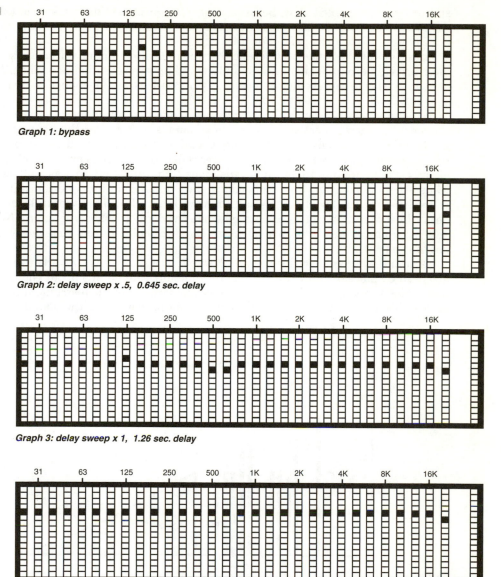

Figure 2-15
Lexicon Super Prime Time $1/3$-octave frequency response charts: direct signal, bypass the delay sweep range × 0.5, and 1× the delay with a sweep setting of 1.5

Graph 1: bypass

Graph 2: delay sweep x .5, 0.645 sec. delay

Graph 3: delay sweep x 1, 1.26 sec. delay

Graph 4: delay sweep x 1.5, 1.92 sec. delay

This coefficient is now called a "Sabin" and a rating of 1 Sabin equals one square foot of total absorption, or according to Sabin's way of thinking, an open window (no reflections). His equation makes one thing very clear: Reverberation time increases with added room size and decreases with added room absorption.

If the reverb time is too long, new sounds will be blurred by earlier ones that linger on still reverberating from the walls, the floor, and the ceiling of the room. At this point any additional speech becomes garbled. Reverberation time or RT-60 is the time it takes for the RMS level in the room to decrease 60 dB from its natural steady state level. This specification should be referenced to some level as per bandwidth.

Once you comprehend the difference between the direct sound, initial early delays, and the later arriving group delay clusters, understanding reverberation is a cinch. It's so straight ahead it's easy to represent graphically. Figure 2-16 shows the change in reverberation level over time.

Without any interference from amplification, the first sound heard will come directly from the source. "Early reflections" are the next sounds to arrive at the listener. These are normally single reflections from the stage boundaries (the side walls, ceiling, floor, and back wall) around the source instrument. "Later reflections" start to arrive as clusters of delayed signals. These are due to multiple reflections from the stage area as well as from the side walls. At some point all these reflections start to build up into one continuous sound with reflections from as far off as the hall's rear wall added in. This overall level then begins to decay. Reverberation time is calculated when the decay reaches a set point. Normally, this is −60 dB from the steady state level, but it can be calculated from time decay level decreases as short as −20 dB.

All of these time periods (early and later reflections) as well as reverberation and its decay, are used by our ear/brain system to determine intricate details about the acoustic environment. The direct sound tells us the distance and position of the instrument in relation to ourselves. While this gives us a good deal of information about the sound source, the early

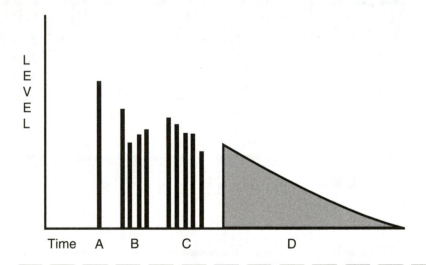

Figure 2-16
This level vs. time graph establishes (A) the arrival of a direct sound. The first cluster (B) of reflections then arrives from boundaries close to the source. The next cluster (C) of reflections arrives from more distant boundaries. As the reflections build they are no longer perceived individually, but become a wash of reverb (D) that fades over time. If the source produces a constant sound (such as when a signal generator is used to feed a signal into a reverberation test chamber), the reflections can build at a steady-state level that exceeds that of the direct signal's level.

reflections further augment this perception. An acoustic guitar, like most music instruments, radiates sound in many directions. Therefore, we can judge its timbre more thoroughly when we hear its full depth as reflected from the boundaries that closely surround it. These reflections also tell us about the size of the stage area (delay time) and the acoustical properties (frequency response) of its surface coverings (as in wood vs. draped fabric). Later reflections start to come to us from the sides, thus rendering details about the width and size of the acoustic space (delay time). Again this information includes clues to surface hardness (frequency content) and irregularities (the number of reflections). Finally the reverberation's decay as per time and frequency content fills us in on the overall size, shape, and surface makeup of the hall or any other type of space.

Okay, say I throw a speaker into a room and put a constant signal through it (i.e., A = 440). The output would spread out and fill the room. The sound would bounce off all the surfaces of the room (chairs, walls, floors, ceiling areas, lamps, books, etc.) not once, but hundreds of times. The actual amount of signal that bounces off would be less than that amount that hit the surface, but because the original signal is still coming out of the speaker, the reflected signal would add to it, causing an increase in level. At some point, this rise in the SPL, or intensity of the sound, would reach a peak where it would level off and remain constant. We have reached the point where the full body of air within that enclosure has been set into motion. If we now disconnect the speaker, the sound will not immediately end but will fade away over a period of time. Common sense, right? It's the measurement of this time, called the reverberation period, which deals with the size of the room, the room's surface area and the absorption coefficients of those surfaces The logarithms, and time factoring, can get fairly complicated. For our purposes, it's enough to say that reverberation is the continuation of sound within an enclosure after the original sound has ceased. Echoes, on the other hand, are distinct repeats of a single sound, like yelling into a canyon. But a series of echoes, or repeats, that decrease in intensity and are spaced so tightly together that they are heard as one continuous decaying or fading sound can sound like reverberation.

However, because an echo *is* a repeat, in order for it to sound like reverb, the original signal must have little or no attack, in the nature of a legato string line or melodic back up vocals. This is due to the fact that echoes copy the attack of the original signal, while reverberation is a grouping of delayed sound images that are smeared together with no attack sound being discreetly discernible. So while echoes are not the same as reverberation, in some circumstances multiple echoes can sound like reverberation.

Echoes are not perceived as repeats unless they occur 50 milliseconds or longer after the original. Once again this time value is dependent on the attack of the sound source. With a drumbeat, delays as short as 30 milliseconds can be perceived as a separate echo, while the sound of an organ can be delayed by 80 milliseconds or more before a separate echo is discernable. Reverberation can begin at any time and it will still be heard as reverb. A millisecond is one-thousandth of a second (not $1/10$, or $1/100$, or $1/1,000,000$); 50 milliseconds is $1/20$ of a second; 200 milliseconds is $1/5$ of a second; and 500 milliseconds is one-half second.

Okay, to recap: Reverberation is a sustain-like effect made up of many, many delays smeared together to sound like one elongated sound fading in level, with no discernable attacks and no limitations as to when it begins. Echo is a repeated copy of the original sound. The delay time of an echo is important, as when it is less than $1/20$ of a second (50 milliseconds) the echo may not be perceived as a distinct repeat. When the original signal has no distinct attack, and the echoes themselves are close enough to each other in time (termed "flutter echo"), it can end up sounding like reverb. Now, let's look at some ways of achieving these sounds. (See Figures 2-17 through 2-23.)

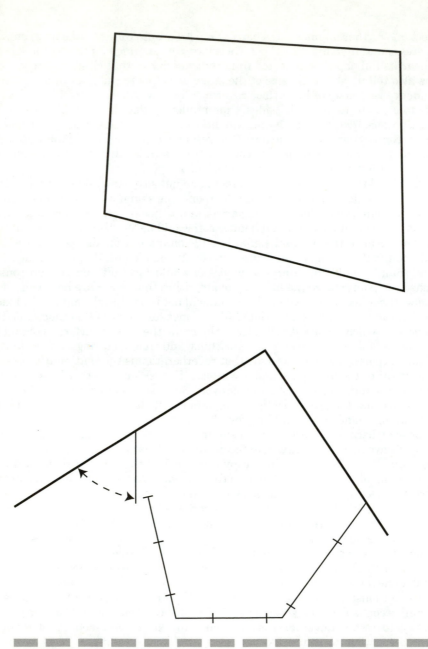

Figure 2-18
An old floor plan of a combination drum booth/reverb chamber that was once built into the corner of a studio. It was used to isolate the drums during basic sessions without having to relegate them to a "dead" acoustic space. During mixdown, it was used as a reverb chamber. It had about a 1.5-second decay time.

Figure 2-19
This is a rough sketch of the finished drum booth showing the fieldstone walls as well as the windows and doors. The ceiling was pitched with its highest point being at the back wall.

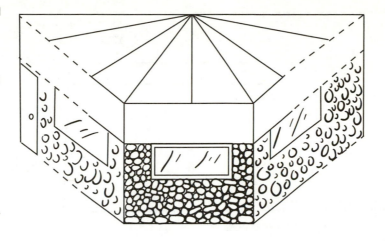

Figure 2-20
A microphone and speaker both placed to face the opposite corner of the reverberant room. Here the mic is behind a speaker that (obviously) has no rear port.

Figure 2-21
This sketch shows how the mic and speaker are positioned in opposite corners. This would provide a longer reverberation time than Figure 2-20.

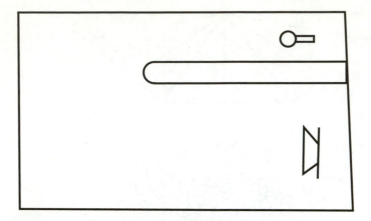

Figure 2-22
This drawing presents a reverb chamber setup that utilizes a partition to cut off some of the direct sound before it reaches the mic. Stairwells not only duplicate this layout, but they can also increase the effect by providing additional (stories) decay time lengths.

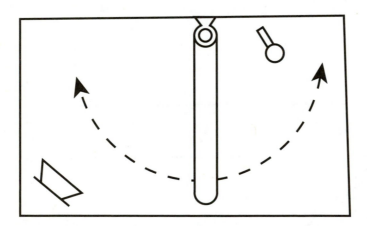

Figure 2-23
This reverb chamber has a moveable partition that is used to adjust the decay time. Positioned closer to the mic, it will increase the decay time; when closer to the speaker, it will decrease decay time.

Echo or Reverb Chambers

Chambers are simply rooms designed specifically for the purpose of rendering reverb. Reverberant chambers run somewhere around 3,200 cubic feet in volume, 18 feet long by 15 feet wide by 12 feet high, and are made up of reflective surfaces. They often have concrete floors and hard, smooth, plaster-covered, 6-inch-thick concrete walls and ceilings. None of the walls are parallel, nor is the ceiling parallel to the floor. A high-quality mic and speaker are placed inside the room, a signal is applied, and the positions of both the speaker and the mic are adjusted for the desired sound. Movable partitions can be added to vary decay times. A chamber's ability to render both reverb and an echo depends on its size, absorption, and speaker/mic placement.

Pure reverberation chambers for professional testing purposes are a little more complex. They often have steel walls. The partitions can also be made into two sections (also of steel) that revolve simultaneously both horizontally and vertically. Mics mounted on rotating booms are common as are dimensions like 19.5 feet long by 13.67 feet wide by 11.4 feet high, and are skewed trapezoidal layouts.

Obviously, neither of these are for the home recording enthusiast or the small independent studio. But by using the principles involved in these professional test chambers, with some trade-offs like less bass response when the chambers are smaller and a little loss at the high end when using walls that are less reflective than steel, you can still get great sound.

At some studios I've worked in, we've used the basement, an alley, an elevator shaft, closets, and bathrooms. Some were great; some not so good. There is, to my knowledge, no proven chamber formula. I'm not saying that you should give up on chambers; just don't invest any money in one. Check out the sounds in empty closets, your basement, or bathroom. Be patient and check out many speaker and mic placements.

Reverberation can also be thought of from the perspective of spaces without reverberation or echoes. Anechoic chambers are used to measure sound pressure levels. They make testing the sound power radiated from a noise source as simple as placing the *device under test* (DUT) and a calibrated microphone inside the chamber and taking a measurement. For "spatial averaging" more microphones are needed to indicate the device's sound radiation patterns. This is because in anechoic chambers, microphones only pick up the direct sound from the source as the remaining acoustical energy will be radiated into and absorbed by the chamber surfaces.

Depending on the bandwidth requirements, accomplishing this calls for the chamber to not only be totally isolated from outside noise and vibrations, but also to have all its inside surfaces covered in absorptive wedges, which can range up to 4 feet in depth. To say the least, anechoic chambers that qualify as test facilities represent a fairly expensive undertaking.

Open field (outdoor) measurements can come close to semi-anechoic chambers (those that do not have absorptive flooring). This "free space" method has very predictable attenuation. The "inverse square law" states that for every doubling of distance between the sound source and the receiver, the level will decease by one half (6 dB). If a level is 50 dB at 25 feet, it will be 44 dB at 50 feet and 38 dB at 100 feet away. In real life the drop-off has more of a curve to it and once you consider the lack of control over noises (dogs, cars, birds, and wind), as well as the weather, you'll understand why outdoor "free field" testing is seldom used. Reverb chambers are less expensive and can be used as substitutes for anechoic chambers.

Reverberant chambers are not only cheaper to build, but in many cases their measurements are more applicable to real-life situations. As some put it, "Why test a speaker in an anechoic chamber when nobody listens to them in that kind of setting?" Test results are also repeatable in reverberant chambers.

Professionally constructed reverberant rooms have highly reflective surfaces so that sound bounces from one surface to the other as easily as light from a mirror. Once a sound generation begins in this type of space, the level builds up quickly until it reaches a steady state SPL. This level cannot be reached or easily sustained in anechoic chambers, open fields, or rooms with open windows and absorptive furnishings.

Reverberant chambers are used to test the absorption and sound transmission loss (sound-blocking properties) of materials, make noise measurements, and to judge microphones and speakers. One of the ways reverberant chambers are used for testing is termed the "reference method." A calibrated (known) sound source is placed in the chamber and its sound pressure level measured. This measured level will be proportional to both the source's output and the room's absorptiveness. Then the subject of the test is placed in the chamber (be it an absorptive panel or an offending noise-making device), and the test rerun. The results will then be analyzed by comparing the difference in sound pressure level in $1/3$-octave bands with the initial test used as a guide for data correction.

The average recordist may not care which manufacturer's appliance makes the least amount of noise, but sound absorption coefficient specifications are also determined using reverberant chambers. The more reflective the surfaces, the longer the tone will last after the original sound has ceased. When testing reverb time, the process begins once the signal has been triggered to cease. The accumulation of data ends when the level decreases to RT-60 (-60 dB or one-millionth of the steady-state level). That's correct; the reverb time or RT-60 is the time it takes the reverberant sound to decay 60 dB, which is a decrease of one-millionth from the steady-state level. For confidence purposes, a number of samples are taken to derive an average result. Most measurements are made according to exacting specifications, which also define the reverberant room's parameters, including dimensions, absorption, and bandwidth.

The lowest frequency the room can handle is dependent on the room's size. The larger the chamber, the better the dispersion of the high frequencies and the easier it will be to excite or cause the build-up of low frequencies. Larger rooms provide better dispersion of sound since all reflections tend to arrive at a particular point more equally from all directions. This lessens the variations in SPL at different locations. Smaller reverberant chambers are not able to produce as much reverberation at lower frequencies. To avoid the added expense of large chambers, rotating reflective panels are sometimes used to break up any standing waves and to even out the sound dispersion within the chamber.

Manufacturers of professional reverberation chambers are concerned with things like "spatial uniformity." The variations at high frequencies even out very quickly so their level is more uniform throughout the chamber, but to achieve ideal uniformity in level at full bandwidth close attention must be paid to low-frequency resonances.

Low-frequency modes that are spaced close together can cause level fluctuations. The frequency of these modes depends on the dimensions of every boundary. Half the wavelength of the lowest resonant frequency will match the largest chamber dimension. This can be a bit confusing so I've spelled it all out so you'll see that it's not all that bad.

Wavelength = the speed of sound (1130 feet per second) divided by the frequency.

1,130 divided by 70 Hz = a wavelength of 16 feet; half that wavelength is 8 feet.

So, 70 Hz is the frequency where its half-wavelength corresponds to a dimension of 8 feet.

The audio community does not get too involved in specifying reverberant test chambers as compared to industrial standards committees. Outside of a few tests, reverberation is either considered as part of the field of acoustics or simply as an effect. On the other hand, professional trade associations and societies spend a lot of time and money researching standards and make their results readily available. Most of this published data can be found in your local library.

The *American National Standards Institute* (ANSI) sites standards for determining the SPL of small sources such as machines using reverberant rooms. The AMCA specifies the use of reverberant chambers to test sound caused by air movement and heating and cooling devices. The ARI has determined the proper method of measuring the sound radiation of refrigeration units and air conditioners. Other organizations to research are *National Machine Tool Builders Association* (NMTBA), *Home Ventilating Institute* (HVI)), National Fluid Power Association, and, more specifically, the U.S. Military Standard MIL-STD-810.

You can completely lose yourself in this kind of research. You'll find some of the things covered are obvious, such as the fact that microphones can be utilized that are capable of picking up low frequencies down to and below 20 Hz, and that sound generation transducers (speakers) must also be suitable for radiation of the lowest frequency required. While other information is just not applicable, such as the fact that some reverberant chambers use nitrogen gas as opposed to air as the medium since its absorption properties are significantly lower (especially at frequencies higher than 10 kHz). I remember one dissertation that examined the degree to which added air moisture increased high-frequency attenuation. I guess this means that you could change the frequency response of a reverb chamber, reducing high-frequency content by adding a common nasal humidifier, or boosting the highs in a damp reverb chamber by using a dehumidifier. Hey, it's a lot easier to EQ the input signal!

I was surprised, however, when I found an engineering bulletin from one manufacturer of reverberant test chambers that included mention of locating speakers in the room corners to increase low-frequency signal content and augment radiation. This is sometimes referred to as the "image-source effect" and is the reason Helmholtz resonators and other bass traps work more efficiently when placed in room corners. Hint, it also makes interesting positioning for subwoofers. The point is, if I hadn't already known this it would have been a big help.

Remember smaller reverberant chambers are not able to produce an equal amount of reverberation at lower frequencies. I found that some professional chambers have "shaped" response curves. Here "higher resonance damping" is introduced to the chamber to decrease high resonant frequencies. This shaping requires an increased amount of input level to correct for the energy that has been lost. In a nutshell, they introduced absorptive materials into the reverberant chamber to lower the high-frequency content. This requires an increased input level to meet the ideal SPL level. This results in getting an increased low-frequency content from a small reverberant chamber.

If I hadn't read this report I would never have thought to take the "go-bos" out of their storage closet and use them to cut down on the high frequencies (and boost the lows) of the studio when I was using it as a reverb chamber while mixing.

The point is that you never know where you'll find that little eye-opener that clues you into something you never would have come up with on your own. So go crazy and research your head off, but don't get overly involved because 90 percent of your findings may come up empty.

Mechanical Reverberation: Springs

Spring reverberation introduces or sets up multiple echos, or repeats, in a network of coiled springs that can be up to four feet long. They were initially designed for use in electronic organs and employed one or more stretched helical (wound or spiraled) springs mounted in a floating framework to help isolate them from external vibrations. Often two springs, which are mechanically different from each other, are used to promote more and different multiples of reflections. To understand how spring reverbs work, you must first understand that they are electromechanical devices.

The audio input is converted to mechanical energy by a moving coil driver, much like a common speaker except coils have no diaphragms or cones as they are designed to induce waveforms onto the spring in the form of vibrations. Springs transmit waveforms at a slower rate than does air. They therefore produce a delay. Each spring may delay the signal by 20 to 30 mS. At the far end of the spring is a specialized pickup, which converts the mechanical energy back into electrical energy and feeds this signal to the output.

The *mechanical impedance* of the pickup is intentionally mismatched to the mechanical impedance of the spring. The effect of this is to cause some of the energy in the form of mechanical motion (vibration) back up the spring. The reflection back up the spring may be reduced in energy, but it's strong enough to return back down again to the pick up. The sound waves can travel back and forth repeatedly depending on the type of springs used, the number of springs, and their tensioning. Weights can even be added to effectively shorten the length of a spring. Different spring types can be connected together. Dampening can be added. All of these variables are used by spring reverb designers to achieve different time decays, numbers of reflections, and frequency response curves.

Some larger devices even utilize springs immersed in oil. Changing the amount of oil affects the decay of the reverberation output. Here, once set, the oil level is usually left alone, as this job can be quite messy.

Another design utilizes springs connected to rotating magnetic rods on each end. On one side, the rod is caused to rotate by a magnetic field created by a coil, through which the input signal is flowing. This rotating rod creates a mechanical twisting waveform in the spring. Again the velocity of the wave's propagation along the spring is subject to the thickness of the spring, the number of turns per inch, and its tensioning. When the waves reach the opposite end of the spring, they cause another rod to rotate, which induces current flow in another coil, thus converting the mechanical motion back into electrical current. As before the wave does not die here but travels back and forth causing multiple reflections and thus reverberation.

While springs are very effective, they are far from perfect. For example, they usually have less than ideal dispersion of the signal through the spring itself. This is due to physical limitations such as the fact that different frequencies travel through springs at different speeds. Because of this resonant (or peak reaction) characteristic, too much input can cause a sound that may be rather unpleasant. Also, unwanted noises can be induced at higher sound levels, due to the microphonics. This noise-induced vibration more commonly occurs when the spring element is physically tapped or vibrated causing it to produce a noise, but the physical vibration of one spring can also be caused by the excessive movement and vibration of another, in devices with low channel separation. With older reverb devices, this is more often than not caused by loose parts.

The average user is not likely to have the equipment necessary to thoroughly evaluate a spring's electrical performance. However, the ear usually does an adequate job in detecting faults, especially those that are mechanical. As with other devices, spring reverbs should be listened to both separately, in order to judge the characteristics of the sound, as well as in the mix, for in-context evaluation. If the spring is to be used on stage in a live situation, it's wise to first point a speaker directly at it to check out its isolation properties.

When judging a spring, listen for hiss and low-frequency rumble. Try clapping your hands or snapping your fingers into a microphone fed to the input to see how sharp transients affect its response. Drumbeats and nylon guitar strings may cause bonging sounds due to their sharp transients.

Springs set up near large transformers may be affected by their magnetic field causing a 60 Hz hum. Also *radio frequency* (RF) interference and buzzes from TVs, SCR dimmers, florescent lights, and AC lines can cause noise in less than adequately shielded spring reverbs.

A flutter or a decay rate that fluctuates rapidly in level rather than smoothly decaying and exponentially down level can be due to the spring not producing enough refections per second or to its having a widely uneven frequency response. This problem is often due to the band that's centered around 300 Hz having an overly extended decay time. Excessively long decay times can be adjusted by changing input and output levels as well as by gating. But in this case having a variable built-in control (EQ) over this parameter is superior.

Plate Reverbs

Ever hear the sound of someone using a violin bow on a saw? What's happening here is that the bow is inducing a vibration into the metal to make it resonate and project a sound. A plate reverb works in much the same way. Instead of a bow, a transducer that operates much like a speaker's coil bit without a cone is attached to a flat, suspended sheet of steel. When a signal is applied, the transducer moves, making the steel plate vibrate.

A second transducer specifically designed to pick up these vibrations is also attached to the plate. When a vibration or wave is induced into the steel plate it travels outward, reaches the end, and moves back again. This process repeats and causes a reverberant sound to be produced.

The decay time is changed by adjusting the position of a soft absorbent pad, often made up of stiff fiberglass. This pad never actually touches the plate, but, as it is moved closer, the air pressure on the plate increases. This causes a decrease in the amount of time that the plate resonates. This damping can be adjusted for different decay times, either manually or by remote control motor drive, often with a meter on the remote indicating the setting. Decay time adjustments should not be made while the plate is in use, as it can cause noise. Plates can be mono or stereo, with the latter having two pickups located at two different areas on the plate.

Plates are convenient because they are smaller than chambers, usually $8 \times 4 \times 1$ feet. And although several can be placed side by side, they must be isolated from extraneous noises above 50 dB. After moving them, they can require recalibration. Some plates are even made of gold foil. These are better sounding, take virtually no setup time, and are smaller ($4 \times 4 \times 1$ feet), but are obviously more expensive.

Electronic Reverberation

Time-processing units that utilize only electronic components fall under two categories: analog and digital. With both, the incoming signal is sampled. Analog units use *bucket brigade devices* (BBDs) or *charged coupled devices* (CCDs). BBDs divide the audio signal into smaller segments of equal length, creating a continuous stream of program samples. The length of each sample is determined by a clocked pulse rate. During each sample period the average voltage value of the signal present during that time period is taken. These samples are then stored or held (delayed) in the BBD. BBDs are made up of charging and discharging capacitors or "storage registers."

Here, the bucket is filled and discharged (or passed along) to the next capacitor. The number of steps in the brigade and the time it takes for the signal to pass from one to the next determines the time it takes for the signal to reach the point where it empties to the output.

Saying the same thing a little differently sometimes helps the understanding. In a BBD, the capacitors are "filled" with a voltage, which is then at some point (depending on the clock time) discharged into the next capacitor. After a sufficient number of clock pulses, the samples begin to exit from the last storage register of the BBD. The waveform is now reconstructed. Although similar to the original signal, it is not smooth, but in steps which correspond to the average voltages of the samples. These steps can be considered as high frequency noise (clock noise), superimposed onto the original waveform. Thus, by putting the signal through a low pass filter, the high-frequency step noises are eliminated, thus smoothing out the waveform.

The result is nearly identical to the original input except, of course, that it is delayed. Through a feedback control, this signal can be reintroduced into the BBD system, causing multiple repeats. Again when the original signal has no defined attack, or is legato or melodic, the effect will be very close to reverberation. A problem with BBDs occurs when using them for long delay times. To produce echos, the clock rate must be slowed down. When this is done, the duration of each sample increases. Now, each sample must hold as larger portion of the input signal. The high-frequency content of the signal will become out of proportion with the low. At the output, the device again will "see" the extended amount of highs, as clock noises and automatically filter them out. Hence, with longer delay times, many units will have a diminished high-frequency response resulting from this additional high-end cut. Therefore, when checking out analog BBD delays, aside from normal listening and specification evaluations, pay attention to the frequency response, not only at short delay times, but also at the longest setting's bandwidth capabilities.

Digital delay units (DDLs) are generally better than BBDs, but they usually cost more. Here the audio input is changed via an *analog-to-digital* (A/D) converter, which samples the waveform and gives a numerical value to represent its voltage, level, and polarity. The signal has now been quantisized and digitized. These numbers are now fed into a shift register, which is controlled by a clock whose pulses cause the signal to move through a shift register one step at a time. Like the BBD, the timing of the clock pulses, and the number of steps in the shift register determine the length of delay. The register can also be tapped into at many points in order to give multiple delay outputs. This output can also be fed back into the delay system, as with the BBD for different effects. Besides shift registers, digital delays can also use RAM. Here the binary numbers from the A/D converter are stored and passed along from one RAM segment to another, again controlled by a clock.

Whether shift register or RAM, the signal retrieved at the output has to be decoded by a *digital-to-analog* (D/A) converter to produce an audio output. The ability of any DDL to handle wide dynamic range and frequency response depends largely on how many digits are used for each number (or stored voltage value). Twelve-bit (12 digits) DDLs are cheaper, but less desirable than 14-bit, 16-bit, or higher-resolution DDLs. Twelve or more bits may seem like a lot, but remember that this is a binary number (base 2) and the polarity, direction of the voltage swing, and actual value must all be documented properly. Some of the problems of DDLs are noise caused by the A/D and D/A converters quantizing and digitizing the audio signal. A mechanical-sounding output is due to low sample rates, and audible clicks are caused by the clock's pulses. The best way to check a DDL is to listen to the delayed signal alone and in the mix. When listening to this signal alone, try increasing the delay time while a chord is sustained. In this way you can more easily hear, once the unit stops reacting to the actual change, if the device has the ability to handle full bandwidth at its longer delay times.

Digital reverberation units are pretty much the top of the line as far as the technology they use. Most have microprocessor control and are software based. Some can even be updated with new effects programs. To understand the processing power required to simulate a real acoustic environment, think of all the separate delay parameters that must be addressed.

Here the source is fed to an equalizer and a limiter within the reverberation device to prepare the signal before it is converted, at a high bit rate, to binary numbers. This digital signal is then fed to a multiple delay processor. Dozens of different delay settings are used simultaneously along with the original direct signal to make up the three main time periods of reverberation. Some of the shorter delays are fed directly to the output along with the direct signal. In most cases both have been equalized according to the software-driven processing of the reverberation program chosen. Later reflections are simulated by longer delays that also must be processed as far as their number, staggered timing, and frequency response.

Group clusters of additional delays are also output. Meanwhile, some of the delays have been fed back to the input and by this time a slowly building amount of reverberation is being output. The number and bandwidth of the refections that make up the full reverberant signal will depend on the program chosen as will be the decay rate. This is of utmost concern as an incorrect (too sharp or slow) decay rate will throw off the whole illusion.

Multiple processors, huge memories, complex software, and extremely fast clock and sampling rates are just common, everyday, ho-hum aspects of today's digital reverberation. It sure beats crawling around inside an old oil tank.

Some of the terms for reverb are wet and live as opposed to a dry and dead sound.

Some Reverberation Devices

A quick review of springs and plates may be called for before tearing into examples of typical electronic processors. These electromechanical devices use specialized vibration drivers or exciters that operate like speakers and pickups that operate like microphones. These transducers are placed at opposite ends of the spring or plate. Some plates use dual pickups for "stereo" reverb. In general, plates sound better than springs because their physical (solid metal) makeup produces reflections with tighter spacing between them. Springs are also more easily overdriven by hard transient signals such as drums. This causes them to produce that "boing" sound. Without proper insolation they are also more prone to interference from outside sounds, vibrations, and *electromechanical* (EMI) or *radio frequency* (RF) interference.

I once had a spring reverb sitting beside me in a control room picking up what was being played through the monitors and shipping reverberation of this signal back to the monitors *without* any input wiring connected to the device. This thankfully is the exception as most spring reverberation designers acknowledge the physical limitations of springs during the design stage.

Admittedly, when used during a mix with both digital and plate reverbs available, I chose the digital for drums as the average digital reverb is simply more versatile. Plates are my personal favorites for vocals.

EMT has to be considered the "king of plates." This manufacturer's plates utilized adjustable dampers (soft, flat vibration-muffling panels) often with motorized operation that allow for remote control of the reverb time. The earliest (and now greatly sought after) versions had vacuum tube electronics, but EMT also put out solid-state versions, and later some pretty wild-looking digital reverberation devices. The original pates were mono or stereo with one driver producing a vibration that sent the plate into a bending motion. These vibrations travel back and forth with their reflections building up over time. The thickness and size of the plate was chosen to set the length, frequency response, and density of the decay. Two pick-ups were positioned to produce a "stereo" effect.

However, my favorite EMT plate was their Model 240 foil plate reverb. This device was almost portable yet still provided remote control capabilities. Its size was a mere 2 feet high by 2 feet deep by 1 foot wide. So why "almost" portable? Well, the foil that they used for this plate was a very thin sheet of gold, a fairly soft and easily damaged material even though it was housed in a protective floating casing on spring suspension. The use of this material, along with very transparent electronics, rendered what can only be described as a sparkly sounding reverberation, making it a pleasure to use on violins, flutes, and especially female vocalists.

Digital Reverberation

This processing is most often achieved by taking multiple time-dependent outputs from a digital delay, adding differing amounts and types of filtering, and feeding some of this signal back into the front end of the processor. What you end up with are many "reflections" of the original signal, some early (less than 20 milliseconds), others late (echo), that when combined with filtering yield real-world room-like sounds.

The acoustic properties of a real room's reverb and decay time are dependent on its overall size while the frequency content is affected by the materials used in making up the room. Reflections are dictated by the front of the room's (stage area) shape. Digital reverberation attempts to simulate these parameters with the use of attenuation, equalization, positive and negative feedback, and delays.

Quantec's digital reverb was one of the first to provide control over the onset of the reverb by adjusting the pre-delay as well as the level of the first reflections. They provided separate decay time settings with a maximum time of almost a full minute for both the high and low frequencies. Equalization will affect the decay frequency-wise much like the variations in an acoustic space when the walls are covered in drapery, the audience is full, and there are many opened windows (Sabin) as opposed to an empty hall with no windows and slab marble walls.

All of these variables form clues that the ear/brain combination uses to identify or picture an acoustic space. Lexicon, with the introduction of the 224 X brought about the complete processing of these acoustic parameters. This unit produced simulations of acoustic spaces ranging from a small garage (great for rock drums) to a full blown Gothic cathedral (a flutist's dream). Additionally, via sophisticated microprocessor control, any setting in between these extremes was achievable along with plate sounds and spectacular chorusing. Even with this very early version I could make two voices sound like eight. Given three or four, it would end up a choir. Things were really popping in the recording industry during the mid-1980s and like MIDI, the 224 X completely changed the way things were being done.

Yamaha later (still in the early 1980s) put out a digital reverb unit, the R-1000, which made this technology affordable for many musicians and most smaller studios. It offered four set reverb times (1.55, 1.62, 2.3, and 2.4 seconds), with apparent in-between decay times achievable by using the direct/effect signal mix control. It featured a three-band full parametric EQ, separate input and output level controls for setting up correct gain staging, a reverberation frequency bandwidth out to 10 kHz, and even a foot-switch-operable bypass function.

Companies like Roland, Tascam, Peevey, Korg, and so on, that manufacture a wide range of professional audio equipment, also offered digital reverbs. This was nothing new as even back in the 1960s when UREI was a major force in professional audio, they provided 10 to 27 band equalizers; notch, low-cut, and parametric filters; as well as compressor/limiters, amplifiers, mixers, and monitors. At that time their equipment was considered top of the line and today many are much sought after and have become quite costly.

The Lexicon 224 X Digital Reverb   The Lexicon 224 X digital reverberation system not only simulated the acoustics of various spaces, but was extremely variable. Like I mentioned previously, you could get a sound that was similar to that of a garage, one more like that of a huge Gothic cathedral, or pretty much anything in between. Since it changed the incoming signal to a digital format and further processing was handled by a very sophisticated microprocessor, the range of sounds it provided was awesome. The unit was made up of two pieces: a 19-inch rack-mountable main unit and a controller. These were connected together by a flat (jacketed ribbon) cable. Once hooked up, the remote was all the operator had to deal with. It had a display that aided in setting up parameters such as decay time, frequency response, depth, pre-echoes, diffusion, and so on. Changing these parameters was accomplished by adjusting fader settings. There were only six faders on the remote, but via a feature called "paging," the push of a button changed the functions controlled by these six faders to six others. There were up to six page settings that gave the operator control over more than 30 parameters. This device was flexible to say the least and this flexibility was what counted.

Let's take a look at one of the 224 X's programs and its variations, all of which were preprogrammed into memory and recalled with the simple touch of a button. Take the program entitled "plates." Variation 1 was a program that sounded pretty much like a standard plate, bright and metallic. Variation 2 more closely resembled a common concrete block garage; 3 had the acoustical qualities of a large stone church and included four pre-echoes; 4 resembled a large Gothic cathedral; 5 was a plate program designed specifically for percussive sounds; and 6, called "church helper," was the same as variation 3 without the pre-echoes and was used with signals that were already spacious-sounding, like organs and other long-tone acoustic instruments. Plainly, the different sounds provided by this single "plate" program were vast, from garage to cathedral. However, we've only scratched the surface of this program. While the variation buttons allowed changes in the program from one variation to another, the six faders mentioned above were used to change parameter settings between these variations, rendering an almost infinite number of different spaces. And this was only one program out of more than a dozen.

However, one point must be made here. The sound of the plate variation itself only resembled that of an actual plate. This unit was not designed to be a substitute for plates. What you had was an infinitely variable supply of other types of reverb sounds aimed at complimenting the studio's other devices. It did this in a way that was second to none. It had halls that actually sounded like different acoustical spaces ranging from classical recitals to very large concert halls, as well as rooms and chambers that were equally real sounding. The 224 X was fully stereo and as such could provide combinations (or "splits") of sounds on its right and left channels. This meant that two different halls or plates, or a hall on one side and a plate on the other, could be used at the same time.

This has to be my favorite chorusing device. It not only rendered six voices for each input signal, but allowed for the control of level, peaks, feedback gains (repeats), and panning of each. Using the 224 X I could turn a single female vocalist into a small choir. Two vocalists become 36. Another program produced resonant chords. Here a single note (even a drum beat) rendered six notes. The levels, pitch, duration, pre-delay, and pan of each were continuously variable. It can actually be programmed to produce a short melody from a snare hit. (I've done it.) Substitute a guitar chord and you begin to understand how mind boggling and powerful this device is. It can also be used as six separate delays (another favorite). Each controls level, delay time (0 to 919 milliseconds), low- and high-frequency cutoff points, pan and stereo control of foldback repeats, and diffusion. Using this mode I can make the aforementioned snare hit a brush stroke.

After setting up all of these parameters, you can store them in one of the 224 X's 36 memory banks that can be switched in and out during a mixdown without any added noise. Unfortunately, these memories, unlike the manufacturer's preset programs, were lost upon powering down the unit. If the studio didn't turn their equipment off there was no problem. In those days some studios actually never turned off the power as the inrush of current could cause console failure and occasionally a loss of substantial amounts of automation memory. I'm a witness.

The Lexicon PCM-60 Digital Reverb   The price of digital reverberation began to fall with introductions from companies such as Dynaco, Yamaha, MXR, and others. The PCM-60 was unique, as it was a cost-effective digital reverberation processor that was produced by a leader in the field, *and* it had the sound expected from a Lexicon device. In the past, owning a digital reverberation device had been a luxury for many studios, and certainly having two such units along with the required plates, springs, and chambers was well beyond the reach of all but a very few facilities. The introduction of the PCM-60 changed this situation. Lexicon developed this cost-effective reverb with the home recordist in mind as well as live applications, but it still deserves a place in any state-of-the-art facility (even one that already owned a digital reverb). Simply stated, the PCM-60's sound is outstanding. It will not give you dual reverb programs as does the 224, and obviously it won't perform complicated processing like chorusing, but it has the famous Lexicon room and plate sounds just like its big brother. In fact, the PCM-60 completely dispels the only two complaints I've ever heard about both the 224 and the 224 X: cost and complexity of setup. Personally, I've never had any problem with any Lexicon processor in terms of setup time and, in fact, have found the variability of its parameter settings to be a big plus.

Even though the PCM-60 is fairly versatile as far as parameter setting, anyone can get the type of reverberation that's synonymous with Lexicon's room and plate programs in a matter of seconds by utilizing only the preset switches. And it will sound good. Yet for those who are willing to do more than push a couple of switches, the PCM-60 does allows for a lot of fine-tuning.

This device easily interfaces with any system. Input (¼-inch phone) can be either balanced or unbalanced, with level being rear-panel switchable from +4 dB to −20 dB. Output is stereo (which enhances the reverberation effect). The left and right ¼-inch phone outputs are unbalanced with the same nominal level switching. The rear panel also houses a bypass foot-switch jack, as well as very convenient effects send and receive jacks that are located in the chain, pre-reverb. Input is set by rotary pot and aided via a five-segment LED display that shows peak and overload conditions. A mix pot controls the balance of dry to reverberant signals, which also affects the decay time. A third rotary pot controls output level. A bypass push-button switch causes direct relay connection of the input to the two outputs (as does the foot switch).

Twelve push buttons control all signal processing, yet they add up to well over 100 different settings. First, either of the two main program settings (plate or room) is selected, followed by one of four size (delay before reverberation) buttons and one of four reverb (decay) time selectors. The combination of these buttons dictates the type of acoustical space simulated, yielding 16 room and 16 plate sounds. The delay times for room are 6, 9, 16, and 37 milliseconds and 1, 2, 7, and 46 milliseconds for plate. Two additional "contour" switches combine for four EQ settings.

As described in the preliminary manual the parameters of these settings were a bit confusing. Memorization of their effects is impossible, since it is different for both plate and room. However, because only four positions are possible, checking each one only takes a few seconds.

I ran a number of signal tests through the PCM-60 to a strip chart recorder. The first graph was of the unit's frequency response with its mix control to direct signal only (unity gain; see Figure 2-24).

This is important (as you'll see) since it allows for more variability in decay-time settings. The manufacturer's specification for this parameter was, "Direct Signal: 20 Hz to 20 kHz, 0.25 dB." I found it to be *ruler flat* from 30 Hz to just about 10 kHz. At 20 Hz it may have been down as much as a quarter dB. At 20 kHz it was more like + 0.5 dB and it was up by + 1.125 dB at 40 kHz.

Remember, this was a cost-effective reverb processor! A second graph (see Figure 2-25) showed the shortest and longest decays of the plate reverberation presets on the same chart with a 1 kHz input. I did this to check the evenness of the decay. No large peaks or valleys were found. Had there been, it would have pointed to processing inadequacies that can cause unnatural reverb sound with some program material. But as you can see, these decay curves were very smooth.

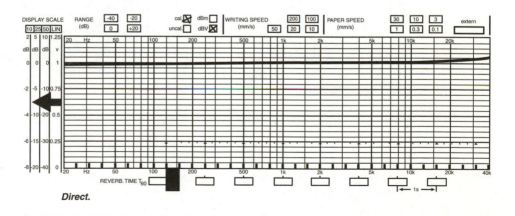

Direct.

Figure 2-24
This graph of the Lexicon PCM-60 dry signal path's frequency response is difficult to make out because it only marginally varies from dead flat (the second horizontal line from the top) at the two extremes of the plotted bandwidth.

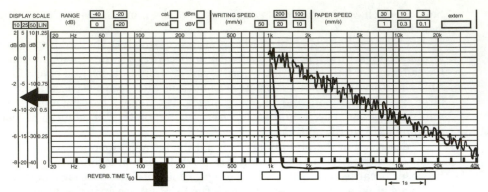

Shortest and longest plate presets.

Figure 2-25
Both the shortest and the longest decays of the PCM-60 plate with a 1 kHz input. Again, both were plotted on the same graph.

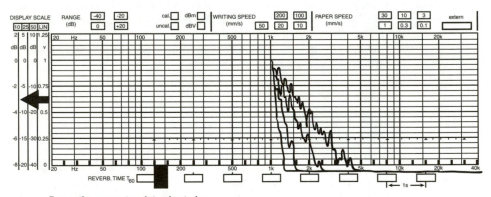

Decay time presets, plate short size.

Figure 2-26
Here I plotted the four decay time presets of the PC-60 "short" plate plotted with a 1 kHz input to see the effect of these settings. They were put on the same graph for easier comparison. As can be seen, they are almost picture perfect in their evenness of decay. This uniformity is remarkable.

    The next graph I plotted (see Figure 2-26) was to see the effect of the four decay-time settings for the plate program when set to the short size. These were picture perfect in their evenness. In Figure 2-27, the actual decay time was not the point but the variability was, as the decay time can be set to anywhere between 150 milliseconds and 4 to 5 seconds.

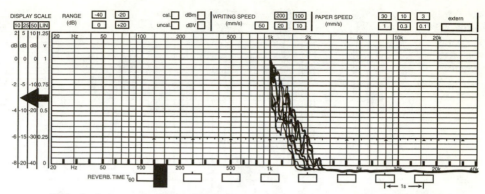

Plate short size, some of the variations between the two mid-decays via mix control usage.

Figure 2-27

(A) This "crazy" graph displays just some of the decay time variations that can be set using the PCM-60's mix (dry to wet) control. The variations shown here are between the two mid-decay time settings of its short plate.

I plotted some of the decay variations between the two mid-decay pre-sets of the short plate. These were made possible by using the mix control. I believe this is important information or else I would not have gone to the trouble of recalibrating the strip chart recorder seven times in a row to produce this graph. It shows that many decay times are available between the two settings by adding the dry signal output and mix control.

Using the patch point for additional delay before reverberation, external EQ, gating, compression, and so on, as well as this mix control plus some gating at the output of the PCM 60, and you've got enough variability to keep yourself busy for quite a while. On the other hand, the sound rendered by utilizing only the preset switches is marvelous. Therefore, even when the operator's setup time was limited, they still had it made.

Figure 2-28 is admittedly an odd graph. This "cost-effective" unit blew away all who heard it, not only in live and home recording situations, but also in a state-of-the-art studio where I used it on drums during a mixdown because I had the studio's 224 X chorusing the vocal. The PCM 60 easily stood up to this comparison. At the time of this review, there was only a preliminary owner's manual available, but it was very complete. The removal of six screws exposed both the top and bottom of the main PC board (switches, pots, and LEDs are on a front panel PC board connected by extra-sturdy ribbon cable), making testing, removal and replacement of almost all components simple. Yet even in this semi-disassembled state, this unit was still very solidly held together. Not only are all part locations and test points clearly labeled, but the PC board is even marked off in identified sections. Main ICs are socketed, the transformer well shielded, and both internal switches and fuses easy to get to. This unit should perform trouble-free for a long time. As far as further judgment of its sound goes, I leave that up to the reader. I have no doubt as to what that opinion will be.

Lexicon also produced several *very* inexpensive effects processors aimed at the musician market. These multi-effects tools had reverberation settings that, although not comparable to their top of the line units, still sounded good enough to be used in professional recording

Figure 2-28
This graph shows the decay times of all the PCM-60 presets with a 1 kHz input.

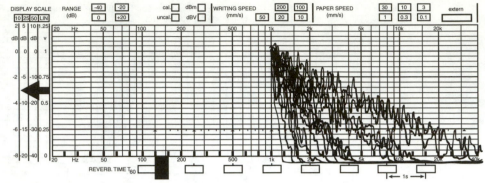

All 16 plate presets.

situations. Some of these were priced at only a couple of hundred dollars. If a truckload could have been taken back in time to say, 1975, the profit margin would have been incredible. Meanwhile, the 224 X evolved into the 480-L (LARC remote), to which was added digital interfacing, thus avoiding extra A/D and D/A conversions. The capabilities were increased beyond that of the 224 to the point where the effects rendered have far more complexity and present a more natural sound. But development continues.

Chambers   Facts are facts, and outside of professional anechoic chambers every space produces some reverberation. Even semi-anechoic free-field environments have at least one boundary (usually the floor or the ground) that reflects sound. Many spaces do not qualify as having any real decay time to speak of (less than a quarter-second), but even one-half second of reverberation decay can be very useful on something percussive like marimbas.

Your best bet is to make it a habit, starting right now, to snap (or pop) you fingers whenever you enter a space for the first time. When I do it, I usually put my hand to my forehead and whisper "Oh, no!" as if I forgot something, just so folks don't peg me for being the nut that I really am. Yet after doing this for countless years, this nut can walk into a tunnel or underpass and with a quick glance tell you what instruments *and* styles of music will work, be adversely affected by, or be in their dream-come-true environment. I know one underpass in Central Park that *any* rock drummer would die for.

Back to studios. Everyone knows bathrooms and empty closets can be used as reverb chambers and the only way to determine the *best* location and orientation for the speaker and microphone used is by trial and error. But I can save you some time as I've set up quite a few of these home-brew chambers in my day.

First, listen to the space by clapping, singing (if you're sure no one's around), or use some kind of full bandwidth signal. In other words, a decent speaker playing programming with at least some fidelity. Hey, a blaster will do for a start, but a mid-priced home bookshelf speaker is going to provide you with more information. You'll need this to aid your choice of positioning the speaker and microphone, as well as any EQ use.

Okay, here are two good starting points: 1. speaker and mic in the same corner, both facing the opposite corner, with the mic behind the speaker (longest delay), 2. both the mic and the speaker facing into opposite corners (most reflections). If you have them, use two of the same kind of mics. You know the drill! Listen to both, move the one that doesn't sound best, listen again, move the one that doesn't sound the best, and so on.

Don't be shy about trying unusual spaces. An empty 50-gallon drum makes a nice little chamber. The end "plug" screw-off opening is just the right size for a small, high-quality (automotive) speaker and a low-impedance mic cable. Wrap that puppy in mud, cloth, sandbags, or drop cloths and it'll even be somewhat isolated. I liked it on a Rhodes piano. Wait a minute, I like just about *everything* on a Rhodes piano. The percussive hit of the hammer on the tine along with their sustain adds up to every audio effects designer's dream instrument. Fifty-gallon drums are also okay on harmonica, sweet on short percussive sounds like a clave, cabassa, or even a tambourine, but they can be a little too short for snares.

At one of my first studio jobs (1967) I was elected to be "lowered" into an oil tank that was no longer in use to set up a speaker and a mic. I was the only one skinny enough (6 feet, 135 lbs.) to fit through the opening.

When they opened the "hatch" and I smelled the fumes (from a small but decades-old oil puddle) I was happy I had spent the time hitting those "physics of sound" textbooks because my mind sprang into "I know *all* about this situation and the best bet is to position the mic and speaker right here below this opening as it will provide the *longest* decay time, which *you* can easily gate to shorten it to *any* length that *you* think will be best and should there ever be a failure or if *you are* required to replace or update something during an *important* session both of the transducers will be right here within *your* easy reach." All this was true, but even if that were not the case I still wouldn't have gone into that tank. You must learn to draw the line. The audio industry is notorious for what it puts its apprentices through, and there's no need for it.

Back to home-made reverb chambers. Much later I worked in a studio that had complete control over the building's elevator after 10 P.M. We used it as a variable depth chamber, so again, I can save you some trouble.

Start at the basement. Locate both the speaker and the mic (again the mic behind the speaker) at the bottom of the elevator shaft. A few employees tried attaching these transducers to the bottom of the elevator, but it became a *horror show* of tangled wiring, things constantly being pulled out of place, and worst of all several heart-stopping *close* "mishaps" around the open elevator shaft.

Hey, if you have total control of an elevator, you have control of the stairwell, too. The latter is an infinitely superior if not perfect reverberation chamber. Think of the turns the walls take as you ascend. It kind of matches the partitions used in professional reverberation chambers, no? How about all those reflectors commonly referred to as "stairs"? I love stairwells. They not only make wonderful chambers but are great places to sing or play in. They're not exactly isolation booths, but if you happen to own the building, a Marshal stack in the stairwell that's being fed by the output of a low-power preamp located in the control room will make the guitarist playing through it eternally grateful.

Remember, the better the frequency response of the speaker, the broader the bandwidth of the reverb. One studio I worked in had an extra pair of JBL 4311 mid-sized speakers in addition to the several pairs of control room monitors that were hooked up. I was forever hauling these to some "new" potentially superior location in that facility, but it was worth it as I had discovered at least four places that had different reverb times. Another trick that works for added decay is to face the speaker right into a baby grand with the lid opened and put some weight (mic stand bases work) on the piano's sustain pedal. This causes the strings to vibrate, which in turn causes the sound board to resonate. Laying towels on certain sections to dampen some of the strings effectively equalizes the decay.

I could go on and on about chambers I've rigged up, but you'd be better off doing your own experimenting. Here's one last note on chambers. Try to be practical. It sounds easy, but once you get involved in a chamber project, it just takes over. I once spent several hours on a discarded refrigerator box. It worked, but everything I ran through it sounded, well, like it was being played in a cardboard box!

Something I never understood is the aversion people have to using two reverbs on the same signal. I don't mean running the reverberated signal through another reverb processor that may sound muddy, distorted, and garbled. As you know, plates can produce low frequencies very well (perhaps a little too well around 300 Hz). Springs, on the other hand, are lacking at the low end but provide the sparkle missing from all but gold foil plates. Because of these idiosyncrasies I always used both at the same time. Try it yourself. Feed a vocal to both and return their outputs to different channels. This gives you the ability to gate (change decay times), EQ (accentuate or reduce particular frequencies), and use the return channel's faders to blend them. For that matter I'm a big fan of psycho-acoustically affecting the output of electromechanical and chamber reverbs.

Other Effects

Panning Using the Console's Faders

Panning changes a signal's image placement within the stereo field. Stereo panning causes the two sides to reverse positions or cross-pan. Panning or having a signal move from one speaker to the other during a stereo mix can enhance the feeling or mood of almost any instrument within a tune. The pan itself is generally achieved by turning a knob (usually marked "pan") back and forth.

During a mixdown, this method is somewhat cumbersome, due to the number of knobs on the board, and the limited amount of space between them. One hand is needed, and even if the pan is done perfectly, some other "move" may be done incorrectly, or its cue missed. At this point you'll very seldom be able to repeat the pan exactly on the next pass, especially if it's a complete run-through. Another method of panning is patching the signal to two separate channels and panning one hard left and the other hard right, and then using the faders of those channels for the pan move itself. One fader is raised, giving you the signal on only the side that channel has been panned to. By lowering this fader and, at the same time raising the other, the signal will move from one side to the other in a much smoother fashion than by turning a knob. But here, both hands are needed, making it necessary to have other people handle all the remaining (often countless) moves required during a mixdown. The explaining and learning of these moves is tedious, to say the least.

Using Console Automation to Program Pan Moves

Along comes computer-automated mixdown. With this you can make a pass through the tune, do your panning, and these moves will be stored into some sort of memory (either on floppy disc or the 24-track tape itself). On the next pass, those pan moves will now be exactly repeated, freeing up your hands for other functions. However, not every studio had a board that was outfitted with computer control as only a few could afford this luxury. I chose to review the Panscan automatic panner, hoping that it would fill this need in those studios that were either too small or did not have the budget for sophisticated computer mixdown. Boy, was I wrong!

The Audio and Design Panscan Automatic Panner   The Panscan was nothing at all like computer memory and control of panning. It was infinitely better. It did more, used no extra tracks, and was extremely musical sounding when performing this function. The unit had the ability to process stereo as well as mono signals. With stereo signals, the Panscan will cross-pan (left to right, right to left) repeatedly reversing the image. This 19-inch rack-mountable unit was not difficult to use at all. It handled an input signal of up to +20 dBm.

The speed at which the signal is panned was continuously variable from an almost instantaneous jump (which, when in the repetitive mode, rendered a great stereo vibrato effect), to a slow and smooth slide-like motion. The pan motion itself was more like a bouncing ball movement. It's very hard to describe, but it sounded much more musical than the more conventional motion of L to R and R to L. The "depth" or width of the pan was also variable. Since you wouldn't always want the signal to be panned to the extreme left and right of the mix, the depth control allowed the amount or extent of the swing of the pan motion to be adjusted to dead center.

When setting this parameter, an image-positioning LED bar graph showed the amount or extent of the pan's travel. When set to the opposite extreme, this control caused the signal to be positioned dead center, for mono. An image offset control was used to change the center point of the stereo image either to the left or right, again continuously variable between the two. Therefore, by using the depth control you could change the amount of swing to, say, half of the full stereo spread. Then you could adjust the image offset to a point that's halfway between the center and almost the extreme right. Your pan would now move from slightly left to almost full right. While this allowed for extreme variability in the positioning of signals within the stereo field of the mix, there was more!

A beat count trigger control, again indicated by an LED bar graph, triggered the pan movement according to a variety of actual rhythms. It counted anywhere from 1 to 10 beats to a bar, with the bar corresponding to the pan move itself. It could pan with the precision of a metronome, and therefore your pan itself became part of the rhythm. When stereo panning was used, only one of the inputs was used to trigger the counts. A count threshold control determined the level at which the beat count control circuitry would be triggered. Say you were panning a background vocal part with the lyrics "I'm doing fine" and the accent was on the "I'm." That word would be slightly louder. By setting the threshold control to a volume just under that level, the word "I'm" would trigger a beat count.

There was a main trigger button that was used to override this circuit. It also reset the count to zero and triggered the count to restart. This was very useful in cases where the program had tempo changes. You could stop the count at the point of a change, reset it to zero, change the number of beats per bar, and restart the new count, all in a matter of a couple of seconds. This count could also be controlled by an "external" input other than those that were being panned. Here a snare drum, for instance, would trigger the count, thereby causing the panning of, say, a rhythm guitar to be in tempo with the tune itself. An interior/exterior trigger switch allowed the use of an external switch where a voltage was sent (i.e., a battery, foot switch, or synthesizer) to trigger the pan. This device also had an image freeze button that instantly stopped the pan, holding the sound in position, in addition to a bypass switch, which defeated the panning completely. All this gave the operator complete control of the panning motion at all times.

The Audio and Design Panscan was simple to set up and easy to use, and the results were wonderfully variable and surprisingly musical sounding. By cross-panning stereo background vocals, they not only moved around the lead vocal, but did so in perfect time. The image center position and speed of movement of any signal is infinitely variable and easily controllable. All in all, the Audio and Design Panscan was a complete gas to use. It changed my whole approach to mixing, as far as panning and placement were concerned. Its unique musical movement added a whole new dimension to mixing that made a song come alive with animated motion.

The Panscan's internal construction is extremely well laid out and put together with clean wiring and soldering. All ICs are in sockets and all components are easily accessible due to both top and bottom cover panels being removable. The PC board has all components part number identifications printed on it.

I just don't understand it. In the mid-1980s, we were still mixing with the final destination of vinyl in mind. This meant that you had to be careful with both your bass content and panning. Bass caused wider grooves to be cut into the vinyl. Phase variations caused wavy-shaped grooves. Panning caused varying groove depths. Lacquer test pressings were being checked by the record companies for tracking capabilities on inexpensive (all-plastic) consumer turntables. With even a slight amount of out-of-phase bass (something I enjoyed using), the turntable's needle would jump out of the track and "skate" across the record. This would both call for a remix and made people very unhappy.

For my own enjoyment, I'd do a separate mix with all kinds of movement going on. Basses walking across the stereo field, single notes of a guitar solo screaming from left to right (an astounding effect with a side guitar), rhythm sections bouncing along with the tempo, and percussion parts that stagger stepped slightly left and right. Myself and many other engineers were constantly trying out ways to push the limits imposed by both tape and vinyl. What I don't understand is why this is not being done today, especially since printing to CDs makes everything that we used to kill ourselves over fairly easy to achieve.

Stomp Boxes in Professional Recording Studios?   Hey, what's this in the bottom of this rack? A "Rockman"? It's hooked up to balancing transformers, but nothing is connected to it. Looks like one of the engineers here is keeping this handy little tool all to himself.

The Scholz R & D "Rockman" was a small (paperback book-sized) guitarist's multi-effects device that was introduced around 1983. It gave the guitarist, or any musician playing an electronic instrument, the full complement of the studio effects common to the day.

These included EQ, ADT, chorusing, and echo/reverb with lots of sustain and distortion, and all in stereo. Variability was limited as the intent was to free the musician from the burden of prolonged dial twiddling to get the sound he or she was after. Limited, yeah, but it sounded good and it was just the thing you'd want to have on hand to help speed up getting rhythm guitar sounds during basic sessions.

I remember one engineer who wouldn't be without his DOD FX50 Overdrive preamp stomp box, that is, until DOD came out with the more radical-sounding FX55 Distortion stomp box.

What's all this talk about stomp boxes? Isn't this supposed to be a book on recording studio engineering? *Anything* that gets the sound the client is after is a professional recording tool. If it gets a specific job done while wasting less time, money, and energy, it is the tool I want to use.

If you're troubled by the less-than-pristine specifications that come with most "foot controlled" effects devices (as some like to call them these days), do something about it! If any effect outputs noise, set the level of the sound so that it masks that noise out. If the noise occurs when there is no signal at the input, gate it.

Many "guitar effects" suffer from the dual powering problem of added noise when levels must be increased due to low battery power, or a 60-cycle hum when inexpensive "wall wart" power supplies are used. Hey, every engineer I knew *always* carried a couple of *new* 9-volt batteries at the ready, and rigging up a couple of blocking capacitors for use with cheap power supplies was a simple affair (see "Capacitors" in the section on electronic components). If this seems like a lot of effort, think again. It will get *the* sound. *You* have it and *nobody* else needs to know that your "secret" weapon is a used $29.95 guitarist's stage effect.

Personally, I always had a Tech 21 Sans Amp on hand. They could be had for under 75 bucks and without using any speakers (talk about isolation!), you could get the sound of a Fender Twin, a Roland Jazz Chorus, a Mesa Boogie, a Marshall stack, or just about anything in between. My little guy was attached to dual capacitors and transformers that dwarfed it in size. The only problem I found was the inability to reset the unit to exactly the same sound on later overdubs. Tech 21 remedied this problem with a MIDI-controlled rack-mounted version of the Sans Amp that even had a front panel numerical readout. It was too expensive for my budget so I opted for the floor model until I got the chance at a used rack-mountable MIDI model. A power supply mod would kill all buzzes and hums, and the sound would be totally resettable. I'll delve into substitute guitar amps further in the recording section.

Multi-Effects Processors

Hey, where's the multiple-effects device system in this place? Oh, okay, I thought something was missing, but they've been loaded into that 2-foot-high rack on wheels that has the multi- and two-track remote controls mounted on top of it. That's a smart move, having the most often used effects processors on hand right there next to the console. Today it's common to find all kinds of multi-effects processors, but they only handle one input or at most dual (stereo) processing.

It all started with guitar processors like the ADA TFX4 that provided many different effects from a single device occupying only a single rack space. Competition was fierce so the number of effects within each device grew and grew. Soon it was possible for the guitarist to afford a single processor that was able to provide compression and limiting for more sustain, a noise gate to help defeat buzzes and hums, a delay that was settable for *automatic double tracking* (ADT), echoes and multiple repeat capabilities, long delays for sample looping, and modulation and pitch change, as well as chorusing, tremolo, flanging, and phasing. Oh yeah, and how about adding in some reverb, equalization, filtering, and "Leslie"-type (rotary speaker) effects while you're at it?

Next came voice processors that were single units that furnished a complete recording chain often including a mic preamp with a pad, gating, equalization, compressor/limiter, de-esser, reverb, harmonizer, chorusing, and ADT. All this could be used so that the vocal output was a polished signal ready to be shipped directly into a recorder or a live sound system.

While these devices were initially designed for musicians, it didn't take small studios, clubs, and touring PA companies long to make use of their extended capabilities. Effects like tape distortion and subharmonic synthesis were later added strictly for studio use while multitap delays and multiple equalization processing was aimed more at the live sound market.

The *orban optimode* is an example of how a multiple effects device was designed to fill a specific bill. This device provided a mic preamp, attenuator/pad, de-esser, EQ, and gate, but its multiple memories made it ideal for use in broadcast audio. I was employed by a facility that had individual processing setups for every one of its "on-air" talents: announcers, DJs, and news reporters. Since only one mic was used, once their particular processing was logged into an individual memory bank, all the talent or engineer had to do was switch the unit on and hit scroll until that person's ID appeared on the display. The processing not only became a simple and fast setup, but its resetability meant that when voice-overs were recorded over several sessions, there were little or no sound discrepancies in the final composite program.

The Yamaha SPX Series

The Yamaha SPX 90 was one of the early multi-effects processors and it gives us a typical scenario as to their development. The original device's compliment of effects included reverberation, delay, chorusing, pitch shifting, compression, gating, equalization, phasing, and flanging, as well as full MIDI interface capabilities. Yamaha continued to improve the sounds of these effects while adding more and more processing power as this device evolved into the SPX 90 II and then continued on through the 900, 990, and 1000 models. The most recent version, the SPX 2000, now also provides multiband dynamics, USB interconnection, and even work clock correction.

But I'm not talking about that kind of multi-effects processor. What I'm talking about are individual effects modules that slide into a rack-mounted housing. These "mothership" rack-mounted, card-cage systems held multiple individual effects device modules—in other words, separate processors that were removable.

I always appreciated the intelligence, cost-effectiveness, small footprint, and versatility of rack-mounted card-cage-based effects systems such as those from Alison Research/Valley People, API, dbx, Audio and Design, Aphex, Vesta Fire, and others. These furnished a built-in power supply and slots for plugging in individual modules whose functions included comp/limiters, expander/gates, EQs, mixers, distribution amps, delays, de-essers, and more.

The biggest advantage of these systems is their ability to group together whatever effects combination may be needed for a particular session into a single rack-mountable unit that provided the power and (rear-panel) interconnection. This was especially useful in multiroom facilities where studio "A" may need six gates and four compressors to handle a basic tracking session, while studio "B" may need a rack unit loaded with equalizers for a mixdown. In any case, the end users could often choose a rack mount that could accommodate from 2 to 10 effects and then load them up with whatever they wanted.

When looked at solely on an operations level, this interchangeability is one of the most desirable features of these rack mothership systems. It makes using them a very intelligent approach to equipment procurement for both the studio owner and an engineer looking to revitalize their selection of processing tools.

This, however, is not meant to detract from the usefulness and flexibility of the more "hard-wired" or "fixed" systems whose modules are not so easily interchanged. These are used in permanent installation situations that usually perform a single set function.

But the rack systems designed specifically with recording studios in mind can give you exactly what you need, anywhere you need it. But you can decide for yourself after checking out some of the following capabilities.

Allison Research

Allison Research was the first of the big-time manufacturers of dynamics processors for professional audio. Their racks of Kepex (or keyable program expander) units and Gain Brain compressor/limiters that had a variable ratio control (a big deal at the time) could be found in just about every recording studio in the early 1970s. Both units incorporated *voltage controlled amplifiers* (VCAs), but the Gain Brain was the device that sold these racks to most of the studios. It included low-frequency cutoff in its triggering circuitry, which gave this unit the ability to avoid the occurrence of false triggering and pumping that the large waveforms (found in this area of the sound spectrum) could cause. Furthermore, the overall degree of processing was based on the whole complex waveform as opposed to the then common level-only tracking. As "Valley People" the company went on to develop a three-band full parametric EQ, the "Maxi-Q," which was also rack-mountable. Valley is now part of the Galaxy audio line, and they've added a de-esser, a peak limiter, and a compressor/expander to the lineup. The specifications of all of these devices are, as always, very respectable.

The dbx 900 Series Rack System

The "900 Series Modular System," offered by dbx, was, and still is, very popular. The mainframe held up to eight modules and furnished the required power with all connections at the rear panel via the screw terminal.

   The 902 de-esser module had an adjustable frequency setting. The 903 had their famous "OverEasy" compression. It also offered "negative compression" (see the section on compressors), which can yield wider dynamics. The 904 gate could both key and expand. The 905 three-band full parametric EQ had peak or shelf switching for the high and low bands. The rare 906 dual delay or flanger module provided pitch sweep. Its control could be via an internal *voltage control oscillator* (VCO), manual (knob rotation), an external voltage, or any combination of the three!

Aphex

What could be better than having a couple of stereo/dual-mono compellers on hand during a basic session? How about a 900 rack with eight of them? Aphex designed their modules to operate in the popular dbx 900 series rack, but they also produced their own self-powered card-cage rack (the 9000R). Besides the Compellor model #9301 and the Aural Exciter #9251, they later added an EQ module to their line.

API

Due to the high demand ($) for gain stage and processing modules that had been "parted out" from their consoles, Audio Products, Inc. produced rack mounts in several sizes: 12- and 10-space card-cage racks for their 200 series modules, a 4-space "Lunchbox," and a 2-slot 19-inch horizontal cage for their 500 series modules.

Some of the 200 series modules include the 205L Direct box, the 212L mic pre, the 215L, which has sweepable high and low pass filters, the 225L compressor/limiter, the 235L expander/gate, the 245L de-esser, the 255L line amp, and the 265L mixer distribution amp. The 500 series includes the 512B mic-preamp/direct box, the original 550 three-band EQ, and the new 550B four-band EQ.

Vesta Fire

At the other end of the user spectrum are the rack "motherships" designed with the musician in mind. While aimed at musicians and home recording studios, Vesta Fire's modular signal processing found a good deal of practical use in the professional recording environment. The Vesta Fire "frame" accepted up to nine modules and provided the required 9-volt DC powering. This was a key feature as any single module could thus be used separately, powered by a simple 9-volt battery. Handy for the musician on the go, right? Their lineup included the MLM-1 limiter, which had a fixed ratio, the MNT-1 gate, which had fixed attack time, the MFC-1 flanger/chorus, the MDL-1 delay line, which provided 50 to 400 milliseconds of delay with a variable amount of regeneration, and the MPR-1, a single-band full parametric EQ with up to 18 dB of boost or cut, and switchable low- and high-cut filtering.

Audio and Design

I have to start the coverage of this next system off by admitting my being totally biased in its favor. While all of the modular rack systems made processing more affordable and easier because their multiple effects devices were in a single centralized location, the "Scamp" rack from the British firm of Audio and Design was the most fun to use. Maybe it was because this unit (especially the 17 module affair) seemed to provide the greatest amount of processing. They offered the usual in terms of a delay module for ADT and flanging, a dual de-esser with a full compliment of controls, an expander/gate, plus a dual gate, a soft-knee compressor with switchable stereo coupling, a three-band parametric EQ that was so comprehensive control-wise it required two module spaces, and a simple three-band straight-ahead sweep (variable center frequency) EQ with 20 dB of boost or cut per band.

It wasn't just all the colorful (wink-eye) switches or LEDs either. Although I must say that the quad LED meter module (switchable between peak and VU reading) was not only an extremely handy onboard addition to any rack, but also spectacular looking. Also handy was the availability of gates with built-in filtering to aid in the elimination of either low-frequency (rumbles) or high-frequency (hiss) noise. Having a two-in/eight-out distribution amplifier module allowed you to send your processed signal to multiple locations without any fear of loading down the output signal, so that everything remained at unity gain.

They also had a dual module with two 10-band (1-octave) equalizers with center frequencies at 31.25, 62.5, 125, 250, and 500 Hz, plus 1, 2, 4, 8, and 16 kHz for use with monitoring systems, or anywhere else it was needed. The addition of their automatic panner module (my favorite) made this package's capabilities truly vast.

However, the ability to load in several transformerless mic preamp modules that were designed to cleanly bring just about any signal to line level meant this system could be used as a portable mixer. Show up *anywhere* with one of these babies fully loaded and all dressed out in patch cable connections and you're "making audio magic." I think the audio community should demand the reintroduction of these systems. What exactly did A + D offer? Both a 17-module and a 6-module rack mothership.

Some of their effects I'd like to see on the market again are module numbers:

S08 2-in/8-out distribution amp

S07 a stereo octave (10-band) EQ (which occupied two rack spaces)

S14 four 12-segment dual-color LED meters (switchable for PPM or VU)

S24 a delay processor that handled ADT and flanging

S23 a stereo automatic panning module

S06 a combination high-frequency filter, de-esser, and gate module

S05 a combination low-frequency filter, hum eliminator, and gate module

S25 a dual de-esser module (which only took up a single rack space)

F300 an expander gate module

S100 a dual gate module

S01 a soft knee compressor limiter (provisions were made so that two mounted side by side could be stereo-coupled)

S04 a three-band full parametric EQ module that incorporated low- and high-frequency shelving filters

S03 a three-band swept center frequency EQ module

S02 a transformerless mic preamp module

Their later additions such as the S30 expanded upon the already versatile functioning of their expander gate by providing 60 dB of attenuation, side chain keying, built-in high- and low-frequency preemphasis, a release range of 25 milliseconds to 4 full seconds, as well as the choice of automatic release plus a 20 LED bar graph.

But they really went all out with the S31, a dual compressor/peak limiter module. The auto attack and release limiter's threshold range was 0 to +20 dB. This module's feed-forward compressor ratio ranged from 1:1 to 20:1 and its threshold range was −50 to +12 dBm. It offered a side chain, stereo coupling, and 20-segment LED gain reduction (0 to −60 dB) indication. These modules were only about 8 inches high and a single inch wide! I was a major fan.

This studio must have a combination tape deck/amp room, because ⅓-octave graphic equalizers and power amps are missing from this control room picture.

Hey, where's that door on the back wall lead to?

The Control Room Monitor System

To consider a recording studio control room an accurate tool for judging the integrity of the music being reproduced and the sufficiency of dialog intelligibility, many requirements must be met.

Among these are having a short reverb time, high sound isolation from the outside sources, including other rooms in the facility, and a minimum amount of ambient noise emanating from the equipment within the control room itself.

The monitor system should be viewed as a complete chain. This would obviously include the speakers and the amplifier along with any equalization used. Most people also understand that the acoustics of the room have an effect on what is heard, as does the cabling between the amp and speakers. Some disregard the interface between the console bus out, the graphic equalizers and the power amp input, thinking this only entails impedance matching and cable selection, but the end result is highly dependent on the correct gain stage setup. Furthermore, there is seldom any concern over the power or utility cable that feeds the amplifiers. To have a monitoring system that is neutral or unbiased in its reproduction of sound, attention must be paid to *all* of these factors as this system, like most, is subject to the laws of the weakest link.

The amp is seldom as much of a problem in terms of performance anomalies as is improper gain stage setup, cheap (16-gauge zip cord) speaker wire, misleading accentuated speakers, equalizer settings that can cause phase and level distortions, or faulty room acoustics. Hey, let's face it. This whole can of worms can be avoided with a good pair of headphones, that is, until the "head in a vise" pain begins.

The Equipment or Machine Room

Hey, what's behind that door in the rear wall of the control room? The door opens outward so there must be some room back there. Check it out; this whole wall is one large diaphragmatic absorber. Its widely spaced mounting studs are sitting on rubber blocks and it consists of $1/4$-inch plywood with black denim wallpapered to the side facing into the control room. I once saw the same kind of approach to eliminating bass buildup using $1/8$-inch sheets of plywood that had been covered with thin industrial-grade carpeting. I measured the room before and after installation and its absorption came pretty close to being even with frequency.

As with Helmholz resonators, there are calculations that help predict the affect diaphragmatic absorpters will have on a room's frequency response and reverberation decay time. These take into account not only the square footage of the panel, but also its total weight. They do not provide absolute results, but can put you in the ballpark as far as the wavelength and resonant frequency are concerned.

I know of a private studio owner who built his control room's rear wall entirely of equipment racks. He left a small, 1.5-foot, walk space behind the racks and covered the back wall with foam, which is much better to lean against while making wiring changes than fiberglass (thank you). He didn't have enough gear to completely fill all the rack spaces so he purchased blank panels. Some of these blanks were predrilled with holes, making them almost screen-like to aid ventilation. The holes gave this guy the idea of using these prerilled panels on his rear rack wall as Helmholz slot resonators. There were obviously no calculations to predict the outcome of this wacky idea, but with a lot of experimentation he actually reduced some of the low-frequency muddiness of his control room. It's amazing what can be accomplished when one has an unlimited amount of time on his or her hands.

I've had success placing small bass trap/equipment rack combinations in the control room's corners, but it's a fact that a well-designed acoustic space shouldn't require any bass traps, reflector panels, or any high-frequency absorption. However, people like the comfy feel of carpeting in a control room. Hung ceilings make perfect air return routes for the HVAC systems. Both absorb more of the high-frequency content than the low. Unfortunately, perfect dimensions and the added amount of space needed for sound transmission loss cannot coexist in the limited square footage of most spaces available in today's cities.

Hey, this rear bass trap must be 6 or 7 feet deep. That's a lot of space to give up, but it does free up a good deal of control room space by not having those two multitracks and two stereo tape recorders in there, not to mention the noise and heat they add to a room. Speaking of heat, these folks had a great idea. They've used a length of flexible duct to tap into the hung ceiling return on the far left and dropped it down to almost floor level.

On the far right they've added a large ceiling vent to complete the HVAC return routing. So for no cost they have added slightly warmed-up air conditioning to their equipment room. And, just as I suspected, they have a video camera with a wide-angle lens aimed at the tape recorders.

You can do everything "by the book," aligning your recorders by first calibrating each output, *then* adjust its meters to read 0 VU, and check that every console return has exactly that same reading. But the first time you hear something that doesn't add up level wise, or if you're a little concerned about how an effects level is going down to the multitrack, *you'll just have to see those multitrack meters.*

A simple switch will give you a bird's eye view right there on the video monitor located between the two main monitor speakers. If you're smart, you'll wire into that monitor's auxiliary audio input so that you can use its little speaker as a mono TV and transistor radio reference.

From the looks of those two structures on the back wall, I'd say they were being used to box-out a couple of windows. So it seems like this machine room is also used to provide additional sound isolation from the outside. This room's depth actually provides enough space to walk around the equipment racks, a full 360 degrees. There's what I came in here looking for. I was curious as to why I did not see any $\frac{1}{3}$-octave equalizers for the control room monitoring system, and here they are.

This studio is using a Klark Teknik DN 30/30 dual 30-band graphic equalizer, but it could easily have been a White, a dbx, or (these days) a Fostex, Rane, MXR, Roland, or any one of a host of graphic EQs now offered. As long as it conforms to ISO center frequencies and ANSI bandwidths while adding no noise or other anomalies to the program material, it doesn't matter. However, every graphic EQ used on a monitoring system *must* be set up using a spectrum analyzer. Do not trust this task to your own or anyone else's ears.

One-Third Octave Real-Time Analyzers (RTAs)

Is it best to design a room for a flat-frequency response or for a set reverberation time? This question can be debated long and hard, but two facts make neither an absolute truth. A perfectly flat listening environment is virtually impossible to achieve and there is no single reverberation time that's perfect for monitoring all program material.

Is it frequency response? Is it reverberation time? It's both!

That's why I liked the Klark Teknik DN 60 real-time analyzer, because it has an RT-60 option that allows it to measure reverb time per frequency. Goldline used to manufacture a complete test system for contractors, which included a real-time analyzer, *sound power level* (SPL) meter, a frequency counter, a sine wave generator, a gated pink noise generator (with timed pulses), a decibel meter, an impedance meter, and a distortion analyzer. The system had 30 memories and printer port output (as well as RS 232), and handled RT-60 and loudspeaker phasing as well as frequency response.

One-third octave analysis takes the broadband input signals and filters them into bands that cover a third of an octave, again with center frequencies set at ISO and widths set to

ANSI standards. These bands are then fed to a display on an oscilloscope, video monitor, and (these days) usually an LED matrix.

If the analyzer's frequency bandwidth only extends to 8 kHz it obviously won't matter if the sound source from the noise generator or the mic used to pick it up extends to 20 or even 40 kHz. A 16 kHz bandwidth is the minimum requirement for an equalizer or analyzer.

The Noise Generators

White noise (the sound heard when a television is tuned between channels) consists of all frequencies at an equal level. Pink noise is similar but adds a filter that rolls off −3 dB per octave. This is done because the $\frac{1}{3}$-octave bandwidths become wider as the frequency increases. As an aside, the names white and pink relate to the light spectrum, wherein red equates to the lower frequencies and a mixture of the whole spectrum produces white light. Due to its high-frequency roll-off (and resulting low-frequency emphasis), it adds red to the white and is therefore "pink" in color. Like white noise, pink noise is a sequence of all frequencies generated randomly or at least pseudo-randomly. Due to this randomness, the operator must average multiple readings to achieve accurate repeatable results.

At one time, dbx manufactured the RTA-1, a computer-controlled real-time analyzer with a very intelligent capability added to its signal source options. It was able to not only use pink noise as the test signal, but also "nontest signals" as well. This RTA made differential measurements so it could use broadband music programming as a test signal. This was obviously a boon to live PA work as a system could be calibrated *during* the performance. In studios it meant the operator could check the system's frequency response without stopping the recording or mixdown in progress.

In live PA applications EQ is most often used to help defeat feedback that occurs when the monitor speaker's sound (picked up by a mic) is loud enough for it to be reamplified. Although I've never tried it myself, another method of adjusting $\frac{1}{3}$-octave equalization is to set up an open mic, increase the preamp level to the onset of feedback, and then drop the level 6 dB. At that point an oscillator is used to excite the system sweeping through the frequencies and adjusting the EQ at points where the feedback reoccurs. This can be a bit brutal on the system's components (and on the operator's hearing), but back in the 1960s systems were often adjusted using a sine wave generator and a voltmeter one frequency at a time! I've performed this kind of testing and the dbx RTA-1 from that perspective seems miraculous. The Alan Parsons and Stephen Court Sound Check CDs $\frac{1}{3}$-octave individual bands at ISO center frequencies do away with the need to constantly readjust a signal generator. When used as a signal source with $\frac{1}{3}$-octave analyzers these CD tracks are more stable, and the "one band at a time" adjustment method is much easier on the eyes *and* ears. Pink and white noise is no fun to listen to especially in situations when its level must be set high to overcome background noise. By the way, the ISO Center Frequencies are as follows:

Octave: 16, 31.5, 62.5, 125, 250, 500, 1000, 2000, 4000, 8000, and 16,000 Hz

One-half octave: 16, 22.4, 31.5, 45, 63, 90, 125, 180, 250, 355, 500, 710, 1000, 1400, 2000, 2800, 4000, 5600, 8000, 11,200, and 16,000 Hz

One-third octave: 16, 20, 25, 31.5, 40, 50, 63, 80, 100, 125, 160, 200, 250, 315, 400, 500, 630, 800, 1000, 1250, 1600, 2000, 2500, 3150, 4000, 5000, 6300, 8000, 10,000, 12,500, and 16,000 Hz

Using ⅓-octave analyzers has become a relatively simple process, but care should be taken so as not to over-EQ. The operator still must proceed slowly in an orderly fashion. If any cut-off filtering (low or high pass) is used, it must be introduced before processing since it will have an effect on other settings. After making an adjustment to the ⅓-octave EQ, the system should be given time to "settle in" or stabilize before another band is adjusted. A final, thorough listening test with familiar program material is most important.

Speakers, microphones, room acoustics, and complete systems are never flat. By flat it is meant that the volume or SPL is the same at all frequencies. So what? Well, operating a live stage monitor system that has a resonant or peak frequency at 2 kHz and using a vocal mic with the same characteristics, or if the acoustical environment of the stage area has a similar response, the performance would be hampered by feedback in that frequency range whenever the stage monitors are raised in level. Or at least to a point that is loud enough for the musicians to hear themselves over the onstage din.

A parametric EQ would help out in a case like this and an engineer with experience and trained ears could zero right in on the offending frequency and bring it under control. However, even the best ears can seldom achieve perfect accuracy. The bandwidth or "Q" of the frequency may be set too wide, thereby causing neighboring frequencies to be affected. Even if the Q is set perfectly, the amount of attenuation could easily be overdone. With this in mind, consider the task of flattening out a recording studio control room's monitor system.

"Ringing out" a PA system is performed by repeatedly raising the microphone's input level until feedback just begins to occur, or "ring," and then lowering the appropriate fader. This is not an option with recording studio monitors as feedback is not the prime concern. Nobody's ears are capable of determining the exact EQ settings for a perfectly flat response. What's needed here is accurate electronic test equipment.

The way it was once done was to use a signal generator to feed a test signal into the system's input and have a decibel meter, or mic/voltmeter combination with the mic positioned so as to pick up the speaker's output. Individual frequencies were generated and the system's output in voltage or decibels read and plotted on a graph. One by one, you'd go through the frequencies until a complete graph of the audio spectrum was plotted. Then, you could set up a ⅓-octave EQ so that its settings formed a shape opposite that of the graph.

To be perfectly accurate, the process was repeated with the equalizer in the system for fine-tuning those settings. This was, and is, called serial or sequential spectrum analysis. The generator used had to be calibrated in absolute values in both frequency and level. If, however, its output wasn't uniform over its entire frequency range, you would have to hook a second meter up to its output and adjust the level output at every frequency in order to get a reading that was consistent. This process obviously required a bit of time but could usually be accomplished between the completion of the stage setup and sound check time. If done properly, sound check was a breeze and everything was ready for that night's show.

Unfortunately, three to four hundred (let alone 10,000!) human bodies can have a substantial effect on the acoustics of any enclosed space. So, by showtime the hall could be a completely different acoustical environment. All those painstaking settings would no longer be accurate.

Then *real-time spectrum analyzers* (RTAs) were introduced. These were capable of obtaining a reading of the whole spectrum in one shot. They are not only able to follow changing signals, but can also give a graphic readout. This was very much faster than sequential or sweep analysis. RTAs are made up of a set of ⅓-octave band pass filters with ISO center frequency bands that should be the same as the ⅓-octave ISO EQ bands.

The outputs of these filters are averaged and scanned in real time, and their levels presented on some form of display for ease of viewing and adjustment. This was a more direct means of performing real-time analysis. The signal was put through the system, amplified, picked up by a calibrated mic (or when performing live venue analysis for feedback reduction, the actual vocal mic being used), and then put through a bank of parallel filters that divided the signal up into multiple frequency bands. These were then scanned and their levels detected and displayed. This was all done very quickly as it happened in real time, hence the name.

The speed at which these results were generated made it necessary to have a display that was capable of constant updating and providing ease of viewing. In the early days, this screen was a cathode ray tube much like the one used in television sets. These were bulky, delicate, and expensive, but compared to sequential analysis they were a joy to use. These early RTAs gave very accurate readings of the audio spectrum.

The same PA system in different acoustical environments may have a good deal of variance, but with a little practice the operator could become very proficient at keeping the response fairly flat. The fact that they no longer had to guess about acoustical changes (with audience arrival) or play with the EQ settings by ear made RTAs well worth their high price.

By the 1980s, with the use of smaller and less expensive components such as ICs as well as less bulky and expensive displays such as LED matrixes, the cost of RTAs dropped radically, even though some of these included a memory function. RTAs have many different features and capabilities.

The Klark Teknik DN 60 RTA   Many people consider these devices to be for one-time use unless acoustical parameters in the environment have been changed. Therefore, many studios rent a unit for only a day or two to take measurements and make adjustments. However, as an ongoing display tool, a spectrum analyzer can be remarkably helpful. This is obvious when performing maintenance or calibration, but I just loved having one all set up while recording and mixing, too.

I can walk over to a 31-band graphic EQ during a live show and pull down the exact fader required to reduce even a slight amount of ring. My ears have been trained over the years to discern frequencies, but in the studio, where my ears may be tuned to one particular instrument, some other sound or, worse yet, a lack of sound may slip by unnoticed.

After having done a good deal of recording with the DN60 sitting right up there on the console's meter bridge, I can understand why some manufacturers outfit their consoles with spectrum analysis capabilities. I can honestly say that its use not only improved my overall sound, but also my ability to perceive spaces in the content. I also found that while clients were at first amused by the "light show" movement of the display, they later started adjusting the frequency content of their instrument's sound as well as toning down their sometimes excessive use of effects devices.

The DN60 is very helpful in laboratory measurement and maintenance work. Reading distortion measurements in numerical form is one thing and looking at the distorted waveform on a scope is also very helpful, but seeing the spectral content across 31 bands can be more meaningful, as it allows you to view harmonic content.

The DN60 features three memories, peak hold, built-in switchable "A" weighting network, microprocessor control, variable response time, and much more. It provides its own signal output in the form of a quality pink noise generator. Output is via an XLR connection and its level is continuously variable (via a rear panel pot) from −6 dBm to +2 dBm.

The nature of pink noise is random, but via true RMS detection and averaged over a long period of time, it can be made to appear flat on a display. The Klark Teknik utilizes three response time settings as well as average and peak detection. By performing the mathematical averaging with a microprocessor, the slowest response time results in a very flat display that's easy on the eyes and allows for optimum speed in EQ adjustment. The thirty-first band shows a composite level for referencing the overall sound level of all 30 bands. I sometimes found this useful for watching peak levels during mixdown.

Once the pink noise has passed through the system under test, it is picked up through a low-impedance balanced mic input that also provides phantom powering. An internal preamp gain control allows for calibration to microphones with sensitivities between 0.25 mV per micro bar (2.5 mV/Pa) to l mV per micro bar (10 mV/Pa). Other input preamplifiers are available for mics not in this range.

However, since it is not only important for the mic to be flat with frequency, but also to have a polar pattern with the same frequency flatness at the sides, it is more advantageous to opt for the mic provided by Klark Teknik, which has that kind of pattern, and when furnished with the unit, the preamp and mic are calibrated to each other. For testing electronic equipment, balanced or unbalanced line-level input connections are provided that will handle a level up to +20 dBm. The unit has a switch for either line or mic input, which doubles as a weighting ("A") filter switch (by holding it in).

An eight-position input-level switch can be used to adjust the input to an optimum range on the display. The display's resolution can be set for either 1 or 2 dB per LED (16 make up each of the 31 bands). The three response speeds affect the attack and decay in the average mode, and the decay only in the peak mode.

A very helpful function for the measurement of short-duration signals is the maximum hold/peak hold, as it leaves a line of single LEDs across the display after the signal has decayed. The three memories store the full display plus the response speed, average or peak, weighting, line/mic, and maximum hold and peak hold switch-setting indications, thus allowing for later comparisons and judgments. All this is due basically to the built-in microprocessor control. Interfacing is also possible with options such as a reverberation analyzer, printer, and data accumulator. Furthermore, error codes are part of the programming and show malfunctions and overload conditions.

Overload means too much input level to either a single band of the display or across the full bandwidth. No damage is incurred or distortion produced due to clipping. The unit is simply warning the user of possible, inaccurate readings. Every time this device is turned on it goes into self-test mode, and if no problems are found the display blinks, gives a readout of "OK," and is ready for use. To demonstrate this unit's ease of use and accuracy, I'll give only one example. I used the DN60 while giving a lecture on parametric equalizers. Most of those who attended had virtually no knowledge of spectrum analysis or of parametric EQs. Using one of the parametric's bands to displace the flat frequency response, I asked each person to reflatten out the display again using a different band of the parametric while I covered my setting. Everyone was able to accomplish this very rapidly. Obviously, the kind of logical and accurate display provided by the DN60 is perfect for EQ setting, but I've used it on every measurement I've done since I received it. It's enabled me to see ground loops and other noise problems, and avoid making inaccurate judgments. I have used the DN60 a great deal while recording and can say that it was not only helpful, but it use actually improved the overall sound quality.

The Optional RT-60 Reverberation Timer Module   The pink noise source in the DN60 is driven by a line amp that is gateable. The gating is controlled by the microprocessor and enables the output turn on and turn off timing to be very accurate. Reverberation decay time is determined by first removing a signal and measuring the time it takes for that signal to drop 60 dB. To do this accurately, the sound source would normally have to first be set at a level that may be very high, so to ease the measurement process; the decay time is calculated from a smaller decibel drop. Through the internal communication card, the RT-60 optional reverb timer can control the DN 60's microprocessor. Once the RT-60 is hooked up, it is included in the DN60's self-test so that it "knows" it is present.

This dual unit is now capable of measuring decay time. It samples the decay curve every 550 microseconds, averages 16 of these samples, and stores them until the cycle is finished, at which point it has 790×8-millisecond samples.

The DN60/RT-60 can measure a decay from 0 dB down to -30 dB, or any range in between, with as little as a 2 dB drop and calculate (extrapolate) a measurement for a 60 dB drop. The parameters of decay can be changed after the measurement without affecting the actual measured value. It is necessary to take many tests with any reverb timer and average these together for more precise readings. This dual unit does your averaging (accumulating) for you (up to 32 readings) and will measure and accumulate the full bandwidth or an individual band. Since the RT-60 is in control of the microprocessor, some of the functional parameters of the DN60's switches are changed. The peak hold switch now shows at which frequency band the measurements are taken. The peak/average controls pink noise turn on. This is important in saving your ears as you can start the noise just before measurement. The RT-60 is also very user-friendly in that it tells the operator via the readout when the unit is properly set up for measurement. It indicates running and calculating, as well as accumulating modes, and more. Once measurement is completed, the display shows a graph of the decay curve. Here the ⅓-octave frequency bands are converted to time columns. The display's time parameters can be changed before or after measurement to three settings: 16 milliseconds/column, 64 milliseconds/column, or 208 milliseconds/column. The curve can also be stored in the DN 60's memory. When recalled, it will show if accumulation was used, how many accumulations, the frequency band measured, and the window and time parameters can still be changed.

Furthermore, the RT-60 accumulator can be used in the spectrum analysis mode to average up to 32 frequency response displays. All this gives you a very powerful tool to work with. I checked studios, control rooms, and all kinds of audio equipment. I set and calibrated plate reverb times, as well as timed digital reverbs, tape reverbs, and chambers, all with great ease.

The DN6000 is the current RTA from Klark Teknik. This one offers both ⅙-octave *and* stereo or dual ⅓ analysis plus a display of the sum and difference with built-in A and C scale filtering. The display can be set to peak or average, and it has a peak hold function. Its internal signal generator produces sine wave and pink noise signals, and has tone burst (for reverb timing) and sweep capabilities. The DN600 handles reverberation timing without the need of an optional add-on. It can store 32 memories and also has a printer port for direct printer output, but the biggest addition is this unit's ability to directly interface with the DN 3600 programmable dual-channel ⅓-octave equalizer. Hold on to your hats, because the combination of these two devices equates to *automatic* correction of equalization settings!

FFT, TDS, and TEF Measurements

While ⅓-octave real-time analysis provides a wealth of information, it leaves out the very important ingredient of the time. It will tell you what the frequency is and how much of it there is, but it will not tell you when it occurs. Remember temporal distribution or a sound's place in time is important psycho-acoustically. *Time-delay spectrometry* (TDS) can give the levels of only the reflections and provide mathematical hints as to where they originate. Thus, more in-depth analysis is provided by these analyzers. *Fast fourier transform* (FFT) transforms the time domain to the frequency domain.

Let's start with a bit of history. In the early 1960s a scientist named Richard C. Heyser was working at the California Institute of Technology's Jet Propulsion Laboratories. Jets were still new then; just ask any old-timer about the experimental aircraft like the X-15 and test pilots who became our heroes as kids. Mr. Heyser was calibrating the frequency and phase response of an open-reel tape recorder and the time lag between the output of the two heads was giving him some trouble with his measurements. So he just went and assembled a bench-load of equipment to compensate for this offset in time. He found that his new test method not only corrected the measurements, but canceled out the harmonic distortion. He realized he had something there and kept working on it and described it in a 1967 AES paper.

Here's my brief synopsis of a *time energy frequency* (TEF) test procedure as used with speakers: a sine wave generator is used to produce the test signal that is fed through the monitor system and a mic picks up the output from the speaker. In addition, the test system utilizes a tracking filter in sync with a generator and a delay module, which offsets the input timing to correct for any delays incurred by the signal such as its propagation through air.

The bandwidth of the filter and the time delay can be varied to allow for a test of only the direct sound, just the early reflections, or both. Even the earliest of these devices were quite informative to say the least. To say that they completely changed acoustic measurement would not be an exaggeration.

Because of this inherent noise rejection of TDS measurements, reverberation decay time can be calculated and displayed even when a high amount of background noise is present. TEF measurements can provide close to anechoic results in *any* room.

A short time segment of a periodic signal can be sampled for a complete or whole number of periods and from the DFT the FFT can be extrapolated using one of the many "windowing" techniques available.

This tiny little area of measurement has acronyms for days, so here's a partial list that will help keep you somewhat afloat.

FFT	Fast Fourier transform
TDS	Time-delay spectrometry
TEF	Time energy frequency
FTC	Frequency-time curves
ETC	Energy-time curve
EFC	Energy-frequency curves
ALC	Articulation loss of consonants
STI	Standard speech transmission index
RASTI	Rapid speech transmission index
RTA	Real-time analyzer
RT-60	Reverberation decay time
NC	Noise criteria
NLA	Noise-level analysis

Back to the history. Mr. Cecil Cable picked up on this measurement procedure's potential and in his 1977 AES paper he described his research into comb filtering caused by early reflections. In 1979, Crown International (the amplifier manufacturer) began manufacturing their TEF system (a TDS analyzer in combination with FFT) that analyzed time, energy, and frequency.

By 1980, Mr. Don Davis was not only teaching "Syn-Aud-Con" (synergetic audio concepts) students how to use a TEF device, but along with Chips Davis, he was designing and building control rooms based on the early reflection/comb-filtering information compiled by this system. The resulting *live end, dead end* (LEDE) control room design had the psychoacoustic ability to make the control room boundaries seem to disappear from the listener's perspective.

The only way to describe the one "true" LEDE control room I've had the opportunity to work in is that your perspective is not one of being in a control room listening to what is going on in the studio, but one of actually *being* in the studio listening to what is going on. The acoustics of the control room do not come between the listener and the sound.

The MLS, or *maximum length sequence*, of signals is used with an FFT to provide frequency response, speech intelligibility, and time analysis. With random signals such as white or pink noise, the signal must be sampled over a period of time in order for the analyzer to generate accurate readings. A pseudo-random signal is made up of a group of sounds that display any appropriate parameters needed to perform a specific test. This signal has a set length that is sequentially repeated.

Level vs. frequency measurements are known to be an important indicator of a system's capability to produce sound adequately, but time or phase vs. frequency measurements are not given the same importance. In fact many people feel that phase inconsistencies are inaudible. Mr. Heyser's answer to this was to point out the fact that with a system's low-frequency driver positioned relatively close to the listener and the high-frequency driver placed a mile back, if the high frequencies were driven with enough level this system could actually achieve a "flat" frequency response. In this case the phase response would give a better indication as to how bad it actually sounded.

In *source-independent measurement* (SIM) FFT analysis the input is fed to the analyzer and the system. The system's output is then compared to the input signal. Thus, the test signal is "independent" of any restrictions other than its being broadband in nature. Therefore, like the dbx RTA-1, it is possible to use a music program as the test signal. However, here the FFT's "windowing" capabilities allow the operator to focus in on the test results down to a single cycle.

By gating the input signal, different sections can be analyzed. The display can show how frequencies change with time. Short gating times provide superior time resolution while longer gates give good frequency resolution. While there may be a trade-off between time and frequency resolution, the 3-D plots produced by these tools are no less than spectacular.

Power Amplifiers

This studio is running dual Crown DC 300s strapped mono, which was a popular configuration as it provided the power, durability, and sound quality demanded for highly accurate audio referencing. But they were used everywhere as they were also extremely roadworthy. Everyone has a Crown DC 300 story, but I've got two witnesses (that are still alive) to mine.

I had helped out a small rock club in NYC with their PA system. It wasn't anything elaborate, but it served its purpose of filling a room with an audience of 150 to 200. It had two stage monitor wedges (Yamaha, 15-inch and horn) powered by the dual 150-watt amp internal to the Yamaha mixer. The main "house" speakers were also Yamaha and had the same drivers but were configured and stand mounted for a forward, ear-level throw and were powered by a DC 300. The use of speakers that were all virtually the same allowed for a single, dual graphic EQ and, due to the fact that the amps were just about matched in power output, the system was an easy one to operate. This suited me just fine as I'm not a fan of panicked emergency repair calls.

One evening around 10 P.M. I get the "Oh my gosh!" call from the club's sound man. Seems a water pipe in the ceiling right above the sound system burst and flooded all the equipment just as that night's band was mounting the stage. I lived close by and, figuring the worst, arrived pronto. But Joe (the sound guy) was on the ball and had powered down all the equipment immediately.

Upon inspection I found that the mixer, EQ, and separate reverb device had only been lightly splashed. They were dried off quickly and the system rewired so that the mixer's amps were driving all four monitors. After explaining to the band that they wouldn't be able to crank up their instrument's amps and still have the crowd hear the vocals (they were amazed that they were going to get to play at all), I set about the task of dealing with the DC 300.

It was positioned off to the side of the rest of the equipment and directly under the flood gates. When I opened it, the poor thing was *full of water!* I drained it, placed it on a side table, and commenced hitting it with the hair drier I had brought with me. Heat shrink guns are no good here as they can melt solder and wire insolation.

I used a lot of freon aerosol; this was before we knew that freon depletes the ozone layer. Too bad, 'cause that stuff was great at displacing water from underneath components, moving little droplets out into the open to be dried up by the hot air. Also, 99 percent pure isopropyl alcohol will do the same thing, *but* it can also remove the connection, component numbering, and values silk-screened onto the *printed circuit boards* (PCBs).

In less than 10 minutes the amp was dry as a desert.

I had the band "take five," reconnected everything the way it had been, and walked out of that place looking like a hero. All because I had convinced the owner to purchase a used Crown DC 300. I can only hope that you end up with the same kind of happy ending "war stories" to tell the next generation of audio professionals.

When the monitor system is viewed as a chain and the weakest link is considered, the amp is seldom as much of a problem in terms of performance anomalies as is improper gain stage setup, cheap (16-gauge zip cord) speaker wire, misleading speakers, frightful equalizer settings (with the resulting phase and level distortions), or faulty room acoustics. The biggest clue to amplifier quality, in terms of reliability, is revealed by simply looking at the manufacturer's warranty.

The professional model amplifiers that are available today have low distortion and noise and more than acceptable frequency response specs. I have a problem with some of these specifications because they were determined on a test bench with the amp feeding a purely resistive (DC) load. For instance, the damping factor isn't specified for high frequencies and any subpar specifications (low values) here can cause these upper-band signals to sound harsh.

Amplifier Design and Types   Low levels can be handled by an amplifier with a single op-amp, transistor, or tube. These are "single-ended" amplifiers. However power amplifiers most often must use two. Which two and how they are configured make for different amplifier classifications:

Amplifiers that always have current flowing through them are "class A"

Amplifiers that only have current flowing through them one-half of the time are termed "class B"

Amplifiers that have current flowing through them less than half the time are called "class C"

Amplifiers that have current flowing through them more than half the time, but less than all the time, are termed "class A-B"

The practical trade-offs are straight ahead. Amplifiers that have current flowing through them less than all the time produce less heat and are substantially smaller in size and lighter in weight.

Class B amplifiers use dual components (tube or transistor) to split up the amplification of the positive and negative sides (rails, cycles, parts, or legs, if you will). This is called a "push-pull" design. While the positive is being amplified, the negative amp is switched off and vice versa. This configuration provides more power from fewer components, but as the components crossover between their on and off stages they produce distortion.

This is the reason all amplifiers classed as anything except "A" have a reputation for being noisy. In the early days of class B it may have been deserved, but that's old history. For example, symmetrical and complementary (matched) NPN and PNP transistors along with filtering overcame a good deal of crossover distortion. By operating solid-state amplifiers as A-B, wherein an amplifier operates in class A mode at low levels where the crossover distortion is most easily heard and as class B when more amplification is required the problem is virtually eliminated. But negatives in the field of audio are hard to kill.

The fact is, by the early 1980s designs were being utilized that provided the efficiency of class B amplifier circuits with very little "notch" or crossover distortion. As you'll see with the Crown PS 200, by the mid-1980s it was no longer anything the audio recordist had to consider.

Professional recording studios have particular amplifier requirements that must be fulfilled as the referencing must be uncolored. They must be reliable under constant use and have protection circuitry to avoid lengthy down time and costly speaker damage, yet not cause those annoying repeated cut-outs as may be found with some relay designs that incorporate automatic resetting. An amp must have enough power-handling capability to preclude distortion at the high levels that referencing sometimes requires. Speaker efficiency also has a large effect on amplifier power output ratings so it's best to start by checking the specs that state the amp's RMS into 8 ohms but also the rating at the load the speaker will present.

Cooling How an amplifier goes about the business of dissipating heat is important as convection fins rely on room air movement currents (hard to come by in a closet). This can be aided by the manufacturer's use of a simple fan that switches on with amp power up or a stepped fan, where the speed is determined by thermal monitoring. The amount of noise produced by the cooling system's fan must be considered. There's an advantage to having the heat blown out the front of the unit as opposed to the rear as it helps avoid heat buildup within a rack of equipment. Having a separate "amp room" may be out of the question, but an air-conditioned enclosed rack or amp closet is a nice remedy to amp overheating and noise.

Ventilation All electronic devices are susceptible to damage from excessive heat. Thermal cycling (periodic heat-up) speeds up the damage. Amplifiers require good ventilation, whether they are convection cooled by the use of metal fins or aided by the use of an internal fan. When mounted in a rack, the space behind and above the back of the amp should be open as this is where the heat sink fins and fan exhaust output will be located.

The idea is to create a chimney effect so that heat can rise, leaving cooler air in contact with the amp. One mistake I've witnessed was the feeding of an air conditioning branch-off flexible duct directly into the top of a rack. This not only pushed the hot air back down around the amps but caused a considerable amount of condensation to develop. Wet amps are not what you want to see when working behind a rack.

This note should not be required but always turn your amplifiers on last in your powering-up sequence as it will avoid the speaker and ear-damaging pops and snaps that will be produced when powering up subsequent equipment.

Bad Amps?   Most problems with modern power amplifiers are due to input impedance and output load interfacing. As always, balanced inputs should be fed balanced signals and unbalanced inputs be fed unbalanced signals or problems may occur.

Take, for example, a source that's feeding 6 volts peak to peak to the amp's input. Each side of the balanced input circuitry would be handling 3 volts. If input as an unbalanced signal, one side of the balanced input circuitry would have to handle twice the amount of level (6 volts) while the other side is fed no signal. In this situation the same source may cause clipping even though the amplifier could easily handle this input level if it were balanced. Forget about modifying the input connector's wiring to "convert" an unbalanced signal into a balanced one. The result could very well be more hum and RF noise from improper shield connection, a lower-level input signal, which diminishes the S/N ratio, or an increase in distortion. By trying every combination you may come up with a method that doesn't cause severe noise degradation, but it's *always* better to use an impedance-matching, balanced-to-unbalanced interface.

Interface Wiring   Most people understand the importance of having speaker cabling that's of a heavy gauge, but few consider the utility cabling (or those lines that feed power to the amplifier). Since the source of the amplifier's power is, in fact, this utility power feed and because all wiring has some inherent impedance that causes attenuation of the voltage passing through them, especially when the current demand is high (see Ohm's law), it is important to utilize power cable sized thicker than the 16-gauge of most "hardware store" extension cords. You cannot expect an amp to produce full power when it's being powered with a 16-gauge extension cord because this cable will lose a lot of power to heat generation and in the process produce a hefty magnetic field.

Here the power that's lost doesn't mysteriously disappear but is generally put to "good use," heating up those extension cables and causing them to radiate a nice 60 Hz field that will do its best to leak into any audio cables that are close. When thought of in this way, the small step of touching to check extension cables for heat generation, if not determining the gauge sizing, seems simple and obvious, yet it tends to be overshadowed by all the attention placed on the speaker wire boutique.

Setting the Monitor System Gain

I've seen amplifier input levels set 6 dB below unity gain to allow for enough extra headroom so that the 10 dB peak levels common in music programming would not cause distortion. Some studios adjust their power amplifier's input attenuators to the full clockwise position.

This is not proper gain stage adjustment as it often requires the console to be set at a lower than optimum output level in order to avoid excessive listening levels. Additionally, these low-level console settings will place more noise (S/N ratio) from the console's output amplifiers into the monitoring system. This low headroom and high background noise situation could have been avoided by proper gain stage setup.

Amplifier attenuation should be set so the volume in the room is at a predetermined "normal" listening level when the output meters of the mixer feeding the amp read 0 VU. This, of course, supposes that the mixers' output voltage has been calibrated to the correct level (adjusted as per a voltmeter reading while connected to the amp). After which the meters' driver amps were adjusted to indicate a 0 VU reading at that level.

Here's the step-by-step process:

1. Set the console output bus levels to unity gain while reading the console's output voltage on a voltmeter with the amplifier (the load) connected.

2. Adjust or "calibrate" the console's meters to read 0 dB at this level.

3. Set the power amplifier's attenuators for what is considered the studio's nominal listening level with the console level set at this new 0 dB output level.

You get less noise and more headroom. It's easier on the ears and makes for longer lasting equipment. Good payoff for five minutes of work, no?

On the load impedance or output side, amplifiers should be connected to loads with impedances that either match its rating or are no more than twice as high. Too high a load simply diminishes output power, but too low a load will not only reduce output power, but may also increase distortion. In some cases low-load impedances can cause repeated triggering of an amplifier's protection circuitry interrupting operation or possibly cause permanent amplifier damage.

It's important to remember that impedance is reactive, meaning that its value (in ohms) changes with frequency. Combining two 8-ohm speakers in parallel would infer a load of 4 ohms, but at some frequencies the actual load could drop to 2 ohms or lower.

Headroom problems caused by load impedance can be avoided by overspecifying (by as much as double) the power-rating criteria the amplifier must meet. Here if the speakers are rated at 4 ohms, the amp should be rated to operate down to 2 ohms. If 400 watts is required by the 4 ohm load, specify at least 600 watts into 2 ohms. This may be overkill, but it's better than requiring more from an amplifier than it is capable of handling, which could result in distortion, overheating, and breakdown.

Fuses If they must be used, fuses should be positioned close to the amp as the amp will not need protection from shorts in the speaker cable that occur between the amp and the fuse.

I'm not an advocate of the use of speaker fusing. Their internal conductors vary in resistance as they heat up and heating up rather quickly is part of their job. I've seen this reaction on a scope and while this distortion is not quite as pronounced audibly as it is visually, it can be heard as high-frequency hash when those near field monitors they are being used to protect are driven anywhere close to fuse-blowing levels. If the speakers *must* be driven hard, and you want to avoid component blow out, just use two pairs of speakers. They'll sound the same, produce twice the level, and be able to handle the abuse.

Bad Specs? During the early 1980s when I took on the responsibility of reviewing equipment for an audio publication, I was required to augment the test equipment I already owned.

So in addition to the two dual-trace oscilloscopes, two multimeters, a stereo (dual indicator) volt/ohm meter, stereo VU, peak, and PPM meters, two signal generators, load resistors, a frequency counter, a variable band pass/notch filter, and a large contingent of microphones, music instruments, amplifiers, and effects I already owned, I bought, rented, and borrowed additional gear. Due to the rented and borrowed equipment all of the following test gear was not available to me on a continual basis, but my test "bench" was at all times during this period, quite formidable as it often included a Leader LAS-5500 system analyzer, Amber 4400A Multipurpose Audio Test Set, Hewlett Packard 8903 Audio Analyzer, Tektronix 500 series mainframe loaded with an AA 501 Distortion Analyzer and an SG 505 Oscillator, Krohn-Hite 5500 Distortion Analyzer, Klark Teknik DN 60 real-time spectrum analyzer with RT-60 reverberation timer, a Goldline model 30 RTA and a Neutrik Audiograph 3300 strip chart recorder with 3322 oscillator and 3312 measurement modules.

This is not a brag; in fact having all this stuff around actually became a pain. But it should help to substantiate the next point. In the half dozen years or so during which I tested several pieces of equipment per month, I did not find one piece of equipment that was not within the manufacturer's rated specifications. When I mentioned this to an "old-timer" his response was, "Of course you didn't. Don't you know the first purchase of every new piece of audio gear placed on the market is made by the competition? And if there was *any* discrepancy, they'd be screaming it from the highest mountain."

Modern professional amplifiers have fine noise, frequency response, and distortion specs. In fact it's actually hard to find an amp made by a reputable manufacturer that has bad specifications. These people have some of the best designers and testing facilities in existence.

Outside of power rating and protective circuitry, other power amplifier performance specifications need to be discussed. I know we've already dealt with specifications in the section on consoles, but amplifiers must meet unique requirements so some of their specifications are peculiar to their functioning. In order to understand the sometimes bizarre nature of audio specifications, one must first understand how they are formed.

Standards and Specifications

How They're Established and What They Mean   The practice of establishing standards dates back to the dawn of civilization with some examples over 5,000 years old. The Egyptian Royal Cubit, for instance, was based on the length of the Pharaoh's forearm, and it was sub-divided into 6 palms and 24 finger widths. This was inscribed on a piece of wood and was actually used in the building of the pyramids. Fortunately, we don't have to carry a hunk of wood around today to confirm that a device is indeed standard 19-inch rack-mountable (6 palms and 21 finger widths). However, at times determining standards can seem almost as bizarre. Doing test reports for an international publication, I found the lack of international performance measurement standards to be oppressive. What do I suggest to alleviate this problem? Let's first take a look at standards in general and how they are developed.

There are two kinds of standards: voluntary and mandatory. Let's discuss voluntary first. While this type of standard does exert some amount of control over quality, it allows for a good deal of flexibility as far as its application goes. Since the measurement of a parameter is not restricted by law, in certain cases the inclusion of, for instance, a weighting network or a band-pass filter, may negate your ability to compare specifications.

How can it be that a "standard" would allow this to occur? Well, it may or it may not. To understand the less than perfect nature of standards, we should examine how they are prepared.

First, a committee is formed. These individuals may be chosen by a company, an organization, or a government. They are generally involved in that particular standard's field to the degree where they have gained wide respect. They will believe in the value of the standard, as they often have to give up much of their own free time during the process of its development. The information-gathering process starts usually with the only authoritative sources available—previous standards (if they exist).

Still, work experience, as well as academic and government research, is equally important. Next committee members are made responsible for specific sections of the standard. They take down everything anyone has to say on their particular subject area, circulate it, argue over it, tear it apart, and reassemble it, only to have the whole process start again. All this is done at meetings that may range from twice a week (small groups) to twice a year. Eventually, the preliminary draft stage is reached and the results presented to the greatest number of people possible in order to get their input.

Federal laws require all voluntary standards organizations to publicize standards projects, while they are in progress, in order to allow those who will be affected by them to be made aware of this preparation and make their views known. To gain committee approval, the draft is now subjected to repeated revisions and rewrites, with ongoing evaluations of returning comments and inclusion of what is considered viable until those involved consider it as good as it can get under the circumstances. It may be understood to be "imperfect," but the consensus is that it's acceptable for its purposes and further revision wouldn't be worth risking the standard's timeliness. Even after all this, however, the process never ends, since new information is constantly flowing in.

Obviously, this is time consuming, expensive, and hard work. You now have some idea of the complexities of standards development. Not as simple and scientific as you might have thought. Now, think of all the organizations writing standards in the U.S. alone. Which one should be universally accepted?

The *American National Standards Institute* (ANSI) is the accepted coordinator of voluntary standards. A neutral organization, it does not itself create standards, but only accredits and distributes them. However, not all standards organizations are affiliated with the ANSI. What's more, many audio-testing standards are not covered in its listing. So, as a "governing" body, it is not conclusive. Furthermore, since testing is not restricted by law, who's to stop anyone from adding in a weighting network or a filter to reach a certain specification rating? Our path, as you can see, is a bit muddied, but when we get to international cooperation on standards it almost turns to quicksand. For example, in 1977 the *National Standards Policy Advisory Committee* (NSPAC) recommended a national standards policy. The ANSI helped with this proposal, which called for an organization much like itself to represent the United States in international standards policies. Some standards organizations, however, felt that the ANSI, as it was structured, could not handle this task. The ANSI has since provided for a new charter the International Standards Council (ISC), but it was, as far as I know, never "officially" appointed to this role. The establishment of a national policy needs to eventually come about, as it is necessary to allow U.S. government and industry to have a voice in international standards. This would mean cooperation between industry, government, and the standards organization.

Countries such as West Germany and Japan are well known for their use of joint ventures involving manufacturers, universities, and government. Furthermore, nations involved in international standards organizations work very hard toward achieving the adoption of international standards that are more favorable to the technologies developed by their own nation's companies. As far as international standards organizations go, there are many. The *Pacific Area Standards Congress* (PASC), for example, includes the ANSI, the *Standards Association of Australia* (SAA), the *Canadian Standards Association* (CSA), the *Underwriter's Laboratories* (UL), the *Japanese Standards Association* (JSA), as well as standards associations of other Pacific area nations. Why only Pacific area nations? I have no idea. What good is an "international congress" that doesn't include the European countries? Recently, I heard of the formation of another organization, to be made up of Canada, the United States, England, and Germany. How can they possibly leave out Japan? The *International Organization for Standards* (ISO), created by the United Nations in 1947, did include the *European Committee for Electrotechnical Standardization* (CENELEC) and the *Pan-American Standards Commission* (PASC), as well as the Soviet Union and China. In fact, today there are almost one 100 nations involved with over 100,000 experts doing technical work comprising over 10,000 documents and more than 3,000 published and revised standards. Yet only a few countries have accepted these as national standards.

In order to further understand the lack of international harmony in terms of standards, let's take a look at a mandatory standard. European Electrical Appliance Safety Standards are obviously very important to American manufacturers wishing to export their products. Even though UL listing and European Standards are relatively identical, aside from variations in terminology and interface, the UL listing has very little value or relevance in European approvals. The *Certification Bureau* (CB) calls for successful completion of testing by two test houses selected by the CB secretariat (from one of the test houses appointed on a rotational basis). This, however, has its problems; its scope is very limited and, being an association of test houses, the CB is not fully representative of the individual country's standard-making bodies. Then, along comes the *European Committee for Electrotechnical Standardization Certification Agreement* (CCA), which calls for testing to harmonized standards for approvals. It is a simple concept, wider in scope, but less binding since it depends on the good will between nations and test houses. The CCA's value may increase with international solidarity to eventually become the main tool for European certification, but as with most generalized agreements it still suffers from test ambiguity and escape clauses that allow for varying interpretations. I do not pretend to know a great deal about standards and the organizations devoted to their development, but I've presented this very simplified outline in order to show the problems we all face in the verification of performance standards. All information on any standards organization or committee is subject to such rapid change that much of the above information is already inaccurate. As you can see, even for safety standards there are no definitive rules. There aren't any international standards to speak of and it will probably be a long time before there are.

What about the manufacturers themselves, can't they come up with some new testing methods that will be acceptable to all concerned? It's not easy. New performance verifications require investments of huge sums of money in equipment and skilled personnel that are well beyond the reach of most equipment manufacturers. Also, once accumulated, the data may not have the acceptance it would if it had been produced by an institution that is both technically credible and competitively neutral. Furthermore, if these new methods are not adopted by the industry in the form of a standard there would be no benefit, not even in terms of prestige.

The end result of all this ambiguity is defined by the term "specmanship." Manufacturers defining parameters by methods that differ from the competition may let them be more cost-effective in their manufacturing process, but they are causing consumers to make direct comparisons between two sets of specs that cannot truly be compared. In many cases nothing is guaranteed, as witnessed by the old escape clause "specifications subject to change without notice." Often, in the manufacturer's printed matter (owner's manual, spec sheets, and service manual), there'll be ambiguities due to dual specifications. Although most manufacturers are straight ahead and honest, all could be suspected of "specmanship," when the consumer begins to regard all specifications as inflated, unimportant, or beyond comprehension.

Manufacturers who are not up-to-date in their testing, both equipment-wise and skilled labor–wise, will fall behind their competitors, both financially and technologically, so that even though they may see their present methods as a cost-efficient approach, this might turn out to be a liability. Eventually, those that exaggerate or misrepresent their products by having less-than-honest specifications will lose face and profits since specmanship only leads to skepticism.

As stated, one of the first buyers of any company's new product is its main competitors. If the specifications are not what they are claimed to be, you'd expect them to announce this fact to the consumer from the highest treetop. Interestingly, the fact is that this discovery will be communicated directly to the other manufacturer. This is due to the fact that it is in the best interest of both to have internal policing among themselves. The result is that I have found no equipment that I have tested not to be within the manufacturer's rated specs (except where shipping damage has occurred). The fact that this whole system of product evaluation even comes close to working at all is due more to the basic honesty and peer pressure between manufacturers than to any outside regulation. So everything was fine for me. It took a bit of work to verify all of those different specifications, but it wasn't impossible. But what about the consumer? Don't they deserve to know what each product is capable of by its parameter verifications being presented in some form of standard terminology?

Expecting international organizations to agree on one universal method is fruitless for the time being. Government involvement won't help, developing new test methods is almost hopeless, and asking manufacturers to change their procedures overnight is unrealistic.

Educating the consumer is tough, as there is a lack of information on standards available even in technical schools. In fact, the only people who really know anything about audio standards are almost exclusively in the industry. I'm talking about you, the audio recording engineer. It's part of your job to be able to explain the technical importance of a parameter's specifications to the client. It's all out there. We've been at this "audio engineering game" for over a hundred years now. You do not have to memorize equations or test methodologies. You just need to have a good enough bead on what's being checked and why in order to be able to relay its importance.

Amplifier Specifications   Some specifications are peculiar to power amplifiers, but others such as distortion are used to rate other signal processors as well. Any production of harmonic content is considered harmonic distortion. This parameter is measured by feeding a pure tone (one with no harmonic content) into the device, filtering it from the output, and measuring what remains. The RMS measurement of this remaining signal is termed *total harmonic distortion* (THD) and noise. If both high- and low-pass filtering is used, the result of the RMS sum of all the harmonics will be only the THD.

Intermodulation distortion is made up of harmonic components caused by an interaction between two signals. Again there are several test methods and they can use different pairs of frequencies. The SMPTE method seems to be the most popular so here's the rundown. Take two signals with frequencies of 60 Hz and 7 kHz. Mix them in a 4-to-1 ratio (60 Hz set 12 dB higher than 7 kHz). When these two frequencies are filtered from the output, the remaining distortion will be clustered in areas on both sides of the 7 kHz band. Why 60 and 7,000 Hz? Well, 60 cycles is easy to come by from any utility AC outlet and 7 kHz was the upper bandwidth limit of the broadcast medium. I swear!

THD measurements at 1 kHz are easy to make and require only a passive "balanced bridge" network consisting of four resistors along with an oscillator and a transformer. This system ignores all the 1 kHz signals common to both the input and output. Therefore, only the distortion will be measured.

THD per frequency is more informative but much more difficult to measure since a filter must track the oscillator's output. Also there can be no delay in the system such as those found when a signal travels through a sound system to the speakers and is picked up by a mic and returned to the test equipment. Yet some modern automatic testers can perform this tracking and output a detailed evaluation right on the factory floor completely unattended.

Cross Talk   This measurement is the voltage ratio, in negative decibels, of the signal on a nonsignal-conducting line as compared to the voltage on the line being driven by the signal. As an example, if the level on the driven line is set at a unity gain level of +4 dBu and the amount of signal picked up on an adjacent line is −60 dBu, the level of cross talk is −64 dB. Cross talk *does* vary with frequency so any measurement should include the test frequency. Graphs plotting cross talk vs. frequency are possible, although I've yet to see one. Cross-talk measurements require the use of bandpass filters set exactly at the test frequency to eliminate it from the noise.

Understanding cross talk can lead to the causes and therefore the remedies of certain problems. AC signals, if large enough, can generate fields that will produce an image of themselves into conductors that are close by. Power amplifiers can put out very large AC signals. Radiated fields can couple with neighboring signal lines capacitively as an electrical bleed and inductively as a magnetic bleed. If the cross talk increases with frequency, it's more than likely capacitive or electrical. Proper shielding or adding a ground plane will reduce the bleed, but it will not do much for inductive/magnetic cross talk. Running crossing conductors at right angles helps eliminate the low-frequency humming of magnetic coupling but will only slightly reduce capacitive bleed. Once you check the problem out you'll have a better idea of how to approach its correction.

Weighting Networks   These are referred to as networks because they are made up of a network of resistors and capacitors that form an equalizer. This pre-set EQ is imposed on a signal during measurement. There are three main (ANSI) weighting equalizers: A, B, and C. They originated in the sound power measurement field, where human exposure to noise is of paramount importance. Human hearing does not have a flat frequency response. We are more sensitive to midband frequencies. The researchers who discovered this phenomena also plotted out this response. The weighting networks utilized these "Fletcher-Munson curves" as a base.

The C scale continuously rolls off in levels above 10 kHz and below 20 Hz at a −6 dB per octave rate. So the bandwidth is down 3 dB at both 9 kHz and 35 Hz and it has the least radical low-frequency roll-off of the three scales. The B scale is down 17 dB at 35 Hz and is also −3 dB at 9 kHz. The A scale is down 38 dB at 35 Hz but only down 2 dB at 9 kHz.

Sound-level meters use weighting networks to derive different responses to noise as a function of frequency. The A scale is a close approximation of the frequency response of human hearing and is used to determine acceptable levels of noise as per ANSI and, more to the point, OSHA and EPA standards. A is also used for speech interference measurements (such as the ability to use telephones in noise-prone locations) and for light noise exposure levels (those below 55 dB). The B scale is used to test noise exposure levels between 55 and 85 dB, while C is used for levels above 85 dB. A and B are used for sound-level measurement and C is used when measuring sound pressure.

I have no problem with weighting networks when used on sound-level meters. I'm just a little apprehensive about their use in specification checks since using an A-weighted scale when performing electronic noise measurements will reduce the level of 60 cycle hum by roughly 28 dB!

Figure 2-29 is a representative drawing of the three weighting scales. The small "x" marks are just some of the exact crossing points that I used to aid my hand drawing of these curves. Who was this drawing for? Myself. I had found a schematic diagram of a resistor/capacitor network that included a switch that made it able to act as any of the three filters. By the way, I found that the A curve works very nicely on drum overheads.

Damping Factor The damping factor is measured to show an amplifier's stability while driving a load that varies with frequency. The rated damping factor can be measured with an input of 1 kHz at a set reference level. However, some manufacturers rate it per a single fre-

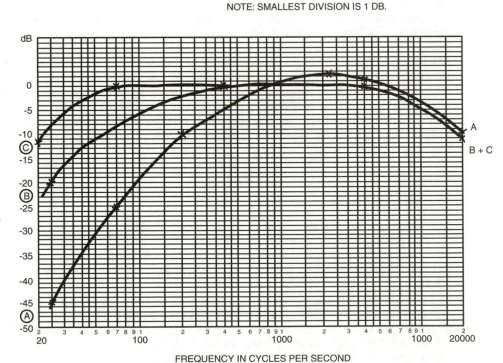

Figure 2-29
This representative drawing shows the three weighting scales. The small "x" marks are just some of the exact crossing points that I used to aid my hand drawing of these curves.

NOTE: SMALLEST DIVISION IS 1 DB.

FREQUENCY IN CYCLES PER SECOND

quency and load, while others use a bandwidth of frequencies (usually at the low end as these frequencies are harder for an amplifier to control) with a specified reference level and/or load.

Due to these test procedure variations, attempting any comparison is fruitless. Another way the damping factor can be calculated is by dividing the amplifier's maximum output voltage while driving a *load* (L) by the product of the amp's *unloaded* (U) voltage minus the loaded (L) output, or L/(U-L). Yet this test method requires driving an amplifier at its maximum power level with *no* load. So this is not something I'd care to attempt. Like slew rate, the damping factor relates to an amplifier's ability to "keep up with" changing signals. These, along with transient response, have elusive definitions. If you really want to know about an amplifier's frequency response, distortion, transient response, and overall performance, look at its full power output on an oscilloscope with square wave inputs of 50 and 10,000 Hz while it's connected to the applicable load. Today the choice of power amplification is more often based on price, protection circuitry, warranty, and power ratings as audible artifacts are rare.

Amplifier Output (Speaker) Cabling Since speaker impedance varies with frequency, the added impedance from the speaker cable can have an effect on the frequency response of the system. This little fact has spawned a whole audio marketing niche designed to alleviate the gullible from their excess cash. The fact is that optimum results occur when heavy gauged speaker cable (less resistance) is used and when the position of the amplifier is as close as possible to the speaker.

Speakers with internal amplifiers (and as far as I know they go back to the 1960s and Phillips) often sound superior not only because of their short amp-to-speaker cabling, but because the amps, electronic crossover, any built-in equalization, and cabinet have all been designed specifically for working in that combination.

Some Facts on High-Power Audio Cabling The resistance of 1,000 feet of 16-gauge cable is about 8 ohms. For 12-gauge it's 3.2 ohms. That takes care of resistance, but there is also series inductance and shunt capacitance to deal with. This means more trade-offs: The inductance of a wire pair can be reduced by having closer spacing between them, but this increases the capacitance; very wide spacing also causes an increase in interference.

To save you a lot of trouble, the fact is that common 12-gauge zip and 12-gauge vinyl-jacketed twisted-pair "speaker" cables are about equal in impedance and make a much better choice than thinner zip cord cables and are often a much saner choice than some "special" audiophile cables.

Large gauge co-axial type cables *do* exhibit lower impedances (lower inductance *and* capacitance) but, due to their cost and stiffness, their use is not always feasible.

In my opinion, it's best to leave specialty cables to the audiophile unless you are offered a trial use with a money-back guarantee in writing! I tell would-be specialty cable purchasers that before investing in *any* "special" cable, they should open the speaker cabinet and have a look at its internal wiring and suggest they start their speaker cabling improvements there.

Passive crossovers are notorious offenders as far as audio quality as they transcend common cabling anomalies, just look at the rolls of wire that form the inductors! Straight out, the best bet is to feed the line-level mix bus into an electronic crossover and bi-amp the speaker system with the amplifiers positioned as close as possible to the speakers. This is pure common sense.

Impedance meters are manufactured that read the combined resistance, capacitance, and inductance of cabling at several different frequencies. The one I owned was made by Sennheiser. It was small, lightweight, easy to use, and accurate. It seems audio people today just don't want to bother with this type of testing. Personally, I like to know when it's time to put my boots on.

I once sat on an AES panel during a discussion on cabling. One engineer (from a small video facility) challenged my "dismissal of specialty cables" by expounding the audible virtues of an esoteric mains power cord. You know, the kind that runs from the wall socket to an amplifier's rear panel. This 3-foot cable ran $250. I almost said, "Hey, if I just spent 250 bucks on three feet of wire, it would sound mighty good to me too." But instead I tried to explain the economic impracticality of replacing every one of the hundred or so power cords in the typical recording facility with $250 three-foot lengths of wire. Economic and electronic facts meant nothing. The gentleman was adamant, so I just had to give up on him.

Hey, let's get out of this machine room. I'm tired of dancing around all the wiring strewn about on the floor. I prefer my cabling accessible but bundled up neatly. It allows for change but avoids damage. Hey, they're using a jacketed pair of twisted 12-gauge cables to drive those 815s in the control room's front wall soffits—a nice commonsense choice.

Monitoring Systems Part II

Back to the Control Room

Before moving on with the discussion of monitoring systems, it's important to get something straight. There is no such thing as a perfect transducer. While nobody would argue with that statement, some folks act as if some products are and a few end-users actually expect a transducer to perform as if it were a short length of wire.

If you actually believe that sound can be perfectly reproduced, try this little experiment. Take the world's best megadollar microphone of your choice. Hey, take a bunch for that matter! Grab an acoustic guitar by its neck right under the headstock, or finger a chord, as long as you can hold it out away from your body. Strum it and record the output. Then play it back. You can use *any* mic, *any* mic placement, *any* recording methodology, *any* amp, *any* cable, *any* speaker system, in *any* room. It will *not* sound the same. Oh, it may be close, but it will not be the same, nor will it ever be. Once you accept this *fact* you are free to deal with all the imperfections in professional audio while totally disregarding the hype. I've been in this position for well over 20 years and believe me, it is not only a much nicer place to be, but my work is substantially better because of it.

Audio Perception

For over one hundred years folks have quested after perfect audio fidelity from their audio equipment, especially microphones and speakers. *Forget it!* No microphone can pick up sound exactly the way our ears hear it. Even if it could, it would not be feeding the information to a brain for decoding and processing.

Although they can do a decent job on trumpets and vocals, no speaker will ever be able to radiate sound exactly the way an acoustic guitar, piano, or any other sound source that disperses sound in multiple directions does.

Our brain processes the audio information fed to it by our ears, and it continuously updates this information, which further aids in our interpretation of the acoustic stimuli from the physical environment around us.

In order for any reproduction system to perfectly match the original sound, it must not only present an unaltered version of the original "direct" sound, but all of the information pertaining to the acoustical environment must be re-presented to our ears as well. This vast amount of data simply cannot be conveyed through a pair of loudspeakers.

Here's another experiment to try. Place a microphone right where you're standing while listening to a music instrument, then go into the control room, and listen to what the mic picks up. It will not be the same. Neither the instrument (the "direct" sound) nor the room acoustics will sound the same as what you heard, no matter what the audio system consists of, even if every piece of equipment in the chain has specs that are superior to those that would be necessary for reproducing that particular instrument. If you don't believe me, then try the same thing with speech from a little transistor radio placed out in the studio.

You can get a good direct sound using mics positioned close in, but picking up the sound from farther away is best attempted with shotgun, hypercardioid, or at least cardioid mics because they help eliminate extraneous sounds. The human ear's pickup pattern is almost omnidirectional, but our brain processes the sound that we hear so that we can discern one sound from another even from a far-off position. Our audio perception also includes the ability to learn about the acoustical environment and the sound source's location and adjust to it so that we are able to pick up only what we want.

The binaural microphone technique is somewhat successful at duplicating the sound of a direct source and the environment it is located in. Some of these methods actually utilize dummy heads with duplications of the human outer ear. But those outer ears cannot reflect, resonate, and feed sound to the mics placed inside the "head" with the same kind of phase shifting as our ears do. Additionally, true binaural reproduction must be monitored through headphones. The information can be reproduced binaurally with the use of speakers that have an added electronic stage (which eliminates a portion of the mono information), but even with this the listener must be located in the "sweet spot" between the two speakers and there's still the limitations of the mics and speakers to consider. At a live performance you can just sit back and enjoy it all because your ear/brain combo takes care of everything for you. You can focus in on the string or woodwind section or blot everything else out while you pay particular attention to the sound of the 12-foot acoustic grand piano.

The Stereo Image   A single pair of mics suspended in front of a sound source should, you may think, render a fairly decent stereo spread. Yet when the recorded program material is played back through a pair of speakers, the stereo left-to-right spread and the listener's ability to locate individual sounds within the front to back field will undoubtedly be inferior compared to what was original heard by our ears. Listening to something that's been picked up by microphones and played back through speakers is a whole different animal as compared to the original live performance.

In our attempt to achieve perfect audio we demand that all our audio gear have near perfect specifications. Yet to get a sound that's even close to the original we must resort to close micing positions, equalizers, and other processing, as we try to match live performances with musicians playing in a deadened studio with added electronic reverberation. Hey, all we need is a simple approach to live recording that is based on the limitations of the components involved and results in a balanced mix that can be both spread or panned out for stereo or summed (combined) to mono so that it can be played on television or by AM radio.

Evaluating Audio

The audio recordist's job entails both art and technology. The technology is not exciting, sexy, or fun, but when used it opens doors to more expressive forms of the art and points out limitations that could detract from it. Either way, knowing the technology enhances the art. Knowing the facts ends up being much more productive than being lead astray by electro-rhetoric, half-truths, folklore, fairytales, voodoo, black magic, dogma, and generalized suppositions.

Although we can discern improvements in sound such as the noise content, most judgments about quality end up being purely subjective since it is seldom possible to correctly compare or A/B the actual before and after conditions. Our ear/brain hearing system has a short memory. Therefore, we should not attempt to perform A/B (either/or) comparisons at different times. Otherwise, one could find improvements that do not exist.

Scientists spend a great deal of time making sure they avoid anything that will "contaminate" the objectivity of their test results, especially subjective influences. But the field of professional audio is rampant with subjective opinions, which are often readily accepted without any objective testing. For some it may be "Let's not rock the boat, but go right along with the program and keep the gravy train of profits going." But for the most part I believe it's due to a lack of understanding of basic audio principles.

It has become almost as bad as it is with consumer electronics where cables that cost $400 are falsely believed to be superior because the audiophile who bought it needs to feel post-purchase gratification with their expenditure. Hobbyist's purchases are no business of mine, but professionals are supposed to be grounded in the real world and not be blinded by subjective illusions.

Of course, the economic realities of our business do not allow us to ignore client demand and the "wow 'em with delight" factor of having the newest craze in pro audio equipment right there enticing clients to shed extra money. As an example, while I honesty liked the sound of EMT's digital reverberation device, it was its R2-D2–like space-age remote unit that led most studios to spring for its hefty cost.

Big-time client thrill factors require no blind listening test that compares the *device under test* (DUT) via A/B switching with a straight wire. One of my favorites terms in audio is "transparency." When questioned, its found that it is not the same as extended-frequency response, ultra-low distortion, slew rate, damping factor, speed, dynamic range, S/N ratio, or headroom, but "all of those are factored in," yet they are not tested for individually.

Another example is cabling. Capacitance, resistance, attenuation, filtering, and induced delay can all be tested. But I once offered to test and review several brands of audio cables and was told, "You won't find any difference using those tests, but you can hear the difference." Time to take out the wading boots.

A/B testing, when properly conducted, bypasses all the hokum and presents us with information that may improve the end product's audio quality. And when you get right down to it, that is really what our clients expect from us.

Okay, as an example of proper A/B comparative objective testing, take two equalizers, set them up on the studio's shop test bench, and feed both of them the same exact output from a signal generator. Using a frequency counter and a decibel meter, adjust their processing (decibels of boost and the frequency settings) to be the same and also set their output levels so that they match each other. Put those outputs through an A/B switcher and feed its output signal to a dual-trace oscilloscope that is also showing the original signal. Now you can actually see any difference in timing caused by the equalizer-induced phase shift.

Do I expect the average recording studio client to care about the results of this test? Of course not, but I would expect the studio owner to want to know about any differences that may cause sonic inferiority. I certainly expect every audio engineer on staff to be highly interested in those results. To be adequately revealing, this testing should include all pertinent frequencies with varying amounts of boost and cut. This may seem like I'm asking a lot, but after using the Neutrick strip chart recorder in phase response mode (see the section on equalizers), I came to appreciate the usefulness of this knowledge to achieving quality sound.

Comparative differences between professional audio amplifiers are seldom audible. You'll find that it's more common that any audio irregularities will be due to improper level adjustments.

Even control room acoustics must be scrutinized. While LEDE rooms are very natural-sounding, our hearing system will compensate for new listening situations brought about by slight changes so that we soon become accustomed to working with them and end up believing a false monitoring reference. In other words, in order to use a monitoring system as a reference it must be subjected to ongoing regularly scheduled testing.

Control Room Speakers

Flat Response   The common goal is to have a system with a perfectly flat frequency response and if this could be instituted and repeated at every studio it would make judging audio quality fairly uniform. On the other hand, a perfectly flat response doesn't sound very good and has nothing at all to do with what people in the real world hear. Personal stereo headphones, automotive sound systems, club PA speakers, radios, television, and home systems all have frequency responses that are far from flat. People wouldn't buy them if they were. This means that the audio recordist must produce programming that can be adequately monitored on these less than perfect systems.

This simply means that you have to understand the difference between the reference monitors that you are using and the playback characteristics of what the "normal folk" are using. I've seen several studios fall into the trap of overly augmenting the monitoring system by boosting the low-frequency content while adding "air" or "crispness" to the highs. This may impress some clients, but a product judged to be acceptable using this kind of "biased" response may pale and thus embarrass the recordist when it is reproduced on normal systems.

If someone came up with a speaker that produced full bandwidth audio with a single driver, referencing would be a lot more simple for the recordist and life would be a real lot more luxurious for its exceedingly wealthy inventor. Don't hold your breath. It can't be done. The physical requirement of reproducing 20 Hz has very little connection to the physical demands placed on high-frequency drivers. Therefore, we must deal with the inadequacies that come with multidriver speakers. These include uneven frequency response, phase smearing, crossover distortion, and power consumption that is less than 100 percent efficient.

Operationally, speakers are fairly simple transducers. The amplified audio signal is fed to a coil, which sits within a magnetic field. The audio being AC causes positive and negative changes in the coil's placement within this field. The coil is attached to a cone, which is held in place by a surround and a lower spider.

Both are pliable and attached to the speaker's frame, or cage. The in and out movement of the coil is limited by the cone's maximum excursion. Cones are often made of paper and the wire used for the coil is about one-hundredth of an inch (0.01 of an inch) thick. At high frequencies, the coil can vibrate thousands of times a second, and at the low-end, excursions can reach as much as a full inch.

Thermal failure is caused by high power heating the coil over long periods until the wire eventually breaks or simply melts. Mechanical failure can result from a single hit of excessive power. Here, the coil can separate from the cone. The cone can separate from the frame or, more commonly, the coil/cone connection will be deformed to some extent. This results in the coil rubbing against the side of the magnet's gap, which means distortion, friction, heat, and a very short life expectancy.

Serious power hits can rip cones, spiders, and surrounds. I've witnessed coils forced completely out of the magnetic gap. They had returned backward off-center so that they ended up stuck outside of the magnet's gap. Besides the fact that they would no longer produce any sound, a coil in this position is exposed to air, has no resistance to power input, and receives no thermal heat sinking from the magnet. Say your goodbyes quickly. Metal dome tweeters can actually shatter by being flexed beyond their limit due to excessive power inputs.

So speaker failure can also be caused by mechanical stress, long-term high-power operation, and excessive peaks such as powering up equipment on the input side of the amp after the amp itself has been powered up. With this in mind the value of sequential power up and delayed turn-on devices becomes apparent.

Speaker specifications are all over the place. I've seen power rated according to peak and continuous operating levels as "program" and "music power" levels. How about a specification like "100 watts RMS continuous power at 1 kHz"? If it is a three-driver speaker system, two of the drivers may only be able to handle 60 watts. If so, operating at the rated level may not cause a failure, but it *will* cause distortion.

I once ran into a pair of drivers that were rated for and could handle extreme power levels. They got my attention because upon lifting a very common two-speaker guitar combo amp, my arm was almost yanked out of its socket. These speakers had giant magnets that caused them to weigh a ton. They could easily handle much more power than that guitar amplifier could generate. Since no modification had been performed on the amp, this meant that it could be driven to distortion without jeopardizing the speakers. When played through, it could get loud, but not as loud as with the stock speakers. In this case the guitarist could have saved himself a lot of money and backache by using a distortion effect with the stock speakers.

Speaker efficiency is rated as the output in SPL with a specific power input. This parameter can also be rated as *sensitivity*. This SPL output rating is referenced to a measurement made at a set distance from the speaker with a specific input level. This is usually a 1-watt input and a distance of 1 meter. Power amps should be rated for at least one and a half times that of the speaker's power rating, so that they can produce the required power and handle peaks without distorting.

The JBL/UREI 815C, used as the main monitors in this 1980s control room, have a power rating of 150 watts. A 100-watt amp may easily be driven to the point of clipping and distortion, and a 300-watt amp, if improperly used, could cause speaker damage, but a 250-watt amp would do the job with an extra margin of safety.

The JBL/UREI 815 C had three 15-inch drivers. One was dual concentric. The passive crossover included three attenuators: one for the dual-concentric's low-frequency driver, one for the dual-concentric's high-frequency driver, and one for the two "low-low" frequency drivers. The specs were

Nominal impedance 8 ohms

Bandwidth 40 Hz to 17.5 kHz (+/– 3 dB)

Sensitivity 103 dB SPL (referenced to 1 watt at 1 meter)

But these were overshadowed by its weight of 250+ pounds and dimensions of 32 by 43 by 21 inches.

The crossover frequency specifications, as well as individual replacement parts, were "not available." This may have had something to do with secrecy over this crossover design's time alignment. These monsters were perfect referencing monitors for dance-hall, rock club, and disco playback, but additional referencing was still required.

Realistic comparisons often called for running a mix out for a listen on an automobile system. Ever wonder why music sounds so good inside an automobile? You already know the answer. They are sealed acoustic enclosures so the low-frequency sound pressure is enhanced. The high frequencies benefit from the added dispersion of the widows and other reflective surfaces so the Fletcher–Munson research on human hearing sensitivity points to this listening environment as being a delightful audio experience. Science may have been proven right again, but it is impractical to bring an automobile interior into a recording studio facility. I know. I tried with a '57 Chevy no less.

For a period I was possessed with accurate monitoring. This was because I was working in a half dozen studios and their monitoring systems varied widely, even though each had three sets of monitors.

Carrying even a small component system around with my (even-response) headphones became a pain. So I switched to a pair of $5 \times 5 \times 7$-inch Visonik David 6000 speakers that had a flat-frequency response from 125 Hz to 18 kHz. To make up for the lack of low end, I added a bass shaker (low-frequency vibrator), which I would gaffer-tape to the console's belly pan (the sheet metal under plating) or some other diaphragmatic surface. But even this mere 10 to 15 pounds of weight was not in the cards on a three-studio, 95-degree summer day.

Next, I settled on a tiny $3 \times 2^3/_4 \times 2^1/_4$-inch (pound and a half each) set of Audio Source LS-6 loudspeakers. They had built-in amplification and when you placed them 8 inches in front of you and spread out 18 inches they rendered absolutely perfect stereo imaging. Due to this closeness, the internal amps and the built-in equalization these little puppies made surprisingly accurate reference monitors.

However, large cones with extended excursions are needed to produce very low frequencies. Small compression drivers such as tweeters and horns are not only better at reproducing high frequencies, but also at dispersing them. Low frequencies tend to be almost omnidirectional while highs tend to beam. It all boils down to the fact that different speakers are better at handling different frequencies. One look tells the story. Low-frequency drivers are large and heavy with concave-shaped cones while high-frequency dome drivers are small and have convex-shaped diaphragms. Low-frequency speakers do not move fast enough to reproduce high frequencies and low frequencies can destroy a domed compression driver. A crossover is needed to divide the program into separate frequency ranges and channel them to the appropriate drivers.

Crossover Networks

Crossover networks, whether passive or electronic, split the full audio bandwidth so that it can be more efficiently handled by a combination of drivers. Passive EQs are commonly enclosed in the speaker cabinet and are positioned between the amplifier output and the drivers. They do introduce insertion loss and can also cause distortion. Active electronic crossovers are usually fed by the console's monitor bus outputs at line level. They divide the full bandwidth signal up into smaller bands and route them to the inputs of separate amplifiers. They introduce less distortion and insertion losses and they provide a greater degree of control over the tonal balance of the monitoring system.

Figure 2-30 is a graphic representation of how the audio spectrum is divided up by a crossover network.

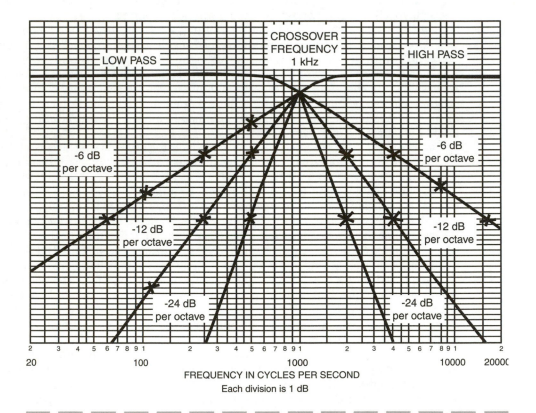

Figure 2-30

This graphic representation shows the audio spectrum divided by a two-way crossover network. As with equalizers, the low-pass filter determines the low-frequency content and the high-pass filter determines the high. The crossover point, here 1 kHz, is where the two roll-off slopes cross over each other. The rate of roll-off (in negative decibels per octave) determines the sharpness of the cutoff. Note that -6, -12, and -24 dB per octave are shown.

Just like equalizers, the high-pass filter determines the high-frequency content and the low-pass filter determines the low. The crossover point is the frequency where the two roll-off slopes cross over each other. The rate of roll-off (given as negative decibels per octave) determines the sharpness of the cut-off.

The highs will be cut by 6 dB per octave (remember an octave is every doubling of the frequency) below the crossover frequency, with a "6 dB per octave" filter. The loss will be cut by the same amount above the crossover point.

Passive crossovers are found inside multiway speaker systems. They are composed of capacitors, resistors, and inductors, and their parameters are generally fixed. They are designed to handle the high-level power amplifier outputs while active crossovers operate at the line levels of power amplifier inputs.

Basic electronics teaches that resistors are not frequency selective, but capacitors and inductors are. They store the signal and release it a short time later. Capacitors can block the lows and inductors the highs.

Active crossovers offer continuously variable crossover frequency control, which allows the system to be adjusted to accommodate specific drivers. They interface well with systems as they do not interact adversely with power amplifier inputs and do not load down (cause insertion loss) the signal.

Bi-amping is accomplished by using different amplifiers for the low and high frequencies. The lower the frequency, the greater the need for power. This is important since excessive low frequencies can cause an amp to clip. This can produce distortion, which has excessive upper frequency (square wave) harmonics that are capable of damaging high-frequency drivers (square waves are made up of an almost infinite range of odd harmonics). By using separate amps, if the low-frequency content causes amplifier clipping just the low-frequency driver will receive the high-frequency harmonics. Now the high-frequency driver's signal remains undistorted while the low-frequency driver is relatively unaffected by the added high-frequency distortion.

The high-frequency amplifier can also have a reduced power output, which is more in keeping with the lower power-handling capabilities of high-frequency drivers while a higher-powered amplifier can be used to deliver a higher SPL at low frequencies. You end up with greater power output, less distortion, and higher reliability. Active crossovers are very beneficial in live biamped PA work. On the other hand, tearing apart a control room monitor system and readjusting its crossover frequencies are impractical, as you'd seldom better a manufacturer's design.

Capacitors can be used to block or stop low frequencies from damaging high-frequency drivers. Their voltage rating should be higher than the amplifier's maximum output voltage and they should be nonpolarized. The best (most durable) are the ones used for motor starting. Their value (in microfarads) is found by dividing 500,000 by pi (3.14), multiplied by the product of the desired cutoff frequency (F), multiplied by the driver's impedance (Z).

$$\frac{500,000}{3.4 \times (F \times Z)}$$

All active crossovers produce some phase shifting, which can impair dual speaker coverage at the crossover frequency. Here the drivers that are a different distance from the listener reproduce the same frequency.

The steepness of the crossover's roll-off is critical to preventing distortion and speaker damage. Sharp high-frequency roll-off requires the use of fast op-amps. This is related to the slew rate and transient response parameters of amplifiers. But unlike the questionable effect speed or the very high-frequency response on power amplifiers, crossover distortion caused by slow-speed electronics is in-your-face ugly.

Mathematicians design crossover networks and it's all about calculus and group time delays. This is real boring stuff that uses weird "vector" diagrams to show the amount of phase shift vs. level.

It once was that crossovers either had a flat frequency response, a shallow roll-off, or had a more linear phase response, but newer "Linkwitz–Riley" fourth-order designs provide outputs that are in phase and provide roll-off rates of 24 dB per octave. The mathematicians are only a spit away from eighth-order designs that will yield 48 dB per octave roll-off rates.

Okay, all this crossover design stuff is all about mathematics, and you're not going to rip open a studio's multiway speaker system to make any changes, so why bring it up? Because crossovers are an essential part of the monitoring system and understanding them can be beneficial. How so? Well, that vibratory bass "shaker" I used to carry from studio to studio as part of my own personal reference system required a crossover adjusted to feed it "subwoofer" frequencies. And at any point in time the recording engineer may have to divide up the frequency content of a music instrument, speakers, or better yet to ship separate bandwidth sections of the same signal to different processors.

Imagine a 10-band graphic EQ that had separate outputs for each band and before they were recombined some of these bands were subjected to level modulation. When the modulation is controlled by the input (envelope following), an accented vocal will initiate a larger modulation. If an outside signal is used as a keying input and it is derived from a rhythm guitar or keyboard, those instruments' level accents would cause a greater amount of modulation. Once recombined, the level modulations of several frequency bands would result in the very bizarre effect that is provided by, you guessed it, a vocorder. The same dividing method is also used for multiband dynamics processing, too, but back to subwoofers.

Using a single bass enclosure with a stereo pair of speakers goes back to the late 1950s. I was still a youth, but from what I've been told, stereo was still in its infant stages so the biggest concern at the time was mono compatibility because everyone was still playing monophonic records. One of the old-timers I knew still had a "Weathers Tri-phonic" three-speaker system in his den (remember dens?). The crossover point was around 80 Hz and the woofer received nothing higher than 100 Hz. The system sounded pretty good, but according to the giddiness of its owner, the real kick was the fact that it was camouflaged. The speakers looked like books on his shelf (hey, it was the 1950s!) and the bass enclosure (due to its having only a 12-inch driver) easily fit under a couch or coffee table. Since low-frequency sound is omnidirectional this split location was not problematic.

The problem with most of the early single subwoofer add-ons was the fact that they tapped power off of the two stereo speakers. Subwoofers have the unique requirement of moving large volumes of air. The large excursion required of the diaphragm means it must be extra stiff (paper breaks down too easily). Additionally, the magnet, suspension, and coil wire must be beefed up to handle that extra strain. But when very low frequencies, down to 5 or 10 Hz are adequately reproduced by a speaker system, the affect (whether psychological or not) is very apparent. Ask the home theater crowd.

Near Field, Close Field, Direct Field, and Free Field Speakers   All these refer to the same thing. This many terms came about because Ed Long trademarked the name "near field" monitors, which meant manufacturers had to make up new names. They were initially used to give the engineer and producer an idea of how the program material would sound on home sound systems. It was found that when positioned correctly, the control room's acoustic anomalies had less of an effect on the direct sound they produce than on the large control room monitors. What kind of anomalies are we talking about? How about reflections (early before 20 milliseconds and late, after 50 milliseconds), boundary effects (added bass boost when positioned close to walls and in corners), and room modes (standing waves or resonances that cause peaks and dips in the frequency response)?

While referencing on the main studio monitors was still necessary for dance hall, club, and heavy metal mixes, the lower power and resolution of these smaller monitors often more readily pointed out small details that may end up being overlooked, resulting in a sound that was lacking on a more consumer-oriented system.

It's important to remember that a small monitor must have an even frequency response from the mid- to the high range. They should also be stood vertically on a stand slightly behind the console with the high-frequency driver aimed at the "sweet spot." (Don't take my word for it. Just try it once.) The system will have improved (a more solid) bass response and clarity, and there will be less smearing, lobing, and comb filtering of the highs and mids. This is caused by the improved dispersion in the horizontal plane and less reflections off of the console's work surface. Exceptions to this are situations where it is important to position the small monitors even closer to the engineer so that its direct sound has more dominance over control room reflections.

Common Sense and Smaller Monitor Speakers

Everyone understands that having trusted reference monitors allows the engineer to accurately evaluate the integrity of the sound. But today the small, close-range monitor is more often than not the primary referencing source and these may not prove adequate for hard-core rock, symphonic, and film postproduction work. The best positioning is to set the small monitors in the two base corners of an equilateral triangle, where the point is located at some distance behind the engineer's position. The "depth" of this point is determined experimentally and is chosen to both allow the engineer a good deal of lateral movement while still remaining in the "sweet spot," and to include the producer's position within that sweet area. The height should be set for a head tilt angle of about 15 degrees as this has been found to be the most natural and comfortable for human hearing.

Damage to small monitors from excessive listening levels is common. Their diminutive size requires more power to produce the gratifying levels equated with some popular music. Many studios turn to fusing to reduce casualties, even though the distortion they produce is clearly evident when viewed on an oscilloscope's display.

You may not notice it while monitoring thrashing heavy metal programming, but that violin solo that explodes in the middle of a ballad may make the hair on the back of your neck stand up. Other remedies include adding a capacitor. Normally, these are used to block low frequencies from getting to the high-frequency drivers. But here values are chosen to stop low frequencies from getting to the bass speaker as well. Passing the full bandwidth through a phase shifting high-pass filter doesn't appeal to me at all.

Common sense tells me that if added level causes damage to the system, back off on the gain. If high listening levels are necessary, augment the system by doubling up on the number of small speakers. Placed side by side (as is common with PA systems), the sound quality will remain the same, but the monitoring levels can be increased without damage or distortion. "But it's still not loud enough for the keyboardist's overdub position at the back of the control room!" Add another set there as well. "Hey, now you're talking about a lot of money." Yeah, like driver reconing and speaker replacement is cheap. This multimonitor remedy makes even more sense when you think about failures at inopportune times (middle of the night) and quick recovery from down time. That's what the world-class studios do. They work at minimalizing down time.

The newest additions to the small monitor system combine the amplifier, electronic crossover, equalizer, and drivers all in the same small cabinet. Taking note of their heat sinks and insuring proper ventilation minimizes distortion and increases life expectancy. I ran into a situation where a studio owner had purchased a $3,000 pair of "self-powered" monitors and was having a bear of a time with listening levels and distortion. This "system" lacked any level control such as an input attenuator or amplifier gain adjustment. With his VU meters at 0 dB, the console's mix bus nominal output of +4 dBu caused the monitors to distort. Backing down on the console's mix level fader reduced the output level and ended the distortion, but during quiet passages the S/N degradation was readily apparent. Mr. Ohm has a simple fix for this problem (refer to Part 1, "The Basics").

Some small monitors are designed to be stand or pedestal "free field" mounted. Placing them against a wall, and certainly in the corner of the control room, will accentuate the low-frequency response. This may seem like a good idea at first, but now the system's response will be nonstandard, susceptible to large variations with incremental position changes, and no longer provide a trustworthy reference. The next step will be equalizing out some of the low frequencies. In smaller rooms this can be disastrous.

These rooms simply do not have enough volume to accurately handle low frequencies in the first place: low-frequency roll-off number one. The wall construction is often less than solid and all openings are rarely well sealed: roll-off number two. Pulling down low frequencies with an equalizer to reduce the boost in the low end that's causing things to sound "muddy" is roll-off and strike three. The system is untrustworthy; it's out.

Linear Phase Response   Everyone knows 180-degree out-of-phase speaker wiring is bad news. One driver is pushing air out, while the other sucks it in. The end result is huge holes in the system's response. This is especially apparent with lower frequencies. Their omnidirectional nature requires that they be reproduced equally on both of the stereo speakers. An out-of-phase condition not only lowers the bass content, but causes it to appear to be emanating from a single point source. It doesn't get much more unnatural sounding.

Phase anomalies that are less than 180 degrees are also problematic. The time smear caused by multidriver placement across the front baffle may or may not be audible, but a crossover with a slope of only 3 dB can place drivers in the difficult position of having to reproduce frequencies outside of their intended bandwidths. Drivers will still add to the overall sound even though their response is down by 12 dB. That −12 dB point is two octaves beyond the crossover frequency in 6 dB per octave crossovers, and four octaves past the crossover point at 3 dB per octave. More roll-off equates to more phase shift. (See Figure 2-30.)

Speaker manufacturers have been trying to solve this dilemma for decades. I remember a dual-concentric self-powered monitor by Philips, from the 1960s no less, that sounded *real*. Its amplifiers were a bit finicky to say the least and since it was designed as a "book shelf" speaker it leaned more toward convenience than high power. So while very accurate, it would be a tough sell for control room applications.

During the mid-1970s, *Bang and Olufsen* (B&O) a company that catered to high-end audio-philes, came up with a novel approach in their Model M-70 speaker system. They tackled the phase response problem on dual fronts. To get around the crossover-induced phase anomalies caused by steep roll-offs, they broadened the slopes. This in turn left a hole at the crossover frequency. So they added an additional speaker fed through a bandpass filter to make up for the lack of content around this center frequency. But they still weren't through. To reduce phase discrepancies caused by varying speaker positions across the front baffle, they broke it up. Here the two main drivers were almost angled into each other. The intended result was to have the two speaker's wavefronts arrive at equal times at the listener.

What about the listener's distance and angle in relation to the multiway speaker? B&O supplied a pedestal that not only adjusted the speaker's height, but allowed for it to be tilted backward and forward for aiming purposes. The gentleman who owned the system I heard was very proud of it, and rightly so, as they sounded very clear and the positions of individual instruments within the stereo spread of an orchestra were a bit more discernable. (His phrasing was "highly discernable imaging.") However, I never saw or heard another pair of these speakers. B&O seemed to have dropped the whole idea and I have no idea if you can get replacement parts. Hey, at least it wasn't a quad system.

The Listening Field   *Mono* is a single audio channel containing all of the sound information. Whether this signal is played through one speaker, a stereo pair, or multiple drivers, it will always be one-dimensional. Depending on the intensity of the sound, the program can only appear to be farther away from or closer to the listener as there will be no apparent spacing differences between the signals.

Stereo has two audio channels of sound information that, when played through a pair of speakers or headphones, provides a two-dimensional listening experience. Stereo creates an audio field that permits a sound to appear either closer or farther away, and to the left, center, right, or anywhere in between, as long as the listener is located in the "sweet spot" between the two speakers.

Expanded stereo was often advertised as "high fidelity" or a "super sound extension." This is actually just a phasing or delay effect. It does have the ability to make the stereo "spread" of the program appear to be wider than the spacing of the speaker positions it's being played through. Most often this is accomplished by adjusting the phase of various frequency bands. This newly introduced timing information tricks the ear/brain into perceiving the signal as being wider. It's an okay effect and an interesting listening experience, but it's not what anyone would call sufficient for critical listening work. Another detriment to the use of expanded stereo is that just as when speakers are physically positioned very far apart, this added stereo spread could cause a reduction in the size of the "sweet spot" area.

Quad, surround, or any other multispeaker arrangement provides more dimensional information by increasing the number of speakers in the system's array. I'm no expert on surround, 5.1, or even Quad for that matter, but having lived through "quadrasonic" audio I can tell you that the viability of any of these approaches depends directly on two factors. First, the control room system *must* be properly calibrated whether the mix is going down to multiple tracks or through some sort of matrixing device. Second, and you can take it from someone who experienced the insanity of Quad playback system variability, the main difficulty is its dependence on proper system setup at the consumer end. Quad was supposed to be so simple that if someone "wasn't down on it," it was because "they weren't up on it." That quote was 1970 speak for "to know it is to love it." While in actuality to invest in it was to regret it.

Binaural

This method of recording and playing back a stereo signal tries to duplicate the way our two ears pick up sounds. Unlike normal stereo that just uses a separate track for the two speakers, the attempt with binaural is to provide each ear with an audio image that is psychoacoustically the same as that which is heard live, including all the audio timing and intensity cues. When played back through headphones, binaural audio can come very close to actually being there. One of the advantages of binaural audio is that these natural sound locator cues are presented to the listener's ear/brain combination.

Okay, here's the deal with good binaural sound. Picture yourself in a large room crowded with people. There are lots of different conversations going on, background music is being played, and there's an occasional public address announcement. You can easily pick out the announcements and the music, but you'll also be able to discern several of the closeby conversations. All of this is due to your two ear/brain systems' ability to both locate and filter sounds. If the exact same audio information is presented to each of your two ears through headphones, you will still be able to pick out the announcements and the music, and you'll also be able to discern several of the individual conversations. When done correctly listening to a binaurally recorded performance can be an awesome experience.

Control Room Acoustics and Monitoring

Although I've put a lot of time into the study of acoustics, I do not consider myself an expert on the subject. I do not design control rooms for a living and never intended to. But I think all audio engineers should understand the principles involved so that they have a better understanding of what they are up against (i.e., you walk into a control room and find its shape is a perfect cube; pull out them trusty headphones, partner). Even though most of the principles involved in room acoustics are based on the unchangeable laws of physics, other aspects peculiar to control room monitoring are a bit more variable.

Here decisions on speaker placement and the reflectivity of surfaces near them come into play. Then there's the possibility of interference from the back wall causing a slap-back delay. It is possible to construct a control room that provides a correct acoustical environment for referencing an audio signal through speakers, but they must be designed for this purpose from the ground up. Add-ons may be helpful, corrections may dissipate some problems, but the overall accuracy of the reproduced signal will always remain questionable in rooms whose dimensions are contrary to tried and true formulas.

But there are so many variables and a whole world of untrained "experts" out there ready to dispense "facts" on control room acoustics. Such as:

Your walls are flexible (nonmasonry) and absorb bass? Add some foam to the room's surfaces to reduce the high-end content and even things out. If you have standing waves and resonances that are causing huge peaks and dips in the room's frequency response? Add some of reflective dispersing panels. Low ceilings and cramped quarters got you down? Hey, move these self-powered speakers that sound just "great" because of their accentuated low- and high-frequency content right on top of the console's meter bridge. Now they're close enough so that the acoustics of the room will have little or no effect on the sound.

But doesn't that all sound just a little too easy? In reality it isn't, and using half-step measures to patch-up faulty acoustic designs does not render predictable results. Your best bet is to learn the real facts and hire a reputable pro to handle control room design.

After writing a short article covering a few of the basic principles involved in correct studio design, I received a letter from a "professional audio consultant" who took me to task for my some of my recommendations. Concerning the construction of control room/studio windows, he felt that two panes of glass were not needed and, if used and angled, the amount of sympathetic vibration would be negligible. Dual panes of glass with dissimilar thickness have different coincidence frequencies. Coincidence in acoustics refers to the phenomena of increased sound transmission by sound waves of certain frequencies striking boundaries with a specific mass at particular angles. Adding the second pane of glass with a different thickness interrupts this event's multiconditions. It's the same principle as in using two different sheet rock thicknesses in wall construction and why heavy masonry walls are good at preserving a room's bass content.

In fact, slightly angling the glass widow panes to avoid the possibility of sympathetic vibrations has been proven to not add to transmission loss, but the practice still plain-out works at providing often needed help in avoiding standing wave reflections between the studio/control room window and the control room's back wall. It also aids sight lines as, instead of needing to duck to "see around" the dual pane's four reflections, they are broken up by the angle. But that doesn't mean that you should go ahead and take the easy route, as in using a pane of thermal glass because it's made of two layers with air between them. Those two panes are very thin, are parallel to each other, and have a quarter of an inch of air space between them at best. They may be okay for temperature, but not for sound.

Next he tore into me for recommending that power amplifiers be located closer to the monitor speakers. He pointed out that the heat dispensed out the front of the amplifier would rise up in front of the monitors and cause the sound from the drivers to be deflected just as light is deflected by heat coming off of a highway. I have no knowledge of any scientific studies on the "deflection" of sound waves by heat, but it doesn't matter because he grossly misunderstood my statement. By "closer" I did not mean right underneath the monitor! It would be crazy to enclose an amp in an unventilated wall soffit. It's almost as crazy as using that positioning so that their fan-driven heat would be pumped out of the amp directly at the mix engineer and the console surface: two items that can normally run a little overheated. I brought this up to show how correct statements about control room acoustics can be made to sound incorrect.

Well, incorrect statements can just as easily be made to sound correct. It's the same with the often-fought battle between the practitioners of the frequency response and reverberation time methods of evaluating room acoustics. The fact is they can be set so deep in the defense modes of their particular test methodology that they both end up being wrong. How can that be? Well, neither test method is complete, but using both can direct you right to the source of many problems.

Their arguments are that the problem with using frequency response is that it provides information on only a single test location and the results hold true for only that one location. The problem with reverberation time is that it provides very little information due to the short decay times of most control rooms.

Okay, single-location frequency response plots will tell very little about high-frequency dispersion throughout the control room, but because the low-frequency content is virtually omnidirectional, a single plot will point to the possibility of frequency imbalance. Multiple frequency response plots have pointed me directly at the source of standing waves: flutter echoes and low-frequency resonances.

In the same manner, multiple (location *and* frequency) reverberation time plots provide absolute confirmation of room/monitor anomalies and can lead to the discovery of the boundary that is the source of the problem: reverb time and frequency response. It just makes common sense to base monitoring decisions on the results of both. Frequency response and decay time are very much dependent on the amount of absorption there is in the room. Adding excessive amounts of high-frequency absorption to correct acoustical problems caused by incorrect room dimensions will result in an unbalanced reverb time. The resultant longer decay time at low frequencies can cause the signal to sound muddy.

Trying to "clean up" mistakes in acoustical design using equalization is just asking for trouble. Comb filtering from the phase shifts it causes can easily result in narrow band notches up to −20 dB. Attempting to compensate for this by additional equalization will not only affect the surrounding frequencies, but add more phase distortion and adversely change the timbre of the sound. Remember that even ⅓-octave EQs have bands that are often broader than the problem bandwidth. In these cases, the ⅓-octave filters may affect surrounding frequencies unnecessarily.

There are no quick fixes in acoustics and no two rooms are the same. Solving acoustical mistakes made during a room's design and construction stages can cost more to correct than the original construction (as now all previous work must also be disassembled). If you're spending any amount over a few thousand dollars to construct a control room, you should start the project by getting the name of an accredited acoustician from either the National Council of Acoustical Consultants (a not-for-profit organization located in Springfield, NJ) or the Acoustical Society of America (asa@aip.org).

Facts on Control Room Acoustics

A spectrum analyzer can only give hints as to phase response, crossover anomalies, speaker inadequacies, and room resonances. Most amps and speakers today provide high-quality sound, but the room they are in can add up to 10 dB of frequency anomalies from mode resonances. Manufacturers know all about anechoic and free-field testing, and as a result flat-response speaker systems have been available since the late 1970s. The main problem now is from the room's boundaries and acoustics. It's a fact that reflections from video monitors and near field speakers can cause comb filtering of the main monitor's outputs. The control room's ambient noise is more apparent today because the electronic circuitry is quieter, microphones have greater sensitivity, and the dynamic range of the recording system has expanded.

Diaphragmatic absorbers resonate at low frequencies depending on their size, weight, thickness, and the amount of air space behind them. There are equations that predict the outcome of constructing diaphragmatic absorbers as well as Helmholtz resonators (both are very effective at curing low-frequency problems), but they are most often built on an experimental basis or from memory of past experience. The math isn't that difficult and the improvement in the outcome is worth the effort. Serious low-frequency build-up problems can be aided with the use of bass traps. Bass traps require a good deal of space to absorb lower frequencies. It takes 3 feet of air space to affect frequencies in the 100 Hz area. A problem down around 60 cycles may cost as much as 5 feet. It's much easier on the wallet to get things right during the initial planning stages.

More importantly this points out that erratic problems can be caused by "accidental" diaphragmatic walls, ones that were intended to be solid but were specified or constructed incorrectly. This not only reeks havoc on a room's acoustic response but can negate all the hard work and cost put into achieving a high level of noise transmission loss.

If noise transmission is not handled correctly at the start of the project, there will be no second chance at adjustment as this aspect of studio/control room construction depends on the accuracy of the structural design and implementation. Good control rooms are close to airtight, completely isolated from outside noise, and structurally solid. That's one of the reasons why basement control rooms always seem to have great bass response.

Ambient noise above 30 to 40 dB can be a problem. Once a quiet background is achieved, any change in the room layout such as equipment additions might cause additional noise. The test procedure that initially qualified the room as acceptable should be repeated often as small changes in background noise may go unnoticed, especially by those who spend a great deal of time in the room.

Sound dispersion in a room can be checked by feeding pink noise into the system and using the ⅓-octave analyzer's display to search out anomalies. If plotted out these will often point to the cause of a problem. I once tracked down the source of a 2 to 4 dB bump centered around 4 kHz to be two structural ceiling beams. They were fairly close together and of course parallel. They set up a standing wave that was adding to the high-frequency content of the signal in the room.

Sound dispersion is normally better at low frequencies. Due to their larger wavelengths, they tend to just go around obstructions. Higher frequencies are more directional and are reflected more easily. If you want to check the evenness of a room's sound dispersion, you can use a spectrum analyzer to check the high-frequency content at multiple locations.

As already stated, resonant or room modes can add up to 10 dB to the level at resonant frequencies. Here the room's geometric shape will cause certain frequencies to be augmented or sustained. There's a lot more to it than lowest resonant frequency and ideal proportions, but people still want to have a go at it without hiring an accredited professional.

The sound transmission of a material is only a function of its stiffness and weight until the acoustic seal is broken. A hole the size of a quarter in a 12-inch cinder block wall will pretty much negate the whole effort. Lead is heavy, but it's also a limp mass that easily conforms to the required shape to seal off openings. But lead sheets wallpapered between two walls will lose some of its sound-blocking qualities. This is not the case with lead-filled or impregnated vinyl materials. These materials add dampening to wall structures when adhered in a staggered array between these two layers.

Things change very slowly in the field of acoustics as truly new developments are rare. When things do change, they are often the result of building on the previous work of many mathematicians, physicists, and acousticians. An example is the superior sound diffusers that are now available.

Since the time of the early Greek outdoor theaters, acousticians have known that a concave surface will focus sound to one particular point (just as parabolic reflectors do with microphones). Convex surfaces are better at dispersing sound, but the outcome is very unpredictable. Highly irregular walls like those made up of field stone masonry are very effective diffusers but still a bit unpredictable. This is because diffusion is dependant on the size (both depth and width) of the irregularities in the surface as well as the wavelength and angle (incidence) of the sound impinging upon it.

The innovation in diffusers started with the physics of optical diffraction no less. Next (during the 1800s) came Karl Gauss's sequential number theory. Later Mr. Manfred Schroeder linked that numerical theory to diffusion. This particular building on past innovations and accomplishments ended up as a new type of panel diffuser that spreads sound impinging upon it across a wider area. Its main achievement, however, is the fact that this scattering applies to a broad band of frequencies and most angles of incidence.

I heard of this in the early 1980s. The folks at Syn-Aud-Con were all over this design, using their TEF systems to analyze its effects. I also vaguely remember an AES paper on the subject, but for me the real head turner was the BBC's use of this type of diffusion in their huge (8,000 cubic meter) Manchester music studio. This diffuser design covered all of the back wall and both side walls for the full depth of the orchestral seating area. The studio ended up only requiring half the normal amount of absorption to even out the frequency response even though it had a reverb time of a full 2.25 seconds! Today you can order them from several manufacturers and UPS will haul them in for you.

More than likely, complete modular control rooms will soon become available. The prospective buyer can visit a sales rep, listen to some programming, select from a list of options, determine size, work out a payment schedule, and make arrangements for delivery. Just like buying a car! This method has already been a common occurrence in the design of broadcast facilities for many years.

The number of individual low-frequency room modes depends on the size of the room. The evenness of their spacing depends on the room's shape. Because the fundamental (lowest mode) is determined by the speed of sound divided by $2 \times$ (the longest dimension), small rooms often need help at the low end. Diaphragmatic absorbers are panels that are flexed by low frequencies and thus transfer this low-frequency energy into another air space. This widens the overall useful low-frequency bandwidth if they are used to reduce the level of the resonant low frequencies. Like Helmholz resonators, panel absorbers are calculated as per the frequency (F) to be affected. The resonant frequency (F) is equal to 170 divided by the weight (W) of the panel in pounds per square inches multiplied by the depth (D) of the air space in inches:

$$F = \frac{170}{W \times D}$$

LEDE Control Rooms If you ever get the opportunity to work in a true LEDE control room, jump on it. The experience will prove that this design method does exactly what it claims. In fact, if you're interested in recording studio control room acoustics at all, *the* place to start your research is with Synergetic Audio Concepts. Look up the name Davis, as in Don, Carolyn, and Chips, or the LEDE design. Their approach to control room design "synergized," or took into account all of the fundamentals of acoustic theory, and used the latest testing methods to prove truths and misconceptions.

By 1980, Mr. Don Davis was not only teaching synergetic audio concepts (Syn-Aud-Con) students how to use a TEF analyzer, but along with Chips Davis, was designing and building control rooms based on the early reflection/comb-filtering information compiled by this system. The resulting LEDE control room design had the psychoacoustic ability to make the control room boundaries seem to disappear from the listener's perspective. The best way to describe the only "true" LEDE control room I've had the opportunity to work in is that your perspective was not one of being in a control room listening to what was going on in the studio, but one of actually being in the studio listening to what was going on. The acoustics of the control room did not seem to come between the listener and the sound.

LEDE designs are based on science. They control early reflections with an absorptive monitor end that provides an anechoic-like path between the speakers and the mix position. The hard rear area diffuses sound reflections in such a way as to add them to the mix position at a time delay that does not interfere with the direct signal. This elimination of the perception of the separate (not temporally fused) early reflections from the back wall is accomplished by paying close attention to the Haas effect. When control rooms are large enough to extend the initial reflection time to a point beyond that of 10 to 30 milliseconds, meaning they would no longer be perceived as part of the "front" signal coming from the speakers but as individual, separate sounds, reflectors are used to shorten the propagation time back to the mix position.

The LEDE's rear-end diffusion also adds to the control room's "ambiance" (reverb time); therefore, there's a balance between the diffusion and absorption. Purely dead studios and control rooms do not work. I know; I've built a few.

A neat trick I heard that Chips Davis used to determine the possibility of comb-filtering anomalies caused by speaker reflections off of the console was to lay a mirror on the console's surface, sit in the mix position, and look into the mirror. If you can see a monitor you've got a reflection to deal with. He also placed the mirror on top of and against the sides of near field speakers to see if the sound from the large speakers would be reflected into the mix position from those surfaces.

As with any control room design, strict attention must be paid to having an inner surrounding structure that is symmetrical with no parallel surfaces. Asymmetrical rooms disturb the stereo image and parallel surfaces can set up resonances and reflections that can result in unpredictable anomalies. Obviously, there must be consistent phase or polarity throughout the audio chain. The power amplifier used must be powerful enough to provide adequate headroom before distortion and have a damping factor that can adequately control the speaker's cone movement. Speakers should be positioned symmetrically.

Some pre-LEDE designs could be termed DELE. Here the wall that the main speakers were soffited (flush-mounted) in, the ceiling, and the side walls immediate to the monitors may seem to have random diffusion, but they were actually designed to project early refections around the listening position to the back of the room. Here the late reflections are absorbed or dispersed so as to not cause late arrival at the listening position. The idea was to break up the very early (less than 20 milliseconds arrival time) and late (more than 50 milliseconds arrival time) reflections, so that they did not interfere with the direct sound. Having experienced some of this method's finer control rooms, I can describe the end result as a situation where nothing comes between the listener and the monitors. In either case, the mix you "walk with" will be the one you heard on the reference monitors, no matter what it's played through. Both design methods work if executed properly, but the LEDE approach has more a predictable outcome.

The Studio Side of the Glass

Studio Monitoring   Instead of room/speaker interface anomalies, studio headphone monitor systems present more down-to-earth problems. Variations in headphone impedances can cause one set of cans to be blaring loud while another's output is minuscule. So it's best when all headphones used are identical. Then there's the imminent danger of the immediate cessation of audio reproduction from one side of the monitor system when someone mistakenly inserts a ¼-inch mono connector into the system. This power outage will be followed by possible amplifier destruction if the system has not been furnished with safety load resistors. These load or build-out resistors are placed in series with each of the amp's outputs to keep the amp from "seeing" a short circuit. The resistor's impedance rating should be the same as the headphones and its wattage rating should be about one-third of the amplifier's rated output. So if the headphone Z is 8 ohms and the amplifier is rated at 225 watts, the system's safety calls for an 8-ohm resistor rated for 75 watts. These guys can run *hot*. I've seen them burn the plywood they were mounted on, so attach them to some kind of metal heat sink, like a blank rack panel.

Studio Acoustics

During the 1980s many studios were designed to be almost anechoic, or at least provide a large, dead area for the recording of multiple instruments. A nonreverberant atmosphere was sought for several reasons. The main benefit derived from multitrack recording was that it afforded the performers the ability to overdub or redo parts they felt were inadequate. Not only can these parts be re-recorded at any time but also additional instruments, vocals, and effects tracks could be added. Music production became wide open. When the number of open tracks was insufficient for planned or unplanned additions, "bouncing," combining, or mixing several tracks down to a single track was used to open up new real estate. Easy as pie. However, track bleed could be a major problem. If the rhythm guitar was picked up by the snare mic, the original performance by the guitarist could still be heard after the guitar track had been re-recorded. Isolation was furnished electronically by the use of gating and direct input interfaces, but the first line of defense was at the source.

Movable panels were constructed to go between (go-bos) instruments. Some were made extra tall for placement around standing performers (horns, vocalists, or whatever). Windows were included so that visual cueing would still be possible.

The studio walls would be made up of absorptive materials. One of the favorites was theater curtains with large folds (air space) over raw fiberglass. Itch city! Next came drum and additional isolation booths to completely separate instruments.

All this gave the engineer and producer a clean, separate, and dry signal to work with. Because of the new-found ability to add quality reverberation along with choices from a huge palette of other effects at any time, things became exciting to say the least. However, some things just weren't right.

The sound of weighted skins and muted drum shells played in a reverberant free environment not only lent itself to the addition of very spacious electronically generated reverb, but it was all the rage for certain forms of music like reggae, hard rock ballads, and disco. But deadened drum skins do not provide adequate rebound in their response to stick hits. This not only slowed down drummers but also caused them to adopt a more heavy-handed playing style. The quick, light touch dexterity of jazz drumming became noticeably absent. Dead rooms did not work on many levels.

During the early 1970s I built a rehearsal studio in a commercial neighborhood. Due to the closeness of the nearest neighbors, sound transmission was a major concern. The room's boundaries included three masonry walls, which was a big plus. The first step was to seal every crevice. This was followed by applying absorptive material. The owner's intention was to provide a space for rock musicians to rehearse "full throttle" as they were unable to do elsewhere. This equated to dual-Marshall stacks for the guitars (one stack for the rhythm and another for the lead), two 18-inch driver bass cabinets (one front mounted and one folded horn), a Hammond B-3 with dual Leslie speakers, a half dozen other smaller "combo" amplifiers for synths, other miscellaneous instruments, a complete dual-horn, double-low-frequency driver, and four separate stage monitor PA systems. The size of this room was only 25 × 30 feet!

With every instrument blaring, the PA levels would have to be set pretty high, so I figured mucho absorption was the way to go to avoid massive feedback levels. It worked! Everyone could set their amp levels all the way up and get the distortion and sustain they craved *and* you could still hear the vocals above the din. Then the problems set in. To say the least, I learned a lot from this experience the hard way. While all the amps, speakers, and drums were on small stage-like risers that were set on vibration isolation pads, I hadn't correctly figured on the massive playing levels of these power-crazed rockers. They were vibrating the building's masonry structure! This displeased our upstairs neighbors no end. Since this business was of the "ill-repute" type, its owners were not happy when their "workers" reported to them that the clientele was disturbed and distracted by the noise. This was explained to me in a very succinct manner.

The 1,500 BTU in-wall air conditioner had been outfitted with intake and exhaust ductwork to reduce the leakage of sound. During the summertime construction stage, it easily chilled the hard-working crew of which I was part. But sound absorption equals thermal insulation. Add in a completely sealed room and a whole bunch of tube amps glowing purple and red while a bunch of sweaty, excited musicians jumped about practicing their energetic stage show and you ended up with a whole lot of heat, humidity, and electronic failure. Equipment was breaking down at an alarming rate! I ended up putting a governor on the PA, which induced the vocalists to plead for more saner amp levels so that they could hear themselves. Luckily, another space became available that was larger and had more understanding neighbors. Central AC was the first investment.

Deadened recording studios didn't exactly suffer the same plight, but some of the joy that comes from recording well-performed instrument playing was lost. Next time you're in a live studio with well-balanced acoustics, have someone play the baby grand. Get a good earfull and then wrap a towel around your head. Electronic reverb will help, but it will not give you back the lost high end. Things just sound better in a large studio whose overall sound is live. Correct acoustic room response is essential for making quality recordings of orchestras and even small string sections. While woodwind and brass sections can be placed in smaller areas, these instrument especially suffer from dead-sounding acoustics as they also project their sounds multidirectionally.

With a large live area, the engineer has the option of recording the drums set up in the middle of the room for a thunderous sound, or they can be relegated to a corner of the room or placed in the drum booth, depending on what the situation calls for as far as the separation needed and the sound desired. This corner area, as well as the booth, can provide various acoustic settings with the use of movable panels or curtains. This kind of acoustic variability also aids the sound of other acoustic instruments as well as amplified instruments.

There's something special about being able to walk out into the studio and flip a hinge-mounted panel next to a Fender twin reverb combo-amp to change the rhythm guitar sound from a biting lead caused by its high frequencies being reflected from the panel's hard surface to a mellower bottom-accented sound due to the added absorption of the high frequencies by the "soft" side of the panel. Acoustic panels can be made in an endless variety of shapes, sizes, and multiple configurations to diffuse, reflect, and provide frequency-selective absorption. There's tons of information out there that will tell you exactly how to construct your own and countless manufacturers offering an almost endless variety of prefabricated panels.

Variable acoustics can be critically important to what microphones pick up. This fact is obviously important to the audio recordist. But to truly understand it, one must first look into the technical aspects of microphones, that is, how they work and the specifications that define their parameters.

3

Microphones

As far as I'm concerned there's no such thing as a bad microphone. At least I can't think of one that I out and out dislike. At the very worst, a mic's range of uses may be limited, but in those few applications it will more than likely perform satisfactorily.

As an example, back in the late 1960s I worked in a commercial recording facility that was, at that time, considered "pro." It had eight-track recording capabilities and the owners were considering an upgrade to a full 16 tracks! Anyway, while rummaging through a box of parts I came across an old Sony pencil microphone. When I inquired about its use, I was told that it was just an old video mic that was once snapped onto a consumer-quality camera. I was told that since it had fallen several feet to a stone floor and then tumbled down a full set of stone stairs it was now "no good." Upon internal inspection (no easy task), I found that this "accident" had caused the thin, metal-coated mylar diaphragm to loosen from its suspension and warp so it had a bit of a wrinkle and a sag.

In the past I had had some luck with supposedly "worthless" audio equipment, so out of curiosity I loaded it up with the required AAA battery and checked it out. While the microphone's noise level seemed unaffected, the frequency response was severely diminished to the point where it had practically no low-frequency output at all. That "no good," worthless thing ended up being one of the best hi-hat mics I have ever run across. It yielded practically no bleed from the kick or the snare, and it made the hi-hat shimmer.

Microphone Types

The kind of diaphragm that's used in a microphone dictates the way in which sound energy in the air is converted into electrical energy. *Dynamic* diaphragms work much the same as speaker drivers do, but in reverse. Here a sound wave deflects a diaphragm that is attached to a coil of wire sitting in a magnetic field. This movement in turn causes the sound to be converted into a small voltage (actually fractions of a volt). They are very durable and the ones made today have a broad frequency response and a dynamic range well beyond those that were common during the 1950s and 1960s.

Controlled magnetic mics are also called "variable-reluctance," "balanced armature," or simply "magnetic mics." Here the diaphragm is connected to, and varies the movement of, a magnet through a coil. These mics are only good at producing midrange frequencies and thus are used mostly for speech communications (i.e., CB radio and paging).

Condenser microphones are referred to as capacitor mics simply because their diaphragms are, in fact, capacitors. Here the received sound impinges upon and moves (very slightly) a diaphragm that is one of two plates that make up the capacitor. This movement changes the spacing between the two plates, causing a change in the capacitance value, which produces the sound as a change in the voltage passing through it. This voltage is so small that an internal amplifier is usually required in this type of microphone. Therefore, some type of powering, either phantom or a separate power supply, is needed to both charge the capacitor diaphragm and to drive the amplifier.

Phantom Power

DC is needed to power the electronics and charge the diaphragms of condenser mics. A handy method of delivering this voltage to the microphone is through balanced audio mic cables. By using the drain wire or shield conductor as the ground or reference and running the same power voltage through both the positive and negative audio conductors, there will be no voltage difference at their output. Phantom power has no effect on the audio and causes no problems when connected to microphones that do not require power such as dynamic mics. However, when used in conjunction with some older tube mic power supplies, an occasional popping noise from a capacitive discharge may result. It is therefore a wise move to turn off the phantom power on any console channels before these microphones are connected to them.

Lately there's been a push toward using larger-sized diaphragms in condenser mics. This was shunned in the past because diaphragms can cause sounds that have wavelengths of an equal or lesser size to come to a complete stop at their front plate. When this occurs it causes an increase in the output level at those frequencies. This starts to become audible with diaphragms measuring over 1 inch in diameter (equal to a wavelength of between 13 and 14 kHz). The old push was to use capacitor diaphragms sized from 0.75 (18 kHz) to 0.5 inches (27 kHz), and this is the reason capacitor diaphragms for ultra-flat response measurement microphones are generally no larger than 0.5 inches, and commonly run 0.25 (54.2 kHz) down to 0.125 (108 kHz) inches.

Most condenser limitations are due to the fact that they require electronic circuitry. All electronics produce some noise and are limited as far as the maximum amount of input they can handle. That's why condenser specs always include noise and maximum *sound pressure level* (SPL) ratings. Yet today's condensers produce a minimal amount of noise and can handle all but the most excessive input levels. While they are often a bit more expensive and can be more sensitive to environmental conditions such as heat and humidity than say a dynamic, they usually handle high frequencies very smoothly and have a superior transient response.

Electret microphones are condenser mics that use a "precharged" capacitor diaphragm, or back plate, and therefore it does not require external powering (for this) but still needs powering for their amplification circuitry. As a general rule, older models had a limited lifespan (as this charge does dissipate) and were not especially good sounding, but current models have a longer life span and can be very natural sounding.

Carbon Mics

These are the oldest type of microphones. Two conductive plates are separated by small carbon granules. Sound waves cause the diaphragm plate to compress the carbon, which varies their resistive value. The current that runs through the carbon is modulated by this changing resistance, which produces the audio output. They have a limited frequency range, have midband peaks, tend to hiss a lot (think of an old telephone handset), and distort like crazy, making them lots of fun to experiment with. That's probably why harmonica players love them and it's why I like to augment the normal front mic positioning by mounting one of them in the back of a R&B rhythm guitarist's Fender twin reverb.

Ribbon Mics

Ribbon microphones work like dynamics in that a magnetic field is utilized, but here it surrounds a thin strip of metallic foil. This metal strip acts like a single turn coil; thus, the diaphragm and the coil are one and the same. This, along with the ribbon's thinness and light weight, makes these mics very sensitive, which equates to their having a smooth frequency response and a natural sound due to their transient response. They add little to no noise to the signal and have excellent low-frequency pickup capabilities. But they can be susceptible to damage from high SPL inputs and wind noise. Because of this, you'll generally see furry blimp-like surrounds applied to them when used outdoors.

Piezoelectric Materials

Piezoelectric aterials transform mechanical forces into an electrical response. The traditional materials used were crystals; then came ceramics, silicones, and most recently polymeric films, which are flexible yet strong and lightweight. This may not seem significant, but in some applications the weight of the transducer can actually change the way in which a music instrument resonates. Piezos all share the same operating principle. By means of mechanical processing, the orientation of their crystallites is controlled to yield a set polarization. When pressure is exerted upon this material by compression, shear, twisting motion, or vibration, it outputs electrical energy. While these elements are commonly thought of by recordists as only pickups, they are also used for microphone elements, speakers, and headphone drivers, switches, and industrial sensors.

Ceramic Mics

Ceramic mics have diaphragms connected to a piezoelectric material. The movement of the diaphragm stresses the piezo material, which in turn outputs a corresponding voltage. Ceramic pickups can have broad frequency ranges that can extend down to DC. Thus, they are highly responsive to mechanical vibration.

Pickups

Pickups have many industrial and commercial applications, which include switches and heat sensors, as well as vibration, acceleration, fluid flow, seismic pulse, medical imaging, position sensing, and security alarms. I know of one manufacturer, Endevco, that puts out a piezo-electric pickup for environmental testing that handles an input of 180 dB SPL! All this is very fortunate for us, as the field of music instrument electronics could not possibly support the extent of research that's been put into the development of these materials.

The result is that today there are a number of pickups available to the recording engineer that faithfully reproduce the sound of acoustic instruments without producing any feedback. Output anomalies due to positioning have been greatly reduced while retaining the signal isolation (instrument separation) inherent to this transducer. Good sound, near-ideal isolation, and the opportunity to run an acoustic instrument through all kinds of effects, including stomp boxes. Why shouldn't viola and cello players get in on the fun? As you can tell, I'm a big fan.

Barcus-Berry makes a popular line of piezo systems (pickups with preamp and EQs) designed specifically for a wide range of instruments. Woodwinds, brass, flutes, strings, and other instruments have the choice of bridge, head stock, or body attachment. There's even a system whose preamp has a belt clip.

I don't know if it's still available, but in the early 1980s a British firm put out a product called a Cducer tape pickup. This was flat strip of material roughly 1 inch wide by $1/8$ of an inch thick and ran 7 or 8 inches in length. Two aspects of this pickup made it distinct. The electronic design utilized both a piezoelectric material and variable (vibration-induced) capacitive functioning to produce its electrical output. As with most pickups, this output was free of phase cancellations and provided remarkable isolation, but due to the strength of their output signal (along with their large coverage area), they were able to deliver even more low-frequency content than normal pickups. These could reach down to subaudible frequencies. I remember being awestruck (and a bit unnerved) by the excessive low-frequency driver excursion of a UREI 813 when I attached one of these to the back of an upright bass. Not a good idea!

Yet even with their large size, they were so light they didn't interfere with the natural resonance of the instrument to which they were attached. These were among the first devices I tried as drum triggers. They worked very well, but I'd caution against mounting them directly to the top of the drum skin. When pickups are struck by a drumstick, the output voltage can reach disastrous levels, as in seriously overloading the trigger inputs gain stage. This is a very different concept (as far as micing instruments) and it lends itself to a lot of experimental fun.

Pressure Zone Microphones (PZM)

But the award for the most fun from experimental sound capturing has to go to *pressure zone microphones* (PZMs). Manufactured by Crown International, this is basically a common capacitor microphone mounted very close and pointed into a flat boundary or plate. Since the mic itself is very close to the boundary, the direct sound it picks up arrives at the same time as those reflected from the boundary. In fact, the Crown PZM is actually a fairly complex system. The distance from the boundary must be very precise so the transducer's high-frequency notch filter can eliminate standing waves, or resonance, caused by those frequencies whose wavelengths match this particular distance.

You see, back in the 1970s, the inventors, Ed Long and Ron Wickersham, were trying to eliminate the problems of conventional microphones like comb filtering, phase cancellation, and the resulting dullness in sound. They not only pulled it off, but they also ended up with a transparent microphone system with a flat frequency response with no phase, filtering, or cancellation problems.

Yet even they could not have predicted how this system would lend itself to ongoing experimentation for the next 25 years. And it's still going strong. One of the latest offerings from Crown, the SASS-P Stereo Pressure Zone Microphone, is a direct result of this ongoing experimentation. So applications for the PZM still seem almost limitless.

Microphone Characteristics

Pickup or Polar Patterns

Polar patterns show the variation per frequency of a microphone's pickup characteristics as per the direction of the sound source. A mic can be termed omni or hemispherical, bidirectional or figure eight, cardioid, hypercardioid, shotgun, or stereo. These specifications can also be listed as simply "dynamic cardioid" or "omnidirectional condenser."

Although these terms shed some light on directional characteristics, this method of specification is inadequate as the off-axis response is significant everywhere outside of an anechoic chamber. For recording use, this parameter needs to be understood in more depth to determine probable applications and point to possible placement problems when reflections may cause comb filtering, unwanted equalization, and phase cancellation.

Because frequency response graphs show all frequencies but not all angles, while polar patterns can give multiple angles but not all frequencies, both are usually provided. But even with both, judgments are still application-specific. For example, a directional mic with an off-axis dip at higher frequencies may seem less than adequate, but it may actually be perfect for recording a vocalist with excessive sibilance if it is placed off-axis to his or her voice.

Directional microphones achieve their pickup patterns through phase cancellation. The difference in the signals at the front and rear of the diaphragm produces the output. Similar signals will have varying amounts of timing differences (phase shift) and cause partial to, in some cases, complete cancellation of sounds when they arrive at the diaphragm from angles that are not perpendicular to the mic's zero degree axis (perpendicular incident).

Cardioid Pattern   It is important to understand that the output of a microphone's diaphragm is equal to the *difference* between the pressures at the front and at the rear. Think of two people leaning on you, with the one in front weighing 250 pounds and the one behind you 175. You're going to be moved toward the rear. Depending on the polarity of the microphone, the combination of the sound pressures that impinge on both sides of the diaphragm determine its output. A microphone will only be affected by sound pressure exclusively from the front when it has no openings or access routes for sound to reach the back side of its diaphragm. Add an opening to the rear, and phase relationships enter the picture. This reality of physics is commonly used to increase a mic's directionality.

"Single-D" cardioid pattern mics have a single rear-entrance port. This provides the sound pressure for a route that is a set distance from the rear of the diaphragm. As with all mics, the closer one is placed to the sound source, the greater input, but with single-D designs that increase is substantially greater at low frequencies. How substantial? Ten to 15 dB is common. This "proximity effect" needs understanding to increase its usefulness.

If you speak very close into a mic, there's going to be some bass boost. It makes common sense that this increase in sound pressure at lower frequencies will have a greater effect on a diaphragm's movement.

It is helpful to understand that test chambers are "filled" with an SPL of 94 dB and this level is sustained throughout the chamber until signal generation is cut off. When 94 dB SPL is uniformly present throughout the chamber, the microphone under test can be located away from the sound source (by a minimum of two feet) to avoid any proximity effect. This is used when measuring *single D*-type cardioid microphones.

It is also important to understand that low frequencies are more omnidirectional in nature, so lows originating from the front will more readily reach around to the back of the mic than the highs.

Different frequencies also travel at different speeds. "Variable-D" mics have several openings or routes to the rear of the diaphragm. High frequencies enter the openings closest to the diaphragm, while lows enter the openings the farthest away, and the mids are in between. Variable-D is a trademark of *Electro-Voice* (EV) and this design reduces the bass-boosting proximity effect by about one-half. It therefore smoothes out a mic's full bandwidth response and provides more even off-axis rejection. EV also patented the "Continuously Variable-D" design wherein the ports are shaped as slots, thus allowing for this effect across a more complete bandwidth. You may be thinking, "So what? Why does the audio recordist need to deal with microphone manufacturing theory?"

Well, since you've stuck it out this far, I'll let you in on one of my best tricks. I like to use it with EV's RE-20, but when I tried it with an Earthworks Z30X, it knocked my socks off. Step 1, get an impregnated piece of lead or weighted vinyl. This is applied to the *front* of the mic by forming it into a cap. It will stop all sound from impinging on the diaphragm's face. Step 2, tape it in place. I like electrical tape because it has the kind of glue that's easily removed with isopropyl alcohol (something that used to be common around recording studios). Step 3, tape around the circumference of the mic and continue toward the rear. You can check out the results of your progress by speaking into the side of the mic (while listening through cans). Step 4, put the mic inside a kick drum and prepare to be blown away. It's amazing what you can come up with once you understand the basic underlying principles of audio technology.

NOTE: For more on microphone polarity, see EIA Specification #RS-221-a, *but* a cautionary note is needed here. You can get away with touching battery terminals to speaker inputs and use the resulting motion to determine polarity, but doing the same to a microphone's diaphragm outputs is *way* out of line, except under laboratory conditions.

Omni Mics

It's easy to understand that a microphone diaphragm that has no openings or access routes for sound to reach its back side will only be affected by sound pressure on its "face." Thus, depending on the way it is mounted, a closed-back diaphragm will react almost equally to sounds reaching it from all directions. Omni, or nondirectional, mics are designed to pick up sound from all directions fairly equally. I've found them very useful as room mics when recording a fine-sounding instrument in a room with good acoustics. They are more "open" sounding and have no proximity effect.

On the other hand, trying to use one placed in the center of a vocal group (as is often depicted in "how-to" books) only seems like a simple process. Even if it were possible to position everyone an equal distance from the mic, they'd still have to self-blend level-wise. This is something that can only be counted on from experienced professional sessions vocalists.

Bidirectional Mics

Bidirectional microphones pick up sounds equally from the front and rear. Since they reject sounds from the sides, they are commonly referred to as *figure eight* mics. Their use with dual vocalists is subject to the same distance and level limitations as omnidirectional mics. I much prefer the level control provided by using two microphones when recording an interview. But, as you'll see, these are unique tools that lend themselves to varied and specialized usage, especially in stereo recording configurations.

All other mic patterns reject sound from the rear and sides in favor of those arriving from the front to a greater and greater degree. Starting with the full heart-shaped *cardioid* pickup pattern, through the narrower (less side pickup), heart-shaped *hypercardioid* to the very narrow, almost tube-shaped pickup pattern of the *shotgun*.

Some shotgun mics achieve their narrow directional polar pattern by using several capsules arranged to face forward and back inside a tube that has a number of slots cut into it, in front of the capsule location. This causes a phase inversion inside the tube that cancels out sounds arriving at the sides. To achieve high directivity some older models were up to a yard long so that their phase cancellation would extend down to low frequencies.

Parabolic Mics

Parabolic mics have polar patterns are almost funnel-shaped in their pickup as the "dish," which reflects sound onto the mic's diaphragm, ignores sounds that occur outside of its tightly focused beam.

Boundary Mics

Boundary microphones, like PZMs, have a pickup pattern that is shaped like a half hemisphere with the flat side equating to the boundary.

Multi-Pattern Many microphones (usually more costly) offer switchable polar responses, often with cardioid, figure eight, and omnidirectional patterns available from the same mic. Dual capacitor-type diaphragms lend themselves readily to this approach. If just the front of the diaphragm is utilized, it operates as an omni. With both diaphragms utilized, it creates a 180-degree out-of-phase configuration of the figure eight pattern. By combining the front and rear diaphragms, less than a 180-degree out-of-phase (and with the help of a little equalization), the microphone can operate as a cardioid.

Microphone Specifications

Yeah, I know first it was consoles specs, then amplifier specs, and now more boring coverage of specifications with microphones! However, many of the specs unique to microphones must also be grasped by the audio recordist in order to be able to get full use out of this, your most important recording tool. Simply put, comprehend the specs; understand the mic. I apologize for the upcoming quagmire that you're about to be exposed to, but no one entity is responsible for insuring the uniformity of the specification test parameters utilized. Thus, the situation can seem a bit inconsistent, arbitrary, disorganized, and confusing. Like everyone else, I had to deal with all this myself, but without the nice, easily understood explanations you're getting here.

First, let me state that I believe that most manufacturers are honest in their specification listings. However, due to the lack of international standards and mandatory testing methods, it is relatively easy to utilize a form of testing that yields better looking numbers. Therefore, I am leery of ratings with reference values not commonly used. I often contrast mics that are new to me with standards that I'm familiar with. When discussing microphone qualities with other people, I generally compare them to the Shure SM58 (see Figure 3-1) as I feel that if there is any one microphone that everyone should be familiar with, this is it. I've found them

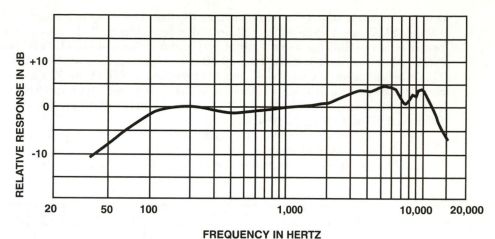

Figure 3-1
Frequency and polar pattern response graphs for the Shure SM-58 microphone

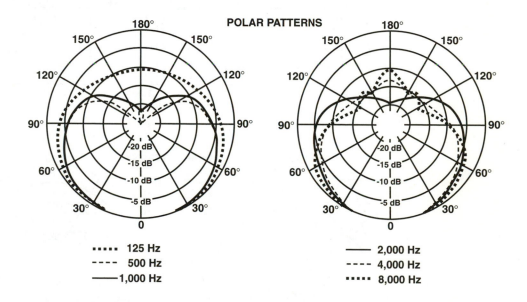

used everywhere from PA systems in the middle of a jungle to state-of-the-art, world-class recording studios. I do not believe, however, in comparing microphones as far as which one is better sounding because too much depends on the individual mic, its history (as in how many times it has been dropped), and the specific application in which it's being used, as well as many other variables.

Since microphone specifications are very helpful in terms of cluing in the recordist on what to expect from an unfamiliar mic, it is important to have a solid understanding of their meaning. People generally don't think about this until they try checking out a few mics by going over their specifications. For example, take the fact that many engineers prefer the sound of a Shure SM57 (see Figure 3-2) to a Shure SM58 for use on an R&B snare drum. The frequency response of the SM57 is down 7 dB at 50 Hz and it has a slight (1.5 dB) bump around 180 cycles, while the SM58 is down 10 to 12 dB at 50 Hz and it does not have the 180-cycle bump. This may not seem like a big deal, but the resulting wallop of the 57's low-frequency fundamental does make a difference. The 57 also has 2 dB more output at 6 kHz, while the 58 has an extra dB of output at 10 kHz. The 57's boost at 6 kHz is 2 dB greater than the 58, but the 58 has 1.5 dB more boost at 10 kHz.

Here's a tip. Look at the marks a drum stick makes on the skin. If they are consistently located and not scattered about, it may be safe to take the metal screen pop filter (it's lined in foam) off the 58. If the hit marks are erratic, just the foam insert can be removed. (Be careful

Figure 3-2
Frequency and polar pattern response graphs for the Shure SM57 microphone

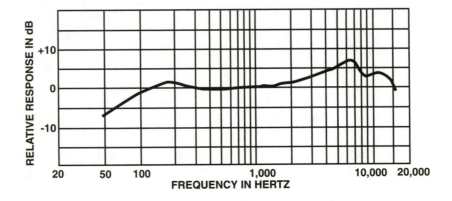

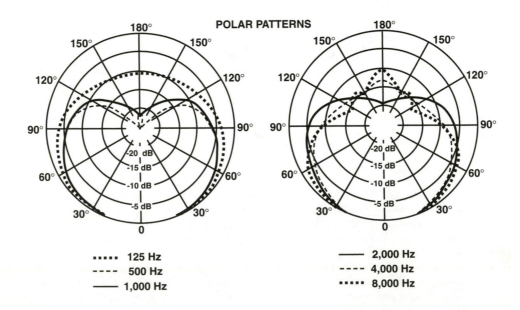

as it can rip.) This brightens up the sound a little further for a great, straight-out pop tune snare. You *will* hear the difference.

Frequency Response   The first parameter generally checked when looking at any set of specs is frequency response. This is pretty much self-explanatory, but a few points should be noted. The extremes of this specification, when given in its simplest form (i.e., 50 Hz to 16 kHz), are the points at which the output level has dropped down 3 dB from the 0 dB level (which is normally referenced to the microphone's output level at 1 kHz). Without additional reference limits like ±.5 dB that give an indication of the response variation between these two extremes, this method of rating is not of much use to the recording engineer. Its like trying to choose a tool that you know must conform to metric standards when you don't have any indication as to sizing. Frequency response in graphic form is much more informative, as all deviations from a perfectly flat response are shown.

These are often viewed as deficiencies, but to the knowledgeable recordist, this information is a clue to the possibility of the mic being used for both helpful corrections and effects usage; it's almost like having built-in equalization. It really all depends on the application. These graphs, however, do not show the deviation from the normal specifications on a per mic basis. This can be important as it is impossible for every production line mic to meet a nominal frequency response characteristic exactly. Yet today, widely differing curves seldom appear during the same production run, and run-to-run variations due to the use of different glues, adjustment routines, or automatic part placement and alignment will make two microphones with the same model number only slightly different because of stringent manufacturing quality control. Yet even a slight difference can negate the whole concept of the nominal frequency response curve. Thus, an individual frequency response graph given for each microphone is best if it does not radically increase the unit's price point. Finally, for directional mics, a 180-degree off-axis response should be included on the same graph, as the recording engineer must know both the off-axis rejection and any frequency response irregularities that occur at this angle.

Again, judging the adequacy of frequency response obviously depends on the application. "Purist" as-is recording requires a flat response, while on-mic equalization can often be helpful and can broaden a mic's range of usefulness, making it a more cost-effective purchase.

In either case, closely spaced peaks and dips are undesirable. Distance vs. frequency response generally deals with low-end boost, which has to do with the difference in pressure at the front and rear of the diaphragm. Ribbon and cardioid polar pattern microphones with a single rear opening are very prone to this "proximity" effect. This can be used to your advantage as long as you know about it beforehand (because you took the time to check out the mic's graphic response charts).

Microphone Output Level or Sensitivity   When a microphone's sensitivity rating isn't high enough, some of the delicate overtones such as those from a tiny bell may not be picked up. This can occur when a heavy dynamic mic diaphragm is difficult to move. On the other hand, too high a sensitivity rating can result in hard or loud transients (such as those found inside a bass drum) to cause distortion. Here a ribbon microphone's overly light diaphragm may be moved to its full extent or possibly stretched out of its normal position. This is the reason dynamic microphones are generally used for in-tight positions on screaming guitar speakers, and why you shouldn't place some ribbon microphones inside kick drums. Knowing a microphone's sensitivity rating is important.

NOTE: *Before we begin, one dyne per centimeter squared (cm²) and one micro bar (μbar) equal 74 dB SPL, while 10 dynes/cm², 10 microbars, and 1 pascal (Pa) all equal 94 dB SPL.*

Sensitivity This specification simply gives the relationship between the level of the acoustic input signal to the amount of voltage *or* power output by the microphone. Thus, this specification defines a microphone's ability to both pick up low-level signals and handle those signals with a high SPL. Here a microphone is exposed to a given input level, usually the same pressure as a nearby speaking voice (74 dB or 1 microbar). The microphone's output level is measured with this input level and the results are given as a voltage (usually millivolts) per microbars, or dynes per centimeter squared, as in "4.5 mV/microbar," which you can read as 4.5 thousandths of a volt output with an input equal to a spoken voice.

An important factor in this test is the load connected to the microphone's output. It must be a minimum of 20 times that of the mic's output impedance to present an unloaded or "open circuit" condition to the microphone. In fact, this test is termed "open circuit output."

Microphone output can also be expressed in decibels. This value must be referenced to some set value or it wouldn't mean anything. Let's start with the lowest amount of SPL that a human can hear, or what is referred to as "the threshold of hearing." This level has been tested to equal 0.002 dynes per square centimeter (dynes/cm²) and this became the 0 dB reference point.

In order to help comprehend this level, normal conversation is rated at 74 dB. This is 74 dB *above* the level of 0.002 dynes/cm², or 1 microbar. An open circuit rating of −60 dB referenced to 1 volt per microbar means that with an input of 74 dB SPL the microphone's output into an "open circuit" will be 60 dB lower than 1 volt (which happens to equal 0.001 volts or 1 millivolt).

You'll find specifications that read "−85 dB re: 1 V (volt)/microbar" or maybe just "−85 dBV." Either way, this means that with an SPL of 1 microbar or 74 dB SPL, the microphone's output is 85 dB lower than 1 volt. A more sensitive microphone will have a smaller negative number such as −70 dB, while a less sensitive microphone will have a larger negative number like −92 dB.

Maximum Power Rating Okay, 74 dB is still kind of a low level when it comes to exciting a microphone diaphragm. By using 94 dB, which equals 10 dynes/cm, 2 or 10 microbars (two microbars = 80 dB), as is done with microphone "power output" ratings, background noise during free-field or anechoic chamber measurement becomes less of a problem.

When the test room or chamber is filled with a sound pressure level of 94 dB, this level is sustained throughout the chamber until signal generation is cut off. Because 94 dB SPL is uniform throughout the chamber, the microphone under test can be located away from the sound source (by a minimum of 2 feet) to avoid any proximity effect. This is important for accurate measurements of "single D"-type cardioid microphones.

This test is performed using a measurement setup that utilizes an input impedance (load) that is the *same* as the microphone's output impedance along with an SPL that is 20 decibels higher than with open-circuit output voltage tests. Here you'll find specifications that read "−58 dB re: 1 mW (1 milliwatt)/10 microbars." This means that the microphone's output will be 58 dB lower than the 0 dB level of 1 milliwatt with a sound pressure level at its input, equaling 10 microbars or 94 dB SPL.

Once again, a more sensitive microphone will have a smaller negative number such as -55 dB and a less sensitive microphone will have a larger negative number like -62 dB. Testing for both open-circuit and power ratings can be performed at either 250 Hz or 1 kHz at the input.

Wouldn't it be great if that's all there was to it?

A microphone's output is equal to the number of decibels as per its output in milliwatts when loaded by a matched impedance when it is exposed to 10 dynes/cm^2, 10 microbars, 1 pascal, or 94 dB SPL as in X dBm re: 1 mW/Pa. Or the number of decibels as per the output in millivolts of a microphone when not loaded by an impedance, but connected to an "open circuit" that is exposed to 1 dyne/cm^2, 1 microbar, one tenth of a pascal or 74 dB SPL as in X dB re: 1 volt/dyne/cm^2). Not a chance!

Since there is no single recognized or *required* international standard for testing microphone outputs, not only can the input impedance of the test setup vary, but also the ratings can be given in milliwatts, millivolts, dBm, or dBV. So not only will you run into power ratings that use matching impedances with an output referenced to 1 milliwatt and a 94 dB SPL input, and output ratings that are related to voltage that utilize a load impedance of at least 20 times that of the microphone output impedance and an output referenced to 1 volt with a 74 dB SPL input, *but* you'll also run into values based on volts per pascal and milliwatts per dyne/cm^2 as well. Thanks!

As if that weren't bad enough, many of the specification sheets are compiled by folks involved in marketing or advertising who are not overly concerned with little things like mW, mV, dynes, microbars, and (I know this is hard to believe) even pascals! (Wait till we get to mOes!)

A couple of real-life examples will bring this fact home. Microphone X's "specification" sheet lists its output level as -57 dB. No reference is given. If that's per mW/Pa or mV/μbar, it's about average. I can't believe it's EIA since the -50 dB that must be subtracted from the test results keep most ratings done to this specification well below that level. So what do we have here? I'll tell you what. A potential customer, looking at this specification sheet and wondering why they should even consider this microphone as a viable purchase option.

This next example is *not* unusual. One manufacturer's specification listing for six of their microphones was printed on a single sheet of paper. The first mic had a "sensitivity" rating of "10 mV/pascal," the second mic had a value for equivalent noise (something none of the other five mics had), but it had no sensitivity rating. The next mic had a sensitivity rating of "-56 dB," and the final three are rated at "-57 dB." There was no reference given and assuming that because the first mic's sensitivity was given in millivolts per pascal meant that this was the reference used for the rest of the mics just does not add up. This is because a mic output referenced to 10 millivolts per 1 pascal would be equal to -40 dB, which would make it a full 17 dB hotter than the others? I tend to doubt this is the case, but what do I know? Nothing, because I'm dealing with what ends up being useless information. Those specs may as have been referenced to "rascals" instead of pascals!

But we also have to take the realities of the manufacturer's situation into consideration. Since testing must be performed by skilled workers using expensive equipment, it can cost a lot of money as a small microphone company outfitting itself with multiple test setups just might not be economically practical.

On the other hand, companies such as Shure and Crown, who provide microphone output specifications with three to four different reference values for each microphone, invite comparisons, which lead to increased consumer confidence.

Comparing the watt value, often called the "effective output" in dBm (re: 0 dBm = 1 mW/1 Pa), to the voltage value or "sensitivity" in dBV (re: 1 V/1 microbar) is not a simple add or subtract, or even a divide or multiply situation. In most cases, the audio recordist is left to struggle with brain-teaser conversion possibilities. I know. I once spent two weeks trying to come up with an accurate conversion for mW/Pa to mV/μbar. I ended up punting because of the differences between the two ratings. Converting watts to volts is okay and comparisons are still possible, when the input levels are 74 and 94 dB. But when I came to the two different ways of dealing with the mic's output impedance I was stumped. A simple conversion equation may be possible, but I'm not bright enough to figure it out. I can't even explain why volts per microbar and milliwatts per pascal often have an equal decibel rating when the microphone's output impedance is 250 ohms.

So don't feel badly if all things audio are not crystal clear. But at least you now know that the relationship between these values is highly dependent on a microphone's output impedance. How dependent?

Well, the volt per pascal level can be seven decibels lower (−7 dB) than the milliwatt per pascal level for microphones with an output impedance of 60 ohms, −4 dB @ 100 ohms, −3 dB @ 150 ohms, −1 dB @ 200 ohms, 0 dB or the same value @ 250 ohms, +2 dB @ 400 ohms, and +3 @ 500 ohms.

So microphone impedance has a great deal to do with sensitivity ratings. This increases the difficulty in comparing voltage per microbar to watts per pascal ratings. It also makes it nearly impossible to compare either of them to EIA ratings.

Electronic Industries Association (EIA) Microphone Sensitivity Standards

The "EIA Standard for Microphones for Sound Equipment #SE-105" dated August 1949 "recommends" a standard for microphone sensitivity ratings with a symbol that "shall be Gm." This test takes the logarithmic value of the microphone's open-circuit output in RMS voltage, divides it by the input level in decibels (SPL in dynes/cm^2), and multiplies the result by 20. From this amount is subtracted the log of the microphone's "rated" impedance multiplied by 10:

$$Gm = 20 \log \left(\frac{\text{mic output}}{\text{input in dB}} \right)$$

$$- 10 \log \times \text{rated imp} - 50 \text{ dB}$$

The result of this, in dB, is then adjusted by subtracting 50 dB from it. That equation again is Gm = 20 log × the output, divided by the input SPL in dB − 10 × the log of the mic's *rated* impedance −50 dB. That's *some* equation you got there, Electronic Industries Association!

There is actually a good reason for all this. Because they take all those variables into consideration, EIA microphone sensitivity ratings can be compared to each other with total confidence. However, comparing ratings as per voltage, or watts/pascal, or even dynes/cm^2 to the EIA ratings, is no simple feat.

Once again impedance figures in the EIA standard utilize a microphone output impedance that is adjusted as per a table that relates the "Magnitude of Normal Microphone Impedance" to their "Rating Impedance" as follows. A "normal" impedance of

19 to 75 ohms	is rated as 38 ohms
76 to 300 ohms	is rated as 150 ohms
301 to 1,200 ohms	is rated as 600 ohms
1,201 to 4,800 ohms	is rated as 2,400 ohms
4,801 to 20,000 ohms	is rated as 9,600 ohms
20,001 to 80,000 ohms	is rated as 40,000 ohms
More than 80,000 ohms	is rated as 100,000 ohms

EV used to make a circular slide rule available that could be used to compare ratings in decibels referenced to milliwatts per 1 pascal, to millivolts per microbar (open circuit), to decibels referenced to 1 millivolt per microbar, as well as volts per microbar and decibels referenced to volts per pascal.

You can also find nomographs that allow for ruler placement across three lines listing output impedance, volts per pascal, and milliwatts per pascal ratings.

My experience is that all of these aids will only put you in the "ballpark," meaning that with the use of a magnifying glass you may get a comparison with a ±2 to 3 dB accuracy. But let's face it. How many audio engineers are going to carry a single function slide rule or nomograph around with them?

I can tell you, however, that since 10 dynes/cm² and 10 microbars both equal 1 pascal, *open-circuit* specifications referenced to *volts* per pascal *multiplied* by 10 will equal ratings referenced to *volts* per microbar and dynes/cm². Conversely, ratings referenced to volts per microbar and dynes/cm² *divided* by 10 will equal ratings referenced to volts per pascal.

Furthermore, a voltage that is 10 times that of another voltage equates to +20 dB. You got it; *adding* 20 dB to the decibel amount referenced to volts per pascal will equal the decibel amount referenced to volts per dyne and microbar. Furthermore, *subtracting* 20 dB from decibel ratings referenced to volts per microbar and dynes/cm² will equal decibels referenced to volts per pascal.

The fact that we have to deal with all this craziness is not due to trickery and deception, or smoke and mirrors being used to convince you to shed your money in one direction or another, but the way audio specifications, which are based on volts, watts, impedance, and decibels, interrelate.

Sorry, there just aren't many easily defined absolutes in audio. But take heart; almost every studio has a Shure SM58 and a Crown PZM, so here's their multiple sensitivity or output ratings if you want to compare another mic's sensitivity, power output, and impedance specifications.

The Crown PZMs are rated at −76 dBV and −53 dBm, have an EIA rating of −148 dB EIA, and a 150-ohm output impedance.

The SM58 is rated at −76 dBV, 0.16 mV, and −57 dBm with its impedance set at 150 ohms.

I once put together a chart for myself that related specifications referenced to volts per pascal, which are usually given in millivolts to their decibel equivalents with 0 dB referenced to 1 volt. Since the millivolt output levels are referenced to a full volt output, all the decibel equivalents are negative values starting with 100 mV, which equals −20 dB. It is accurate to only 3 dB. By the way, negative 20 dB, even when referenced to pascals (a 94 dB SPL input), happens to be a very hot output as far as microphones go.

100 mV	−20 dB	5 mV	−45	1 mV	−60
55 mV	−25	4 mV	−48	0.55 mV	−65
40 mV	−28	3 mV	−50	0.4 mV	−70
32 mV	−30	2.8 mV	−51	0.25 mV	−75
22 mV	−33	2.5 mV	−52	0.1 mV	−80
20 mV	−35	2 mV	−55	0.75 mV	−85
12.5 mV	−38	1.6 mV	−56	0.45 mV	−90
10 mV	−40	1.4 mV	−57	0.001 mV	−100
8 mV	−42				

Another trick I used that's also not dead-on accurate but, like the above chart, will get you to within ±3 dB when comparing decibel ratings referenced to 1 mW/Pa to EIA ratings. Add −96 to the dBm value. For example, −51 dB re: 1 mW/Pa + −96 results in an EIA rating of −147 dB EIA. However, I've found that this conversion method can be off by as much as 7 dB simply because it does not take the microphone's output impedance into consideration.

Right about now even the most die-hard audio enthusiast starts wondering, "Is dealing with all this really worth it?" *You bet!* You think I made up charts and tried to figure out equations for comparing mic specs because I had nothing better to do? Think again!

Once sensitivity is understood (including whether it's referenced to either 250 or 1 kHz) and used in conjunction with a microphone's frequency response graph, it opens the door to a vast amount of information pertaining to ways in which you can use a mic. Add in the polar response chart and you'll *know* the tool you're about to work with. How so?

Let's take a Shure SM57 and SM58 comparison. Since this will be somewhat subjective, it's best to stick with a single manufacturer. Shure gives its microphone sensitivity ratings in dB and millivolts "Output level (at 1,000 Hz) Open Circuit Voltage re: (0 dB = 1 volt per microbar)" per its output impedance, and in "Power Level" in dB re: 0 dB = 1 milliwatt per 10 microbars. Thank you.

The SM57 is rated at −75.5 dB (re: 1 volt/microbar), 0.17 mV, and −56.5 dB (re: 1 milliwatt/10 microbars) with its impedance set at 150 ohms.

The SM58 is rated at −76 dB (re: 1 volt/microbar), 0.16 mV, and −57 dB (re: 1 milliwatt/10 microbars) with its impedance set at 150 ohms.

The polar patterns for these two mics are just about identical. The frequency response graphs show the 58's low end to be down 10 to 12 dB around 40 to 50 cycles, while the 57 is down only 7 dB at 50 cycles and it has a slight 1.5 dB bump around 180 cycles. The 58's output is +5 around 6 kHz and +4 dB at 10 kHz, with the 57's being +7 at 6 kHz and +3 at 10 kHz.

In the early 1980s I heard a microphone being used by a vocalist on a television variety show. It sounded clear and crisp with a real even top end, which was apparent even over the cheap portable TV speaker. I just knew two of them would make for great drum overhead mics, so I researched like crazy until I found it was a Shure SM85 (*now called* SM87). When I tested them in a recording studio setting, they proved to be even better at this application than I expected. Why was this the case?

The specs tell it all. First off, it's a capacitor mic so the *common assumption* would be that its lighter diaphragm will be more sensitive than an SM58. It is, but only by 1 to 1.5 dB with its output of −74 dB (0.20 millivolts/μbar). The frequency response graph shows it to be down 15 dB at 50 cycles and to have a peak of +7 dB between 7 and 8 kHz and 5 dB at 10 kHz. Pretty similar to the 57 and 58, and in fact I found that it provided just a little more high end when used on a snare.

The difference in polar pattern, however, was much more pronounced. The 57 and 58's maximum off-axis rejection occurs generally around 140 degrees, while the SM85's maximum rejection occurs at 180 degrees and is more even with frequency.

This is exactly what's called for when placing a mic pointed downward from above a drum kit to avoid phase cancellation and comb filtering caused by reflections from the ceiling. Removing this mic's wind screen/pop filter made the cymbals sizzle.

Why were those words "common assumption" italicized? Well, a few years later, along came the Shure Beta 58. This mic is dynamic (like the SM58 and 57) so you may assume its sensitivity would be a little lower than that of the SM87, which is a condenser. Guess again.

This is audio. The land where *no generality is always true!*

The Beta 58 utilizes a neodymium magnet, which gives it more output. How much more?

The Beta 58 is rated at −71.0 dB (re: 1 volt/microbar), 0.28 mV, and −51.5 dBm (re: 1 milliwatt/10 micrbars) with its impedance set at 150 ohms.

Its output is not quite double (+6) the voltage of the SM58 (being only 5 dB higher), but it is more than double (+3) the power output at +5.5 dBm. This is a big difference, but there's more.

The Beta 58 has a hypercardioid polar pattern. This means that the frontal pickup is narrower than the normal cardioid and it has a larger pickup "lob" at its 180 degree, or back position. As with many hypercardioids, the maximum rejection occurs just behind the diaphragm. In the Beta 58's case, *all* frequencies are attenuated by at least −20 decibels 120 degrees behind the front of the diaphragm.

Its frequency response is not much different than the 58 and 57, being down about 12 dB at 50 cycles with peaks of +5 dB at both 5 and 10 kHz, and I would discourage the removal of its windscreen because I'm not sure how well a ceramic magnet would survive multiple drumstick hits.

So what does all this boil down to? Yeah, I was right about the SM85/87 for overheads, but the Beta 58 is now my favorite snare mic. Not so much because of the higher output (though it helps), but like the SM85/87, the pickup pattern is exactly what you want in this situation. To get near-perfect separation of the snare from the sounds of the kick and hat, all you have to do is place the Beta 58 pointed downward and aimed at where the stick hits the snare with both the kick and the hi-hat facing the 120-degree point on the mic, which is just behind the pop filter. Place it like that and simply move right on to the next mic you have to set up.

Take a look at Figures 3-3 and 3-4.

As stated, know the specs, know the mic. This section contains other polar and frequency response graphs to check out. There's another thing to be aware of as far as sensitivity and power output is concerned. With microphones that require powering (condensers, ribbons, etc.), the output level can vary with the amount of power supplied.

Take Telex's "Cobalt" series SE 60 condenser mic, for example. It was designed for both live and studio use, so they made it operable by phantom (48-volt) and AA battery (1.5 volt) power.

Their specification for sensitivity (0 dB = 1 V/pascal @ 1 kHz) is −50 dBV with phantom power and −52 dBV when powered by battery. No great shakes. But there is another specification that is dependent on variations in powering. There are situations (especially in live performance) where it is very important to know the *maximum sound pressure* for each powering condition. With the SE-60, it's 141 dB for phantom and 137 dB for battery power. A 4 dB

Figure 3-3
Frequency and
polar pattern
response
graphs for the
Shure SM85/87

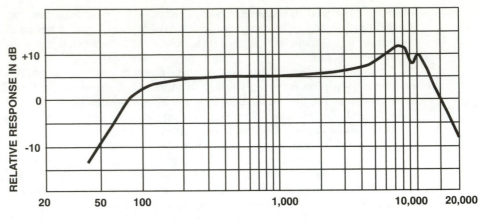

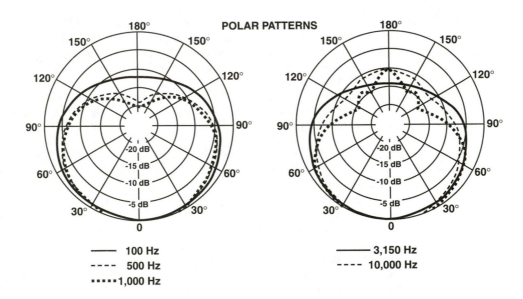

difference in the SPL range is normally of no concern to the audio recordist, but it's important to know when micing guitar amps in live venues.

Although manufacturers are pretty up front with these facts, it can still be a little tricky.

Sanken's little omnidirectional lavaliere, the COS-11s, comes in three versions to handle differing power capabilities. The COS-11sBP, which can be powered by either a 1.5-volt battery or 12- to 52-volt phantom power, only has one sensitivity rating of −44 dB ±3 dB, and 6.0 millivolts re: mV/Pa. Yet while the manufacturer lists only one output for the two powering

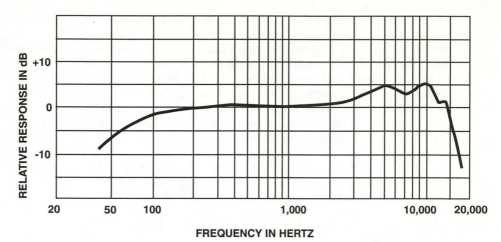

Figure 3-4
Frequency and polar pattern response graphs for the Shure Beta 58

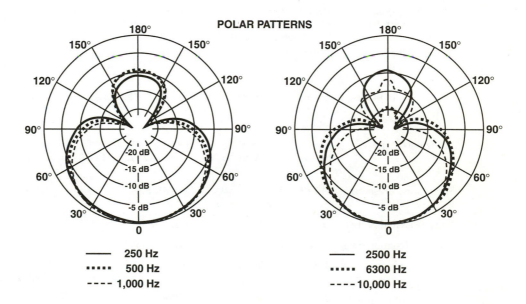

possibilities of the BP version, because of the other specifications given, such as current consumption, which is designated only per the battery usage, you can *assume* that the BP's specs are with battery power. I hate assumptions!

The COS-11sPT (pig-tail connection) takes 3 to 10 volts of power and puts out −41 dB ±2 dB and 8.9 mV/Pa.

The COS-11s is the standard version and it is powered by 48 volts ±4 volts phantom and has a sensitivity rating of −35 dB ±2 dB and 17.8 millivolts per pascal.

Aside from the fact that the standard (48-volt phantom) version puts out twice the level of the pig-tail version (3–10 volts), again the SPL-handling capabilities depend on variations in powering. The dual-power BP (again I have to assume with battery usage) will handle up to 120 dB, while the PT is rated at 123 dB and the phantom-powered version can handle 127 dB of SPL. It takes a little work to figure all this out, but all the information is provided by this manufacturer on a single sheet of paper (see Table 3-1).

Audio Technica's 813R is one of my all-time favorite mics. I say that up front, because I'm going to use it as an example of how sensitivity ratings per power source can be a real challenge. There are four different versions of the AT813 if you include its replacement, the ATM31a. The "R" in the original version's model number stood for remote or phantom-powered only (12–52 volts). The 813 version could only operate via a 1.5-volt AA battery. Few people know about the 813 "a" version that could be powered by either a 1.5-volt AA battery

Table 3-1

The Sanken COS-11s specification listing

Specifications	COS-11s	COS-11sPT	COS-11sBP
Transducer type	Back-electret condenser	Back-electret condenser	Back-electret condenser
Direction pattern	Omni-directional	Omni-directional	Omni-directional
Frequency response	20–20,000 Hz (±2 dB)	20–20,000 Hz (±2 dB)	20–20,000 Hz (±2 dB)
Source impedance	Approx. 180 ohms	Approx. 700 ohms	Approx. 200 ohms
Sensitivity at 1 kHz	17.8 mV/1 Pa	8.9mV/1 Pa	6.0m V/1 Pa)
	(−35 dB ± 2 dB)	(−41 dB ± 2 dB)	(−44.5 dB ± 2 dB
Sensitivity at 1 kHz (Red Mark)	6.3 mV (−44 dB/Pa)	3.2 mV (−50 dB/Pa)	2.1 mV (−53.5 dB/Pa)
S/N to IEC 179	>66 dB	>66 dB	>66 dB
Equivalent noise level 179)	Less than 28 dB SPL (0 dB = 20 μPA, IEC 179)	Less than 28 dB SPL (0 dB = 20 μPA, IEC 179)	Less than 28 dB SPL (0 dB = 20 μPA, IEC
Max. sound pressure for 1.0% THD	127 dB SPL	123 dB SPL	120 dB SPL
Max. SPL (Red Mark)	136 dB SPL	132 dB SPL	129 dB SPL
Total al. dynamic range	95 dB min.	94 dB min.	92 dB min.
Power supply	48V ± 4V Phantom	+3V to +10V	1.5V lEC R6 battery or 12V–52V phantom
Current consumption	2 mA	0.3 mA	0.5 mA (battery)
Output connector	3-pin XLR	Connectorless	3-pin XLR
Connections	Pin-1 = ground	Shield = ground	Pin-1 = ground
	Pin-2 = audio hot	White = audio hot	Pin-2 = audio hot
	Pin-3 = audio cold	Black = DC power	Pin-3 = audio cold
Dimensions	4.0 mm × 16.1 mm	4.0 mm × 16.1 mm	4.0 mm × 16.1 mm
Supplied accessories	WS-11 windscreen, RM-11 rubber mount and HC-11s holder clip		

Table 3-2
Different Outputs

Model	Power	Open-Circuit Sensitivity	EIA Sensitivity	Maximum SPL at 1% THD	
TB AT813R	12 to 52 volts	−70 dB 1 V/1 µbar	0.32 mV	−143 dB	141 dB SPL
AT813	1.5 V AA battery	−71 dB 1 V/1 µbar	0.28 mV	−149 dB	125 dB SPL
AT813a	Phantom	−44 dB 1 **V/10 µbars?**	6.3 mV	Not given	Not given
	Battery	−45 dB 1 **V/10 µbars?**	5 i.6 mV	Not given	Not given
ATM31a	One specification for phantom and battery powering	−44 dB	1 V/10 µbars	63 mV	Not given

The bold type 813a sensitivity ratings are because no reference was given, but with these output levels "1V/10 microbars" would make sense.

or phantom power (9–52 volt). This is because these three microphones were intended for different end users.

At the time (late 1960s to early 1970s), most live mixing boards generally did not have phantom power capabilities. So you could use the "813" battery version (it was even supplied with a ¼-inch phone plug output cable), while the "813R" phantom-powered version found a happy home in professional applications. Later when musicians wanted something they could use live and for home recording, the "813a" fit the bill. An updated model, the ATM 31, replaced the 813 mics. After 20 years the "ATM31a," an update of the original dual-powered 813a, is the only remaining version.

So here's what I could dig up in my files on all the different outputs (see Table 3-2).

The point is that even when you try your hardest to do your job, you can still be left with a few blank spaces. But hey, I have a pair of original AT813R mics and they've been my go-to mics for almost anything for over 25 years. Knowing the specs means you end up with great mics, so it is worth the effort *and* the aggravation.

Microphone Impedance Ratings

While a microphone's output impedance rating may be listed as 150, 600, or 2,400 ohms, the actual impedance can range from 75 to 300, 300 to 1,200, or 1,200 to 4,800 ohms. This is not something the audio recordist should worry over, but a microphone's output impedance *is* important for maximizing output level, minimizing unwanted signal pickup, and avoiding high-frequency loss.

As in all things electrical, having the correct impedance between the microphone and the preamp (whether it be internal to a console or a separate unit) makes this interface less susceptible to noise pickup, decreases the possibility of frequency response degradation, and maximizes signal output strength. Its really not such a big deal because the rule of thumb here is that the preamp's input impedance should be 10 times that of the mic's output impedance, and most mic preamps have very high input impedances.

However, impedance "matching" should not be taken too lightly as you may end up with interface problems as bad as those that occur when you connect a high impedance microphone (50 kilo-ohms or more) through unbalanced cabling to a low-impedance balanced preamp. In this situation, the cabling will more readily pick up extraneous *radio frequency* (RF) and *electromagnetic interference* (EMI) from things like taxi cabs, walkie-talkies, or elevators, motors, and power-handling switching devices that can induce buzzes and hums into the system.

This kind of poor interfacing can also result in a degradation of the signal's frequency response due to the combination of the inductance, capacitance, and impedance of the cabling in addition to a loss of level due to the increased loading down of the microphone's output caused by the impedance mismatch between it and the preamp. Don't worry too much about this, but should you hear excess noise and notice the lack of high-frequency content, best check the microphone/cable/preamp interface.

Test engineers, however, must know the mic's actual impedance in order to determine its output level. Under open-circuit microphone test conditions, the output of the microphone is noted. A variable resistor replaces the "open-circuit" load and is adjusted until the mic's output level decreases by one-half (drops 6 dB). The resistor is then measured and its value at that −6 dB level will equal the microphone's output impedance (as per Ohm's law).

NOTE: *This means that matched load specs will be lower than open-circuit specifications by 6 dB.*

Signal-to-Noise (S/N) Ratio   This is sometimes referred to as S/N level. Here the output of a microphone is measured with a specified input (usually 1 pascal). This input is then cut off and the residual noise output from the microphone is measured. These two values are compared and the difference ratio between the signal and noise levels is given in decibels. Here a larger dB rating is better.

Output Noise   This specification is usually given as an amount of dB SPL and it's generally stated if any weighting curves were used. Here a lower decibel rating is better; however, only specifications that utilize the same weighting curves can be compared.

Dynamic Range   This rating is usually the amount in decibels between the mic's own self-noise and the output at the highest sound pressure that it can handle before distortion. Even though the average studio background noise of 30 dB will generally mask out a mic's self-noise, this spec is important as the practice of close micing can cause a mic with a moderate dynamic range to clip and thus distort.

Common ratings should be around 120 dB and many microphones can withstand levels in the 140 to 150 dB range. This rating may also be given as "clipping level," but more likely this will be the same as maximum SPL. It should be referenced to a certain amount of distortion and noise with a given load.

Hum Sensitivity This rating is pretty much self-explanatory and is generally given in decibels referenced to 1 mV/Oersted or SPL/mOes (milliOersted). There are those mOes again! The term actually comes from an 1800s Danish physicist named Oersted (Hans Christian, no less) who discovered a way to measure current flow through wire and the resulting magnetism it induced by noting the amount of deflection it caused in a compass needle.

Hum pickup, such as a 60-cycle hum or its harmonics, can be induced into a mic's voice coil, transformer, or casing. The ability of a mic to reject this pickup is commonly rated in dB SPL per a magnetic field of 1 mOe, which is about equal to that which is found in the average studio. You may also run into this specification referenced to a 1 milligauss field. I believe they are the same and, if not, I don't know if there's any conversion routine for comparing the two because I never really bothered myself about it. You see modern-day microphones do not have hum pickup problems because no radio, television, film, live venue, or recording studio can tolerate having this kind of noise injected into their sound system at a point just prior to where the greatest amount of amplification takes place. If a microphone were susceptible to this kind of interference, its manufacturer would be stuck holding on to a lot of product.

Transient Response While often talked about, transient response is actually rated by only a few manufacturers. General statements like "since a condenser's diaphragm is lighter than that of a dynamic, it therefore renders a better transient response" do not take into consideration the acoustical impedance on either side of the diaphragm and the diaphragm's inherent voltage-generating characteristics. Transient response is best shown by an oscilloscope view of a microphone's output when excited by a spark gap generator (a controlled discharging of a capacitor). This view gives rise time vs. amplitude, called an *energy time curve* (ETC), and shows any overshoot. This response should ideally be performed off axis as well. Yeah, sure!

I've seen the results, used them, and actually performed this test procedure. They work, are accurate, and provide a wealth of information, as do spark gap tests done on speaker systems. Yet the number of people who would pay attention to these results often does not justify the manufacturer's expenditure of time and money.

Conclusion The use of words like "can," "often," "generally," and "should" exemplify the possible variations in mic testing methods. This variability, along with the fact that all of the specs covered and many other ratings such as *total harmonic distortion* (THD), difference or *intermodulation* (IM) distortion, phase response, and environmental operating conditions may or may not be provided, makes microphone comparisons via the rated specifications very difficult.

Yet since most audio recordists do not have access to adequate testing facilities such as a Bell Labs' specification, judgments must be based on what is provided by the microphone manufacturer. So due to the average recordist's inability to verify these specification ratings, microphone judgments must be based on subjective listening tests as well. This is okay because the most important aspect of a mic is its sound quality as rated subjectively by the end user.

One more point, do not be suspicious of dynamic microphone specification listings that leave out ratings such as hum and *radio frequency interference* (RFI) sensitivity, output noise, and clipping level, or maximum SPL. Dynamic microphones are less susceptible to these problems than are condensers, ribbons, and other microphones with active electronics. These generalities do not mean that passive dynamic circuitry is unaffected by these problems, but in most cases the effect is negligible. Active microphone circuitry is more inclined to overload (or clip) and thus to distortion. Active electronics are also more at risk from hum, RFI interference, and noise because of their amplifiers. Any noise output from a dynamic mic will usually be caused by impedance matchup problems.

Micing Music Instruments

Different tones are often produced from different areas of a musical instrument. These combine to make up its full sound. No matter where on an instrument a tone originates from, it will radiate in all directions. Generally, all the different tones combine together into a single sound at some distance from the instrument. However, this combined sound does not automatically appear at all locations that are equally distant from that instrument. This is best demonstrated by first standing in front of, then to the side of, and finally in back of an upright acoustic bass. It's easy to notice the huge differences in sound radiation patterns with this large instrument. You'll find it a little more difficult with saxes and flutes, and close to impossible with trumpets. The "book" on trumpets is that *all* the sound is radiated out of the front bell. Not true. Ask a trumpet player, or better yet, try placing an additional mic pointed into the off-hand side of the horn. The output will be a lot lower in level than that projected out the front, but when soloed and brought up to the monitors, you'll find it to be a very pleasant sound. When mixed in, it plainly enhances the horn's main sound.

Obviously, placing a microphone close to an instrument will emphasize the sound from the area that it is closest to, but it will still pick up sounds from all sources. These "secondary" source sounds travel a longer distance to the microphone. When they arrive (after being reflected by a nearby boundary), they can be delayed enough to be out of phase to some degree.

Therefore, an instrument's proximity to boundaries within the studio must be considered along with the acoustic characteristics of those boundaries. While reflections can present problems, the absorption properties of some panels may also be too frequency-dependent to augment the instrument's full sound. Because of these factors, panels used to disperse the sound can sometimes be the safest bet. The farther away from a sound source a microphone is placed, the more ambient sounds in the form of ambient room noise, or "bleed" from other instruments, will be picked up.

A long studio reverb time can cause conflicts when recording dialog. The additional sound can not only causes smearing, but can also actually garble the talent's words. You would never think that a short room reverb time could disturb the tempo of a piece until you actually hear it. Variable acoustic panels along the walls of a large room solve a host of potential problems for the audio recordist.

Critical distance refers to the position where the sound source and the reverberant field will be equal in level. A mic positioned at any point farther than this distance from the source will render a sound that can lack definition, sound blurry from excessive reverb, or sound muddy due to phase cancellation.

A Rule of Thumb   One "rule of thumb" is to place a directional (non-omnidirectional) mic within one-half of the critical distance:

1. Listen to the mic's output level at a close position.
2. Move the mic farther and farther away until you reach a point where the output seems to stop getting lower and you've found the "critical distance."
3. Move the mic at least one-half that distance back toward the sound source.

Try this method just once and you'll agree that it's a natural. You can count on it to always work because it is not based on anyone's subjective opinion, but on the laws of physics. Pretty soon you'll be able to judge the correct distance without all the stop, listen, and go movement.

It's faster and easier to listen to the source while you move away with one ear blocked. But unfortunately, half the critical distance may be too far away in terms of other room noise, *heating, ventilation, air conditioning* (HVAC), or instrument bleed.

It is fairly easy to place go-bos around acoustic instruments, but often excessively loud combo amplifiers are best set up in the corner of the room that has absorptive surfaces. In some cases the amp will need to have an enclosure of go-bo panels assembled all around it, or be placed in an isolation booth to prevent its output from "bleeding" onto other instrument's tracks by leaking into their mics.

Reflections can cause what sounds like utter havoc at the microphone's diaphragm. This is because the direct sound can combine with the reflective sound to produce any or all of the following three distinct and undesirable results. If two sounds arrive at exactly the same time, there will be a doubling of the output. If two arrive out of phase in timing by 180 degrees, they can cancel each other out. Anything in between these two extremes can result in lower or higher output levels at narrow frequency bands, the cancellation of certain frequencies, the smearing of frequencies with low frequencies being affected more at greater distances than high frequencies, tube- or tunnel-like sound effects, muddy sound, or just about anything else. The difference between all of these results can equate to as little as a one-quarter inch change in the mic's position. So using two or more microphones on a single instrument, or in a reflective environment, can be bad news.

Another Rule of Thumb Called the "3-to-1" Rule   This says that when using two omnidirectional microphones it is best to keep the second mic at a distance from the first that equals three times the distance the first is placed from the sound source. This means that if the primary mic is 1.5 feet from the instrument, you must place the second mic 4.5 feet from the first one.

Okay, the problem when multimicing an instrument is obvious. I like to place a mic 1 to 2 feet in front of a saxophone's bell. This is because I don't like a sound that only comes from the bell. From this position, I can aim the mic higher up toward the neck and pick up all the sound coming off the upper sound holes while still getting some of the raspiness from the bell. But it is also essential to mic woodwind instruments at the player's right-hand side to pick up its lower frequency content. The 3-to-1 rule presents no problem when micing a baritone, but for a curved soprano it's plain ridiculous.

It could equate to a reduction in phase cancellation problems, but the mic that's more distant by three times will have an output that's 12 decibels lower. Even though this distance can be lessened for directional mics, it exemplifies how the addition of a second microphone can introduce a whole set of unpredictable problems.

Combining two mics together that are picking up the same sound from different distances will always result in some amount of phase cancellation.

The 3-to-1 rule does not apply to microphones placed very close together (a coincidence pair), as sound waves will reach both mics simultaneously. The 3-to-1 distance can also be reduced for microphones with cardioid pickup patterns if the two mics are pointed away from each other (near coincidence pair), as this will not expose them to the exact same sounds. Spaced omnidirectional microphones make wonderful environmental ambience pickups.

Luckily, there are other ways around this problem besides the 3-to-1 rule. Using a directional mic and taking advantage of its off-axis rejection characteristics (as per its polar pattern chart) to eliminate pickup from unwanted sound sources is a recording engineer's mainstay, especially around a drum kit.

If you mic the back as well as the front of an open cabinet guitar amp (like a Fender twin reverb) to boost the low-frequency response, the two signals will be close to 180 degrees out of phase with each other. The same goes for dual micing a kick drum, a snare, and even double-skinned mounted toms. So hitting the console's (180-degree) phase reversal switch may rectify the situation.

You can locate 180 degree out-of-phase positions on things like Leslie speaker cabinets by summing the two mics to a mono output and moving their positions until the sound just about cancels out. Now hit the console's phase reversal switch and the signal will jump from nothing to twice the output of the single mic.

The Inverse Square Law   Doubling the distance of a mic from the sound source can halve the output (-6 dB), and four times the distance can decrease the output by as much as 12 dB. However, the amount of leakage or bleed that is picked up by the mic will be just the opposite. Twice as close (half the distance) may equal half the amount of bleed (or room noise pickup).

That's a lot of "cans" and "mays," but I have run into comb filtering problems that were caused by reflections off of music stands. The science (art) of correct mic placement is never completely predictable.

Concave surfaces are a nightmare because they focus or reflect beams to specific areas. These exact locations are unpredictable because they depend on the location of the sound source. If a microphone happens to be located in one of these "hot spots," its signal output will be severely affected. Hint: Adding a few dispersion panels to the sides of amphitheatre band shells has proven to not only remedy this focusing problem, but also to facilitate a more balanced and even blend of the stage sound from the musician's perspective.

Want to know about the studio's acoustics? Put on a pair of cans, hook up an omnidirectional microphone, and walk around the room snapping your fingers or use castanets (they work better). What you hear is what you have to work with, unless variable acoustics are accessible.

The Dilemma   A single mic placed at a distance from an instrument can more readily pick up its full sound. But it can also introduce more room noise and bleed from other instruments. Move the mic in closer and it is no longer exposed to the full sound. Add a second mic in close to augment the sound, and you open what seems like a Pandora's box of phase problems.

At one point in my career I went at this problem head-on. *Not* because I was any kind of crusading "correct all things audio" fanatic, but because I was afforded the luxury of having plenty of studio time and the equipment to attempt to find a way to make my job of microphone placement a lot easier.

Okay, back in the 1980s my time was divided pretty equally between engineering, teaching, writing, and maintenance work. To perform preventative maintenance (my technical specialty) wherein you are repairing problems *before* they cause breakdowns, you need to have a lot of off-hour control room/studio time. Reviewing professional audio equipment meant having a test bench at the ready. Since my reviews (as well as tutorial articles) contained a good deal of practical user information, the review equipment had to be placed into professional studios.

This was just fine with the studio owners I knew and they were more than willing to provide me with ample studio time to run the equipment through its paces myself.

Most often I'd also gather several engineers and musicians together to get their input. All found no problem with this as it gave the engineers a chance to check out some of the newest equipment available first hand, and the musicians knew they would walk away with a great-sounding demo tape. It also afforded me the opportunity to experiment with different micing techniques while being scrutinized by the most critical of onlookers: fellow recording engineers. The task of repeated microphone repositioning, monitor level adjustment, and multiple subjective evaluation note-taking was eased by my class full of assistants during teaching hours. Finally, the majority of my engineering work during this period came from other engineers who were overbooked. They were often involved with several projects at the same time so their being present at every recording "date" was not possible. This is how staff assistant engineers pick up their first full-fledged engineering work. I was overdubbing all kinds of instruments. Recording overdubs are easier than basic sessions as there's a lot less setup involved. The atmosphere is more relaxed, there's less pressure from time constraints, and you're usually working with well-maintained instruments played by ace musicians. Once I had matched the new part to the old sound, and laid all the punch points into the multitrack's memory, I was pretty much free to listen to the results of all the other microphone positions I had set up. Their positions were based on the theories that I had been experimenting with during my maintenance, teaching, and reviewing hours.

Musicians are not shy about letting engineers know how they feel about the sound of their performance. If the overdubbed sound doesn't cut it, there will be no additional session work from that same engineer.

If there was enough real estate on the multitrack tape, I would also record my "experimental" mics to a separate track. I'd include a note to the engineer/producer to ask them for their opinion and possibly suggest that they try blending in some of "my" sound to augment the overdub track. This was appreciated by almost everyone and quite often my tracks were the ones used during mixdown. This was especially true when I recorded solos. It was a nice life.

Positioning Microphones

Placing a mic close in on an instrument increases separation and reduces background noise, but limits the tone's ability to fully develop its natural sound. This is especially true at the lower frequencies. Yet with some mics it may produce an unnaturally "tubby" bottom-end due to the proximity affect.

With vocalists, a 3-inch distance may produce popping, and a closer positioning will make it worse. Using a side-address positioning will limit this proximity effect as well as the frequency response. Hey, those pop filters do work and if it's a single vocalist overdub, back the mic off, or use an omnidirectional mic as they are less susceptible to low-frequency breath pops since they have no rear-entry port.

Any microphone "technique" used will depend on the following:

- The instrument that's being recorded
- The tempo and pitch of the piece being played

- The musician's style of playing and whether it requires a more noninterfering mic placement
- The characteristics, especially proximity effect, of the available mics
- The sound of the room's acoustics due to its size, shape, reflections, and reverberation time
- Any other instruments being played such as when the sound must be blended with the other members of a section
- The amount of bleed from other sound sources
- Whether or not you are using the microphone position to achieve a purist or natural sound, or to change or affect the instrument's sound in some way
- And on and on and on

What I'm getting at here is that it's impossible to have any hard and fast rules. So all suggested positioning should be looked at as only a starting point for your own further experimentation as too much depends on the nature of each particular situation. Besides, in addition to knowing the standard mic placement procedures, as a professional you need to develop your own methods. If the standard configuration works, great. If not, you'll know what to do about it.

Years of experience and simply listening to the source will bring you a long way, but you also need to understand the way microphones, music instruments, and room acoustics function together in producing sound.

It's obviously important that the instrument should be playable, hold tune, and produce an adequate sound, and that the mic used can handle the full frequency content and dynamic range. Less obvious is how the acoustic properties of the recording environment will affect a sound. That's because you're hearing the instrument in that acoustic space with your dual ear/brain combination, which works to compensate for most of the sound irregularities that you encounter. Microphones, on the other hand, only have one "ear" and no brain so they pick up *everything* and that's what you get. This will include the direct sound from the source, reflections from nearby structures, plus sounds from more distant sources such as reverb, HVAC noise, and a bleed from other instruments, all picked up from that one position.

The difference between sounds picked up by your hearing system and that of a cardioid microphone is remarkable. If you don't believe me, stick a finger in one ear and give the world a listen. Better yet, use a *soft* in-ear hearing protector instead of your finger, but remember to keep that open ear at a distance from loud sound sources. Hey, I always wanted to stick my ear close down to a snare or right in on a guitar speaker too, but that's like a video color corrector staring at the sun. It's not for someone who has any long-term professional goals!

Or, as I like to put it, "You got to protect your money-makers!"

Most microphone placement works when positioned from the perspective of the audience, or listener, that is, 3 to 6 feet in front of a music instrument. This makes sense as it takes into consideration the way an instrument was designed to project sound. Even though it will pick up information from reflections off the floor, walls, and other surfaces surrounding the instrument, a single mic only has one "ear" so it cannot provide the spectral information obtained by listening with our two ears.

Dual mic usage may be able to make up for this, but that's not guaranteed and, as explained, this is often impractical as they can be cumbersome, introduce comb filtering, and add leakage from other sound sources.

The Musician's Perspective   Try looking at the way a musician's ears are aimed at the instrument: the violin family, a seated cellist, an upright bass player, or someone sitting at a grand piano. They are satisfied with the sound as they are hearing it, and while there certainly may be better positioning as far as instrument timbre is concerned, I have found this to be a very good starting position. However, you must remember that body contact with an instrument provides a lot of low-frequency information from the vibrations.

My Big Theory   Eureka, I'll combine the two using microphone placement from the musician's perspective and get the added low-frequency instrument vibrations by using contact pickups. *Piezoceramics* (PCs) produce an electrical charge when subjected to a force. That's the way an accelerometer produces an output voltage when exposed to pressure. The opposite is true of alarms, which, when activated by electrical energy, mechanically resonate sound by expanding and contracting in one or more dimensions.

PCs are produced during a polarization process. The prepolarized material starts off with randomly arranged electrons. By applying heat and a strong electromagnetic field to the ceramic, the "dipoles" reorient themselves to be parallel in the field. The material is cooled while the field is still applied. What results is a material with a negative charge at one surface and a positive at the other. Okay, don't get all disturbed by the term "dipole." It simply means something that has a pair of oppositely charged poles. Magnets are dipoles! But when PC dipoles are subjected to mechanical stress, they produce an electronic voltage. This is called a direct piezoelectric effect.

Industrial applications predominate their usage. Switches made of PC materials convert forces such as changes in air pressure into an electrical charge. This charge can then be used to turn a mechanism on or off, completing the application. Shock or vibration pickups can be used to monitor rotating engines and signal a controller to change the speed or timing to reduce wear, breakdown, and failure, thereby reducing expensive equipment downtime. Industrial usage fuels most of the research into PC and other accelerometers. Acoustic emission sensors have also benefited. These are used to measure and signal the approach of mechanical failure (i.e., the more frequent and stronger the emission rate, the closer a pressure vessel is to rupturing). Acoustic emissions sensors are also used to measure gravity.

I've used accelerometers from the measurement field, which are normally used as solid-state industrial monitors to pick up flow, pressure, and acoustic emissions. Big money goes into research and development in areas like the measurement of skin surface pressures of high-altitude aircraft.

This ended up being just fine by a couple of bluegrass guitarists whose instruments' sound benefited from their usage. Like those high-intensity flight measurement accelerometers, the corresponding on-ground sound pressure microphones have frequency responses that are flat within plus or minus 1 dB from 2 Hz to 4 kHz. They handle levels up to 180 dB SPL and have stable (1.5 dB) outputs within temperature ranges of −65 degrees to +500 degrees F. I opted for ones with a broader frequency response at 1 Hz to 7 kHz, but others can easily extend down to 0.1 Hz. How about a miniature (3/16-inch diameter by less than 3/8-inch long) high-intensity piezoresistive measurement microphone that uses a micro-machined silicon chip (as in IC material) that will handle 190 dB SPL! They exist.

Piezoelectric accelerometers may be able to measure down to 1 Hz, but piezoresistive accelerometers are used to sense the acceleration of the earth's gravity. Now *that's* low frequency.

Attaching Acoustic Pickups to Music Instruments

Wax can be tricky as it forms an elastic interface, which can dampen vibration. But by using a very thin coat on *flat* surfaces, wax can provide the same frequency response as fixed mechanical screw or stud mounting methods.

Increasing the damping of any material that interfaces the sensor with the vibrating surface decreases the resonant frequency of the interface and thus its bandwidth. Here the interface material acts like a spring. Since the system's frequency response is determined by the combination of the transducer's mass, the size of the contact area, and the interface material's damping factor, the method used in attaching pickups can become a big deal.

I found a small amount of beeswax to work very well; however, even that can require the use of rubber cement thinner for complete removal. You would have to be out of your mind to apply a liquid chemical solvent to any piece of wood, let alone a music instrument. I don't want to have to answer the question, "Hey, how come the finish is flaking off my two hundred-year-old irreplaceable viola?" I've heard of using rubber bands, but most of the musicians I've dealt with were a bit too active while playing to allow for such a non-"fixed" mounting scheme.

Pickup Mounting Experimentation   As far as methods of attachment, the best results I obtained came by adding three elements to the system. First was a flexible ruler I purchased from an art supply house. This was just a strip of lead coated in vinyl. You bent it into shape and it stayed there. Next was a small sheet of cork $1/64$ of an inch in thickness. (Where you gonna get cork $1/64$ of an inch thick? Try your local woodwind repair shop and, while you are there, make friends and take a couple of business cards that you can pass on when you run into a player having trouble with loose pads, weak springs, or noisy keys. Okay, the cork goes between the pickup and the instrument. At $1/64$ of an inch, there's not much of a damping factor involved, it protects the instrument's surface and is flexible enough to conform to any shape. (This is important with piezo film materials as the best results depend on the size of their contact area.) The flexible (vinyl-coated lead) ruler is then shaped and positioned on top of the piezo film material. The third element was a couple of 3-foot strips of double-sided Velcro (one side has the hooks and the other fuzz) that I picked up at a florist shop. These are normally used by cutting the Velcro with scissors and wrapping it around a plant stem and a wooden support stick. Several strips wrapped around the instrument holding the "sandwich" of cork/pickup/flexible ruler against it ended up being secure enough to hold its positioning regardless of the musicians' movements, yet was unobtrusive to their playing style. More important was the fact that its output level was greater than that of the average mic, especially at low frequencies. The frequency response goes so low you have to be careful about how much of this overpowering signal you use, especially when applied to places like the back of an acoustic upright bass.

C-Ducer Tape Pickups   I found the same exact thing was true when I used the "C-ducer" tape pickups that were available back in the mid-1980s. The large sizing of this pickup made it less sensitive to placement variations and provided a more even response across large instruments. Acoustic pianos with two of these mounted on the sound board had no holes in the output across the whole keyboard (as could happen when using other smaller multiple pickup arrangements). This large coverage area also meant they were able to pick up very low frequencies. Some of this content would normally have been termed subaudible. In fact, excessive low-frequency content is a common problem when using contact pickups.

Even with their large size, the C-ducer pickups were so light in weight that they didn't interfere with the natural resonance of the instrument they were attached to.

These were among the first devices I tried to used as drum triggers. They worked very well, but I'd caution against mounting them directly to the top of the drum skin because when they are struck with a stick, their output voltage can reach levels that overload the input trigger's gain stage. This was a very different concept as far as micing instruments and it lent itself to a lot of experimental fun.

Next came polymer piezo films or sheets. They are also flexible and light, have low mechanical and acoustical impedance (meaning a high output), and can easily mate with any shape by simply cutting and forming them. Some of the first applications for these materials were for speakers and headphones. I remember an omnidirectional high-frequency speaker made by, I believe, Pioneer, which utilized these materials bent into a cylindrical shape. The ability to size and shape these thin materials to fit the intended instrument meant that I no longer had to be content with pickup sizing and positioning that had proven to be inadequate.

You see, I felt that commercial pickups that were designed to be mounted to a stringed instrument's bridge had a metallic, or electronic sound, and since I was using the pickups to accentuate the warm low-frequency sound of wooden instruments, that was not what I was after.

On the other hand, some commercial pickups worked rather well. A company called Frap (or flat response audio pickup) made pickups mounted inside a cylindrical cork that could be press-fit into the end of a flute with no muss or fuss, and the sound was not only full bandwidth (as it was not aimed at the player's mouth embouchure) but it was also almost impervious to feedback.

I wouldn't have anything to do with drilling holes in instruments or mouthpieces to mount pickups or miniature mics. I thought it was best to leave that to someone with the proper insurance coverage.

One of the best sounds I was able to get came from using a 6 × 8-inch piece of piezo film placed right against the back of an acoustic guitar. The film itself is nonmarring. On top of this I placed a flat piece of lead-impregnated vinyl that was normally used for acoustic dampening and held the two in place with a crisscrossed double-sided Velcro (with the hooks facing in). The guitarists could play in the normal style, with the back of the guitar right up against them. In fact this added contact pressure actually accentuated the instrument's depth.

My final experimentation was with a combination of a film piezo pickup mounted to the bottom of a PZM microphone. I was trying for a linear phase response from this dual unit. To understand this, you must first understand the operating principles of the PZM. This mic was designed specifically to eliminate phase cancellation. Remember that when multiple sound waves, carrying the same signal, arrive at a microphone diaphragm at different times, they can cause comb filtering. The result of this can often be a harsh or muddy sound. The amount of distortion this causes depends on the different path lengths that each wave must travel. If the path lengths, and thus the time delays, are very short, the sounds will (according to the Haas effect), actually reenforce each other. Because a PZM receives no direct sound, all sound picked up is via its reflective plate. The very close spacing between the microphone element and the plate equates to a very short path length. Nothing but short path lengths equates to a phase-coherent output.

One factor that must be considered when using PZMs is their low-frequency limitations. A 5-inch square reflective boundary will not handle frequencies much below 500 Hz on its own. That's why this plate is considered the "primary" reflector and PZMs are mounted on larger, flat surfaces such as floors, suspended panels, and piano lids. But this is not ideal as far as separation is concerned and 4-foot square panels can be difficult, even when they are made of clear acrylic. The resulting low-frequency roll-off of a free-standing PZM is known and understood.

The point at which low-frequency pickup starts to decline (by 6 dB per octave) is determined by dividing 61.7 by the boundary dimensions in square *meters*. You know the drill. Get out the calculator and always check my (and everybody else's) work for mistakes.

Five inches is about 130 millimeters, 1.3 centimeters, and 0.13 meters. So 61.7 divided by 0.13 = a roll-off that begins at 475 Hz.

Ten inches is about 255 millimeters, 2.55 centimeters, and 0.255 meters. So 61.7 divided by 0.255 would halve that to 242 Hz.

In order to get down to 20 Hz, the boundary would need to be 3 square meters (61.7 divided by 20 Hz). One foot is equal to 0.3084 meters; therefore, 3 square meters would be 3 divided by 0.3084 = 9.73 square feet! That's a bit too much to haul around, but that "no phase cancellation" aspect just wouldn't go away.

What we have: A PZM has no phase cancellation but can be lacking at the bottom end. Piezoelectric pickups provide every ounce of low end and only pick up the direct signal.

I couldn't come up with the right configuration. Yes, the pickup responded perfectly to the instrument it was attached to, but the half-sphere-shaped polar pattern of the PZM ends up facing away from the instrument. Starting in 1990s, I had to deal with an illness so except for a little part-time maintenance I had to stop working. Therefore, I never got to finished that project, but I still believe there's something there.

You need to understand the underlying principles behind a problem before you can attempt a remedy. The problem of combined delays from two microphones naturally pointed me in the direction of pickups because they have no phase problems and can provide full low-end bandwidth. Now the mic selection process is narrowed down to a polar response rejection of unwanted sounds, an adequate frequency response of the upper bandwidth, and a not too excessive proximity effect, which shouldn't be too difficult! Now that you understand the way mics operate, we can go over a few examples.

The Shure A15 *Tone Generator* (TG)   It happens to everyone, maybe during sound check while doing live PA work, just before air time in broadcasting, or while getting sounds before recording. You bring up a fader and there's nothing there! The fault could be a bad cable or connection, or perhaps the snake, possibly the mixer's channel, or how about the microphone itself. The normal troubleshooting procedure calls for someone to first check the mic connection to make sure it is plugged in properly, then comes cable replacement, followed by trying a different mic or even switching to a different mixer input, all after several switches on the console have been checked.

For the most part, these common faults can be corrected before very much effort is exerted or time is consumed, that is, if there's a crew involved and the communication between them is adequate. However, should the engineer be alone, or the crew starts to disconnect and check the sections of the audio chain mentioned in a random fashion, that fairly simple problem can grow astronomically.

With a Shure A15TG (see Figure 3-5), the fault will normally be located in no time at all. This device is a portable, self-powered signal generator that measures only 5 $^{17}/_{32} \times {}^3/_4$ inches in diameter. This sizing, along with an XLR end connector, allows it to be plugged directly into the audio system at various points to confirm operation. The overall design of this device is as intelligent as its configuration. Its tone output is roughly 700 Hz, which is a distinctive yet not

Figure 3-5
The Shure
A15TG

unpleasant sound. The level when terminated in 150 ohms is approximately that of a micro-phone. Not only was battery choice that of a common and easily found hearing aid-type, but the circuitry's current draw permits a battery life of 1,500 hours under continuous use and, when used about an hour a day, the same as the battery's shelf life. Power on/off is via a slide switch recessed in the housing assembly. Granted, this is no precision instrument (see Table 3-3). Voltage output can vary, as can frequency, depending on battery drain. However, fixed level and either settable or selectable frequency output would have increased this unit's list price. Instead, Shure provides a very durable (steel-cased) unit that's perfect at performing its task. It is not affected by phantom powering and disassembly will not be a problem in the unlikely event of the A15TG needing service.

I received my first A15TG as a gift from a friend in broadcast audio maintenance, and dur-ing the time I've had it, it has saved me many, many hours of troubleshooting time and it's more convenient to use than my much higher-priced function generator.

In the previous situation, you simply divide the mic/output connection/cable/snake/console input connection/mic preamp/preamp output insert normaling chain in half, and plug the A15TG directly into the questionable console channel.

If you hear tone, the problem is before that point; no tone, it's after that point. Next, divide the remainder in half. You can track down the cause of a problem very quickly using this method and it does not entail unplugging numerous cables that end up in a knot. The U.S. Navy came up with the divide-in-half method of electronic troubleshooting after much research. When a ship's electronics are out of commission, the fault must be found quickly to avoid becoming a floating target.

If you really want to know how to solder electronic components, look to the U.S. Air Force's publications. The vibrations in early jet propulsion caused electronic intermittencies and com-plete failures that ended in crashes, so the people who were responsible for outlining proper soldering techniques were not playing around.

Table 3-3
A15TG specifications

Frequency output	700 Hz. 20%
Output level (150 ohms)	−41.5 (±4.5) dBV, 8.4 mV nominal
Battery	1.4 volt (Eveready E 675 or equivalent)
Size	5 $^{17}/_{32}$ inches long × $^{3}/_{4}$ inch diameter
Weight	4 ounces

The AKG D-330BT Dynamic, Cardioid Handheld Vocal Microphone The need for phantom powering can cause grief when the live PA board doesn't have it built in, or if your batteries always seem to run low at the wrong time (usually from forgetting to turn off the mic's power switch or, worse yet, turning it off after sound check and not turning it on before the performance, causing the first phrase to go unheard). Sometimes a dynamic mic is the best way to go. One of my favorite handheld dynamic vocal mics is the AKG D-330BT. I used this mic in both live and studio work, and it never failed to give me both a good sound and amaze anyone who was not familiar with it. The AKG 300 series offered three handheld dynamic mics that were rugged live vocal mics, which were also studio-quality vocal mics.

The D-310 was the basic cost-effective model, while the 320B features a hum-bucking coil, a bass equalization switch, and an easily removable capsule assembly. The D-330BT is the deluxe version (see Figure 3-6).

This mic incorporated two diaphragms (one aimed forward and one aimed to the rear), which were wired out of phase to provide a cardioid polar pattern that virtually eliminated onstage feedback. It had a hum-bucking coil that canceled magnetic field noise, an onboard bass, and presence EQ. The equalization had two 3-position filters, which provided up to nine different response curves. At the top end it was not that variable (only a couple of dB), but the cut at the bass could be drastic. This was helpful, as this mic had a good deal of proximity effect.

The bass and presence EQ increased this mic's range of usefulness. The main problem with using handheld vocal mics in the studio is handling noise. AKG's use of dual transducers in the 330 was more than helpful. This mic did the best job of dealing with this problem than any other mic I know of. Ask anyone who has ever used one. You can literally drop this mic on the floor with your level set for vocal gain and only get a very quiet thud. It's amazing! All of the 300 series mics have reinforced plastic caging in front of the capsule, in addition to the steel wire mesh-and-foam windscreen. This, along with dual isolation points between the capsule and other housing, adds up to trouble-free operation. But should the 330 or 320 ever break-down, simply remove the one-piece internal electronics assembly and replace it. This is one of the few mics that can be serviced by anyone.

Figure 3-6
Frequency and polar response graphs for the AKG D-330BT

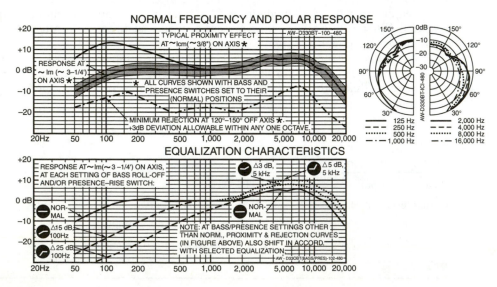

Used live, these mics are right on the money. Their 8 kHz peaks yield a clear, crisp yet not harsh sound that helps the vocal cut through the onstage din. With their flat midfrequencies and bass proximity effect, they can also give a very warm, intimate sound. Their cardioid pattern starts to narrow to a lobe at the high end where most feedback problems occur, thereby providing them with more isolation as in higher gain before feedback.

In the studio, these mics obviously will handle vocals with no problem at all, but here the 330's ability to vary the pickup response becomes an important feature. It allowed me to do several background harmony overdubs by either the same or different vocalists and, via changing the EQ, capture every nuance. High-hat pickup is quite good due to their natural 8 kHz boost, with isolation from the snare being excellent. On the snare itself, the signal rendered was a great dynamic sound. They were also very useful for micing guitar amps. If you want bright or intimate horns, all you had to do was position the D-330BT back (or up close) and let its proximity effect and high-end peak do the rest. Strings were captured very nicely, and they made recording percussion a breeze. I'd hand the D-330BT to any vocalist in any top recording studio and not have to apologize. For onstage use, this mic was just about perfect (see Table 3-4).

The Electro-Voice 667A Dynamic Microphone This mic has long been pegged as a broadcast-only (television/film) mic due to its boom mounting and its ability to almost completely eliminate wind noise problems. However, it is also one of the best dynamics available for all-around studio use. As a vocal mic it is completely impervious to even the worst cases of pops, sibilance, and proximity effects. It has a built-in, pin-switchable, color-coded programming board that is accessible via a removable cap at its rear.

This enables the engineer to obtain up to six different frequency responses right at the mic itself. I found this helpful in multimicing situations as the 667A could be readily adapted to blend with the other mics used. Not having to use an outboard, or console EQ, helps cut down on phase distortion. Its a virtual tank of a mic and is not affected by temperature, humidity, or shock.

Off-axis and especially rear rejection is optimized via the use of EV's "continuously variable-D" principle, which uses a matched pair of slotted tubes coupled to the back of the diaphragm. The acoustic length of the tubes varies inversely with frequency. This configuration causes almost all off-axis sounds to be canceled out via phase inversion. The result is an exceptional front-to-back pickup ratio and a rejection pattern that is symmetrical and smooth all around.

Table 3-4

AKG D-330BT specifications

Frequency response (with bass and presence switch set at flat)	50 Hz to 20 kHz
Sensitivity at 1 kHz	Open circuit: 1.2 mV/Pa (−58.4 dBV)
Maximum power level	−58.5 dBm re: 1 mW/10 dynes/cm^2
EIA Gm	−156 dBm
High impedance output	−48.4 dBv at 1 Pa
SPL at 1% THD (40 Hz, 1 and 5 kHz)	128 dB
Hum sensitivity (in a 1-milligauss field)	−142 dB

The output of this mic is pretty hot at −51 dB (putting it in the same ballpark as the Shure Beta 58). However, its size pretty much precludes its use around the drum kit. It is not the best kick drum mic for hard rock, yet it is acceptable for lighter rock, and some country music. As an overhead, it is on the money when set to 3 and A positions (see Figure 3-7).

As a matter of fact, the ability to change the response curve makes the 667A a studio workhorse.

Although maximum input is not specified, it easily handles guitar amps and loud trumpets when positioned right at the bell. This is such a versatile all-around studio mic that I'd set it up on a rolling boom stand and leave it there. I found myself using it on almost every session either by itself or to augment a favorite micing technique.

Where the 667A really shines, however, is with problem vocals. Pops and sibilance will be a problem at one time or another, as well as those low-level, whisper-like vocals when excessive preamp gain setting causes mouth and background noises to be moved to the forefront. Then there's the breathy "mic eaters," who can trigger excessive low-end rumble due to proximity effect. With this mic, the operator can feel confident taking on one or all of the above as the end result will be a warm, clean track without using any equalization except for the 667A's built-in passive equalizer (see Figure 3-7 and Table 3-5)

Table 3-5
EV 667A specifications

Frequency response	Uniform 40 to 10,000 Hz
Poler pattern	Cardioid
Impedance	50, 150, 250 ohms, selected at rear of case by moving one pin. Shipped set at 150 ohms.

Output Level

250 ohm tap	−51 dB
150 ohm tap	−51 dB
50 ohm tap	−52 dB

(Ref. 1 mW/10 dynes/cm², with response selector in the A-1 position)

EIA Sensitivity Rating

250-ohm tap	−149 dB
150-ohm tap	−145 dB
50-ohm tap	−150 dB
(Response selector in A-1 position)	
Switching:	Response and impedance selection accomplished with gold-silver-plated pins
Hum Pickup Level:	−121 dBm (ref: 001 gauss field)

Figure 3-7
Frequency and
polar response
graphs for the
EV 667A

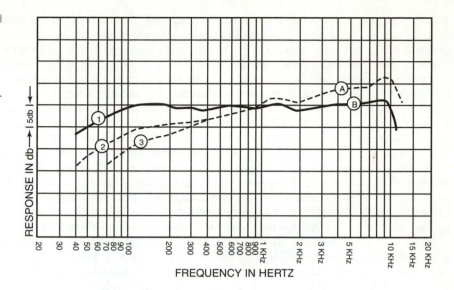

Figure 3-7
Frequency and
polar response
graphs for the
EV 667A

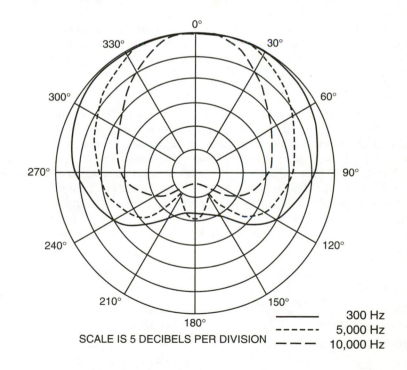

SCALE IS 5 DECIBELS PER DIVISION

———	300 Hz
- - - - -	5,000 Hz
— — —	10,000 Hz

The Sennheiser MD 402 U Dynamic Microphone   The MD 402 "K" microphone cost less than $80. It was specifically designed to replace the common crystal mics provided with the average consumer tape recorder. The "U" version's balanced low-impedance output made it useful in professional recording applications as well. One look at the frequency response graph shows that it may be less than ideal for some applications. Additionally, its super cardioid polar pattern curve shows a somewhat odd shape, with a midband indentation at one side from 90 to 150 degrees. This may also seem to be less than adequate. However, the way I looked at it was that the MD 402 U had a built-in roll-off to one side, as well as high-end boost. These two features, along with the end user's willingness to try a little experimentation, can make this mic very useful to studios operating with a moderate budget.

As far as drums, it is best to forget about kick and toms altogether. Yet positioned under a snare, this mic yields a nice high-end sizzle that, when added to the sound from the top mic, provides that urgent crack. With a bit of rotating trial and error, the 402 becomes an excellent hi-hat mic with the ability to do a very good job of canceling out quite a bit of snare leakage by aiming this mic so that the side with the indentation faces the snare.

I did a lot of playing around with this mic and found some unusual applications. By placing the 402 just outside a kick drum and aiming it to the side, I was able to place it in the mix so that it was out of phase with the main kick mic, thus canceling out almost all of the snare and cymbal leakage. In fact, in most instances of multiple micing, the 402 will be an asset. By placing it in front of a guitar amp, you gain the ability to either boost or cut the high end by using the channel's fader and phase switch.

If you should ever want a real raspy sax sound, just drop an MD 402 into the bell. On the other hand, when used in conjunction with other mics on acoustic instruments like piano (attack) or guitar (string sound), it can be used to add realism or presence. The point? All mics are useful, and this one sounded better and was a lot easier to interface than a telephone mouthpiece mic (see Figure 3-8 and Table 3-6).

The Shure SM7 Dynamic Microphone   The SM7 is a studio-quality, dynamic moving-coil mic. It has a smooth, flat, and wide frequency response and is commonly found in use by radio announcers. As far as I'm concerned, it's also a great studio mic. It sounds clean and has a fairly tight polar pattern that's even across its full frequency range. It provides a good deal of rejection (cancellation at the sides is approximately 6 dB, while at the rear it ranges from 15 to 20 dB). This adds up to very little coloration from off-axis sound pickup. The mic comes with a yoke mount, which has a captive nut for stand mounting. This yoke also has two hand-tunable side screws that set it at almost any positioning. The SM7 is solidly constructed with the diaphragm suspended within the body for protection.

Table 3-6
Sennheiser MD 402 specifications

Frequency response	80 Hz to 12.5 kHz
Directional characteristics	Super cardioid
Output level	−53 dB (open circuit at 1 kHz)
Impedance	200 ohms

Figure 3-8
Frequency and polar pattern response graphs for the Sennheiser MD 402

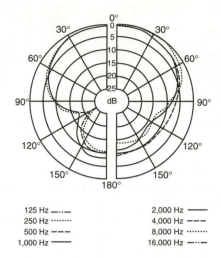

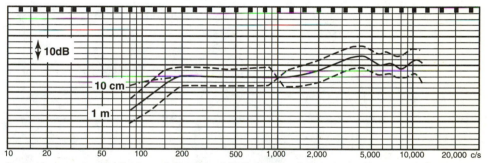

Standard frequency response with tolerance limits and close field characteristics

This mic has hum-rejection circuitry and incorporates a fitted windscreen-type pop filter to help with close-up vocal micing. Furthermore, there are two onboard frequency response switches with graphic displays of the selected response settings. The first switch, for bass roll-off, provides either a flat response or a gradual low-frequency roll-off (see Figure 3-9).

The second deals with the high midrange, or presence area, and can be set flat or used to boost these frequencies, resulting in a crisper sound. (It is often used to help announcers cut through background noises.)

It works great on the bass drum and, depending on the sound you're after, it can equal one of the standard mics (the EV RE-20) for this application. For bass amp and bottom-end Leslie micing, it works well because of its low-end response and built-in windscreen. It can be helpful on a muddy-sounding upright bass, as it allows the operator to adjust the sound to the instrument using the bass roll-off switch or add more punch to the mids and boost its high-frequency range. On other acoustic instruments such as guitar, strings, violas, violins, and cellos, it renders a very smooth response that is either low and warm or bright and midpeaked, depending on the filter switching. Although no SPL rating is provided by the manufacturer,

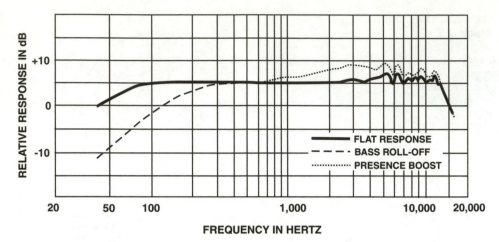

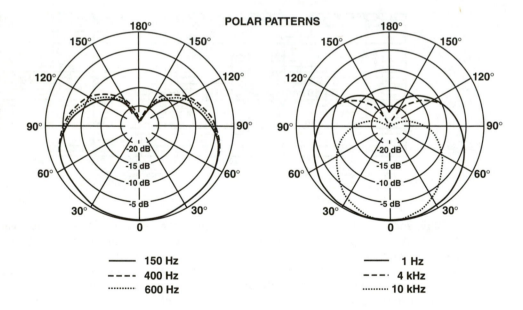

the SM7 can handle brass with no problem at all. I used it to mic all kinds of horns, including trumpets, in real tight without any sign of signal breakup.

On instruments like vibes and piano, where you're generally double micing, this mic's off-axis response provides a good deal of separation and therefore less coloration and phasing problems. But on vocals, it shines as it easily controls sibilant vocal sounds and again, due to the versatility of its switchable frequency response, it's helpful on diverse vocal styles. All in all, the SM7 (like the EV 667A) is a great all-around studio mic, a real workhorse, and one that will find use on most dates due to its fine sound and extreme versatility.

Shure SM7 Specifications   Outside of the polar pattern and frequency response graphs, the only specifications given for the SM7 are as follows in Table 3-7.

What's the idea of using microphones that do not normally coincide strictly with recording studio usage as examples in a book on recording? The A15TG was included for obvious reasons. I'm such a big fan, I purchased a second one. I was even going to add a resistor to its power supply line so that the frequency output would be different, thus making the pair applicable to stereo troubleshooting.

It's true that the AKG D-330BT was designed as a live performance handheld vocal mic, but there's no law against using it on an instrument. Also, because of its variable frequency response switching, it does make a fine studio utility mic. However, when put to its intended use it really shines, even when used in the studio. Quite often vocalists who are more accustom to live work just feel better when they're using a handheld vocal mic so they often end up performing better. I found the AKG D-330BT especially useful in this application due to its ability to handle high SPLs, reject all handling noise, and its switchable frequency response.

The EV 667A is more readily found in television and film work. This fact is exemplified by its huge windscreen, but television has turned more to using the much improved clip-on lavaliere and shotgun mics that are currently being made. Video houses close down quite often and at the auctions folks are not all that interested in dynamic microphones. This means that used 667A mics can be pretty affordable.

Granted, the Sennheiser MD 402 was designed as an affordable mic to be used with consumer cassette recorders, but it can also be put to good use in state-of-the-art multitrack recording facilities. The point is that not even an inexpensive (less than $80) consumer mic should necessarily be viewed as an inadequate recording tool, particularly one as useful as this.

The Shure SM7 found its niche in radio broadcasts as it is impervious to the "mad-cap hi-jinks" of excitable DJs. Most on-air talent now get miced using condensers. With the resurgence of ribbon microphone use in radio due to the availability of more resilient models such as those from Royer Labs, the SM7 is now often ignored. If you get a chance at one for a good price, jump on it!

I go back to my opening statements on microphones: There are no *bad* mics. The recordist just has to know how to take advantage of their characteristics.

Judging a microphone's sound performance from frequency response graphs and polar patterns can be a little difficult, but you can use the following examples as a guide. I've pointed out the features that are important to the recording engineer and explained how they affect the way microphones are used. With a touch of patience and a magnifying glass, you'll be able to "Sherlock" your way through most mic specs in no time.

Table 3-7
Impedance (150 ohms) and output level

Open-circuit voltage	−79 dB re: 0 dB = 1 volt per microbar (0.112 mV)
Power output	−57 dB re: 0 dB = 1 mW per 10 microbars

Polar Patterns

Reading polar patterns is fairly easy once the parameters involved are understood. The first polar graph (Figure 3-10) was left blank to make the numerical designations of the response levels for each ring more readily apparent.

The mic's main point of focus, and thus the angle it will be most sensitive to (perpendicular incident), will correspond to the 0-degree point. These graphs cover 360 degrees within a single plane divided into 30-degree sections. The pickup sensitivity is designated by positioning the response on the graph rings. The level decreases as the response curve moves inward away from the outermost 0 dB ring.

Actual polar pickup will vary depending on the sound source and its location in relation to the mic within a specific environment, but mics generally conform to the pickup pattern they were designed to cover.

The polar response graph in Figure 3-11 is a depiction of an ideal omnidirectional response. In reality, a plot this even would only occur below 1 kHz. The polar pattern of Figure 3-12 is more realistic in that higher frequencies (designated by a dotted line) tend to arrive directionally even when the mic is omnidirectional.

Cardioid patterns are the most popular mic patterns, but seldom will one have a polar pattern as symmetrical as the graph in Figure 3-13.

More often there will be a greater amount of high-frequency pickup at the 0 degree point and the pickup of low-frequency information will extend well behind the 120-degree point as in the polar pattern of Figure 3-14.

Wide cardioid patterns (see Figure 3-15) are not too common, but they do have applications (such as ensemble recording).

Figure 3-10
Microphone
polar pattern
graph

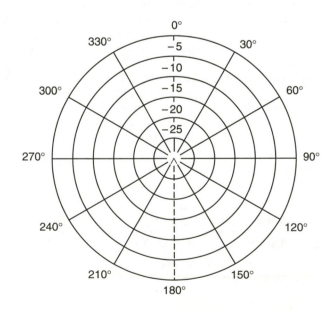

Figure 3-11
A polar pattern
of the ideal
omnidirectional
mic

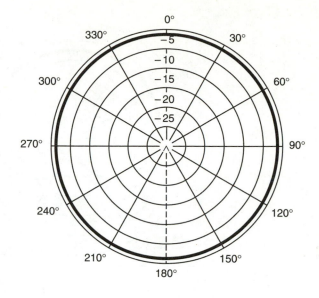

Figure 3-12
A more realistic
polar pattern
of an omni-
directional mic

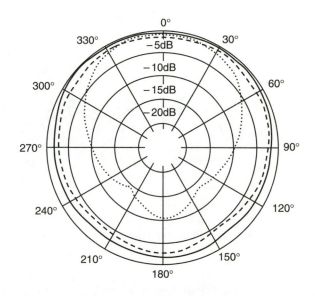

Figure 3-13
A polar pattern of an ideal cardioid mic

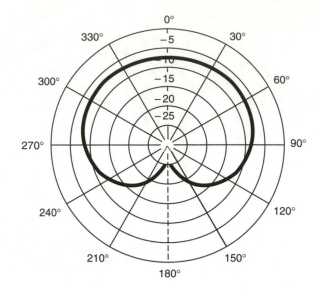

Figure 3-14
A more realistic polar pattern of a cardioid mic

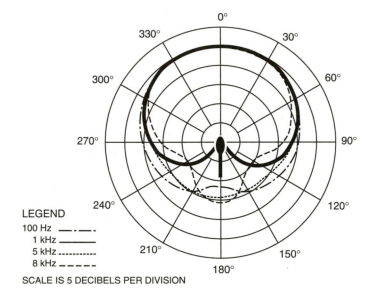

LEGEND
100 Hz —·—·—·
1 kHz ————
5 kHz ············
8 kHz — — — —

SCALE IS 5 DECIBELS PER DIVISION

    Here low frequencies "sneak into" the rear at a wide angle, while a lack of high-frequency rejection will cause the rear to have a "lobe"-like projection. Wide cardioid patterns, as in Figure 3-16, introduce another variation in polar pattern displays.

    Here, in order to discern the response of several individual frequencies, their polar response curves are split between the two halves of the graph to either side of the 0-degree point.

Figure 3-15
An polar
pattern of the
ideal wide
cardioid mic

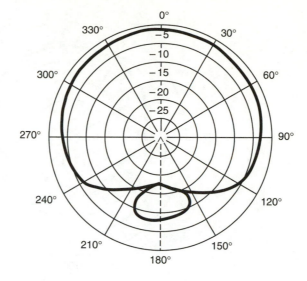

Figure 3-16
A more realistic
polar pattern
of a wide
cardioid mic

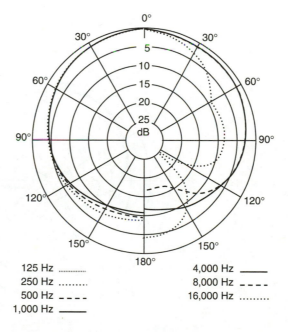

125 Hz	4,000 Hz
250 Hz	8,000 Hz
500 Hz	16,000 Hz
1,000 Hz	

There's a bit of disagreement when it comes to super cardioid (see Figure 3-17) and hyper-cardioid (see Figure 3-18) polar patterns.

In general, hypercardioids have a little more side rejection than super cardioids, along with much wider rear lobes. The polar pattern of Figure 3-19 could be termed a hypercardioid, even though it is called super cardioid by its manufacturer. Since manufacturers generally do not

Figure 3-17
A polar pattern
of an ideal
super cardioid
mic

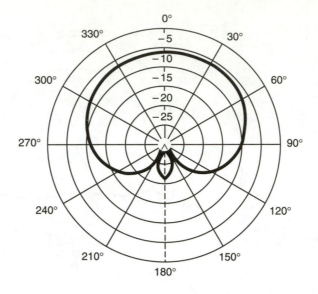

Figure 3-18
A polar pattern
of an ideal
hypercardioid
mic

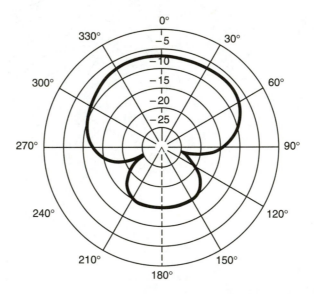

make one of each, and they both have equal side rejection, it's not something to be overly concerned about.

Figure eight or bidirectional polar patterns (see Figure 3-20) are often compared to dual but opposite-facing cardioids.

Figure 3-19
The polar
pattern of a
hypercardioid
mic that is
called a super
cardioid mic
by its
manufacturer

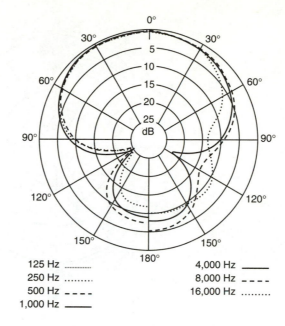

125 Hz ....
250 Hz ....
500 Hz - - - -
1,000 Hz _____

4,000 Hz _____
8,000 Hz - - - -
16,000 Hz ....

Figure 3-20
A polar pattern
of the ideal
figure eight or
bidirectional
mic

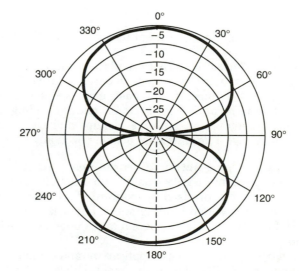

However, their pickup response is in fact closer to dual-wide cardioids as they have less rear pickup. As indicated by the polar pattern of Figure 3-21, figure eight mics can sound a little bright because high frequencies tend to be accentuated at the sides.

In some cases, as with ribbon mics, the lack of a back plate requirement lessens the diaphragm's resistance to movement. This increases their sensitivity, which makes them sound more transparent.

Figure 3-21
A polar pattern
of a a close-to-
perfect figure
eight mic

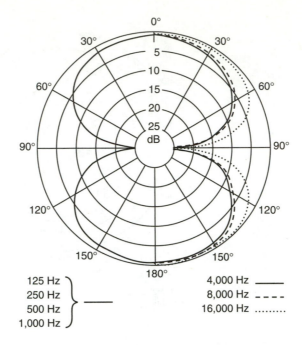

125 Hz ⎤
250 Hz ⎬ ——
500 Hz ⎥
1,000 Hz ⎦

4,000 Hz ——
8,000 Hz – – – –
16,000 Hz ·········

Shotgun mics come in two general flavors: physically short (see Figure 3-22) or long (see Figure 3-23).

They all have slotted tubes that are placed in front of their pickup elements. The slotted openings allow sounds from the sides to enter the tube. This sets up what can be termed a phase cancellation chamber. In the past, increased side rejection as in Figure 3-23 meant these tubes had to be fairly long. Shotgun mics extending 2 to 3 feet were not uncommon. Today, "short" shotguns use multiple pickups and electronic cancellation circuitry, which enables them to have a large amount of side rejection.

Some Examples Using Actual Microphones   The following collection of polar pattern and frequency response charts was included to help acquaint you with how the sound of a microphone can be surmised by examining specifications. Nobody expects the recording engineer to sit around reading specs, but with a little bit of practice you will get to the point where giving a mic's specs a quick once-over will clue you in on many possible applications.

After you try out many of these applications and examine the results, your thought process will be readjusted. You'll end up with a mental library of microphones that seem to magically "spring to mind" whenever the need arises.

A good example of a multipattern microphone is the Neumann U 67, which had omni, figure eight, and cardioid settings (see Figure 3-24).

Each frequency response graph also shows the effect of switching in the high-pass filter. The cardioid frequency response graph also shows the off-axis rejection. The polar patterns are shaped as expected. They show high frequencies refusing to behave in an omnidirectional manner, while the low frequencies will not be forced into tight directional patterns. Of note is

Figure 3-22
A polar pattern of an ideal short shotgun mic

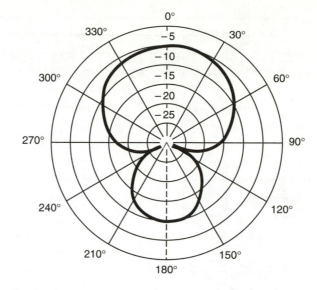

Figure 3-23
A polar pattern of an ideal shotgun mic

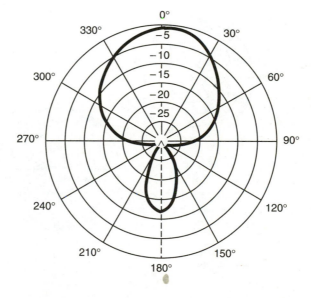

the fact that the manufacturer rates this mic as having a frequency response of 40 to 16,000 cps. This conservatism is typical of leading microphone manufacturers. The fact is that the −3 dB points are actually 30 Hz and 18 kHz. No, that's a *big* difference.

Neumann continued these conservative ratings with the subsequent U 87 FET version (see Figure 3-25) which is *the* microphone to which all others are compared. It can handle an addi-

Figure 3-24
The polar
pattern of the
Neumann U
67 mic

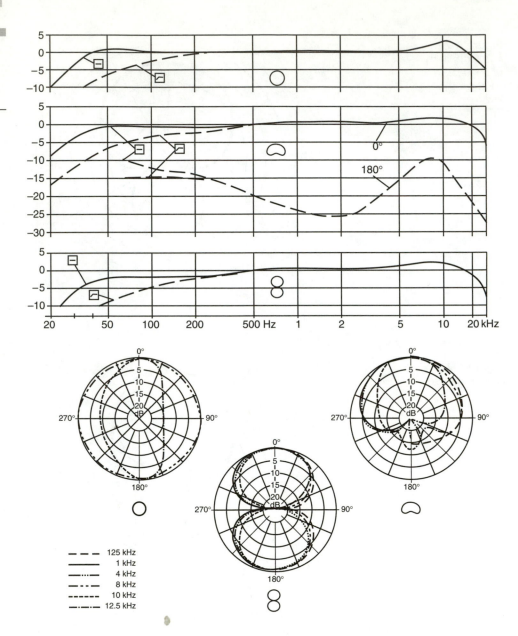

tional 4 dB of SPL (120 dB as opposed to the 67's 116 dB re: 0.5% THD) and can be battery operated (dual 22-volt). But if you're not going to continually record for a straight week, don't forget to turn off the power switch, because that's how long these batteries last.

While the two mic's capsules are identical, the U 87 has a 2 dB increase in its frequency response around 10 kHz. This is more than likely due to the increased transient response (speed) of the FETs and I suspect this is the reason many feel the U 67 sounds "warmer."

Figure 3-25 The polar pattern and frequency response graphs of the Neumann U 87 FET mic

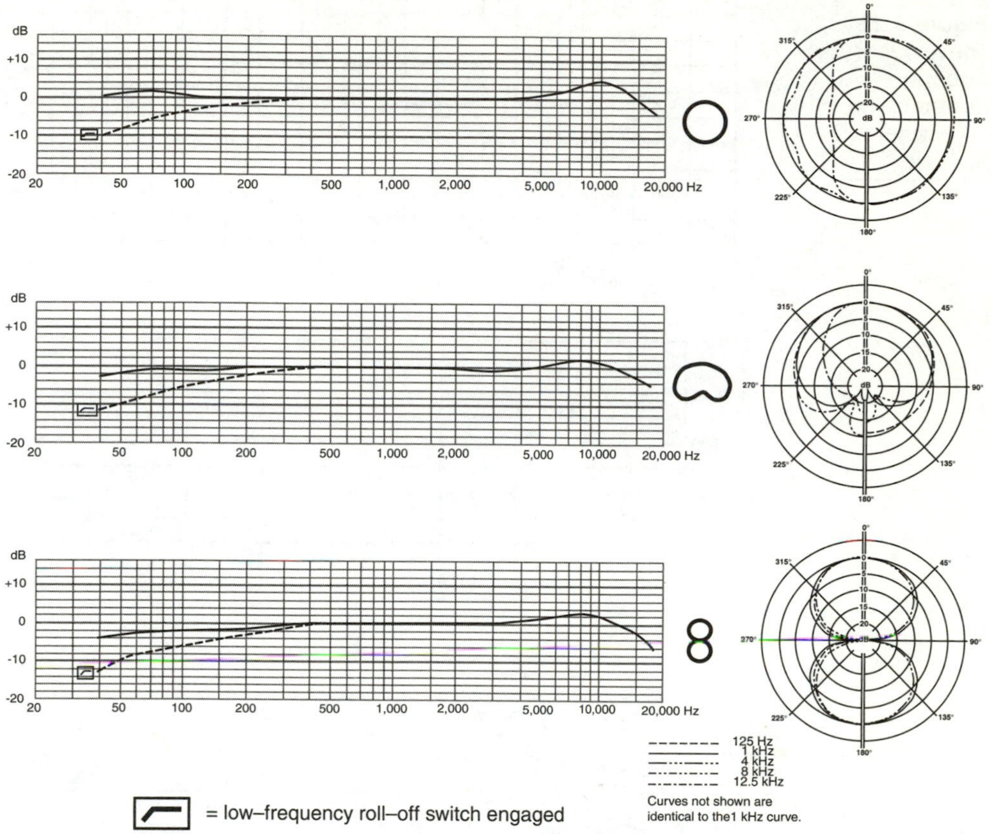
dB
+10
0
-10
-20
20
50
100
200
500
1,000
2,000
5,000
10,000
20,000 Hz
0°
45°
90°
135°
180°
225°
270°
315°
dB
125 Hz
1 kHz
4 kHz
8 kHz
12.5 kHz
Curves not shown are identical to the1 kHz curve.
= low–frequency roll–off switch engaged

The frequency response graphs in Figure 3-26 provide a great deal of information in a very small space. The top (Sennheiser MD 421) gives the frequency response with a reference level set at 1 kHz as well as the response tolerances.

The center (AKG C 414 EB), in addition to its amazingly flat and extended frequency response, provides the distance at which the response was measured. This accounts for the lack of proximity effect as well as the minimum deviation with frequency often caused by sounds picked up from the rear as shown by the off-axis 180-degree plot. The 50 dB range as marked on the last (Milab LC-25) frequency response graph points to the fact that the far left level designations should be used. Its frequency response and rear rejection do not seem as remarkable as the previous mic (C 414 EB's), until you consider the fact that this mic was designed to function as a handheld vocal mic.

Continuing with this same Milab LC-25 (see Figure 3-27), its cardioid polar pattern is perfectly symmetrical and its proximity effect is very predictable.

At 100 Hz, the boost is +3 dB at 0.5 meters (about 20 inches); at 0.2 meters (around 8 inches) the boost is +9 dB, and at 1 centimeter (under one-half inch) the increase is +14 dB. No wonder this mic is known for following the voice so well.

Figure 3-26
The frequency
response
graphs of the
Sennheiser MD
421 mic

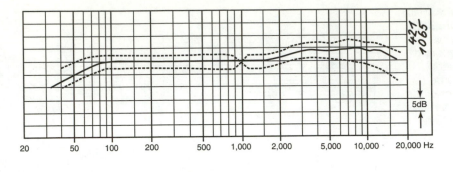

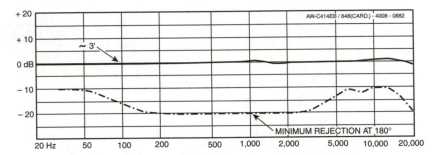

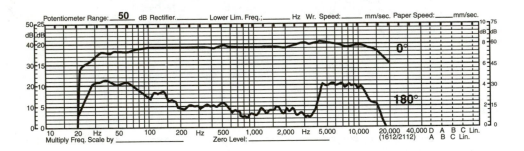

While the next two mics' (Audio Technica's AT813 and AT813 R's) polar patterns are identical (see Figure 3-28), their frequency responses are not exactly the same.

The phantom-powered bottom mic (AT813R) has a slightly smoother response than the battery-operated version (AT813). However, the main reason these graphs are included is because of their unique use of the kind of dotted and dashed lines normally used to define different frequencies in polar patterns. Here they are used to show the mic's proximity effect per distance.

Figure 3-27
The polar
pattern and
frequency
response
graphs of
the Milab
LC-25 mic

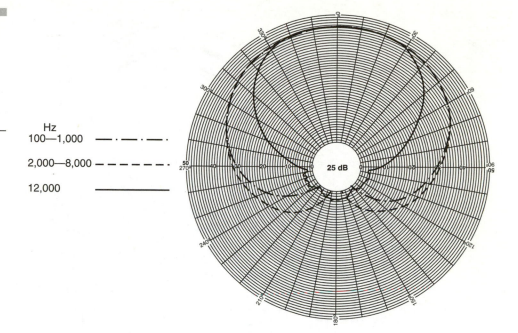

Hz
100—1,000   — · — · — ·

2,000—8,000   — — — — —

12,000   ——————

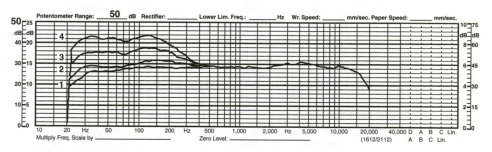

1: Free field
2: 0.5-meter distance
3: 0.2-meter distance
4: 1.0-centimeter distance

    In addition to the very comprehensive information given in the AKG D-330BT mic's flyer (see Figure 3-29 and Figure 3-6), the manufacturer outdid itself in the owner's handbook.

    These two graphs show frequency response, off-axis rejection, high- and low-pass filtering, and proximity effects. Everything is clearly labeled, the layout is uncluttered, and no magnifying glass is required.

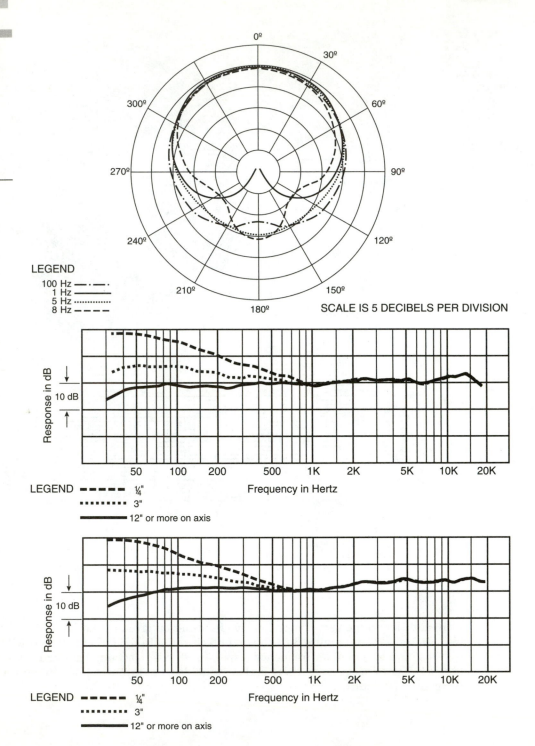

Figure 3-28
The polar
pattern and
frequency
response
graphs of
Audio
Technica's
AT813 and
AT813R mics

LEGEND
100 Hz —·—·
1 Hz ———
5 Hz ·········
8 Hz — — —

SCALE IS 5 DECIBELS PER DIVISION

LEGEND — — — ¼"
·········· 3"
———— 12" or more on axis

Frequency in Hertz

Response in dB

10 dB

LEGEND — — — ¼"
·········· 3"
———— 12" or more on axis

Frequency in Hertz

Response in dB

10 dB

Figure 3-29
The frequency
response
graphs of the
AKG D-330BT
mic

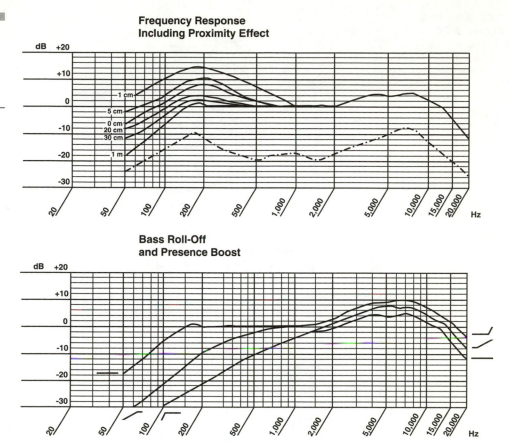

The polar pattern of the bidirectional (M 88RP Fostex ribbon) (see Figure 3-30) mic clued me in on something as far as using this mic in an MS stereo configuration. When looking at the manufacturer's logo (located at the 0-degree point), the side that is to the left should face forward. In this position, the sound from the stage will be more even. The irregularities on the opposite side, when picking up the rear's diffused ambient field, will not cause any audible anomalies. However, the real reason this mic was included here is due to the way its bottom end is depicted in the frequency response graph.

Is that the proximity effect or is it the cut of a high-pass filter? Because of the dashed lines and the way they are lower than the main response line, it could be cut. You could simply look at the mic, but the switching may be internal, too small to notice, or hidden (behind a ring), or the legends may be worn off. Your answer will almost always be found within the specifications. Here, it was a low-frequency roll-off switch designed to offset the proximity effect when used for speech.

Figure 3-30
The polar
pattern and
frequency
response
graphs of the
M 88RP Fostex
ribbon mic

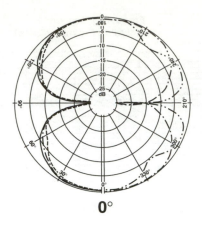

0°

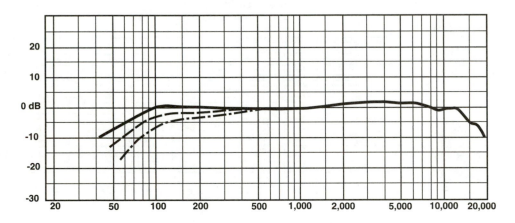

   The Shure SM81 provides shovelfuls of information (see Figure 3-31), especially when you compare its R104A omnidirectional capsule to the cardioid version.

   An omnidirectional mic's frequency response will always be a bit smoother at the high end. This is because while this frequency area is more directional, an omni mic is also "open" to more high frequencies that arrive from the back and sides. The SM81 polar patterns illuminate this point perfectly.

   The low end will, because of its nature, arrive from all around the diaphragm, so here too an omni will have a more even pickup pattern than a cardioid pattern mic. Cardioid mics are more susceptible to low-end increases due to proximity effect. This is clearly illustrated in the frequency response graphs. Shure includes information on the high-pass filters within their frequency response graphs, but then goes further by adding a proximity effect chart that also gives you the results of using the high-pass filter. The distance at which these measurements were taken is also included. While all of this may not seem all that important, in actuality it's critically important to the professional recording engineer.

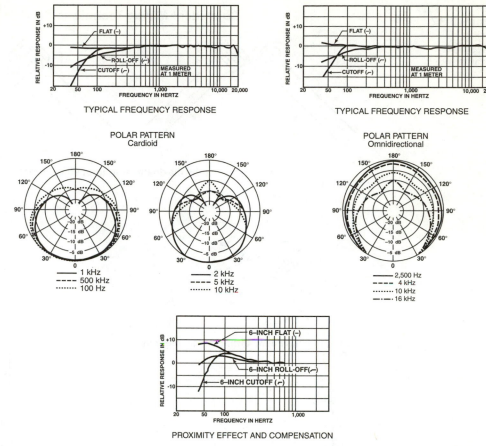

Figure 3-31
The polar
pattern and
frequency
response
graphs of the
Shure SM81
mic. Also
compared are
the R104A
omnidirectional
capsule to the
cardioid mic.

Microphone Positioning Techniques for Stereo Recording

When it comes to stereo micing, it's a good idea for the recordist to know as many microphone placement techniques as possible (see Figure 3-32). That way, should a situation change, you know how to adjust the mic setup to match it.

Live Stereo Recording The choice of whether to pan mono signals from multiple microphones into a semblance of a stereo event, or to try capturing the event so as to exactly recreate it, is usually based on either the "purist" desire to preserve the integrity of the performance or a "realist" need to provide as much audio information as possible by improving the balance while controlling both the reverberation and audience ambience, including HVAC, and other noises. This section is about using microphone selection and placement to capture an event, record an instrument, and gather audio information on location. The choice

Figure 3-32
Micing
placement
techniques

Coincident Micing Technique

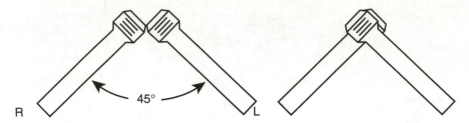

Coincident at the maximum setting of 90 degrees

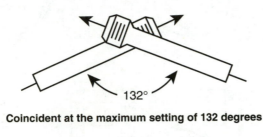

Coincident at the maximum setting of 132 degrees

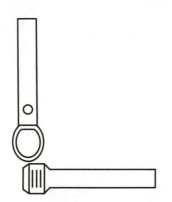

**MS Coincident with a bidirectional mic facing the side
set at 90 degrees to a cardioid mic facing forward**

**The Blumlien Array, two
bidirectional mics offset
90 degrees from each other**

Figure 3-32
Micing
placement
techniques
(continued)

Near–Coincident Micing Techniques

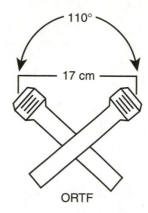

110 degrees and 6.5 inch spacing

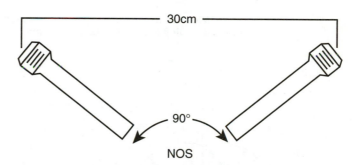

90 degrees and 11 3/4 inch spacing

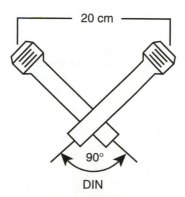

90 degrees and 7 3/4 inch spacing

Spaced Pair Micing Technique

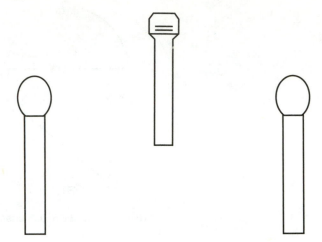

**Mics can be cardioid, figure 8, or omnidirectional.
Here a cardiod is added to accentuate the center stage area.**

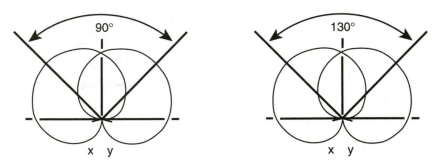

**The effect of mic angle on the stereo spread. These two drawings
depict how cardioid microphones in an XY configuration can be
aimed to narrow or widen the width of the stereo field.**

Figure 3-32
Micing
placement
techniques
(continued)

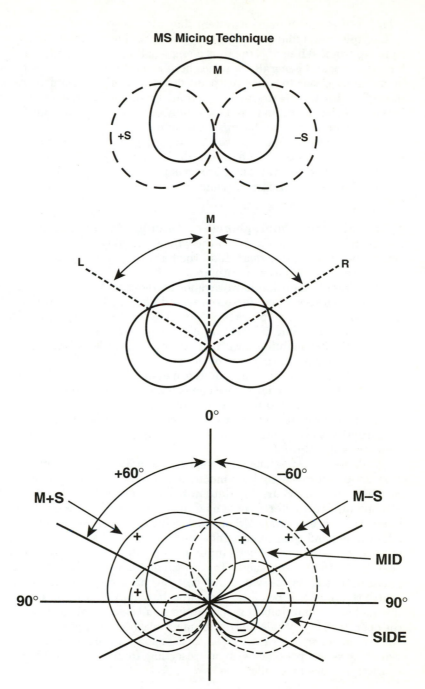

MS Micing Technique

This layout, showing width adjustments and the crossover and null points, gives some indication as to this stereo micing method's variability.

then becomes one of using a system that combines two or more diaphragms in a single mic or multiple individual mics. To be perfectly fair I must admit to my own personal prejudices right up front. While it's true that a single, all-inclusive stereo mic is more convenient, easier to set up, and generally provides better mono compatibility because the diaphragms can be mounted closer together, my point of view as an independent recording engineer is that multiple mics are a better way to go than a combined stereo mic. If I had been more involved in live, television, or film recording, it may be a different story, but for the work I was involved in, I was much better off having a pair of mics.

No matter what type of stereo micing you go for, mono compatibility is a priority second only to sound quality. Even in the studio when using stereo micing techniques to try and create the widest acoustic guitar or broadest piano in the world, it's important to avoid phase distortion. Phase distortion and mono compatibility are the two sides of the same coin called time.

Mounting Stereo Microphones on Stands   Any shifting of a stereo multiple mic setup can negate hours of work. Often the whole assembly will have to be repositioned; therefore, ridged mounting is necessary. If all the hardware is firmly joined together, you will not have to deal with a midconcert nightmare.

The king of mounting hardware for stereo mics has to be Schoeps. While this company does not exclusively deal in stereo recording, they have it fully covered and they have the mounting hardware to prove it. Their UMS 20 "Universal Stereo Bracket" and their model number A22 dual mic holders simplify the recordist's job immeasurably. Microtech Gefell's Model SH93 X/Y mount is a work of art and Shure's Model A27M is elegantly simple yet provides comprehensive coverage of most of the dual mic stereo positioning possibilities. It's important to check out many stereo mounting bars and other assemblies, including the adjustments they provide before one is decided upon. Requirements will change with differing situations. Get pictures or make drawings to help you decide what you need and like. If you find the price tag (often in the neighborhood of $500) out of your range, try presenting the situation to your local machine shop. Bringing your own mic clip attachments and stand adapters will help.

While we're on the subject of making life a little easier, a big-time aid when you need to monitor the stereo signal for mono compatibility (even while on location) comes from the Swedish company Pearl. In addition to their line of high-quality studio microphones, Pearl also manufactures stereo condenser mics and the Model MSV 15 MS matrix. This is a very handy transformer-based MS network, which is mounted in a small metal box. So what? Well, follow along and then contact Independent Audio, Pearl's U.S. contact. The two MS mic outputs are connected to the box's input and its output is connected to the recorder. This is no big deal, nor is the potentiometer that's included for adjusting the stereo width. It has a headphone output, a bypass switch that circumvents the MS matrix network, and a locking toggle switch that makes it possible to monitor the signal in stereo or mono. This is all very nice, but it would be unremarkable except for one small detail. This little box was designed to clip onto the operator's belt. While this device may have been designed with location recording in mind, with a pair of sealed cans (closed to the open air), monitoring mono compatibility at a live venue becomes possible without resorting to the often difficult and always attention-attracting use of an oscilloscope.

Stereo Micing Techniques   Let's start out with some straight laws of physics. Low frequencies are nondirectional. They go right around barriers like our heads and appear at both our ears pretty much simultaneously. Therefore, clues to the location of their source within the stereo field can only come from the level intensity differences at the receiver. On the other hand, high frequencies beam, diffract, and often take more than one path to the listener's ears (and a mic's diaphragm). Therefore, our stereo localization of the source of high frequencies within the stereo field is determined less by intensity and more by arrival times or the phase differences between our two ears (or mics). Every stereo microphone technique must take these two facts into consideration.

Stereo microphones are usually configured as coincident pairs, spaced pairs, near-coincident pairs, or omni pairs with baffling between them. All these methods have their advantages, but the use of any one of them will result in some amount of compromise. This *is* audio, remember? Some examples follow.

Spaced pairs use timing difference clues to place or locate instruments within the stereo field and thus they take advantage of the human ear/brain combination's ability to recognize these positional clues. This method's accuracy is highly dependent on having the center of the stereo field located equally distant between the positioning of the left and right microphones. Furthermore, localization during reproduction through stereo speakers also requires the listener to be positioned exactly halfway between two speakers that should be pointed toward the center at an angle of 60 degrees.

Coincident (XY and MS) methods place the left and right microphones so close together that the signal reaches both diaphragms at basically the same time. Thus, the two mics have little to no phase differences. This provides less in the way of timing clues, so these methods depend more on level differences for stereo localization information.

The Short Form Rundown

Coincident: narrow field, good imaging, mono compatible

MS: adjustable stereo spread, good imaging, mono compatible

Near coincident: accurate localization, not completely mono compatible

Spaced pair: This method usually uses two cardioids or two omnidirectional mics. These are faced forward and the spacing between them can be set at 3 to 10 feet or more. A spacing beyond 3 feet provides a somewhat exaggerated field. At 10 feet, this method delivers good, "warm" room ambience coverage, but side instrument localization can be poor. Often a center mic must be added to enhance localization.

Coincident methods can utilize two cardioids, or figure eights, but not two omnis because when they are closely spaced and combined the result is mono. With XY the two cardioids are usually set at a maximum angle of 135 degrees. Ambisonic or "stereosonic" coincident uses two bidirectional mics set at 90 degrees. In order to be mono compatible, the coincident micing technique requires the use of directional mics. Since these two mics come close to occupying the same space and therefore receive no timing information, they are aimed in different directions. The localization capabilities of coincident micing is more important to film than TV because of its ability to match the audio to picture. When used on classical orchestras, this method provides the listener with aural information as to the layout of the instruments on the stage.

The MS micing technique uses a cardioid and a figure eight (bidirectional) mic with the cardioid facing forward while the bidirectional picks up the sides. MS provides superior imaging and has more central focusing due to the mid-aimed mic. It is very flexible in that later adjustments of the side levels can be made to increase or decrease stereo spread. Since the two side mic signals cancel out when summed to mono, the remaining mid-mic presents a phase-coherent, full frequency mono signal with little room ambience.

- MS can use a hypercardioid mic for tighter (105-degree) front pickup than would be provided by a super cardioid mic.
- Hyper = more rear pickup and maximum rejection at 110 degrees from the front.
- Super = little pick up from the rear as the maximum rejection occurs past the 120-degree point.

Combinations Often combinations of stereo micing methods are used (i.e., a spaced pair made up of a center cardioid and two figure eight mics, or an MS configuration at the center with the addition of two spaced figure eight mics). Sometimes close and distant spacing is used to capture the direct and ambient sounds. The distant mics can be placed 20 feet back and their phase response should not interfere with the front signal because the delay of their signal will still be under 18 mS.

Additional processing, such as adding delays to close mics, can get mighty academic and is not always helpful or practical in the field. Adding more mics can cause phase problems, so their effect should be monitored with a scope in XY mode.

The Negatives It seems that whenever stereo microphone techniques are discussed, their deficiencies get the most play. Therefore, I feel an even, unbiased, and undistorted presentation of all the negatives would be beneficial.

Coincident The close positioning of the two capsules allows for no time differences; therefore, the imaging at high frequencies is poor. Additionally on-axis or center signals now arrive more at the mic's sides. Due to the response characteristics at the sides of directional mics, the reproduction of these sounds arriving at this location will have a less than ideal frequency response. This method can be lacking in stage "breath" and room ambience.

Near-Coincident This method, like full coincidence, colors the centrally located sounds because they arrive at the directional mics off-axis, while the signals arriving from the sides and rear are attenuated, which reduces the pickup of room ambience. Furthermore, due to the added distance between the two capsules, this method is not completely mono compatible.

Spaced arrays Spaced arrays use large distances between mic placements. This results in nonuniform level differences, which can cause blurred focusing and poor localization. This distancing also makes for large differences in timing between a signal's arrival at the two microphones, which results in a phase response that is not mono compatible.

Binaural The frequency response of this method is not flat or mono compatible and it does not translate well when reproduced on stereo speakers.

The Blumlein System Sometimes referred to as "Ambisonic," the Blumlein System is composed of two bidirectional (or figure eight) mics crossed at 90 degrees to provide a sharply focused stereo field, but its localization qualities vary between the high and low frequencies. Electronic circuitry can be added to equalize or decrease the highs, but this colors the response.

This configuration puts severe limitations on placement, since flawed positioning results in a less than ideal stereo field and inadequate phase response. Its 360-degree pickup can end up providing excessive reverberation and it can be difficult to position so as to exclude HVAC and other noise sources.

Depending on the situation, every one of these methods can be very useful. For example, the Blumlein method, because it provides equal pickup in all directions, may be just the ticket in a great sounding hall. So it's best to keep your mind open to all micing methods. Understanding all the pros and cons of every micing technique is part of every recordist's job, but it seems to fall a little heavier on those involved in live recording. Thinking of live recording reminds me of a letter published in a nonaudio magazine that was replying to a article on classical CD collecting. The author's opinion was that these recordings were nothing more than collector fetishes. The letter compared the enjoyment of listening to a classical recording to being present at the actual event. The author had used the term "tremor of drama" to describe a live performance and the reply agreed wholeheartedly with the "tremor" of coughing, paper crackling, cell phone using, and chattering in the audience all around him. He concluded that he was thankful to have the opportunity to really hear the performance by listening to the recording. My reaction to this was twofold. I was proud of the job we recordists do in trying to make great music available to as many people as possible, but I also thought, "You think *you* had it bad?" Try recording an event while people throw things at your microphones! Why does everyone *need* to pat a binaural model on the head?

That's enough of the stereo micing small talk and generalities. Now it's time for some details. You may think the following coverage of stereo microphones is a bit drawn out and overdone, but I purposely left out most of the 5.1 and surround sound techniques just to reduce the *reader mortality rate* (RMR).

Coincident Pairs   The "XY" method refers to the use of two directional microphones (usually cardioid) that should at least somewhat match each other in characteristics (the same make and model number will do). They should be placed as close as possible to each other to avoid differences in the sound's arrival time at both diaphragms as this helps to avoid phase anomalies. To facilitate this, the diaphragms are often placed one on top of the other with their two sides touching. When set at a 90-degree angle they make up the side and bottom of an "L" shape. This angle can be varied, such as when aiming the mics to pick up the full spread of the stereo field.

Since directional mics are more sensitive to the sound arriving from a "perpendicular incident" (from directly in front of them) and have an increasing amount of rejection as the sound arrives more off-axis, the level of the signal they are pointed at will be louder. However, using large angles to capture a wide area (such as a broad stage) means that more of the center signal is picked up by the sides of the mics. Some feel this overly diminishes the effectiveness of the XY method because sound pickup from an off-axis incident (at the side) by directional mics is very frequency-dependent and thus not ideal.

No matter what angle is chosen, the center or the "crossover" point of the two mics should be physically positioned at the "center stage" location. This accomplishes two important requirements. It keeps the two mic's angles symmetrical so that the center stage sound is picked up equally by both mics. Later during reproduction, the dual mic pickup of the center location will be added together to the imaginary mono or center position between the two speakers. This results in the lower pickup level from this area becoming more proportional to the other levels within the stereo field. So remember that the physical center position of a coincident pair is very important to accurate image placement within the speaker-to-speaker spread.

Some folks complain that the XY micing method provides a narrow field. Hey, how can any positioning be expected to provide a stereo spread that extends to the full "breath" of the stage and pick up all of the room's ambience when two directional mics are used?

Since the sound may end up a little more centrally located between the two speakers, different polar patterns are often used, but trying to use the XY placement with two omni mics does not work because both mics will pick up the same thing and the result will be a mono signal.

XY placement with two figure eight patterns set at 90 degrees renders an equal amount of front and rear pickup, but some feel it provides too small a recording width. Shifting this configuration's angle by 45 degrees results in an MS configuration.

Using wide cardioid pattern mics in an XY pattern produces a stereo field that's wider, but since the rear rejection of these mics is increased this configuration will provide less low-end content.

XY placement with two super cardioid pattern mics will obviously be useful with narrower stages, but any instrument to the side can come across sounding unfocused and more like part of the ambient sound.

Coincident XY micing using two regular cardioid patterns provides a focused center, less sound from the rear, a good stereo spread, and its setup is almost foolproof. Coincident micing allows the use of two of the same type of mics, and the stereo signal is available without the need of further processing. However, with coincident micing there's no fix in the mix so what you hear is what you live with. This method can be mono compatible to the point where the frequency response will be the same in mono as it is in stereo.

Near-Coincident Stereo Micing   This method also uses a pair of directional mics that are located close together, but here the diaphragms are not touching and are angled away from each other. In some cases the two diaphragms can be up to 12 inches apart. Yet it only takes an inch or so of added spacing for coincident mics to begin picking up more of the room's ambience.

Because of the additional time arrival (phase) differences, this method is more like a combination of coincident and spaced-pair stereo micing techniques. Therefore, the same rules for both apply: very wide angling provides an unnatural or exaggerated stereo spread and tight spacing results in a narrow, closer to mono, spread.

Wider Coincident   Some configurations are set wider than coincident and angled farther apart. These can render an almost quasi-binaural response that translates well to stereo speaker monitoring.

The French Broadcasting ORTF stereo micing method is one of my favorites. This near coincident method places two directional (cardioid) mics 17 centimeters (6³/₄ inches), or a head's width apart, and angled away from each other by 110 degrees. This setup is simple and mounts are readily available. I've taped band rehearsals to cassette decks with a pair of SM58s set up this way and the result served the purpose of self-evaluation very well, and the localization was well defined, even on speakers.

Being partially spaced, the ORTF method ends up with phase cancellation that causes a dip with a center frequency around 1 kHz. This is due to comb filtering. But this loss of information is partially made up by the increase in the signal content from using two mics and it can actually help decrease any "bunching up" of midfrequency levels in the center or mono position between the two speakers. This method does not require an electronic anticross talk circuit, it is fast and easy to set up, and it uses common, readily available (even handheld vocal) cardioid mics. Because it's just about foolproof, it's easy to get good results using this method.

The Dutch have a slightly different take on the same method. Their NOS method uses 12-inch spacing between the capsules and a 90-degree angle. This results in the phase-induced dip being lowered from 1 kHz to around 250 Hz. The German's DIN method tightens the spacing back to 7⅞ inches but retains the 90-degree angle. *The* method to use depends on the situation, but it doesn't take much more than a few minutes to try out all three of these near-coincident methods before you make a choice, so dog-ear this page!

ORTF (French)   6 ¾-inch spacing; 110-degree angle

NOS (Dutch)   12-inch spacing; 90-degree angle

DIN (German)   7 ⅞-inch spacing; 90-degree angle

Near-Coincident Micing Using Omnidirectional Microphones Often referred to as "the baffled omni pair" stereo micing method, two omnidirectional mics are positioned much like the ORTF method in that the distance between the two is usually head width, or 6 to 7 inches. Since we're dealing with a pair of omnis the angle is nonspecific and varies with manufacturers and operators alike. This dual-mic method provides low-frequency localization information from level differences and high-frequency information from arrival time variations. The baffle blocks some of the high frequencies from appearing at both diaphragms and increases separation while decreasing comb filtering from phase anomalies. Called the *optimum stereo signal* (OSS) by its originator, a Mr. Jeckin, his "disk" system sets a pair of omnidirectional mics again positioned much like the ORTF method but with a baffle between them. Thinking about it for a moment reveals the logic. Two pickup patterns that are very close to our own are used, with a baffle as a substitute for our face. The result: omnidirectional low-frequency response with some of the high-frequency cross talk blocked. These systems sometimes utilize dual boundary mics, but versions made up of two mics mounted on opposite sides of a foam-covered 12-inch diameter disk with clips allows for increased studio usage from these two, high-quality omnidirectional mics.

Spaced Pairs This method of stereo micing places two mics aimed directly at the stage and spaced some distance from each other. The greater the space between their positioning, the wider the stereo spread. Omnidirectional mics are normally used, but any polar pattern is possible. Why omni? Well, this method depends on both proximity and time differences to convey the stereo field. Sounds closest to a mic will be louder and arrive sooner. The same sounds will also be picked up by the opposite omni mic, but they will be both delayed and lower in level. The difference in level alone will cause a signal to seem to appear louder on the closer mic's speaker and the delayed signal on the opposite speaker reinforces this perception. Large microphone spacing (I've heard of people using separations of up to 35 feet) can provide an unnaturally exaggerated stereo spread, but even 10 feet of separation can result in a center stage that sounds empty. Center stage positions should be picked up equally by both mics. The use of omnidirectional mics helps, but wide spacing can require the addition of a third mid mic to fill in the central area. Stereo recording with spaced microphones readily provides the stage breadth and room ambience lacking in the coincident pair method. When omnidirectional mics are used, the low-frequency response is more extended and a "warmer" sound is provided. Because the sound pickup of omnidirectional mics is more open, the instruments on the stage are blended more smoothly with room reverberation.

But because this method depends so heavily on phase differences, sound source locations often end up blurred. Therefore, it can be difficult to achieve the accurate localization of instruments within the sound field using only spaced mics.

Mono Compatibility *Multichannel television sound* (MTS) demands that the audio be mono compatible and to consistently provide correct image localization. This equates to little or no phase cancellation. In this case, spaced pairs must be set up so that the distance from the main sound source to both mics is equal. This results in there being no time difference between channels while still fulfilling television's need for room ambience or crowd noise. Some folks argue that the French 110-degree, near-coincident method of using a hypercardioid mic is better than the 135-degree spread you often get when normal cardioid mics are used. But let's be practical, the stereo spread of TV is pretty narrow, often more in the area of 10 to 20 degrees from the viewer's perspective of the screen. This means that most dramatic audio-imaging motion will not "play" on television like it does in film. In fact, the coherence between television audio and video works best when all dialog is placed dead center.

The only wide space audio imaging that would remain coherent to the television viewer would be ones that are similar to theatrical offstage sounds. Even when the program material is music and broadcasted in stereo, you'd want the lead vocal to remain in the center position to retain its correct audio imaging within the room's ambient sound. Keeping the spread of the stereo field within practical limitations also greatly improves the possibility of mono comparability.

Mono compatibility is important to television and AM broadcasting, but what if you're recording the event in stereo for CD or a live FM stereo broadcast? Well, wouldn't it be nice if the end product were mono compatible when checked by other radio stations for later rebroadcast? Most stations do in fact check program material for mono compatibility. My experience has been that it's quite common to find either a phase meter or a vector scope in most audio production and broadcast control rooms. Why should stereo FM broadcasters worry about mono compatibility?

Almost everyone has experienced an interruption in a stereo broadcast caused by interference from wind, rain, and especially lightning. Listeners will not usually put up with the resulting noise, but if they are interested in hearing the programming they may try switching their receiver to mono in the hope that this will at least partially remedy the problem. However, even additional retuning will not help if the program is not mono compatible and the broadcast's frequency content suffers from audible holes due to phase cancellations. This results in the listener moving on to somewhere else on the dial.

Picture yourself as a radio station's chief engineer who has just been told by management that every advertiser that had purchased air time yesterday between 4:00 and 6:30 p.m. now wants a refund or credit because the rate they were charged was based on a set number of listeners that was drastically reduced during that period as the station was broadcasting gibberish. This scenario, while fictitious, is not without foundation. Ask anyone involved in broadcast about "dead" air or "black" picture problems and they won't exactly shudder as there's not much you can do about nature. But it *is* relatively easy to ensure that the program being broadcasted or recorded is mono compatible.

Vector scopes, waveform monitors, and phase displays operate much like oscilloscopes. Normally, a scope displays level on the vertical axis and time on the horizontal, but the XY mode is used for displaying phase. The resulting "Lissajous pattern" relates the amplitude and phase of one signal to another. Here, the left signal is fed to the vertical axis and the right to the horizontal. When both inputs are fed identical, pure tones, the display will be a single-line readout that rises from the lower left to the upper right at an exact 45-degree angle.

This will rarely be seen (outside of test bench setups) as normal audio inputs show up as more like a pulsating brillo pad. The best you can do is estimate the phase response as per the amount of deviation from the in-phase cluster around the 45-degree line. Audio vector and phase scopes improve upon the common oscilloscope by tilting the cathode ray tube 45 degrees so that the response rises from the bottom to the top at the center of the display. Either way, if you see what looks like a brillo pad having a real bad hair day, you probably have phase problems.

Some audio vector scopes also include level meters for the two channels, as well as sum and difference levels. This combination of level and Lissajous pattern displays makes for a more easily noticed warning of potential problems.

Then there's good old-fashioned mono referencing. The best bet is to have this capability, which requires a summing amplifier, or isolation transformers, to combine the left and right signals to mono. This, in turn, is fed to a make-before-break switch (to avoid loud switching pops) along with the normal stereo output. The switch's output is routed to the monitor speaker amp, or a smaller amp connected to a set of headphones when it's essential that the normal monitoring not be interrupted, as when it's being listened to by the producer or musicians. With this capability it becomes a fairly simple process of checking for differences in frequency response between the stereo and mono signals. If you look closely at photographs of older recording control rooms, you'll often see a solitary Auratone speaker atop the meter bridge. Most recording consoles used to have small mono speakers built into the meter bridge or the producer's desk. Hey, at that time we were cutting singles, as in mono 45s for AM radio play.

The Mid-Side Stereo Microphone Technique   MS offers a great deal to television production. Recording the (prematrix) two-mic feed allows the producer to later open up or close down the stereo spread at will. Here a narrow or mono setting, depending on the forward facing mic's polar patter, can be used to exclude ambient room sound should there be a risk of it causing confusion when listening to dialog. Meanwhile, the main sound arrives at this forward-facing (front) mic on-axis, which equates to better fidelity. With this micing method the stereo does not diminish with distance and the two-mic assembly is easy to deal with.

In television, audio and video points of view must be coherent, yet mic locations and levels can not change with every video perspective change. Music programming may have many video close-ups, but if the mic levels were to also change, it would just confuse the viewer. Sports coverage, on the other hand, may lend itself to dual stereo and mono sound pickup. Here, stereo can be used for wide-angle sounds such as crowd noise, while mono pickup can be used to isolate sound sources.

MS and Mono Compatibility   When summing MS mics, the noise levels will be 3 dB lower than when combining other coincident pairs. This is due to the fact that combining signal's normal adds the noise from the two mics. With MS, since the two sides or "difference" signals are exactly 180 degrees out of phase with each other, when they are summed their signals and noise will cancel out. Additionally, because the side mic signals cancel out when summed to mono, the mid mic remains intact to provide a phase-coherent, full-frequency mono signal with little room ambience. In other words, the mono signal from summed MS is free of ambient signals and comb filtering. Furthermore, this mono compatibility is reliable.

When used for stereo, the close proximity of the two mic's diaphragms results in better phase coherence. Since the main pickup is by a mic that is front-facing, it means that the sound arrives at its perpendicular angle of incidence. The MS method's superior imaging and flexibility is due to the central focus provided by this forward-facing mic.

Another benefit provided by MS is that unlike other coincident methods very little, if any, of the directional mic's side is exposed to the sound source. Remember, the frequency response at this area is more compromised with directional mics. Additionally, since figure eight microphones are less frequency dependent, the overall frequency response is more even and natural.

The Mid-Side Setup   For MS, one cardioid and one figure eight (bidirectional) mic are positioned in a tight, coincident spacing with the cardioid diaphragm aimed forward and the bidirectional aimed to the side.

An MS setup always utilizes at least one figure eight microphone that is rotated 90 degrees from the front or mid mic. This places the bidirectional mic's side null point on axis with the front-facing mic. The pickup of a figure eight mic at the null point is usually 20 to 30 dB lower than its front and back pickup. It thus provides little front and back information, but the sounds from the sides will be picked up in a close to half hemisphere pattern.

An MS setup with an omnidirectional mic in the center or mono position provides both front and rear coverage. With the side levels set low, the output will be mono, while large increases in the side levels will greatly diminish the front and rear coverage.

MS setups with a wide cardioid as the mid mic provide a wider front sound, but there is little pickup at the rear. Once again, increasing the side levels will diminish the central coverage. MS with a cardioid mic also eliminates much of the rear pickup. Additionally, signals outside of the array's area of focus come across as vague ambience.

MS with a super cardioid mic is much like that of a normal cardioid in that it eliminates most of the rear signal. The front spread will be equal to, or only slightly less than, when a normal cardioid is used, but here signals outside of the array's focus may be completely eliminated. However, MS stereo micing requires the use of a matrixing network.

An MS Matrix Setup Using Console Phase Switching   Start by feeding the mid mic to a console channel, pan it to the center, and bring up the level. Then feed the side mic to two channels and switch one out of phase. For consoles that do not have phase-switching capabilities, a plug or a patch cord that has been phase-inverted will do. These *must* be marked as such or all kinds of problems will someday (more than likely very soon) take a bite out of your butt.

Now raise one of the two side mic faders and determine if it is the left or the right feed using visual cues from the positioning (during the recording process), or as notated (when dealing with the program in the post-production stage). Otherwise, without any phase inversion, the front signal will increase a bit more when one of the sides is raised. This will be the front-facing side, which needs to be phase-inverted. If the sides are reversed, repanning is faster and easier, but repatching will avoid near-certain confusion. All the guesswork can be eliminated; just make sure you always *face the bidirectional mic's front side to the left*.

The front of a bidirectional microphone (the side where a singer should face) will always have the manufacturer's logo on it. When a sound impinges on this side of the mic, it produces a positive electrical output voltage. When sound hits the opposite side of the bidirectional mic, the electrical output will be negative.

The two, close to half-hemisphere lobes that make up the figure eight pattern will cross-over into the sides of the forward-facing cardioid mic. When the right side of the forward-facing mic is combined with the out-of-phase right-side output of the figure eight mic, all signal content common to both will be cancelled out. Depending on the polar pattern of the forward-facing mic, it will now pick up the sound from the center point out to some angle to the right. At this angle the right-facing side mic begins to pick up the signal and this coverage extends around the right side. Because the left output is in phase with the forward-facing mic, its output must be reversed in polarity.

Finally, pan the two side channels hard left and right, set up an equal mix of the two side faders, and bring the center mid mic up for a medium blend. The left is to the left, the right to the right, and the center is spread between the two.

Now the fun begins. Adding more of the center channel will tighten or "collapse" the stereo field. This adjustment has the effect of focusing the width of the whole pickup pattern. Adjusting one side fader up, while lowering the other, as is done when cross-fading, will effectively allow you to "aim" this narrowed pickup pattern across the width of the stereo spread. This is a wonderful tool. When the two microphone outputs have been sent directly to the recorder, bypassing the matrixing circuitry, the operator can vary the perspective during post-production by adjusting the mid-levels to increase or decrease stereo spread and by adjusting the side levels to aim the refocused pickup pattern across the stereo spread. I can't remember the model number, but many years ago Sony offered a tube MS mic that allowed this function to be remotely controlled at the power supply. I could put this mic out in the studio, facing into the window, and then, back in the control room, aim it around until I found the "live" ambient sound I was looking to add to the rock drums that were set up and miced in the middle of the studio. When heavily limited this became an awesome audio experience.

MS is about as simple as it gets, but on the other hand, it is important to have matching microphone characteristics so with this stereo recording method the operator is faced with the task of combining two matched mics that also have different polar patterns. Don't anguish over this as in some situations unmatched mics work out better.

Binaural Recording

Once you've heard an accurate demonstration of this method you'll never forget it. "Stepping inside" those cans will actually seem to put you into the original environment. Localization is pinpoint accurate. Questions of placement get answers like "5 to 6 feet from the right rear at about 270 degrees," and they're dead-on accurate.

True binaural recording is accomplished with two mics set into the outer ear part of a model of the human head, or at least spaced the same as our two human pickups, about 6 or 7 inches apart with some sort of baffle between them. Is it better to use a model of the human head? The first problem becomes, how wide should the head be? There could be less than perfect spacing. Some folks demand that the model have a face. How big should the nose be and what about hair? How long? Torso? Fat or thin?

I swear people get adamant about this stuff. It's too bad, because it's actually an elegantly simple recording method that provides astounding results. When done right the listener will perceive localization clues from a full 360 degrees. Even vertical clues will be included.

Binaural dates back to 1881 and a Mr. Clement Ader who used the method to broadcast the Paris Opera through two telephones lines!

Microphones, audio electronics, knowledge of psycho-acoustics, and the testing of the human auditory system has progressed greatly since then. While no one would state that any link in the audio chain is now "perfect," many get fighting mad about which particular binaural method is the best.

Scientific studies have shown that the outer ear, or pinna, affects not only the frequency response from the front, but also attenuates the highs of sounds arriving from the back. So using an outer ear-shaped baffle with binaural recording does in fact augment its 360-degree pickup.

Today, some mics are available that look much like "ear bud" consumer headphones that fit into the wearer's ears and thus use the wearer's head, hair, nose, and pinna. Outside of comfort, I have a couple of problems with these. They are not in fact headphones, so you can't actually hear what they are picking up while you're wearing them. Secondly, when wearing them you cannot monitor anything else (the live sound, other cans or speakers) as they will interfere with what your ears would normally hear. Finally, recording with this method requires no head movement by the wearer, which is near impossible. (I tried it a few times with a headphone-type rig.)

However, I don't believe their size compromises the sound quality as there are flat-response microphones available that are much smaller. Measurement microphones are often most accurate when the diaphragms are $1/4$ inches in diameter. So audio quality is not the question here. One of the original models of this type of in-ear stereo microphone was the Sennheiser MKE 2002, but this set was different in that it came with a dummy head.

What's the big deal about being able to listen to the sound through the headphones as well as pick it up? Ever do any live mixing where the mix position is relegated to side stage, a corner booth, or in another room? Having a "head" out there in the sweet spot to monitor the sound from does not require that it have built-in speakers. However, being able to wear and monitor the output through them while you hunt for this so called "sweet spot" is invaluable, unless you'd prefer to run back and forth between the pickup and monitor location until the head is positioned "right." Either way, binaural recording is great for reference during later sound evaluation (live mixer's homework).

I once had extended access to a pair of JVC Model HM-200E headphones that were fitted with omnidirectional electret mics. These cans were closed (not open air) and fitted to block most of the outside sound from the wearer's ears. The outside molding included a pinna addition and inside were a pair of fine-sounding headphone speakers. They were battery powered and provided stereo high impedance, unbalanced outputs to recorders, *and* a monitoring connection for use with headphone amps. They came with a Styrofoam head that was coated with a felt-like covering to simulate skin. This head also had a front shaped like a face with a nose, chin, and lips. Since it was designed to use a mic stand mount for recording, it came with multiple microphone stand adapters.

Using this system was easy as pie, was inexpensive, required no additional interface hardware, and provided endless hours of pure audio fun. If they were still available, I'd buy a pair in a second. Maybe if everyone who reads this writes JVC, they'll make them available again for all our enjoyment.

But you must wear cans! Much of the localization properties of binaural is lost when played back over speakers. This is due to the fact that each ear now receives some of the opposite ear's signal. For true binaural, total separation between the two channels is necessary, all the way to the listener's head.

Cross talk is not as detrimental with music programming as it is with speech. Localization is not as bad as when using a single omnidirectional room mic, but the dialog will appear to have been recorded "off-mic." There have been attempts to reinstate channel separation when using loudspeakers by adding electronic circuitry that equalized, delayed, and phase-shifted some of the information from each side and mixed it with the other. It worked pretty well *if* the listener *and* speakers were positioned precisely. This area is still being investigated now by using *digital signal processing* (DSP).

One fact is often overlooked by proponents of binaural and other stereo recording methods that have extended localization properties. While important to classical (orchestra and especially chamber) music and some jazz recording, localization clues are virtually nonexistent in popular music (especially rock). For this reason I find the main problem with binaural recording to be the initial approach. How can one rely solely on a system that can be thrown off by something as little as pinna size? Remember, binaural recording is a one-shot deal. There's no fix in the mix and adding microphones to augment or "ensure" coverage can add phase and other distortions. I know, I know! Stereo recording can make you crazy!

Combining Stereo Micing Methods Microphones must be backed off from the sound source to capture the full frequency content and room ambience, but this positioning can provide a weaker stage signal and a loss of central localization. This is especially apparent with equally spaced stereo micing. Meanwhile, coincident micing often lacks room ambience; therefore, it's common practice to utilize combinations of stereo and mono, or stereo and stereo techniques. Some commonly found combinations include the following:

- Two spaced omnidirectional mics with an additional center, forward-facing cardioid mic.
- A center cardioid mic with two spaced figure eight mics.
- An MS configuration in the center position with a pair of spaced figure eight mics.
- Close and distant stereo spacing for picking up the direct and ambient sounds.

Here if the distant mics are kept within 20 feet of the front, the two will not conflict because the delay will be less than 18 mS.

Things can get mighty academic when folks start talking about adding things like delays to mics that are positioned close in. On top of being impractical, these approaches do not always help when applied in the field. Since additional mics can cause phase problems, their effect should be monitored with a scope in XY mode.

However, when the tracks are available I like to add a separate binaural setup because it provides documentation, which during later comparisons can help to ensure accuracy. They may also accentuate something that the main stereo mic setup ignored.

My Stereo Setup

By now you've probably got a good bead on my approach to this profession, so my "Cover Your Butt" method of stereo micing should come as no surprise. The best way to describe my "take" on this whole stereo micing situation is for me to go through my mic collection and explain how they were used for live recording. This will also give the novice some insight as to how engineers go about building an arsenal of microphones.

As stated, the reason I chose not to purchase any single-unit stereo mic was due to both economics and functionality. As an independent engineer, my budget was limited and since I engineered in many studios it was important that I use the same mics to achieve continuity. Even though many stereo mics with combined diaphragms have better mono compatibility, multiple mics were just more in line with what I was doing. If I had been working in a single facility or venue these combination stereo mics may have been the better way to go.

With that out of the way, let's get right into it. I always liked using the French (ORTF) setup. It's not perfect in terms of mono compatibility, but it reproduces well when played back through near field monitors. The German (DIN) and Dutch (NOS) configurations aim their mics a little tighter and thus come across a bit narrow sounding on the wider spaced main control room monitors. The French method also seems to translate a bit better when monitoring with headphones. A pair of mics can be repositioned into all three of these setups very quickly and easily for comparisons.

The end results are always reliable. The first pair of mics I purchased were often used for this purpose. The Audio Technica 813R unidirectional electret condenser has an even, predictable polar pattern; has a wide, flat-frequency response; handles high input levels; and provides a fairly hot output with low self-noise. I guess that covers it all except that they were also relatively inexpensive and performed exceptionally well on just about every instrument in the studio. Utility was the key.

The next pair had the exact same studio utility except that they were a little less useful on vocals. However, with the pair of optional R104A omnidirectional capsules, the Shure SM81s were my main choice for spaced-pair stereo micing. Furthermore, stories about the durability of this mic rival those of the Crown DC300 amp.

Very early on in my career I got a taste of using a Sony C-37A tube mic. It was actually made by Fostex for Sony back in the early 1950s. This mic's electronics consist of only a power supply and a couple of band-pass filters. In fact, within the mic itself, there's only a 6AU6 tube and three resistors. This translates to the necessity of having a great-sounding diaphragm. This mic was purchased and used for even-order harmonic warmth (distortion). I actually kept three tubes in its case marked "good," "bad," and "ugly." Ready to accommodate acoustic instruments, vocals, and even amps (with the mic backed off by at least 6 feet, of course).

This mic, along with my Neumann U 67 tube mic, comprised an MS pair that could mellow out the brightest of venues. I swear I could get good sound in a high school cafeteria with that pair.

I've had a number of C-37A mics and I have always traded up whenever I found one that was better sounding. My current one was acquired fairly recently (within the last 15 years). It was in pristine/new condition and in the original box. A Midwest radio announcer had purchased it in the 1950s. He used it a couple of times, but because he didn't always remember to power it up and was confused about the use of the "non"-directional screw adjustment on the capsule, he put it in his closet and went back to using his other mics to record his radio spots.

Forty years later it was still in the box, with all its papers, taking up valuable space, and I jumped all over the chance at it regardless of the exorbitant price.

But the story of how I got the U 67 is even better. It was originally purchased form Gotham Audio in 1962 by the Julliard School of Music in NYC and hung in one of their recital halls. Twenty-five years later, the new owners decided to change the mic to a newer Neumann FET stereo mic as part of a system-wide renovation. Considering that the distance from the mic location to the "control room" was over 500 feet, this was a good idea. I was hired by the equipment supplier to handle the installation. I was the individual who stood tiptoe on a ladder atop a 1-foot-wide catwalk, 35 feet above the tenth row, to take down the mic. They still had the original box and papers and, with the cost of a pile of money and a bunch of mics, as the song goes, she was mine! Since then I've compared it to every other U 67 I've come across, including two that were newly manufactured from original parts by Sennheiser after they took over Neumann. No question, it was always better sounding (or else I would have traded up). It now has every mount Neumann made for it, two spare Neumann (Amperex EF 86/6267) tubes, a spare capsule that could use resputtering and extra elastics for the suspension system, all housed in its very own flight case. Once you've done a little audio repair work, you want to have every spare possible.

During the mid-1980s Fostex manufactured a line of "printed ribbon" microphones. Their Mylar diaphragms were coated with conductive material much like most other condenser mics, but Fostex had etched out some of the conductive material, leaving a coil shape behind. When moved by sound, this coil/diaphragm changed position in relation to a magnet. Remember, the operation of the diaphragm in a ribbon mic is as if it were a single coil. This combination of diaphragm and coil increases sensitivity. The Fostex printed ribbon microphones were *not* as sensitive as single-coil ribbon mics, but they were more robust.

It's kind of like the difference between titanium and aluminum. The rap is that titanium is stronger than steel and lighter than aluminum. The fact is that titanium is lighter than steel and stronger than aluminum. That's a huge difference. Fostex was very up front about the design details and capabilities of their printed ribbon mics.

They were certainly more sensitive than the dynamics available at that time, and more robust than the ribbon mics available at that time as well. But they also *sounded* great. I could put my M11 RP cardioid-pattern printed ribbon mic in front of any horn and just hit record. It is the best and most natural-sounding horn mic I've yet to come across. Strings, keys, percussion, you name it, the sound is right there.

My M88 PR bidirectional printed ribbon mic is as wide open and clear sounding as they come. It's great on vocals but a complete knockout when recording dialog. This mic was designed with dialog in mind, and it includes a three-position (flat, −3 dB and −6 dB per octave) bass roll-off adjustment switch to offset its proximity effect. The sensitivity is such that I'll often give a screaming vocalist a handheld mic, but record from the M88 backed off about 6 feet. From this position the M88 would still get it all. Close in for speech recording it doesn't distort or break up and, like the M11, it just sounds natural.

Oh yeah, Fostex did make an MS printed ribbon mic, the M22 RP. I could have saved some money by purchasing the MS mic, but I just didn't need an MS mic as much as I needed a pair of all-around studio workhorses. Plus I could just combine them and they'd do the same job.

In the mid-1980s while I was testing and reviewing professional audio equipment, an editor asks me to check out a new mic that was made in Sweden by a company called Milab. I had already done a review of this company's variable pattern DC-63 studio mic that was a complete knockout, but this was a handheld vocal mic. However, the LC-25 turned out to be much more than just that. It was a cardioid-pattern condenser that required phantom powering to drive its dual FET amp. I fired it up and noticed that it had a black chrome finish and exquisite machine work, and that it was a demo sample, was etched in gold right on the mic. Within one sentence I also realized that it was one of the most transparent mics I had ever heard. It also followed the voice perfectly. If you ever get a chance to see films of some of the great old jazz singers, watch how they move the mic in close to accentuate low notes (making use of the proximity effect), and away to smooth out notes that must be sung louder in order for them to "make" a certain pitch. A mic that can do this well is said to be good at "following" the voice.

The LC-25 is exceptional at voice following. I discussed this property with the folks at Milab and they thankfully did not attribute it to the fact that the circuitry was transformer-less. It was the result of each individual microphone's amplifier being adjusted via parts changed out to match its diaphragm.

The fact is they would not sell a spare diaphragm or a preamp separately. You *had* to buy a matched set. At the time this mic's price was too rich for my blood. I sadly waved goodbye but never forgot it. Many years later, while cruising around a pro audio auction, I saw it! I knew it was the same mic because of the "demo" etching.

Since everyone else was ignoring that lowly "live" handheld vocal mic (while eyeballing big ticket starship-bridge-like effects devices), I was able to get it for half its original retail price. The first thing I did was order a spare diaphragm/preamp combo from Sweden. This mic is so special, my will states that it goes to the jazz singer Nancy Wilson. She'll know what it's all about.

In the late 1980s I was trying to explain to a studio owner/operator that his digital tracks sounded "extra bright and harsh" when he used his (original silver) AKG C 414 EB because it didn't roll off above 16 kHz like his dynamic mic did. He would have none of it, because some-one told him AKG 414s were all "top heavy" and "extra bright." He was going to sell that $1,250 mic for $1,000 so he could buy a "vintage" reribboned mic from this individual. I had done a bunch of repair and engineering work at this studio and he had owed me about a grand for several months. I tried once more to explain that the 414 was the better mic, with four polar patterns, flat frequency response, great sensitivity, and lots of headroom. He didn't want to know, so I walked with a great mic in perfect condition at a bargain price. The silver AKG C 414 EB and black chrome Milab LC-25 mics make up my crystal-clear MS combination.

The last mic I'll go over didn't get a lot of use in stereo recording, but since this whole run-down is also meant to show how engineers go about acquiring mics, its story has something to add. During the 1990s I was the main maintenance tech at a multiroom broadcast facility and my boss was the VP of engineering. This place had 10 production suites up and running, and they had decided to add another. I was told they would need to purchase another Electro Voice RE 20 Dynamic cardioid mic. I decided to try an experiment and told the VP I would get them one. A pro audio equipment dealer I had done a lot of work for had recently closed shop. Three of its salesmen had moved on to become the heads of pro audio sales at different local audio retailers. Since "local" here means NYC, these were big establishments with lots of inventory. I spent the better part of a Saturday going through the broadcast facility's 12 RE 20s to choose the one I thought sounded best. Then it was off to visit three old friends. At my first stop, they had six RE20s in stock and the one I had with me sounded superior to them all. The next outfit only had four, but one of them was a perfect match to the one I had brought with me. I was tempted, but got myself out of there quick before I strapped myself with a $1,200 expenditure ($450 for that mic, $450 to replace the one I had with me, and $300 for two yoke shock mounts). Microphones can make even the most stable among us go loopy, so for a dedicated mic aficionado, this was very dangerous territory. I made it to the last stop without spending a dime and was pretty proud of myself even though I was thinking of the impossi-bility of finding another matched pair.

This place had 22 RE 20s in stock, so I returned the next day. After four hours of side-by-side testing I found no other matched pairs; however, one of the mics they had was head and shoulders better sounding than that "best in the facility" mic. I walked out proudly, carrying both winners of this "best of 44" contest. Does this mean that RE 20s are not uniform sound-ing? Of course not!

Hey, when I did a vocal session in a top, world-class recording studio, I'd have them bring out six or eight of their U 87s to try them out as one will always sound at least a little better in some situations than the others.

Meanwhile, management at the broadcast facility decided not to go ahead with adding another production room. What was I to do? Return the mic after putting a day and a half, and every connection I could muster, into finding it? How you acquire mics is often unpredictable. But back to my stereo mic setup.

A pair of cardioid coincident mics, a spaced omni pair, and three MS combinations should be more than enough for anyone. But not me!

First I tried adding a front-facing hypercardioid/shotgun mic, figuring this would allow me to aim an extra-narrow MS pattern across the stage and pick up individual instruments that were low in level. *Not a chance!*

While the pickup of better shotgun mics can be down by 25 dB at 90 degrees, it will be down only 5 dB at 30 degrees. Many are only down 15 dB at 90 degrees. This means that from 12 to 15 feet out, by the time your "narrow" pattern reaches the second violin section, your aiming at it will not only be joined by the violas beside them but the trumpets behind them, as well.

I would also add a rented binaural head when the budget would allow it. What was I going to do with all these microphone outputs? First, because I was recording to an eight-track deck, limit them to eight. This meant that if the binaural was used, I only had six tracks left. But it also eliminated the need for the ORTF near-coincident pair. The forward- and rear-facing MS sets took up four tracks, leaving two tracks open for the omnidirectional spaced pair.

I would choose the MS mics during rehearsal. While this would only give me a slight indication of what the hall would sound like fully populated, I could nail down the forward-facing MS selection (bright stage, tube mics, dull sounding, semiconductor-powered condensers, and medium or naturally good sounding printed ribbons). I have been told that rear-facing MS setups are unnecessary, but remember, this is my *cover your butt* (CYB) stereo micing technique.

True story, small orchestra, 1,700-seat hall, bad weather, and a lot of no shows. Everybody upgrades their seating by moving forward, except for a group of 200 or so friends, family, and members of the same local lodge who stuck together in the right rear section. The hall's reverberation time was more than slightly affected by the gaping hole in the population of the left-side. Yet later in post-production, I was able to electronically re-aim the rear facing MS setup to a point where the reverb didn't splash about seemingly out of control. But covered!

Yet I still wasn't through as I had to spend a whole bunch more to facilitating my "big plan." You see, I figured that my best bet was to work directly with conductors. These folks act as the producers, talent co-coordinators, and arrangers all rolled into one. I'd record the event and then let them have the final say on the instrumentation blend, depth of the stereo spread, and the amount of audience ambience. To achieve this, I designed and built my own recording and mixing board. Its eight mic preamps consisted of a dual ATI unit, four of the final MCI 600 series mic pre-upgrades, the ones that utilized tiny Neutrik metal, encapsulated transformers, and a pair of cost-effective MXR tube mic preamps that could be overdriven for just the right amount of added even-order harmonic content whenever that type of warmth (distortion) was required.

The faders were all Penny & Giles, of course, and the metering you already know about, but it follows. A stereo (dual-needle) VU, a pair of 29-segment LED Peak/VU meters, and a single PPM meter. Three MS decoders were also built in. One I made using precision Penny & Giles (real stereo) pan pots. If you have any difficulty clocking this move, go back to the description on using a mixing console's phase switching, pan pots, and faders to size and aim MS configurations. The second was composed of switches with fixed resistors. This was modeled after MS modules as found in old Neumann consoles, but my resistor values (5 percent), were not as well matched as theirs (1 percent). The last matrix was a store-bought Audio Engineering Associates MS 38. While it only has control of width (no directional adjustment), it does have something that is unique for an active MS matrix: It's a two-way street, MS in/stereo out *and* stereo in/MS out. Or should I say binaural or ORTF in, MS out, which was then put through another MS-to-stereo matrix to adjust its aim and spread just to squeeze every ounce of information I could get out of these mics.

Finally, 10 Jensen JE-11P-9 line in (+26 dBu max) 1:1 ratio isolation transformers allowed me to do anything I wanted, phase- and level-wise, to a signal without affecting the original feed. The end result was twofold:

1. It actually worked. Oh, I couldn't "pull up" individual instrument levels, but I could bring-up a whole section to some degree and easily blend the sides with the stage and room sounds. I could make up for some acoustic problems like uneven reverb decay (aim the pickup pattern to where the decay was more even), dead-on match the exact stereo field of the stage, warm up a bright hall, and to some degree brighten up a dead one.

2. Nobody cared! That's right. It was as if I were peddling baked potatoes on a hot August day. Well, I'm exaggerating, but the amount of work I "picked up" was minuscule compared to my close but not quite $25,000 total outlay.

There's a lesson here, but I'll save it for the end of the discussion on stereo micing techniques.

With the onset of ill health I had to part-out most of that mixer. The Jenson transformers and the mic preamps went first. I still have most of the faders, the AEA, home-made Penny & Giles MS matrixes, and all the mics, but in retrospect I would have saved a lot of money buying a Soundfield "stereosonic surround" mic.

The Soundfield Surround and Stereosonic Microphone System   I'll start this coverage by stating that true three-dimensional surround sound requires, at a minimum, a four-speaker playback system. But that doesn't mean that you can't use this system for stereo recording. In fact, Soundfield offered a less expensive stereo-only SPS 422 studio stereosonic version, which was followed by the ST 250 stereo version that is capable of battery operation. Like other all-inclusive stereo micing systems, this single stereo mic is more convenient to use and easier to set up.

However, even the stereo versions of the Soundfield mic go far beyond what is considered normal stereo pickup. Here, the mic's controller provides the operator with two (left and right mic) outputs, whose polar patterns can be continuously adjusted from omnidirectional through hypercardioid to figure eight. To understand how this is accomplished, it's necessary to take a closer look at the microphone itself.

The first Soundfield mic I dealt with was the CM 4050, and like all that followed, the pickup patterns are based on its *four* closely spaced diaphragms. The diaphragms are matched and this, along with their close spacing, results in less time-delay (phase) and frequency-response errors. The electronics within the control unit are also matched to each microphone. (They both have the same serial number.)

The controller takes the mic's four separate outputs and processes them into three figure-eight-like polar patterns that cover the up/down, the front/back, and the left/right fields. The control unit adjustments actually read "azimuth" (or rotation) of the front- and rear-facing horizontal planes, "elevation" (head tilt) for vertical height changes, and "dominance" for forward-to-rear (zoom-in) control. On the other hand, this ability to turn, twist, tilt, and move the mic electronically does not preclude good commonsense mic placement.

The "height" information improves the three-dimensional quality of ambient and directional localization when using normal two-dimensional horizontal stereo speaker setups. The importance of vertical information may seem far fetched, but I assure you it is not. It is worth your time to investigate the *listening environment diagnostic recording* (LEDR) monitor tests that came out of the Psychoacoustics Department of Northwestern University.

The sounds this recording produces not only travel up and down, but also across and *over* your head when listening to a properly aligned stereo monitor system. It's enlightening to say the least.

The Soundfield mic's stereo spread can be monitored on standard stereo speakers from any listening position. It seems like people need to complain about something, so they would point out the fact that there was no "onboard" metering for checking mono compatibility, stating that, "At $15,000, I expect to at least get a phase meter."

The newest version, the Mark V, allows the operator to set the controller's pattern to omni and the angle to 0 degrees (both are located in the controller's headphone section), and monitor the mono signal. Its "azimuth" control is adjustable through a full 360 degrees and the four-band metering is switchable to show either the outputs of the four capsules, or the stereo MS *and* XY levels.

The SP 451 Surround Processor unit, along with the software-driven "Surround Zone" (and other plug-in options) make the Soundfield Mark V capable of not only 5.1 (four "corner" channels, a front center channel, and subbass), 6.1 (an added rear center-channel output), and 7.1 (front plus left- and right-center channels), but also a preset for eight channels (front left, center and right, right-side center, rear left, center and right, and left-side center), but no subbass output. I guess that would be 8.1! Hey, how about a nine-speaker surround system (eight-way satellite and a subwoofer)! Every home should have one. Like I said, folks just naturally get carried away to extremes with this whole stereo/three-dimensional/surround sound thing. So watch out!

People still complained about the Soundfield's price, and the fact that to really get the most out of this system its preprocessed outputs should be recorded to four tracks!

In Figure 3-33 the Soundfield stereophonic microphone system's frequency response graphs and polar patterns are representative of the SPS 422 Studio Stereo System (top) and the Mark V full surround system.

The frequency response graphs show the response to be flat from below 50 Hz to above 20 kHz (re: 1 kHz). I specifically included both because, while they at first look very different, they are in fact just about identical. The surround version is flat to 35 Hz, while the stereo version is flat down to 40 Hz. The polar patterns illustrate the fact that the stereo version's two "mic" outputs can range from straight cardioid, through figure eight, all the way to omnidirectional, and the surround version can deliver everything from a mono to a full four separate-output feed.

The Crown Stereo Ambient Sampling System (SASS)   By using dual boundary microphones, this micing method provides phase coherence and total mono compatibility with good imaging and stereo localization, even when monitored through stereo speakers. The two boundaries are spaced a head width apart (17 cm as in ORTF) and aimed similar to near-coincident mics.

Figure 3-33
The polar
patterns and
frequency
response
graphs of the
SPS 422 Studio
Stereosonic
System (top)
and the Mark V
full surround
system
(bottom)

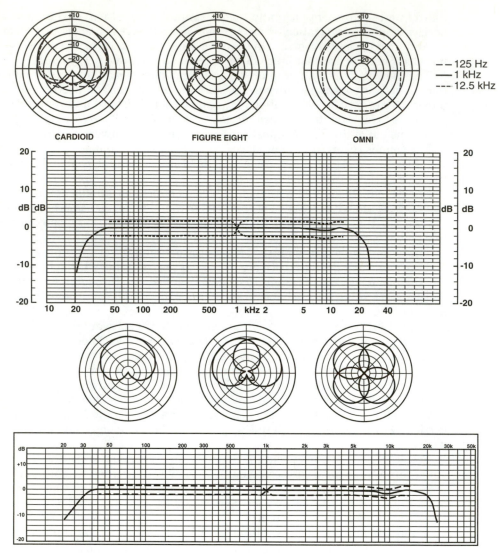

This system also utilizes a baffle to deflect the high frequencies. It not only lessens the amount of cross talk, but increases the level differences of the high frequencies between the two channels. It provides an even low end because boundary mics are closer to omnidirectional. Its use also eliminates the possibility of any proximity effect or any off-axis coloration.

Since the low-frequency content is picked up omnidirectionally, it is equal in both phase and level at the two microphone outputs. The pickup angle is set at 125 degrees to increase the amount of ambient room signal. These facts negate all restrictions as far as placement within specified distances.

In order to understand the significance of this system's approach, some facts about human hearing and localization response must be reiterated:

■ With human hearing, sounds that are lower than 100 Hz are not localized.

■ We use phase or time arrival differences to localize sounds between 100 and 700 Hz.

■ Humans depend on a combination of phase and level differences to provide localization clues in the high midrange area of 1.25 to 2 kHz.

■ For high frequencies above 2 kHz our localization clues are mainly from level differences.

Now for the SASS's method of providing localization clues:

■ At frequencies under 500 Hz, the direction clues can only rely on delay between each channel since the levels in this frequency range are roughly the same.

■ From 500 Hz to 2 kHz, the directional clues are provided by timing differences between the two channels.

■ From 2 to 4 kHz, directional clues are from both phase and level differences between the two channels.

■ Above 4 kHz, localization clues are provided by the difference in level between the two sides.

Additionally, by giving the two boundaries the same spacing as our two ears, the stereo imaging ends up being more natural-sounding. This makes it is easier to later focus in on individual sounds, and because the two half-hemisphere pickup patterns are aimed at 125 degrees, this method does not result in a stereo field with "a hole in the middle."

Due to a barrier between the mics, which helps prevent cross talk, the SASS outputs are completely mono compatible, meaning that the stereo and mono frequency responses are the same.

So the SASS represents a significant design achievement in that its response parameters are perfectly in tune with the characteristics of human hearing response. *However,* even though the SASS does in fact have "good imaging, superior low-frequency response, less off-axis coloration, and good mono compatibility," I've heard some folks downplay this stereo micing method as "sounding dry and lacking any character."

I remember the same things were said about pressure zone microphones (PZMs) when they were initially made available to a wide audience. I have always been under the impression that a good microphone is not supposed to have any character. So as far as I was concerned, the PZM represented a major technological breakthrough in microphone design. The SASS is a continuation of this logical design and it works because it is based on scientific facts.

It can be battery operated (9-volt) and has built-in, low-frequency roll-off switching (one setting for when phantom powered and another when battery powered). The SASS comes with a windscreen, a swivel mount that can be set to any angle, and adapters for use with European (³⁄₈ –16) mic stands and photography stands (¹⁄₄ –20). It's no accident that the purchasing price also includes a hand grip for *electronic news gathering* (ENG), sports, and other "field" recording work. Figure 3-34 shows its frequency response to be wide and even, while the polar patterns show the two boundary mic's outputs to be very symmetrical.

Figure 3-34
The Crown
Sass-P mic: The
polar patterns
of the two
sides are about
as symmetrical
as it gets. The
frequency
response
graphs show
the response to
be wide and
even with a 12
dB per octave
high-pass filter.

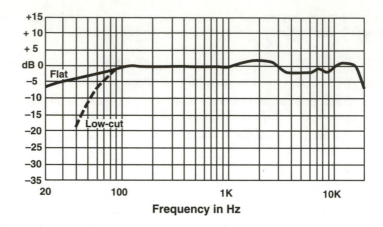

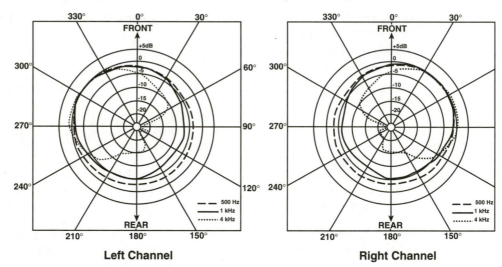

Left Channel

Right Channel

If a recordist is working in a single location, it makes sense to take the time to experiment with all of the stereo micing techniques within that environment using the mics that are available.

Novice as well as professional audio recordists also owe it to themselves to check out some of the current "ready rolled" stereo mic packages. Lately there has been a marked improvement in these mics due to large investments in research and design budgets initiated by television's embracement of stereo audio. I would like to say it had something to do with the audio music recording as well, but more sales of stereo mics are made to video, sampling, sound effects, "gaming," and field recordists. If you don't believe this, check with the mail-order houses.

You may get better pricing from a video distributor than from an audio house simply because their turnover of these items will be far greater. Although most of the newer stereo microphones were designed for on-location use, many (as with the Soundfield and SASS) certainly have a place in studio recording. So once again, the audio music recordist has benefited from research expenditures brought about by other areas related to our field.

The Shure VP88 Stereo Microphone System   The VP88 consists of a pair of diaphragms (placed in an MS pickup pattern) and its matrix housed in a single microphone package designed to be used for location recording, ENG, and studio recording. Since it has single-point (coincident) pickup, its phase response is more time-coherent than would be provided by using two separate microphones. The use of cardioid (front-facing) and bidirectional (side-facing) pickups in an MS configuration within a single mic is nothing new, nor is its built-in matrix network, or even its onboard adjustable stereo spread. But the fact that this is a handheld system that's rugged enough for field news gathering opens up its use to many applications.

For example, while producing a live broadcast of a sporting event, an announcer is commonly sent out onto the field of play with a cameraman to conduct interviews. The mic used will be for the interviewer and the interviewee. The person handling the audio mixing duties would normally have to unmute the interviewer's mic, bring up its level, and mix in some of the background ambience (crowd noise) from another mic.

This works fine if the interviewer's mic is directional and, as long as the crowd noise from the ambient mic matches what's going on in the picture, from the interviewer's location. But how about a more exemplary solution?

The Shure VP88 is a coincident (single-point) MS stereo microphone with a front-facing cardioid that operates just as a handheld vocal mic would. A second figure eight capsule is mounted perpendicular to the cardioid and faces the sides. For all intents and purposes it can be used exactly like a handheld vocal mic in that it is simply positioned close to, and aimed at, the sound source. Therefore, there is no learning curve for the talent. It has accentuated pickup at the front with ambience at the sides. What could be more convenient?

How about the ability to output the two signals independently, or as a fully matrixed stereo pair. The VP88's onboard stereo matrix electronics have a three-position switch that controls the stereo "spread," or width. It can be battery powered (6-volt) or phantom powered, has a built-in switch for high-pass filtering, and comes with all the necessary stand and cabling accessories. It can use a regular SM58-type mic clip, but since it's a bit longer (around 11 inches overall), when positioned on a camera, a right-angle mount with double clips (much like those used for rifle scopes) is used. Both mounts are included in the accessory package.

The studio recordist can position this mic, feed its dual outputs to the mixer, and adjust the front- to side-levels to change the width and the left-to-right levels for aiming, and with its frequency response of 40 Hz to 20 kHz, there is little to no compromise involved. It's also rugged enough for road use and 100 percent mono compatible.

This mic's sensitivity specifications are different for each element with the side being 1.6 dB higher, so it should come as no surprise that the output levels would also differ and affect the clipping and noise levels. The output level of the side element (in relation to the forward-facing element) is also dependent on the stereo width selected. At "low," or the narrowest setting, the side pickup is 6 dB lower than the front. The side output at the midway stereo spread setting is lower than the mid by 1.9 dB. At the widest stereo spread setting ("high"), the side output level is greater than the mid by 1.6 dB. This means that the side is being boosted by almost 4 and 8 decibels to achieve a wider stereo field. This information is not all that earth-shattering on its own, but there are additional clues to this mic's performance (see Figure 3-35).

The separate side- and mid- (forward-facing) frequency response graphs show the mid-element to have a slight 2 dB dip at 5 kHz, while the side frequency response has a 2 dB bump

Figure 3-35
The Shure
VP88
handheld
stereo MS
microphone
system

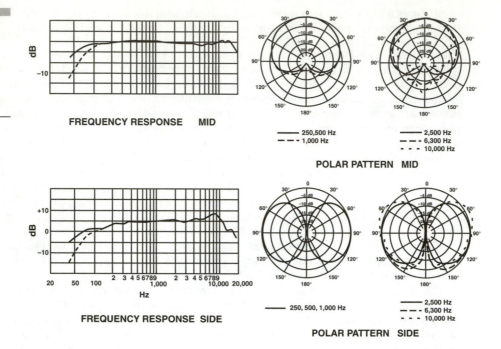

FREQUENCY RESPONSE   MID

—— 250,500 Hz
- - - 1,000 Hz

—— 2,500 Hz
— — 6,300 Hz
- - - 10,000 Hz

POLAR PATTERN   MID

FREQUENCY RESPONSE SIDE

—— 250, 500, 1,000 Hz

—— 2,500 Hz
— — 6,300 Hz
- - - 10,000 Hz

POLAR PATTERN   SIDE

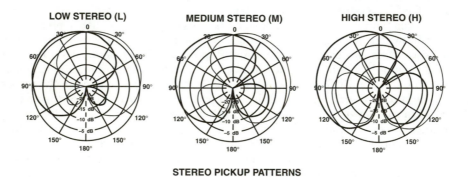

LOW STEREO (L)   MEDIUM STEREO (M)   HIGH STEREO (H)

STEREO PICKUP PATTERNS

at 6 kHz. However, at 7.5 kHz the side's response is up by almost 8 dB. There is an additional clue from the polar patterns. Here, the mid's output is shown to have rear lobe bumps at 6.3 and 10 kHz. The side's 10 kHz response is enlarged by 3 dB at 60 degrees, and its rear pickup at 6.3 kHz is increased by almost the same amount.

It all really comes together when you view the stereo polar patterns. As the stereo width is increased, the side lobes extend beyond the 0 dB point of the polar pattern chart to almost +5 dB at the 60- and 300-degree points. This holds true to some extent in the MS mode, as here the sides extend past the 0 dB point at 90 degrees by about +2 dB.

So what? All that information could have been had from knowing any single one of those facts. Not necessarily. I now know that when the stereo spread is increased, there is a high-frequency boost at the sides. This boost will be in the area of 7 to 10 kHz and will occur at 60 and 300 degrees. According to the polar patterns, this increase in the high-frequency content will be by as much as 3 dB.

I have never worked with the Shure VP88, nor do I know anyone who has. But I bet that when using the VP88 as a drum overhead mic, I would be able to position it down closer to the toms so as to increase their pickup content while at the same time diminishing anomalies caused by ceiling reflections. Lowering overheads down closer to the toms would normally negate some of the cymbal pickup, but the VP88 side reception's high-frequency boost should compensate for this loss.

I also believe that it would work out well in a live acoustic music setting, where all the instruments are positioned to the sides of the lead vocalist. Narrow down the spread, back off the positioning, and aim it at the vocalist. Nothing beats well-done minimalist recording techniques. *It's all in the specs.*

The Sanken Model CSS-5 Stereo Shotgun Microphone The Japanese company Sanken helped make on-location ENG and film recording a bit easier with the introduction of their CSS-5 stereo shotgun microphone. With its pistol grip, it looks a lot like a Ruger Mark II semi-automatic and its balance is also about the same. Easy to handle and point, but in this day and age you better keep it in its Rycote blimp-shaped windscreen and wear readily seen identification when using it outdoors. The CSS-5 can also be stand-mounted.

So what's the big deal? Well, the fact that the forward-facing element is a true shotgun is operationally important. (It's rejection is −10 dB at the side and −20 dB at the rear.) How this directional shotgun polar response is achieved with a mic that's less than 12 inches long is also of interest. Sanken used multiple elements like other shotgun mics, but this mic is composed of five directional condensers in a front- and rear-facing array. The mic's pickup is equal in the horizontal and vertical planes, but with the addition of some EQ the sound picked up by the front-facing element is brought forward to accentuate the sounds from the center.

Remember, this mic was designed for film and television location work where (like other shotgun mics) it is pointed at a primary subject, but here the recordist also gets the benefit of a realistic pickup of the surrounding ambience, if wanted. "If" because this mic also has switchable modes of operation. In "mono" mode, the CSS-5 operates as a straight shotgun mic. It is important to note that it operates as a true shotgun and that this function is not compromised because of its other capabilities. In "normal" mode, the CSS-5 puts the primary subject in the forefront while adding a blend of the surrounding stereo field around and slightly behind it, yet it retains excellent localization. "Wide" mode provides a pure stereo output. The pickup angle at this setting is a full 140 degrees, so the designation "wide" is no exaggeration, yet due to the close proximity of the CSS-5's diaphragms, this output is still highly mono compatible.

This design configuration also alleviates the normal extended proximity effect normally found in shotgun use. This makes the response more even when distances are varied. The specs tell the story: high sensitivity (−30 dB per pascal), a maximum input of 120 dB (1 percent THD @ 1 kHz), and a noise level that's below 18 dB.

This mic may have been designed for location use, but it's quite enough to be used in any studio, sensitive enough to pick up a light brushstroke across a ride cymbal's bell, and I wouldn't be afraid to find out what those front- and rear-facing capsules did to the sound when placed *inside* a kick drum, or a folded-horn bass speaker, and when placed underneath a driver connected to a guitar amp. From my experience, these could be ideal studio applications to add to the CSS-5's normal functions. It would be a real treat for attendees at trade shows if Sanken had one set up with a pair of cans for monitoring.

The frequency response graph in Figure 3-36 shows this mic to have an even response between its −3 dB roll-off points of 100 Hz and 16 kHz.

Because of the nature of this mic (a shotgun), the off-axis rejection per frequency is given for a 90-degree incident as well as 180 degrees. The polar pattern, in addition to showing the pickup to be very even, points out the fact that low-frequency pickup at the rear is not as accentuated (being only a small lobe) as is common to all directional mics since low frequencies are omnidirectional.

The Josephson Stereo-Q Microphone   Even though describing this stereo micing system is going to be a little difficult, I just had to include it. Imagine a microphone body with two flexible donut-shaped rings on top of it. Picture those rings as having the ability to be rotated to any position in relation to each other. Now imagine that you could slide microphone capsules into those rings. The result would be aimable XY and near-coincident configurations. Mounting two out-of-phase capsules in a single donut would result in a figure eight pattern, making MS and Blumlien techniques possible.

Now add the choice of omnidirectional as well as hypercardioid to the list. Pipedream? Wishful thinking? No. In fact, this microphone has *four* separate outputs just in case the recordist wants to record a full four-capsule pickup pattern (surround) or feed them to this system's matrix unit for a stereo output. You see, the Stereo-Q comes with four matched-cardioid capsules and an XY/MS matrix network that allows the operator to record stereo in either MS or XY configurations, all four of the separate outputs for stereosonic recording, or for stereo processing later during post-production. Setting the capsule's rotation angle is aided by detents at 60, 90, and 120 degrees and the "donuts" they sit in are ringed with foam. In addition to providing a little added shock absorption, this defuses the sound around the capsules so that their bodies do not cause the sound to diffract onto another diaphragm, causing a change in its frequency response and polar pattern characteristics.

This is all very versatile, but it wouldn't rate much more than an aside comment except for the quality involved. This company's products are designed for "critical" recording. Their condenser microphones are manufactured by MB Electronics in West Germany, which has over 40 years' experience in making microphones to the exacting standards required by German broadcasting and classical music halls. I've included the frequency response graphs and specifications for their KA-10 and KA-10 D omnidirectional capsules along with their KA-20 and KA-40 cardioid capsules to illustrate their propensity for technological detail.

The KA-10 D has about a +6 dB (re: 0 dB @ 1 kHz) presence boost centered at 6 kHz. The output specifications for these capsules vary by $^2/_{10}$ of a millivolt. This makes the output of the KA-10 D 2 dB hotter than the KA-10, while it also decreases its noise level by 2 dB (15 to 13 dB) A-weighted. Although the two cardioid capsules have close to identical frequency responses, their other specifications vary a bit more. The KA-40's output is a one-half of a

Figure 3-36
The Sanken Model CSS-5 stereo shotgun mic. The CSS-5 frequency response graph shows an even response between the steep roll-offs 3 dB/octave below 200 Hz and above 16 Hz. Since this is a very directional mic, the rejection response is given at 90 degrees as well as 180 degrees. The polar pattern is very even with a rear low-frequency lope that's common to directional mics, as low frequencies are nondirectional.

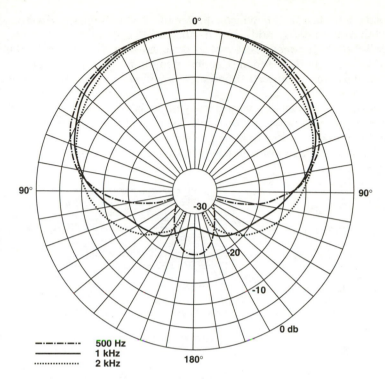

Polar Pattern Mono Mode

500 Hz
1 kHz
2 kHz

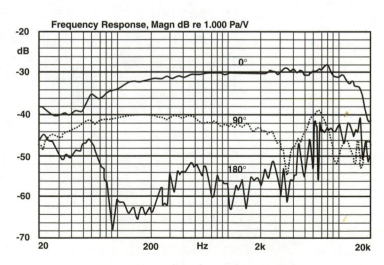

Frequency Response Mono Mode

millivolt hotter than the KA-20, which equates to +5 dB. Thus, the KA-40 has a 5 dB lower noise level and can handle an additional decibel of SPL.

The real difference between these two cardioid capsules is in their off-axis rejection, with the KA-20 providing 10 dB of added rejection of midband frequencies. I can't remember having ever seen these kind of minute differences provided by any other manufacturer (see Figure 3-37A).

While these capsule specifications are fairly impressive, one may wonder if details this precise are necessary. Maybe not, but it certainly boosts confidence in their quality. Furthermore, this penchant for detail resulted in polar pattern graphs of the stereo-Q outputs that ended up being perfect visual examples of stereo micing techniques. I'm impressed. (See Figure 3-37B.)

Figure 3-37A
The Josephson KA Series condenser microphone capsules. Note the even polar patterns, the flatness of the frequency responses, and the detail of the specifications.

Omnidirectional

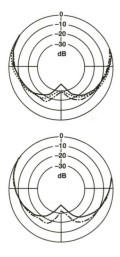

Cardioid

.... **300 Hz**

———— **1 kHz**

–·–·–·– **5 kHz**

– – – – **10 kHz**

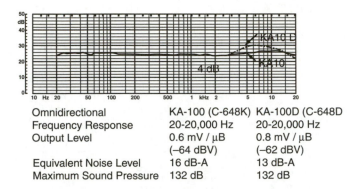

Omnidirectional	KA-100 (C-648K)	KA-100D (C-648D
Frequency Response	20-20,000 Hz	20-20,000 Hz
Output Level	0.6 mV / µB (−64 dBV)	0.8 mV / µB (−62 dBV)
Equivalent Noise Level	16 dB-A	13 dB-A
Maximum Sound Pressure	132 dB	132 dB

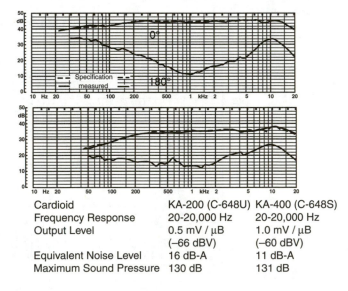

Cardioid	KA-200 (C-648U)	KA-400 (C-648S)
Frequency Response	20-20,000 Hz	20-20,000 Hz
Output Level	0.5 mV / µB (−66 dBV)	1.0 mV / µB (−60 dBV)
Equivalent Noise Level	16 dB-A	11 dB-A
Maximum Sound Pressure	130 dB	131 dB

Figure 3-37B
The Josephson
stereo Q
system capsule,
matrix output,
and polar
pattern graphs
illustrating XY
and MS stereo
micing
techniques

XY with two cardioid capsules

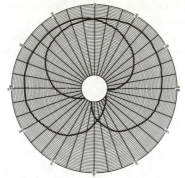

Direct and matrix output

MS with three cardioid capsules

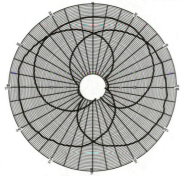

Capsule outputs

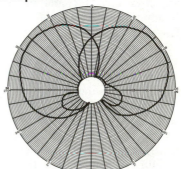

Matrix output

MS with two cardioid capsules and an omnidirectional midcapsule

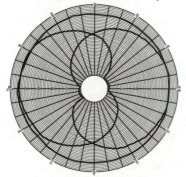

Capsule outputs

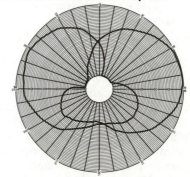

Matrix output

The Royer/Speiden SF-12 Microphone   I included this mic mainly because of the fact that it is a ribbon mic. However, its capabilities go beyond what would normally be expected from microphones operating on the common ribbon velocity principle. Remember that when a sound wave impinges on a thin metal strip suspended in a magnetic field, its velocity causes this metallic ribbon to move some distance, at some speed, depending on the ribbon's weight (inertia). This movement causes the generation of voltage. Because ribbons do not need a back plate, they are naturally bidirectional. The lightness of the ribbon material not only translates to an excellent transient response, but can also mean a disproportionate low-frequency output and possible susceptibility to damage from high SPL. One of the reasons I'm such a fan of the Fostex printed ribbon mics is the fact that they can handle high SPLs. With that said, we can move on to a discussion of the SF-12 from Royer Labs.

This is a dual-element stereo mic that has two bidirectional ribbon capsules set at 90 degrees to each other. It has no ready means of angle adjustment, but this is not unusual as this positioning is required for both of the stereo patterns (Blumlien and MS) that utilize bidirectional microphones. The SF-12 can fulfill these requirements with the use of a single microphone. This is no big deal except that the two elements are identical and thus provide two figure eight patterns whose outputs are proportional to each other. Now consider the fact that ribbon mics have the same impedance at the front and rear of the diaphragm so that their bidirectional response is uniform and results in four lobes of equal size and shape.

The frequency response has -3 dB down points at 30 Hz and 15 kHz. This seems like no great shakes, until you eyeball the chart in Figure 3-38 and see that this response is just about flat from 40 Hz to 14 kHz.

A sensitivity rating of -52 dBV (re: 1 volt per pascal) is nothing to write home about, that is, until you consider the dynamic range this mic is capable of. This ribbon mic's maximum input level rating is 130 dB. Royer is a small company, so I forgive them the omission of self-noise and spark-gap transient response specifications. But I have heard the Royer ribbon mic, so I know it has the smooth response, accurate sound quality, and the sensitivity expected from high-quality ribbon velocity operation. I also know that a 1.5- to 2-microns-thick ribbon (less than $^1/_{10,000}$ of an inch) that weighs one-third of a milligram makes for very low inertia. I'm also fully aware of the additional flux density gained from the use of neodymium, rare-earth-ceramic magnetic materials.

What I cannot understand is why Royer Labs does not make these facts evident in their literature so that those without my experience would be able to understand it, too. Hey, all this *and* a maximum SPL of 130 dB makes this a very big deal to anyone who understands the underlying technology involved.

However, it is fairly easy for everyone to understand that these ribbon mics are robust as Royer's warranty states, "Lifetime to original owner (repair or replace at Royer's option)."

Conclusion   This quick review of four stereo mics is obviously not intended to be all inclusive. Hey, Sanken and Josephson offer other equally impressive stereo mics, which I encourage you to check out. But hopefully you now have a better idea of what to look for in an all-inclusive stereo microphone. Taking the time to check out the various capabilities of many stereo mics (like any other audio tool) puts the recordist in a position to make more accurate judgments, and thus better purchasing decisions. There's tons of info out there since all of the major microphone manufacturers have stereo versions in their line. So check out AKG, Beyer Dynamic, Sennheiser-Neumann, Audio Technica, and Sony, as well as smaller outfits like Schoeps, Microtech Geffell, Milab, Pearl, Rode, and Brauner. Then contact local businesses that rent audio equipment and note what they carry. You never know when you'll have the

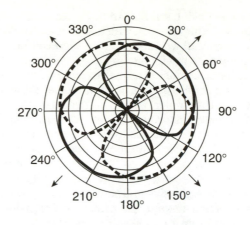

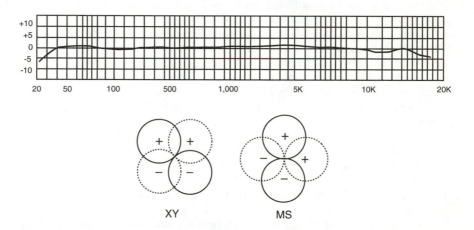

Figure 3-38

The Royer/Speiden SF-12 stereo ribbon microphone. The SF-12 has the typical full, flat frequency response expected of high-quality ribbon mics. The XY and MS illustrations demonstrate the fact that when you rotate a dual, bidirectional (Blumlien) mic setup 45 degrees, it becomes an MS configuration. The polar pattern is also instructive. Taking the time to "eye" the two bidirectional lopes shows them to have symmetrical traits, which are inherent with ribbon mics. Furthermore, because the two diaphragms are of the same type, all four lopes are just about identical. All of these qualities are very advantageous, especially for stereo microphones.

need (and the budget) to make this expenditure appropriate. And don't forget retailers. Unlike a picture, a demo is worth much more than a thousand words.

This discussion of stereo micing could go on all day. Hey, there are whole books out there covering just this one aspect of audio recording, and many others devoted to the wide-ranging subject of microphones. I suggest you read them all.

But now let's go out to the studio and have a look at the acoustic panels they're using. It's right this way, through this corridor, which acts as a control room/studio sound lock. Here you go, a real studio, almost all wood, and a nice high ceiling. You don't see many of these being built anymore, real-estate pricing and all.

Uh-oh. It looks like the musicians are arriving for the next session so we better get out of the way. But let's stop in the sound lock for a moment. I want to show you a clever idea. They used the space within the support beams and in between the isolation mounts underneath the speaker soffit for two metal storage cabinets. They're locked now, but I can explain the setup. Each one is divided into multiple, foam-lined compartments to hold the studio's microphones. They are all metal and lockable, but the fact that each compartment is clearly labeled with the make and model number of the microphone that should be occupying it is more important. Because of this, you can tell with just a quick glance if any mics are missing. This can be critical as microphones are expensive and they tend to "grow legs" very easily. Losing a mic that's an old friend leaves a hole in your gut. I know; it's happened to me more than once. It makes sense for recordists to protect and keep tabs on their tools.

I think a session's about to start, so we better get out of the control room. Hey, there's the studio owner; give me a second to give him my regards.

Guess what? He says his engineer called in, notifying the studio that he's going to be late. The assistant engineer recently quit to go on the road with a rock band, *and* he asked me if I know of anyone who can step in. Well, buddy, do you want to go for it? Don't worry; I'll be with you every step of the way. Just read on. All this information can cause a recording engineer to become so finicky about mic selection that he or she could end up blowing the musical feel of a session. As an old friend and a great jazz recordist ("great" because he was an expert at letting the music happen) once put it:

> People are always raising the question: "What do you like to work with?" Really, it's apples and oranges. If the music's good and the situation's right, personally I don't care.
>
> David Baker, from an interview I did with him during the summer of 1983

Of course, *he* knew all the basic theoretical principles involved and was also an expert at mic selection and placement.

The Recording Process (The Job)

Some of this section was written two decades ago. So what? Sound is still subject to the laws of physics, and mics still operate on the same electromechanical principles. The only reason I'm mentioning it is because some of the microphone models used to describe positioning techniques may no longer be available. This is not a big deal as manufacturers do not drop popular products from their line; they replace them with better ones. For example, Audio Technica no longer makes the AT813 condenser mic, but their AT4051 is more sensitive, less susceptible to the proximity effect, includes a high-pass filter, and its cardioid capsule can be interchanged with ones that have omnidirectional and hypercardioid polar patterns. Granted, it was designed more for instrument pickup than for vocals, but Audio Technica (like every other microphone manufacturer) offers a wide array of vocal mics. Therefore, if you contact a manufacturer and ask them for something that would be equivalent to any microphone used in the following setups, they will have something to fit the bill. More to the point, since every recording situation will be different to some degree, microphone models and physical positioning are only part of the story. Mic A may be more sensitive, but due to this, you might have to position it a little farther back than, say, mic B, from a loud sound source to avoid distortion. However, in this location there may be more leakage picked up from other instruments. It's audio, and every choice involves trade-offs. Take a standard combination like a Shure SM57 on a snare drum. A condenser may break up when placed in close, but in some situations you may want that kind of distorted sound. A purist, pristine omnidirectional measurement mic is going to add all kinds of bleed to the track unless it's positioned very close in, but that sensitivity may be exactly what you want on a jazz date where the drummer's using a lot of brushwork.

Making position choices in terms of a microphone's response capabilities is really the key, so if you don't have access to any mic I've used as an example, don't sweat it. The principles are valid for any comparable mic and I made the effort to add most of these mic's polar patterns and frequency response graphs in Part 3, "Microphones," so you have something to compare to those of other mics. This should cut down your research time considerably.

This was designed to be an overview of the methods and techniques used in professional recording. While coverage of recording, overdubbing, mixdown, and editing will be useful and informative to the layman, musicians, and engineers alike, keep in mind that any written piece can only provide insight and is no substitute for experience. I urge you to be inquisitive to the point of being a pest and to experiment as often as possible as this is the only true route to mastering this art.

A Basic Session

Hey, you're in luck. This gig I got you is a basic session, so if you stick with it, you'll get to follow the project from start to finish.

Okay, maybe jumping into a full-blown basic setup is a bit much to ask, kind of like taking a first swimming lesson in the middle of the Atlantic Ocean during a storm, but I'm not going to hold back on any info. I'll be right with you, giving you all the basic information you'll need, including any necessary wiring and interface details.

From an overall production point of view, basic sessions are the most critical.

What's most important here is that the players lay down the rhythm or backing tracks with the correct tempo and feel (the "pocket"), as this will affect all the additional recording that follows. The feel will come not just from the drums, but from the full rhythm section, including the bass, guitar, and keyboards. Even if some of these instruments are to be overdubbed later, if the pocket or groove they lay down becomes well defined during this initial process, everything else that follows will go down much more smoothly.

It is also very desirable during basics to get a good, complete drum take since overdubbing and punching in a complete drum kit can be a difficult chore for drummers and engineers alike. Cymbal and drum sounds decay over time, making punch-ins more noticeable. The liveness of the room will also be different during a less crowded overdub session, and even if this is not noticeably different when compared to the sound of the original tracks, it's never easy to get all the mic placements and drum tunings to exactly match. Furthermore, the drummer's energy level will not be the same during an overdub as it was when it was all going down to tracks while they were playing along with the whole rhythm section. Therefore, *during basics, drums are the priority*.

But first, we must discuss the people involved in the recording process and their responsibilities. The producer, after hearing a tape of the band, or seeing them perform live, will have some idea as to the type of sound they are after. Their function is to capture the style and sound of the artist(s) while making it understandable and enjoyable to as many listeners as possible. If they feel that a certain piece of equipment is needed to achieve that sound, it's up to them to make sure it is available at the session. The artist is expected to have their material, arrangements, and parts somewhat together before entering the studio. However, some (with the budget) may prefer to compose and arrange while actually in the recording studio. Either way, they must perform with emotion and proficiently execute their parts to the best of their ability.

The engineer is responsible for the technical running of the session. They must act as an intermediary between the technical and artistic facets of recording. They're a second set of ears for the producer and expected to contribute their expertise to the session via impressions, observations, and *possibly* an occasional idea on production.

Previous to the session, most engineers check out all the equipment. Cleaning and demagnetizing is rarely necessary with today's equipment, but the engineer must still verify the alignment of the recorder being used.

A minute or two of tones (usually 1 kHz, 10 kHz, and 100 Hz) is printed on all tracks for reference, especially if the tape will be taken to other studios for additional work. Often this job is left to the *assistant* and it can also entail striping (recording) any control or data tracks necessary to the recording format being used. If possible, now is the time to lay down SMPTE, MIDI, or any other time code tracks that might be needed. Their levels will be determined by the engineer and will usually be lower than the 100 Hz, 1 kHz, and 10 kHz tones that are recorded at 0 dB (or "unity") reference levels. Additionally, I always like to lay down a couple of minutes of A 440 Hz at the head (beginning of the tune). But the best bet is to get the keyboard player to lay down a sustained chord in the song's key for you. These can then be used by the musicians as a tuning aid during later overdubs.

Digital tapes must be preformatted. Here the "control track" must be printed from head to tail. This information will be used by the recorder to sort out the digital information and will feed tape location information to the remote. Like SMPTE time code, it has numbers and a machine will "flip" over a 24-hour crossing to 0 hours. Therefore, like time code, starting things off at 10 hours is a real safe bet.

If you're not using a *new* tape, or one that has been adequately bulk-erased, you may be in for a hard time. Many machines do not actually erase the control track but record over the old one. There's nothing like having a "ghost" control track along for the ride.

The control track is *not* SMPTE, so if you expect to interface the tape with video decks, or some brands of console automation (SSL, Neve, etc.), it's best to stripe any required clock information as well. Just because a machine is digital, some people don't think that it is necessary to print tones at the head of the tape. I print tones on *everything*. To me, it's like having a copy. You may never need it, but . . .

Bookkeeping is also part of the assistant engineer's responsibility. It's important to start a session log that includes all the information about levels and locations of the code and tone tracks.

When the musicians start to warm up during the final setup stages by playing the tune, it's a good idea to take note of the song's sections while referring to the music charts if they have been provided by the producer or arranger. Later the assistant will begin keeping a running log of SMPTE time code readings and their relation to these song parts to speed up moving to these locations during the overdubbing and mixing process. Part of this record-keeping should include a list the mics used (serial numbers are helpful too), along with any effects patches and their control settings as an aid to matching sounds later. Since it is often difficult for the engineer to leave the control room during a session, it is the duty of the assistant engineer to be is or her eyes, ears, and hands in the studio.

The basic, or rhythm, session is the initial work and usually consists of drums, bass, guitars, piano, and possibly scratch or reference vocals. The studio is set up in a specific order starting with the drums, followed by bass and guitar amps, keyboards, and other instruments.

Mic stands are placed in approximate yet out-of-the-way locations, mics attached; then music stands and chairs for the musicians are set up. Baffles are then positioned so as to isolate the instruments from one another. Yet they should allow visual contact between all musicians and the control room (I've used closed-circuit video cameras and monitors when necessary). Then, any AC wiring that may be required for electronic instruments, music stand lights, etc., is run so that it's completely out of the musician's way. This is followed by the mic and line cabling, which connects the instruments to the recording console, often through mic "snake" boxes and *direct input* (DI) interfaces. Finally, the cue lines are run and headphones hooked up.

Recording Drums

Drum setup is performed by either the drummer (if present) or the assistant engineer. When set up by the assistant engineer, it should be done loosely. By this I mean that although the positioning is pretty much straight ahead, all clamps and height settings should be left semi-loose since the drummer will invariably need to readjust them to his or her own liking. Mics should also be placed to leave room for this repositioning in order to avoid their being hit by metal hardware. Drum keys, tape, and tissue paper should be left within easy reach of the drummer, and headphone lines run from behind so as not to inhibit their playing motion. If the drums do not belong to the studio, the assistant has the option of either helping with the setup or moving on to the bass amp. Once set up, the drummer begins to tune the drums. Drumheads should have equal tension all around the rim so that when tuned to the desired pitch the drum will produce an equivalent tone, even if not always hit in the exact same place. This is achieved by hitting the skin close to and in front of the lugs around the rim, adjusting each lug so that all areas of the head have the same tension. Double-headed toms are more time-consuming to tune since the top and bottom heads interact in producing the pitch and therefore must be tuned to each other. However, once tuned properly, double-headed drums produce a more resonant tone as compared to single-headed drums.

I find that the best way to deal with the kick or bass drum is to remove the outside skin and its retaining hardware. Place two folded blankets on the bottom of the shell with a mic stand base in between. This sandwich of absorbent material is then adjusted against the inside head, near where the beater hits. The kick's tone can then be varied with blanket pressure from a resonant boom (loose) to a dull thud (tight). For reggae sounds, I would dampen the resonant shell sound by putting sandbags across the top of the kick.

The toms should have separate mounting as opposed to the standard mounts that are attached to the kick drum. This will prevent the transfer of vibrations, which can cause unwanted leakage. Toms can either be tuned by ear or to a piano. If you use a piano or other instrument as a reference when tuning the drums, remember that a 14-inch tom is a full step lower than a 12-inch, whereas a 15-inch tom is one and one-half steps lower than a 12-inch tom, and a half-step higher than a 16-inch. To save you a lot of effort, I've found that toms can only be readily tuned about a half step up or down.

The snare drum, being double-headed, must have its skins tuned to each other. The tension can be fairly tight for that gunshot crack, or loose for a low, soggy thud. **Caution:** Loose heads are very difficult to tune and will also not have enough tension for quick stick rebounds. This can cause the player to lose speed and agility.

The snares themselves must lie flat, but they can be adjusted for a sound that is long (loose) or short (tight) in duration. The choice of drum skins will have a lot to do with the sound of the set. Thin, clear heads will have a higher pitch as compared with hydraulics, with the thick clear heads (like black dots) lying somewhere in between. No matter what the choice, all toms should have the same type of heads.

Every drum should then be listened to individually for rattles or squeaks from hardware, and for rings, or buzzes, from the heads. In live recording this is not as important as amps and other instruments tend to cover up these noises.

Ringing from the drumheads can be brought under control by various means called dampening. Masking or cloth tape around the edges or across the head of each drum should help. If more dampening is needed, tape tissue paper or paper towel to the sides of the heads. On double-headed drums this may be beneficial when done to the bottom skin. Also, try taping a square piece of foam right behind a black dot for that low thunder-drum sound. Another engineer I know tapes coins to the skin. If the taping is tight enough you'll get a deepened drum sound that's trouble-free. A drummer who hits the drums very hard will need more hardware dampening as this style of playing causes more ringing. Hardware rattles can be cured by either tightening the adjustments or taping loose sections. Here, gaffer or duct tape is most efficient. Simple oiling will end bass pedal squeaks. Hi-hats and cymbals can also be tuned, and their length of decay adjusted by using masking tape across them. A clean cymbal is a bright sounding cymbal, so keep some brass cleaner on hand.

It is part of the assistant engineer's job to check the kit between takes to ensure that dampening and other work done to the drums remains intact.

After tuning and dampening are completed, mics are placed in the positions desired by the engineer. The engineer's micing technique will be dependent on many variables. The size of the kit, the type of recording equipment, and how many instruments (including overdubs and vocals) they are planning to record will all affect the number of tracks available for drums. If only one track is to be used, the drums will be mixed down to mono during the recording session.

This limits the variability of the drums during the final mixdown to simply one overall level control, while recording to two tracks enables the engineer to pan them left and right for a stereo split of the kit. When three are available, the kick or bass drum can be assigned to a separate track, while the rest of the kit remains in a stereo split. As more tracks become available the snare and hi-hat can also be assigned separate tracks. With seven tracks, the kick, snare, and hi-hat are recorded on individual tracks, leaving four tracks for two separate stereo splits of both the toms and overheads. To record every piece of a standard drum kit separately, eight tracks are needed with three toms assigned to individual tracks. With more toms and the use of ambient or room mics, the number of tracks used for the drums can increase further.

Mic placement is also dependent on the sound the producer is after. For a "tight" sound, a close micing technique is used. If a more live or open sound is sought, the drums will be placed in a more reflective area of the studio and fewer mics will be used. When the number of tracks available permits multiple close mic placement on the drum kit and a pair of ambient mics are backed off in the room, the sound will be augmented when the overhead mics are brought in closer than normal. The end result is a sound that's open and live yet retains its intimacy.

Individual mic selection is done by experimentation, or from experience. Directional mics, because of their polar or pickup patterns (which are less sensitive to sounds arriving from the back or sides) are good for multiple micing the kit. Whether micing from above or underneath a skin, they should be no more than an inch or two back, and positioned so as not to pick up sounds from adjacent drums, cymbals, or other instruments. Mics should be pointed at either the center or the outside perimeter of the skin, since these are both different sound sources, but placement in between can cause unpredictable cancellations.

Three variables come into play when micing the kick. Micing the middle of the skin close up will give more of a hard, beater sound. Close, but off to the side, will give more of a skin sound. When backing up, the mic will render a boomier drum or shell sound.

I use two mics on the snare drum. One on the top head, placed an inch or so above the rim at the side farthest from the drummer, and pointing to where the stick hits the skin. A second mic is placed under the drum in between the stand and pointing at the snare itself. Depending on tuning, dampening, and mic selection, this method gives me the ability to get most any type of snare sound I want. Toms are miced with the same principles in mind. Because of the probability of more dampening here, the mics should be positioned and aimed so that the dampening will not interrupt the sound path between the mic and where the stick hits the skin. Overhead mics are placed and aimed to pick up all cymbals and/or toms equally. The hi-hat is miced from above, aiming at the midpoint between the bell and rim farthest from the snare. Ambient mics, which are used for adding live coloration to the overall drum sound, should be placed farther away from the kit and near reflective areas of the studio. As with all mics used for a live sound, these should be more omnidirectional in order to pick up more of the reflected and delayed signals.

Once the drums are completely miced, the assistant writes down the drum and the corresponding channel it's plugged into. The engineer then labels all his input faders accordingly. He assigns and mixes these down to the output faders, and notes these. Both these labels, as well as a track sheet, will be kept with the tape itself for future reference. Starting with the board's equalization flat, each mic is checked out separately. Levels are checked and the mic preamps in the board are adjusted to prevent overload and distortion. If a tight sound is sought, the assistant now covers the rear of the kick drum with a heavy packing blanket. The engineer begins to set the board's EQ to obtain the drums' sound and placement within the spectrum. By using the console's EQ, along with an outboard equalizer and limiter/compressor, the engineer can make a light-touched jazz drummer doing a rock session almost sound like he's using baseball bats for drumsticks. If the sound of any drum is not acceptable, different mics, positioning, redamping, or retuning should be tried. Some experimenting will almost always be required.

Via soloing, each mic should be checked for leakage, phase anomalies, and any extraneous noises. The drummer is then asked to play a basic rhythm, starting with the kick, snare, and hi-hat. Tom fills and cymbal accents are then added until the whole kit is up and running. Other outboard gear can now be used along with other effects. Different reverb or echo devices can be tried out. I always liked to use separate delay units for the drums and vocals, especially in live work as it avoids muddying up the vocals. I prefer to hold off on gating drums until the mixdown session because when used during recording, lightly hit sounds may not be above the threshold setting and might therefore not be recorded. Big mistake!

Now some particulars.

I like a dynamic on the kick, especially when I'm positioning it inside to increase separation. Here, I'll opt for the standards like an RE-20, SM7, or, if I can get one, an old AKG D-12. These mics can take the *sound pressure levels* (SPLs) without breaking up and they add a nice boost to the low-end sound. I *believe* you'd get a mighty sound from a Royer ribbon mic inside of a kick and I'd love to try it. But when I can, I'd prefer backing up the positioning and using a directional condenser, or a *pressure zone microphone* (PZM) mounted just outside the shell. This can yield a full, karate-kick-to-the-chest bass drum sound, but it lacks some of the high-end beater against skin sound that disco thrived on. In this case, I'd add a mic to the drummer's side of the kick, mounted on a short stand and pointed directly at where the beater makes contact with the skin. A couple of cautions are in order here. Point out the mic to the drummer and make absolutely sure the drum pedal doesn't squeak. Two more tips: Wooden beaters sound best; when everyone started using plastic or went back to felt, I started carrying my own wooden beaters with me. Believe it or not, the best mics at picking up that high-end beater "tack" sound are those dynamics designed for speech communications. You know, the kind of mic that's normally attached to a live mixing console by a gooseneck.

I also use this type of mic under the snare. The application for which they were designed requires them to be fairly directional and to have a frequency response that accentuates speech (2 to 5 kHz). I carried an AKG D58E, but most will serve this purpose well and you'll appreciate the maneuverability afforded by the gooseneck mounting.

Standard fare for the top of the snare is an SM57, but Beta 58s, condensers that can take high SPLs, and exotic mics all render different amounts of "punch," so it comes down to which mics you're willing to put at risk. I know my Milab LC-25 would sound devastating on a snare, but there is no way I am going to take the chance, even though I'm pretty sure it can survive multiple stick hits. You should see some of the 57s and 58s that are sitting in recording studio maintenance shop drawers.

I've heard the sound of a Sony C-37 on toms and, yeah, it was just what you want, but their capsules are not replaceable. Standard fare here are Sennheiser 421 dynamics and the sound is straight-out predictable. On the other hand, I've used Shure SM81s, Audio Technica ATM 11Rs, and other condenser mics on the outside of toms getting great results from their more intimate sound. However, I find condensers to be a bit too "boomy" when placed inside drums, so if I need the extra separation this positioning affords, I'll hands down stick with a dynamic.

I love the sound you get from a shotgun mic on the hi-hat, not to mention the separation and the newer "short" shotguns eliminate the obstruction problems caused when using the older 2-foot long models. But for the most part, I'll go with a condenser with a tight polar pattern unless I have an ace in the hole like a damaged pencil mic or a Sennheiser MD 402 dynamic available.

In general, condensers are used for overhead positioning, but it really all depends on the sound of the kit. If the set includes three crashes, two rides, a "china cymbal," *and* a couple of splash cymbals, you may want to tone down the high end a bit. Using dynamic mics will help a bit, but in a case like this, the console's low-pass filter may need to be pressed into duty.

The standard for overhead micing is a pair of Neumann U87s set to a cardioid polar pattern. For a more open sound, the omnidirectional setting can be used *if* there is enough ceiling height and its surface is absorptive. The placement of overhead cardioid-pattern mics will depend on the size of the kit as well as whether or not the toms have been individually miced. If they haven't, and you're going for a single stereo spread of the toms and cymbals together, coincident or near-coincident spacing, as well as lower height settings may be used to bring out the toms more. I love the live sound of a few mics backed off from the kit, but individually micing each piece of the kit *does* have its benefits.

Click Tracks and Drum Machines   Occasionally, it's necessary for even an experienced drummer to be fed some sort of click track in order to aid them in achieving an unwavering beat. One method of doing this is to mic a metronome.

Whether this is of the electronic or common pendulum type, it must be isolated so that the mic on it doesn't pick up any other sounds. This method also gives the engineer more control over the signal. The simplest way to accomplish this is to make use of one of the many road cases, which are always lying around the studio during basics. The assistant engineer should first check the battery, or spring tension, to ensure that it will make it through the tune.

The producer, or one of the musicians, after either listening to a tape of the song or discussing the tempo with the other members of the band, sets the metronome. The assistant then puts the running metronome inside a road case, places a mic directly in front of it, wraps some sort of foam or cloth around the mic cord for protection, and closes the lid. If no road cases are available, an unused isolation booth (or closet) will work. I once put one under the cushion of the studio's couch and the isolation was sufficient. The mic is then plugged into the system and bussed to the headphone cue mix, a track, or both. A jack can also be mounted onto an electronic metronome to make a direct output available.

A less monotonous type of click track can be recorded from an electronic rhythm machine. Although these are endless in variety, for our purposes I categorize these into two basic forms: those with single outputs and those with multiple outputs. The multiple output variety is generally more expensive, as they employ designs that provide separate outputs for each sound that makes up the overall rhythm. Here, the bass drum, snare, hi-hat, toms (which may or may not be tunable), cymbals, and other percussion sounds like congas, bongos, hand claps, claves, and cow bells can be plugged into different channels of the board. If a single-output type of rhythm box is used, you can create a situation that's approximate to multi-output units by bussing the one output signal to several channels. Each sound is then separated from the others via equalization.

If, for instance, you wish to have separate control over the bass drum, a low-pass or parametric EQ can be used to effectively cut off all the frequencies above the kick's sound. Now when you raise or lower the output of that channel, the bass drum's sound will be affected to a greater degree. Cymbals are brought out by using a high-pass filter, while parametric EQ will generally be sufficient in accentuating the dominant frequencies of the other sounds. At this stage, each sound can be delayed, modulated, gated, limited, compressed, or whatever, to give you the sound you are after. By delaying the snare, for example, you can double the number of beats. Modulation increases the number of beats and also changes pitch, which is an effect that does wondrous things to bongos, congas, and claves. Gating can be used to eliminate any noise and compression, or limiting can make the drums sound bigger than life. If too many beats of any sound are being synthesized, switching the console's mute button on and off will eliminate the ones you don't want, but having the drummer reprogram it is a better bet. In fact, when it comes to programming drum machines, I kept a list of drummers I could call in to handle the task. The small fee they charged often saved a lot of time, money, and aggravation.

When this substitute kit sounds the way you want, it can then be mixed down to the number of tracks desired. This method can be so effective that it is often used in place of real drums. Drum-drop records (now CDS) can also be used for the same purpose. Here, you generally have a stereo recording of a real drummer. These two tracks can also be bussed to several tracks and split up as was done in the previous rhythm machine example. In the past, it was often first necessary to EQ out the noises caused by scratches and mars on the record. Okay, enough on percussion for the time being. I'll get back to it when we discuss overdubbing. Now it's time to move on to the bass and guitar setups.

Recording the Rest of the Rhythm Section

Bass and Guitar The most important thing when recording these instruments on a rhythm date is isolation. The last thing you want is to have the guitar all over your tom tracks, or the bass volume to increase when you raise the cymbal tracks. To avoid this problem, these amps should be positioned as far away from the drum kit as possible. Sound-absorbing panels should box out the speaker cabinets and, if needed, I'll even drape several packing blankets over the mic/cabinet combination to eliminate the transmission of the sound as much as possible.

I almost exclusively use and even demand that a small guitar amp like a Fender Twin or a practice amp like Princeton be used by the guitarist on a rhythm date. If a loud rock 'n' roll Marshall-type sound with plenty of sustain is needed, wait for overdub, where you'll get a better sound and it won't bleed all over the drum kit.

At this point in the project, you have to be more concerned about getting good clean drum tracks and personally I prefer redoing bass and guitar tracks, if needed. This does not mean that the rest of the musicians should have to try to play their best while putting up with wishy-washy, no "guts," tiny sounds. No sir, and if they didn't happen to bring along any of their effects pedals, I always carry a couple of modified little stomp boxes with me that are absolute ringers in the recording studio. I'll divulge these secrets when we get to overdubbing.

The hookup of the amps is left to the assistant engineer. Direct boxes are either of the transformer or electronic type. Basically, both are used to match the high (600-ohm) impedance and level of a music instrument's output to the lower (150-ohm) impedance and balanced mic level of the console input. This is why both 1/4-inch phone and XLR connectors are on the box. The first is transformer-based while the latter utilizes battery- or phantom-powered op-amps. Direct boxes (DIs) enable you to get the sound directly from the pickup uncolored by amplification and with less noise. When the EQ control on the guitar is at its midway setting and the volume control is positioned all the way up, DIs faithfully reproduce almost any sound. In fact, they can often reproduce more lows than an amplifier, but combo amps usually provide more of a mid-bass punchier sound. But, it all depends on the choice of amps and mics.

The block diagrams in Figure 4-1 are provided to illustrate the different types of DI hookups.

The first (A) is a basic two-channel direct box output and mic hookup. The guitar is plugged into the DI first. From here, two outputs are sent, one to the console (via XLR), and the other to the amp (via ¼-inch phone) and the amp gets a microphone placed in front of its speaker. The second set (B1 and B2) show two ways to double-track when effects are used. In the first of these examples, the effects are on the DI track, while the other shows the effects going to the amp. In Figure 4-2, C1 shows the way to get the direct, effects, and amp sounds separately, utilizing two direct boxes and thereby feeding three channels.

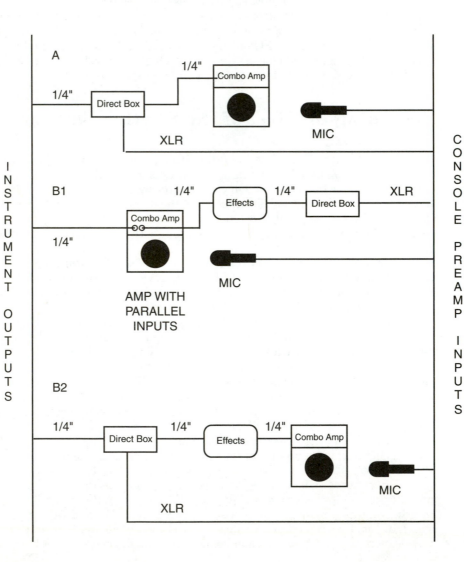

Figure 4-1

Three methods of using a direct box to record an electronic music instrument. (A) This is a basic setup. It is possible to use only the feed from the direct box, but this can be a little sterile and "lightweight" sounding. The sound is best when a little direct signal is used to clarify the amplified sound. The two (B) drawings add the musician's effects to the overall sound with B1 after the amp, or B2 before the amp.

Figure 4-2

This illustrates two methods of recording an instrument's direct output, the effects processors, and the mic/amplifier sound to three separate tracks. The second method (C2) results in a track that has only the effects as well as one that routes the effects through the amp and microphone. The difference between these two can be huge.

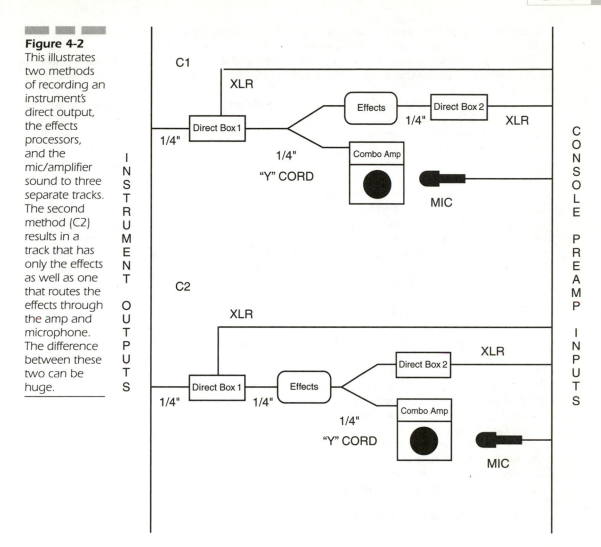

The next (C2) shows the direct and the effects being taken separately, with the effects sound also through the amp.

These channels can be either mixed down to one or two tracks or dumped onto the tape as they are to be mixed later. I always let players use any effects they bring so that they can feel more comfortable. By using the last method, anything can be saved and used later, or erased without any loss of material. This can be a very consoling fact.

I have to make one point clear. Keep those interconnecting unbalance cables as short as possible. The pickup of noise and the degradation of high frequencies increases with the length of unbalanced cables.

Another point about effects boxes. These devices are generally manufactured assembly-line style and therefore may be incorrectly wired. On more than one occasion I've found that the two (direct and effect) output connectors were wired 180 degrees out-of-phase. I use phasing quite a lot to get interesting sounds, but when it is unintentional it can be horrendous, especially on bass.

After having rewired at least a half dozen of these units, I started carrying a cord that was intentionally wired out of phase with me for just these occasions.

DIs can also be taken from an amp head if it has a preamp output, or right off the speaker leads as long as you pad down the signal. These methods, however, do add color that may or may not be desirable. I used a common 100-watt lightbulb as a dummy load (see "Guitar Overdubs").

One of the main priorities here is to eliminate all buzzes and hums caused by ground loops. This can sometimes be accomplished simply by flipping the amp's ground reversal switch. Most often it'll take a bit more effort. I used to add a standard two-pronged ground lift plug to the amp's three-pronged plug. While I was never stupid enough to clip off the ground pin of an AC plug, I was guilty of flipping the way this plug is inserted into the AC receptacle, that is, until it was graphically explained to me that electrocution can occur from as little as a few dozen volts of alternating current. So now it's *my* duty to pass the word on to you.

Treat the electricity powering the music instruments and other studio equipment as if it were totally separate from the control room's electricity. It should be identical, but it's very easy for things to change, and you can never be sure. *Ground*, whether signal or electrical, is there to offer stray electronic noise such as RF, EMI, and capacitor leakage a nice low-resistance path out of the system. If you have a noise problem, something is wrong with the ground system you have set up. Do not let the pressure of even a full-blown basic session cause you to take a willy-nilly, hit-or-miss approach to its correction. *Lives may be at stake here!* So a levelheaded, straightforward, logical approach to problem-solving is what is required.

This is going to be an overly simplified explanation to help you get the point across when you explain it to someone else. Thinking of it this way may help. The electronic circuitry within the guitar amplifier generates some noise. (They *all* do). The amp's ground system picks up this noise and tries to conduct it away from the audio signal just as it should by dumping it out into the utility AC power's ground conductor.

Try looking at a ground loop like this:

utility ground > to amp > to the audio output > to direct box > to mic cable > to console input> to console power supply > back to utility ground.

This kind of loop can keep the noise within the system and could actually increase its level. To break ground loops and get the noise out of the system, you must first find the source of the problem.

Dividing the system in half will help you find it more readily.

1. The halfway point here is the direct box. Lift its ground via the switch. If the noise ceases at the amp and in the control room, you had a common ground loop. Leave the ground switch in the lifted position.

2. If the DI's ground lift switch has no effect on the noise, try bypassing or swapping out the DI.

   If the noise now stops, you had a defective DI (either its switch or internal wiring). Turn it in to the maintenance shop.

3. If there's no change in noise, disconnect the input to the DI.

   If the noise ceases in the control room but remains at the amp, then the noise problem is at the amp end.

4. Check the amp's AC connector for a cut ground pin. If removed, replace the power cable or amp.

   If the pin is intact, *do not* lift or "float" the amp by disconnecting its utility ground connection.

5. Disconnect all pedals and effects connected to the amp. If the noise ends, the source of the problem will be obvious.

   If it's still noisy, disconnect the audio input to the amp.

   If it's still noisy, you may have a faulty amp. Try swapping it out.

   If the noise ceases, it may have been due to the high gain of a guitar's built-in preamp (notorious). Turn it off.

   If you still have noise, it could be a faulty guitar cable (very common). Try a different cable.

   If it's still noisy, it may be due to an improperly shielded guitar (quite common). Try connecting a different instrument to the same setup.

Why You Should Not Float an Amplifier

Fault #1   Floating the ground connection of *any* piece of electronic equipment means that there is nowhere for the noise current to go. This increases the potential amount of voltage on the ground system. If the amp's chassis is connected to the ground system, this increased potential may also appear on the audio ground.

Fault #2   A microphone has been improperly wired so that its metal casing is connected to the audio ground.

   The audio ground is normally connected through the guitar cable directly to the guitar's ground system.

   The guitar's bridge *and* strings are part of this sytem.

   The player's left hand is in contact with the strings.

   The player's right hand reaches out to adjust the position of the microphone.

Faults #3 and #4 Occur at the Amplifier

Fault #3   A short circuit has developed due to the wearing away of some insulation material, and one of the current-carrying AC conductors becomes connected to the amp's chassis.

Fault #4   The circuit breaker on the utility power has become defective or the current value of the fuse being used greatly exceeds the proper rating.

   The electricity on the amp's chassis flows through the guitar cable to the guitar, out the strings, to the player's left hand, across the player's heart, and from his or her right hand to the mic and ground. This could spell *electrocution!*

   "Yeah, but all those other faults also need to happen on top of having a floating amp."

   If the amp were not floated, it would have either blown the overrated fuse or started to smoke due to fault #3.

Faults #3 and #4, floating the amp, could easily raise the voltage potential to, say, 30 volts, but the real problem is from the amount of current. The human heart can have its fibrillation process interrupted by as little as 100 milliamps.

Each year in the United States, more than 30,000 people receive permanent injuries and 1,000 others die from electrical shocks. Meanwhile, thousands more get hit with stronger jolts of electricity and walk away as if nothing happened. To comprehend this, you must first understand how your heart works.

The pumping of blood by your heart is triggered by an electrochemical reaction. The chemicals involved are potassium, which has a positive charge, and sodium, which has a negative charge. They pass right through the cell walls in opposite directions when the heart expands and contracts during its pumping cycle.

The voltage this chemical movement generates causes first one side of the heart to contract, and then the other. After these two stages, the heart waits for the potassium and sodium to return to their original states, and when they do, it triggers the next contraction cycle.

If an external voltage flows through the heart during the wait period, it may try to beat prematurely. However, with the potassium and sodium out of their triggering positions, the heart may not be able to complete a full contraction. Instead, it begins to quiver, or fibrillate. At this point, even 400 watt seconds of DC from a defibrillator (which forces the potassium and sodium ions back to their reset positions) does not guarantee the heart will begin to beat again normally.

The heart's cycle is susceptible to this outside interference for only about one-sixth of each heartbeat cycle, which is why you can get a shock from 120 volts to 15 amps one moment and have nothing happen, and you can get hit with 30 volts AC at 500 milliamps the next second and make your big exit.

I'm glad I got the chance to pass this info on.

Recording the Bass Guitar   During basics, the bass is set up exactly like the guitar except I would often completely eliminate the use of an amplifier and speaker. During overdubs, or when the use of only a DI is frowned upon by the bass player, you can choose either a tube amp, which gives a fat sound, or a transistor amp, which renders a more punchy sound. Depending on the kind of sound we're after, I prefer to use a smaller, front-mounted speaker such as a 15-inch as opposed to an 18 or 20-inch. Sometimes I'd plug the bass head into a 4 × 12 bottom. This is done for two reasons. First, isolation; it's a lot harder to control the transmission of the lower frequencies that emanate from large speakers (18 or 20 inches) mounted in a folded horn enclosure. Secondly, I often didn't need the low end as this was picked up quite efficiently by the DI. What I often needed was more mid-bass punch, which is exactly what you'll get from, say, a 4 × 12 cabinet hooked up to an amp powered by transistors.

For mics, I would use either a warmer-sounding tube mic like a Sony C-37 or a Neumann U 67 (if I can get one). Or for more midrange definition, a handheld vocal mic with an additional EQ cut between 200 and 400 Hz. The volume control on the bass guitar itself should be at maximum to improve the *signal-to-noise* (S/N) ratio, and the tone control starting at midpoint should be adjusted up or down for the desired sound.

On the other hand, I could get downright radical in the heavy bottom direction when it was called for. I got a real kick out of blowing away my reggae bass-player friends with this next method. (I said I was not going to hold back, so hold onto your hat and follow along.) You know from the section on mics that the two sides of figure eight or bidirectional mics are 180 degrees out of phase with each other and the two diaphragms are as close (coincident) as it gets. I would face two bass speaker bottoms into each other with my Fostex M88RP bidirectional printed-ribbon mic between them. I tried a lot of combinations, but settled on an Acoustic folded horn cabinet with an 18-inch driver and an Ampeg cabinet with two front-facing 15-inch drivers. For power, I would use a 250-watt Ampeg tube amp and reverse the polarity of the wire feeding the speaker aimed into the back (nonlogo) side of the mic. Finally, I would heavily go-bo around (and place many packing blankets over them) to improve the isolation. The sound of this setup can scare you.

When recording bass, the engineer must get down as much of the player's dynamics as possible. They must keep all of the signal content audible and yet avoid distortion from too much level. In order to do this, a compressor/limiter is often used. Here the compression is set to keep the signal's dynamic range to within, say, 6 dB. The release setting should be pretty fast so as not to affect any subsequent notes, which may not be as loud. And a slightly delayed attack time should be set up, causing the limiter to kick in after the string has been plucked, leaving you with a punchy sound. The output is then set to the level desired.

Oftentimes on R&B or funk dates, where the bass player is using a finger-slapping style of playing and where the slaps and pops may cause distortion, I'd roll-off the top and the low-bottom, and go for a tight, punchy sound around 100 to 150 Hz. But wherever you situate the bass via equalization, be sure to avoid conflict that can happen when it and the kick drum occupy the same frequency range.

More Guitar   The choice of an amp for guitar, outside of the small size previously mentioned, will depend on the guitarist and the guitar used. Often a guitarist will bring his own amp and even if it is only barely acceptable, I'll go with it for his comfort.

Aside from that, an amp's frequency response and tone must be considered, for they will determine whether or not it fits into the particular orchestration of a song. For example, if both a keyboard and a guitar are used for rhythm, they should not clash, but rather interact together. This is greatly aided if their sounds do not mask out each other because they are occupying the same frequency range.

Once chosen, I'd elevate the amp off the floor or tip it back about 35 degrees in order to avoid any reflection off the floor, which could cause comb filtering at the mic's diaphragm. A dual-speaker amp would often get two mics, usually a dynamic like a Shure SM57 or a Sennheiser MD 421, and a condenser like an AKG 451 or a Shure SM81. I can use a combination of both of these mics plus the DI or switch off between them for different sounds on different sections of the tune. Remember that your direct line will be extremely clean sounding. Even when combined with the mics, it may not sound that great when soloed, yet the sound may end up being perfect when mixed with the rest of the track, so continually check both. Try adding a little direct signal to a heavily distorted dynamic mic sound for added definition. Mixing a little tube mic warmth with the direct will ease some of its extra high-end cleanness.

Adding digital delay to one of the signals, and panning them both for a stereo split, will render a nice fat sound that moves slightly from one side to the other. As with everything else, experiment.

Keyboards During Basics   Unless the studio has their acoustic piano either in an isolation booth or has some sort of isolation built up around the back of the instrument (high enough and soundproofed enough to keep the lid wide open), I'd rather record it during an overdub. This is because I personally prefer to record an acoustic piano without the sound bouncing around off its very reflective lid. This is especially bad when the lid is almost fully closed. For this reason during a rhythm session, if the drummer and other players need a keyboard to get the feeling of a tune right, it makes more sense to use an electronic keyboard.

This may be an electric grand like a Yamaha CP, a Fender Rhodes, or a synthesizer. These can be taken direct with any effects placed before the DI, and as either a stereo or mono feed, depending on the song. With a little EQ, panning, and delay, you can make a mono output Rhodes sound very close to stereo.

Again, if you're using a synthesizer that has reliable A 440 tone, it is wise to put down a few measures of it on a track for tuning purposes. Also, continually check the tuning of older synths as they can drift. Whenever other musicians tune up to this, or anything else, they should tune their instruments while listening to their amplifier's output, as opposed to head-phones, as the latter can be very misleading and therefore making tuning difficult.

Communications   These players (bass, guitar, and keyboards) should also be given vocal mics for communication purposes. But if mics are used to pick up the room's ambient sound, they can also be used for this purpose between takes.

If there is to be a scratch or reference vocal on the date, it should be done in an isolation booth, if possible, to prevent instrument bleed onto the vocal tracks should the singer lay down a take that's too good to replace. There's nothing worse than a singer laying down a great take and you having to tell him that he must do it over because there are drum or gui-tar sounds all over it.

When everything is set up, tuned, and all the sounds are acceptable to the producer, the tape is rolled, the record button hit, and you're off. The drummer should count off the time clearly, often hitting his sticks together with the count. This will be very helpful later when trying to overdub instruments onto the intro.

During the actual take, the producer will be listening to the overall tempo and sound of the tune. It is therefore the engineer's job to listen for mistakes, tuning faults, and other problems. Remember here, that at elevated listening levels, an instrument that is in tune may sound out. The recording process should be stopped *only* when the producer tells you to stop. The assistant engineer should stick close by the control room in case any problem develops so that, upon instructions from the engineer, they can be remedied.

In between takes, playback can be sent to the musicians either through their headphones or through the studio monitor speakers by switching the recorder from input to playback mode. When a take is a keeper, it should be notated by the assistant along with its SMPTE start and stop points.

Synchronized Monitoring and the Overdubbing Process

From this point on, all recording will be done as overdubs. In order for this process to work, the musician must be able to hear the tracks that were previously recorded while at the same time hearing the part he or she is presently playing. This point brings to light one of my main prejudices. With *all* the digital recorders that I'm aware of (and I have been keeping very close tabs on this technology since 1982), there is no way to monitor what is being recorded *while* it is being recorded, that is, hearing the actual sound that is being recorded while the record-ing process is taking place. Previously recorded tracks are played back in perfect synch with the tracks being recorded; however, the recordist is limited to only being able to monitor the tracks being recorded as they appear at the recorder's *input*. I don't mind if the signal is delayed, as long as it is being monitored from the *output* of the recorded media. You can blame it on my "protect-your-butt" attitude, but I came up having that assurance and I like it. You see, with analog tape recorders, the recording engineer has the capability to momentarily switch the monitoring position of the tracks being recorded from the record head to the reproduce head. *This should not be done* while actually recording a session, as the sounds being recorded will now be out of synch with the rest of the tracks and throw everyone's timing completely off, but it was my common practice to pop into the repro mode for a second while "getting sounds," just to be absolutely sure of the outcome.

You may think that this is not important, but how would you like to find yourself sitting in a state-side, world-class recording facility, just back from Vienna where you had *the* boy's choir (no less) add backing vocals to your tracks, only to be told by the tech on duty (me) that the distortion you were hearing on these digitally recorded tracks was not the fault of the machine they were being played on, but that they were distorted on the tape. This is a true story, but I'll never tell which internationally known superstar this happened to.

Therefore, I stick to using at least a backup analog recorder when doing live recording. Here, none of the musicians are monitoring from the recorder, so I can keep that deck in repro all during the performance and know *exactly* what the end result would sound like because that's what I'd be listening to.

In order to overdub, analog recorders have the ability of recording in synchronization (or synch mode). Here, the input signal is fed to the record head while the playback information is reproduced by that same record head. Without synch, the input being added would be heard as if it had been performed sometime after the original tracks that are being reproduced by the repro head, which is spaced some distance ahead of the record head. (The time delay will equal the distance between the record and playback heads divided by the speed of the tape.) This would obviously be unacceptable since the purpose of overdubbing is to have all the tracks sound as if they were played, even if not at the same time, at least *in* the same time. To avoid this problem, the recorder is put into the synch mode, where the previously recorded tracks are played back from the record head to the musicians through their headphones. The signal being added is fed to a track, which is put into record mode or "armed," and fed from there to both the cans and the record head. Thus, both the signals that are being played back and those being recorded appear at the record head at the same time. Later, during normal playback, all signals arrive at the reproduce head at the same time and all will be in perfect synch with each other. Since record heads are designed specifically for the purpose of recording, they do not have the playback fidelity of reproduce heads. For this reason, excessive bounces can degrade the sound.

However, the playback capabilities of record heads are quite adequate when used for reproducing cue-signals through headphones.

If (and when) the hard disks used in digital recorders start to incorporate a second, read-only head that allows the engineer to monitor information while the recording process is in progress, I would consider coming out of retirement. Someone has to demonstrate the butt-kicking outrageous possibilities afforded by the fidelity of 24-bit 96 kHz (and higher) sampling rates on those portable eight-track location recorders that are the size of a standard dictionary. I have been waiting my whole life for this. You're hearing this from someone who, during the 1960s, cut holes in the wooden side panels of a pair of two-track, open-reel tape decks and connected one of the capstan drive motors to both deck's capstans with a vacuum cleaner belt so that four tracks could be recorded at once. I couldn't rewind or fast-forward them in synch, but once both tapes were repositioned to the beginning grease pencil marks, they played back in synch. One of the old-timers who witnessed this process got such a kick out of what this "crazy kid" was up to that he took me under his wing and taught me a bit of radio, audio, and electronics technology.

I tried to explain to a couple of manufacturers that three of these new eight-track digital location recorders have the potential to replace analog 24-track decks at half the price if they could be synched together, and they were uninterested.

Just to help you comprehend what can be possible, take this one scenario of a rockabilly project. First, dump a rough mix of the tunes to one of the three portable eight-track decks. Then grab it, the singer, a mic, a couple sets of cans, and head out to Central Park. Next, go into one of three underpasses (depending on the tune's tempo) I am familiar with that have absolutely perfect slap-back, delayed echo and reverb for rockabilly and print vocals. The resulting vocal tracks will not only sound great, but since the acoustical space around them will sound different than the other tracks, they'll end up having more separation in the final mix. Now imagine recording a complete project this way.

Multitrack Overdubbing   When doing overdubs, the problem of leakage from other instruments does not exist since generally only one instrument will be recorded at a time. However, care should be taken when setting up headphones since anything other than a tight fit, as in the use of "open air" headphones, or feeding the cans too much level may cause leakage into the mic being used, bleed onto the track, and possibly painful feedback into the musician's ears.

If the piece being overdubbed is not continuous, the musician must try to be quiet when not playing to avoid unwanted sounds on a track that would later have to be erased as it can be a real pain to hear a cough come out of nowhere during a final mixdown.

Many recording studio consoles are made up of *input/output modules* (I/Os). These are used so that every single channel has the capability to process signals either from the recorder or from the studio, at any given time. When overdubbing, the console is put into playback mode. The individual channel or channels through which the overdub signals are coming are put into input by hitting a single switch on each channel, or a main console-wide control, which is usually marked monitor input reverse or program. The level of the overdub track should be made louder on the control room monitors for the purpose of scrutiny, while the cue or headphone mix is adjusted as desired by the player. In fact, I prefer to have the musician with me in the control room when I set up the cue system mix.

The section of the song to be overdubbed can now be recorded over and over until the musician is satisfied with his or her performance. If a part, such as a solo, is particularly difficult and there are several open tracks available, more than one take can be saved. In this way, a performance that is relatively good can be tried again without the loss of the original take. After recording, say, three or four takes of a part, a composite of the best sections from each take can be put together to render a more perfect rendition. This is accomplished by bouncing or "ping-ponging" the tracks. Here, while in the synch mode, all the overdubbed tracks are mixed and assigned, via the console, to a single track that is placed into record. The track being recorded onto should have at least one open track between it and any of the tracks being recorded from, since if adjacent to each other, high-frequency oscillation may occur.

The section is then played back, levels set to match, and the overdubbed tracks are switched on and off via the console's mute switches so that only the best sections of each overdub are recorded onto the new track. After the new composite track is checked, the old overdub tracks can be erased, freeing them for other instruments.

When an overdub is not during just one section of the song, as when re-recording an entire bass track, punching in and out of record may be more appropriate. If a mistake is made, the tape is rewound to a point where there is some clue as to where you are in the song. The play button on the recorder is hit, the musician plays along, and the record button is then punched in at a convenient place (in terms of spacing between notes) somewhere before the mistake. This process is repeated throughout the song. Another method is that of letting a performance go all the way through, and then going back and punching in and out to correct mistakes. My experience has been that the results are often better when you let a musician "go for it" by continuing the recording process after a flub is made; however, both methods are common practice.

Additional Pointers on Overdubbing   Another engineer asked me to sub for him, as he had two sessions booked at the same time. This is fairly common and the session I was to take over was a routine hi-hat and drum fill overdub. No problem. The producer was also not able to attend, so I questioned and was filled in on exactly the type of sound they were going for. On the date, I set up and tuned the drums, chose the proper hats, and adjusted the EQ while the drummer played until I got the designated sound. Everything was done to the letter. Once all was set, the drummer and I discussed which parts he was to play on, as well as what he needed to hear in his headphones. Because I had never heard these particular tracks before, as soon as I finished setting up his cue mix, I raised up all of the faders in order to get a feel for the tune.

Just as I was about to punch the deck into record, the drummer said to me, "Don't forget to stop." We hadn't discussed this, so I stopped the tape and asked him what he meant. The drummer looked at me like I was nuts and told me he didn't understand what I was talking about. Chalking it up to a communication problem, I started the tape again, only to have the same thing happen. Again I questioned the drummer only to find that he was getting irritated because I was stopping the tape just as he was about to play, which was interfering with his concentration on the timing of the song. Then it dawned on me. I asked him to come into the control room and I played that section of the tape again. Now we both heard the statement "Don't forget to stop," which was obviously already on the tape, but was unheard by the drummer because it hadn't been brought up in his cue mix.

It seems that during the basics, the producer was directing the band via a mic through their headphones. Those verbal cues were also recorded but never erased. It was not a big deal for the original engineer. He knew where they were and could drop them out when they weren't needed. Because it was intended to be erased, it was never written down on the track sheet and therefore I had no clue as to what was going on. The result wasn't that bad (a wasted half hour), which at a hundred bucks and my fee of $25 per hour, came to $62.50. Couple that with the fact that it was my first experience with that particular drummer, who at this point was not all that impressed with me, and you may be able to imagine that I didn't feel too happy.

The final mix of that tune had to, for some reason or another, be scheduled for that weekend. Both the original engineer and myself were previously booked, so the chief engineer of the studio where it was to be mixed took it on himself. I made a point of warning him about the "voices" and pretty much forgot about it. But, when I ran into the chief engineer later, he questioned the capability of the original engineer, telling me he had to spend a great deal of time cleaning up the tracks before he could start to mix. Besides the "voices," it seems there were also bits of outtakes, coughs, tuning up, and other unwanted sounds all over the place. The original engineer was a very competent one, but he had gotten into the bad habit of not cleaning up his tracks.

Any sounds that are not supposed to be on the tape should be eliminated immediately. In this business, if you don't do your job properly, it will not only affect everyone else involved, but come back to you as well. He may not have lost those clients, but I know they got an ear-ful during the mixdown from that chief engineer.

Just to be fair, this next example was totally my fault. I was booked for a session by a studio on the spur of the moment, so I showed up for the date with no knowledge of the band or the material. This is not uncommon for freelance engineers. There were more than a dozen people in the rather small control room. Even though only six to eight of them would be performing, all had comments about the production. The players were well rehearsed and itching to go. Apart from the vocals, a hi-hat, and some percussion, all the tracks were done in the control room (digital drums and multiple synthesizer keyboards); in other words, everything was laid down as an overdub, one instrument at a time. This meant that there were always half a dozen people impatiently waiting to do their thing. Now, add to this the fact that we were working on a 10-minute song and you've got a situation that calls for speed on the engineer's part just to keep the emotional flow going. Those guys got uptight when I got up (after four straight hours) to go to the bathroom! We completed the recording in two days (22 hours total time), which is fast, considering we laid down between 50 and 60 tracks, bounced them down, and stacked them (placed two or more instruments, whose playing times do not overlap, on the same track). Remember, we're taking about a 10-minute tune in analog, where going from the intro to the tag takes a while, even in fast forward. So, as any engineer can tell you, speed is very important. In this case, due to the band's finances, it was an absolute necessity.

Now, there are many times, such as when doing lightning-quick punch-ins and -outs, that you can't look at your meters. If, in addition, the client wants the mix loud in the control room so that he can hear his cues, your ability to make sure a track is laid down correctly (undistorted) becomes tougher. This was the situation during a tom-fill done with a synthesizer, where the artist not only changed the output level, but also the octave in which he was playing. This led to recording levels that were a bit too hot and therefore distorted. Here, the pressure of the momentum got to me and I allowed them to continue on to the next overdub. Anyway, because of the nature of the sound (white noise added to varying pure tones), I knew I could fix the tom-fill in the mix.

A week later, I was back in the same studio for some reason or another, and that client was picking up his tape to take it to another studio where he had gotten "a better deal on time." All I could think about was that synthesizer tom track. I pulled the guy aside and gave him instructions for the engineer who was to do the mix as to how to correct the track. For three days I walked around kicking myself in the butt, thinking about some other engineer's reaction to my distorted tracks! As it turned out, I ran into this same client, in another studio, and I ended up knowing the mix engineer very well. We discussed the problem and everything turned out okay. Lucky me.

What should I have done? Should I have taken a chance of breaking up the emotional flow of the session by having the tom-fills re-recorded? Of course! Additionally, any time you can't hear the instrument being recorded well enough because the control room mix is being tailored to the player(s), give them a set of headphones. Those monitors are for the engineer.

Overdubbing Drums   The first instrument overdubbed will usually be the drums. This is required only if there is a problem that will throw off the timing. In this case the overdub will be needed in order to keep the subsequent overdubs in tempo. Whether the problem is a single missed beat, or something like the drummer's sticks hitting each other on a tom-fill leading into a bridge, the overdub should be done right after the basics have been completed. This is because the traps are already set up and the tones of the kit will match perfectly with the sounds already on tape.

Overdubbing can be very difficult for drummers because their timing must be right on. It is therefore necessary to discuss the punch (in *and* out) points with the drummer beforehand. The engineer should start the playback at the point that gives the drummer enough time to get into the groove, or feel, of the tempo.

Since an early or late punch will cause the erasure of part of the sound's attack, or even a whole beat, it is critical that the engineer also have the correct timing when punching in and out of record. However, it is possible to record the overdub on another track, thereby affording yourself the later opportunity to practice, via the console's mute switches, the punch timing before bouncing the tracks down without any risk of unwanted erasure. Once the drum track's timing is such that it will not hinder subsequent overdubs, if there are other parts to be added, I prefer to let it go until later. Why put the drummer through the strain? It's usually been a long, tiring day and subsequent drum overdubs will come out better when done later when everyone is fresh and you're only recording a partial kit.

Overdubbing the Acoustic Piano   If there is to be an acoustic piano on the song it will usually be the next overdub done. Whether all, or any, of the original basic tracks (aside from the drums) are kept or re-recorded, just as the drums will set the tempo for the rest of the instruments being recorded, the acoustic piano will set the tone, helping determine the rhythm as well as the placement of other instruments within the audio spectrum.

Many articles, books, lessons, and descriptions that concern microphone placement can be vague. I personally believe in experimentation as even the most standard mic setups will need to be adjusted slightly from application to application. However, I feel that telling someone to "just experiment," or to set up two mics, listen to both, choose the better, move the worst, listen to both, choose the better, move the worst until the two sound good is not very helpful. The average professional recording studio has at least 20 different types of mics. Experimentation is fine if there's nothing else to do that day. This is rare indeed and, if you're on a budget, time spent trying out different mics and placements may not be affordable.

Therefore, from here on out, I'll be giving examples of overdubbing many different instruments with both the exact placements and the actual mics I used. These placements and mics are by no means the only choices, but each have been set up by me personally in a recording situation with better than just acceptable results. With this in mind, we move on to some of the various ways to mic an acoustic piano.

Most state-of-the-art recording studios should have a pair of Neumann U 87s or AKG 414s. These are high-quality, expensive condenser microphones. For a classical or full concert acoustical piano sound, these will be the ones most commonly used (see Figure 4-3). They should be spread across the piano to cover all notes, and located about 8 to 12 inches above the dampers (where the hammers strike). Since each mic will pick up both high and low notes, some adjustments may be needed to increase the stereo spread.

The mics can be tilted or cantered slightly (the left clockwise and the right counterclockwise). Note that the lid is wide open. This is intentional as it gives the low-frequency wave forms a chance to develop, and it limits the phase cancellation that can occur from the sound bouncing off this very reflective area. Again, if the same signal reaches the diaphragm from two directions that have differing path lengths, the resulting time difference can cause frequency response anomalies (such as comb filtering).

Figure 4-3
Two high-quality mics located 8 to 12 inches above the dampers spread and cantered to pick up the sound from across the whole piano

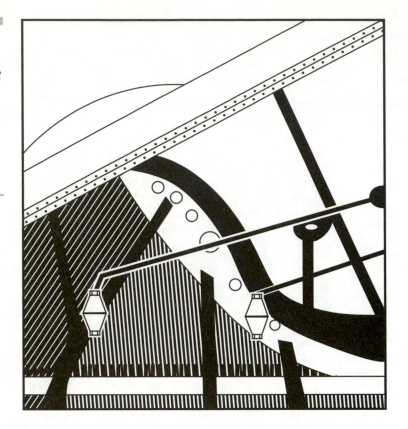

Simply put, because the mics are tilted to pick up lots of sound coming off the soundboard, as well as the hammer strike on string attack sound, this setup will give you a rich but detailed piano sound. If you're not lucky enough to have any of these top-shelf microphones available, it shouldn't be a problem as there are many fine microphones being manufactured, both condenser and dynamic, that will handle this job almost (if not just as well) as these expensive mics. Audio Technica's AT811, Shure's SM81, AKG 451 condensers, and Sennheiser 211 and 441 dynamic mics are good examples. Because of their cardioid pickup pattern (as shown in Figure 4-4), I'll go with an XY positioning that has a setup that's closer to near-coincident.

I would like to say that the acoustic piano micing technique in Figure 4-4B was the result of deductive reasoning based on my knowledge of the fundamentals of polar patter response, but that is not the case.

The fact is that I was checking out some acoustic piano micing techniques while it was being played. The musician was following a set score and I marked all the position changes I made on my copy.

Later during playback the "standard" XY or coincident setup (as depicted in drawing A) resulted in an accentuated midrange. There was just too much overlapping of the two mics' coverage so they "doubled up" the coverage of the mids.

Figure 4-4

Two versions of XY coincident micing as would be set up above the strings where they are struck by the hammers.

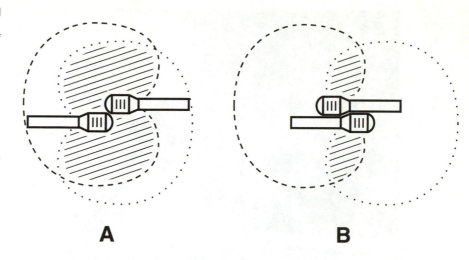

A   B

NOTE: *While this is not something that would normally work for a classical piece, it may just be the ticket for a rock recording where the sound of the piano would otherwise have difficulty being heard in the mix.*

On the other hand, the response from the setup depicted in B was smooth and even across the whole keyboard. While the doubling up of the midrange coverage still occurred, here it was beneficial as it made up for the deficiency of the mic's side pickup.

During the 1980s there was always time available for recording engineers to experiment like this, and the sound quality of the recordings from that era verify that this time was well spent.

If these types of mics are not available, other methods of micing an acoustic piano will work out as well. The recordist can make use of more common dynamics like Sennheiser 421s, as shown in Figure 4-5. They can be mounted on boom stands and placed above the strings by about 6 inches. One should be placed at the rear of the piano where the bass strings are attached, while the other is located nearer the center where the middle- and high-register strings are attached, at the rear of the metal "harp."

Crown PZMs can also be used to get a good acoustic piano sound. I normally use them in conjunction with another mic, but have had good results with them when mono micing pianos, too. They can be positioned all over the piano, with the choice of location dependent on the sound you are after. From underneath the piano placed on the floor, aimed up at the sound-board, they provide mostly a low-end response. (Someday crawl under a grand piano and listen.) When taped inside the piano at the curve, they produce a nice, high-end sound. Attached to the lid, they can give you a surprisingly true sound. The amount of isolation does increase with the lid closed, but things can get a little "boomy" in there. I've always found

Figure 4-5
Two dynamic
mics 6 inches
over the piano
bass with
upper register
strings
attached

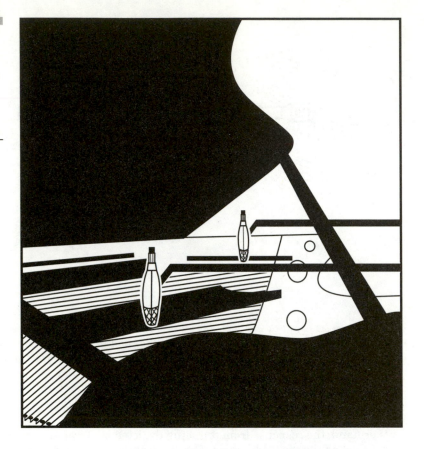

stereo PZM positioning difficult, whether in a V-shape for drum overheads, side by side on wide instruments like pianos and vibes, or spread out on instrument setups like percussion groupings. Affixing PZMs to dual panels attached to mic stands so as to be sturdy, aimed correctly and out of the musician's way, is something you have to be practiced at. This is why the new SASS PZM is a major improvement over "home-made" PZM stereo positioning rigs.

Hey, but this is an overdub session where you get to take the time to set everything up just the way you think it ought to be. For me, this meant removing the piano lid (you'll need *at least* two other people to help with this), and hanging a single omnidirectional mic 6 feet above the center of the piano. As far as I'm concerned, nothing surpasses the sound of a fine instrument placed in the center of a large, wood-paneled room with a quality tube mic (in this case a U 67) picking it up. That's right *mono*. It's a lot easier for those brainless microphones to handle mono recording. Don't get me wrong; I'm a fan of stereo micing, too, but sometimes when the situation is right, a single mic just "says" it all.

Figure 4-6 may seem ridiculous, but a binaural head will give you a sound that'll have many professional engineers asking you how you did it. JVC, at one time, made a set of headphones with a mic located in each can. This unit not only picked up and reproduced the signal, but it also came with a styrofoam head with a mount on the bottom. This head, headphone,

Figure 4-6
Using a
binaural head
to record an
acoustic piano

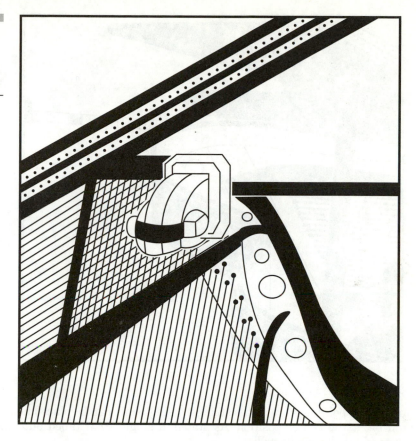

and dual-mic combination recorded sounds binaurally, that is, exactly the way you heard it. In the case of piano micing, all I did was put it on a boom and place it dead center, face down, 8 inches above the strings. The result was a sound that was most acceptable, but the kicker was the stereo spread. In the control room, the piano sounded like it was 20 feet wide. It blew me away. The price? At that time, a little more than one hundred bucks! The same exaggerated stereo spread can be achieved using two condensers in a XY coincident setup with their diaphragms touching at the side and aimed outward at a 45-degree angle like the sides of the letter "L." This rig should be placed dead center, over the strings, 2 to 3 feet behind and facing the hammers.

Another single mic solution (see Figure 4-7) makes use of a shotgun mic. Any one of the many common examples from AKG, Sennheiser, Audio Technica, and almost every other manufacturer will do nicely. Here, we're taking advantage of their tight directional polar pattern. Isolation is increased with the lid on the short stick and the mic positioned in close. In a noise-free environment, the mic can be positioned back farther. In either case it will be aimed at the bottom of the lid.

Here, we're using the reflectiveness of the lid to capture the piano's sounds. If the mic you have is not as directional, it should be brought in closer, positioned above the point where the

Figure 4-7
Mono pickup
of an acoustic
grand using a
shotgun mic

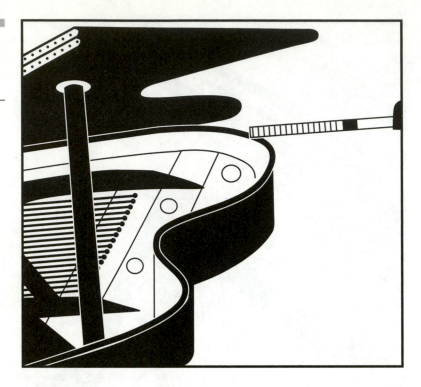

piano curves, and pointed more at the strings. A single mic can be placed near the hammers, but should be at the center position. As stated, a single PZM taped onto the middle of a slightly opened lid also works well.

What about an upright piano? In this case, a mic can be placed at the back to pick up the sound by aiming it at the soundboard or, for that matter, a PZM can be attached to the wall behind the instrument. If you get too much bottom end, try using a second mic over the top with the lid opened to pick up some added highs. More midfrequencies can be had by micing the inside of the piano pretty much dead center.

With live recording, because of the inherent leakage virtually none of the techniques previously described will guarantee you success. Therefore, when doing live work, I always demanded an electric grand piano like a Yamaha CP-70 or a Kawai. If your demands are ignored, threaten, then beg. If an acoustic piano must be used, try a piano pickup; at least you'll get isolation. If that's not in the cards, there's only one thing left for you to do. Grab a couple of handheld vocal mics (anything like Shure SM58s or 57s, AKGs, Beyers, or EVs), whatever's available. Mount these somewhat inside the holes in the metal harp above the soundboard. The smaller holes will give more of the high end, while the larger will give you more of the bottom. The angle at which these mics are inserted is very important and should be adjusted to produce the least amount of ringing. The lid should be on the short stick and the remaining opening should be covered by draping heavy curtains or packing blankets over the piano. At this point, check again for ringing, phase cancellation, or any imbalance in pickup, and then readjust the mic placement.

These examples of micing acoustic pianos could go on and on, but the way I see it these descriptions will enable you to get a very good acoustic piano sound in just about any situation. More importantly, they provide the basis for developing your own techniques and sound.

Guitar Overdubs   Guitar tracks from the basic session are rarely kept, unless an unusually great take was recorded that for some reason just cannot be reproduced. On the other hand, if the performance was excellent but the recording was not up to par, several means can be taken to "fix up" the track. One way is to feed the guitar from the recorder back into the studio through an amp and re-record it to get the desired tonality. Other common methods involve the utilizing outboard gear, which we will discuss later, because re-recording the guitar tracks is the norm.

Look at it this way; during the basic sessions the guitarist had to wait around a long time while the drums were set up and their sounds gotten. Due to this, as well as other distractions, the performance may not have been the guitarist's best. Additionally, because of the restraints placed on the amp's level due to the need for added isolation, the sound of the guitar was probably not at its optimum. So generally (90 percent of the time), the guitar will be overdubbed in order to go for a better sound and/or performance. Rhythm and lead guitar overdubbing techniques are pretty much the same, so there's no need to cover them separately.

At this point the song is beginning to take on a definite sound and feel. The drums, bass, piano, scratch vocal, and even the old guitar tracks already recorded will help in determining how the guitar parts should fit in with the rest of the instruments. But the recording engineer needs to have a bead on the producer's intent. The guitar's frequency content will be adjusted depending on whether the guitar will be a part of the rhythmic background or dominate the overall sound. While the final decision on the guitar's sound, as well as where it will be physically placed within the stereo field (by tone, panning, or the use of reverb and delay), will come during the mixdown process, it is never too soon to start preparing for that stage.

Therefore, whatever guitar sounds you're going for, a major consideration will be orchestration. This is important since, for example, the guitar and keyboard rhythm parts will often occupy the same frequency range. These two should be complementing and interacting as opposed to clashing with or masking out each other. Blend is the key here. After an important point, we'll continue with this instrument, which is often a lot of fun to record.

Communications   Whenever possible during overdubbing, I prefer to avoid the communication hassles inherent in the recording process. I am referring to control room/studio communications. Because both are soundproofed, you must depend on the one way at a time, talk-back button, cue system, headphones, and room microphone hookups, which can be a pain. It's much easier to have the player sitting next to you at the console. He's released from having to wear headphones, the producer/engineer/musician communication is never cut off (or, worse yet, started at the wrong time), and the player gets to listen to the actual sound that is being recorded through the reference monitors.

There are two ways of accomplishing this. First is to supply the guitarist with a long cord (up to 50 feet) or a wireless hookup. I'm not a fan of either of these methods because of their susceptibility to high-frequency degradation (due to the long length of high-impedance cabling) and noise pickup. Taxis seem to have a habit of passing by during takes that utilize wireless interconnections.

An alternate method is to patch the guitarist directly into the console either via a modified patch-bay cable or through a direct box. The signal is then "brought up" to an I/O module or channel, and its cue send is used to deliver the signal to a headphone box in the studio, which is then patched into an amp. In this way, you'll have more control over the sound, but caution should be taken so as not to overload the amp's input to the degree where it is damaged. In either case, the amp's controls should be set after everything is connected since this hookup will invariably change the output from the guitar to some degree.

However, there is a method that is superior to all of these. Just use a guitar preamp situated in the control room. The studio may not have one, so I would bring my own. I had a Mesa-Boogie, tube-based preamp for a while, but ended up with an inexpensive MXR microphone preamp and a heavily modified, original rack-mounted Sans Amp guitar effects processor. While these line amplifier outputs *can* be run to remote power amps and speakers, and I could get *any* guitar sound you'd want with this setup, I'd often opt for the guitarist's own rig.

My personal feeling is to generally let a player use his or her own guitar/effects/amp combination and check them out as is. These individuals have often spent years developing their own style and sound, so who am I to try to force something on them that they may not want? You should consider the guitar-amp combination as a single instrument. Only after I've heard their setup will I make suggestions or experiment. On the other hand, if the producer is looking for something specific, the player brings in a record or tape with a special sound, or if the equipment they have just doesn't cut it, then it's up to the engineer's experience to put it all together.

Like all instruments, guitars should be recorded with as much frequency response and dynamic range as possible: bright, yet not brittle, and warm on the bottom without being muddy. It is also important to get a good sound without radical console equalization or before you start to use outboard effects. In this way you know you have something substantial, so that when you do turn to your outboard and special effects gear instead of using them to try and fix up the negative aspects of the sound, you can use them strictly for enhancement purposes. How is this accomplished?

First is the choice of amplifiers. For everyday pop, jazz, and R&B, small practice or combo-type amps fit the bill perfectly. If you want a cleaner, crisper sound, something like a Roland jazz chorus amp is going to be right up your alley. However, if it's a loud rock 'n' roll cone-ripping rhythm or lead part with lots of sustain that you're after, it would be more in the vein of a Marshall-type amp opened up all the way.

As previously described in the basics section (in the method on setting up DIs and effects boxes), all buzzes and hums should be eliminated first. Let the player warm up with the track. Listen to his or her style, sound, amp, and the room itself. Don't be afraid to move the amp around. Stick a finger in one ear and walk around the room, and make a mental note as to where it sounds best. To avoid unwanted phase cancellation from reflections off the floor, either raise the amp onto a chair or stool, or tilt it so it aims upward.

Now that you've chosen the amp and its location according to the studio's acoustics, different microphones and positioning techniques must be considered. I regard spending a large amount of costly studio time experimenting with different mics and EQ settings as foolish. It is much faster and simpler to go through three different mics placed in front of a speaker at the same time than to check them out one at a time. I would put a high-quality mic at the middle of the speaker, something like a Neumann U 87. Care must be taken as too much volume and/or too close a mic positioning can cause some microphones to distort, but for the most part it will be okay. Also, by putting some amount of pad or resistive load (attenuation) on the mic, the sound should be quite nice. If you're going for a warmer sound, try an old tube mic like the Sony C-37 as their sound can be fantastic, especially when placed farther out from the speaker to let the low frequencies develop. In cases where none of these mics are available, I would run the signal through an old Pultec tube limiter to warm it.

A high-quality mic that is very real sounding is the AKG 414. To each side of this mic I would place another mic. One would be a dynamic. Some examples are Shure SM57s, the Sennheiser 211 or 421s, and the AT41s. A dynamic mic can easily handle the dynamic response of something like a guitar amp, which more often that not will put out a high SPL, so you can use them to get a real close-in sound. Weak amp sounds can be beefed up by using something like a Shure Beta 58 because it has a high output. A dynamic mic's diaphragm acts almost as a limiter. Here, when the sound hits it, it is moved to the extent of its travel fairly quickly. The result of this can cut off certain subharmonic frequencies, which arrive at the mic after the note. The only way I know of to describe the sound of a dynamic that is tight in on a guitar speaker is to say that they have a slow response time. This can mean a mellow sound with a large diaphragm dynamic mic like an SM7, RE20, or a 667A. On the other hand, the normal frequency response of a dynamic handheld vocal mic does not equate with a laid-back guitar sound. So, in short, rhythm guitar sounds benefit more from large diaphragm dynamics, while handheld vocal mics add more bite or cut to a lead guitar's sound.

On the other side of the AKG 414 will be a condenser like an AKG 451, ATM 11, or a Shure SM81. Condensers react to high SPLs differently. They are not self-limiting to the extent of dynamics so that more bottom end or subharmonics will often be transmitted by them. The old line on them was that due to their mylar diaphragm's light weight, they were prone to distortion when subjected to high input levels. This is why maximum SPL ratings are always given for condenser microphones, but today's capacitor mics easily handle SPLs of 120 decibels or more.

Back in the control room, I would now have three mics to choose from without running back and forth. More importantly, some combination of the three may work well when blended together. At this point, I may need to pull back the level of the two outside mics in order to make the center, clearer mic more prominent. Furthermore, since each mic has different frequency characteristics before adjusting the sound's equalization, I can use the console's phase-reversal switches and play around with the phase relationships between the mics to cancel or accentuate different frequencies. In this manner you can check every possibility within a few seconds.

Sometimes for rhythm guitar overdubs, when using a dual-speaker cabinet like a Fender Twin, I would put a mic on each speaker such as a dynamic on one and a condenser on the other. By doing this, especially if working on more than one tune, I can change the sound quickly, without effort. You never really know what type of sound will fit a particular tune until it is actually brought up to the control room monitors along with the other tracks.

Another trick I use with these open-backed enclosures is to mic the rear of the speaker as well as the front. I based this method on an old speaker design called the push-pull cabinet. These had one speaker facing out while the other faced in. The speaker that faced in rendered more bottom-end signal because the center (where much of the highs radiate from) was blocked by the magnet. The two speakers were wired out of phase so that, although aimed in opposite directions, their diaphragms moved in the same direction. Since you can't wire the same speaker both in phase and out of phase at the same time, it is necessary to hit the phase switch on one of the mic's channels to make one mic out of phase. This straightens out everything and gives the guitar a nice, warm sound. If you don't believe me, just listen to a twin from both the back and front. Coles ribbon mics are especially good in this application.

Most cabinets, especially large 4 × 12 enclosures, are designed to project. Therefore, the sound from them doesn't really open up or become complete until they are some distance from the speaker. For this reason, an additional mic is needed, say, 10 to 15 feet back. Using any fairly good mic like an AKG 414, U 87, AT811, or Shure SM81 will capture the fullness of the sound. It is also important to take the guitar direct. This hookup was previously explained. DIs have certain characteristics that you should be aware of, such as the fact that they are more subject to string noise and that occasionally their outputs can be affected by microphone preamps that have low enough input impedances to load down the signal.

Depending on the guitar used, the direct input will generally be cleaner and richer in the midfrequencies. If I was dealing with a loud rock 'n' roll lead with mucho distortion, I would add a touch of direct signal. Not enough that you would notice that it was there, but just enough to give the notes a bit more definition, clarity, and punch. All this will give you a clear, clean, and tight sound. Even with the mic 10 to 15 feet back, mixed in with the basics, with all that leakage and ambience, it can be pretty sterile sounding. If that was not what the song needed, I would also add an additional mic to pick up ambient or room sound. Here, an omni-directional mic, possibly 20 or 30 feet back in the room, will be used or maybe something more exotic like a shotgun pointed into a reflective corner of the studio, or, more commonly, a PZM taped to the glass on the studio side of the control room window. This signal should be recorded to a separate track. Later during mixdown, it will be positioned between the drum overheads. Since some guitar would have been picked up by these drum mics during the basic session, this method makes the guitar overdub sound more natural and it fits better into the basic tracks. If done correctly, the guitar and drums will not sound as if they had been recorded at the different times.

Five mics and a DI may seem a little extreme to you, but my record usage is seven mics and two DIs. On that particular occasion I actually ended up using only a single SM57 for the guitar sound needed. Foolish? Excessive work? Waste of time? No way! All the work was done before the client arrived and, on this occasion, by using the board's mute and phase-reversal switches I was able to give them every possible combination in a matter of seconds. It just happened that the sound of a single SM57 was best for what was being done. Most of the time it is not even close to that easy. You never know exactly what will be needed and a little work beforehand will make you look good when you display an almost endless number of different sounds extremely fast.

The guitarist should have his volume control set all the way up as this improves the S/N ratio. If the guitarist is using any effects of his own, it is better to put them through the amp so he will get to hear the exact sound he wants. Now, aside from a separate track for the ambient signal and one for the mix of the mics being used, I would give the direct signal (without any effects on it) a separate track, too. This is so that later we are not locked into keeping those effects as they are and also to have the ability to pan the two opposite for a fatter sound. For example, by panning the direct hard left and the amp sound with effects hard right, you can adjust the two volumes for the desired sound to change the guitars, the apparent position in the mix, or both.

Okay, you have the ambient sound, the direct sound, and the combination of mics sound. Everyone is happy and they all think that you're the greatest. No problem. But what if it still isn't magic? Well, don't sweat it as you've only used the studio. You still have all your outboard special effects gear left and with this arsenal you will be able to make almost anything sound good.

Say, for example, that the amp has some sort of weird noisy buzz on it. (This may be caused by an inexpensive, pedal-effects device that guitarists often bring with them.) If you put the signal through a noise gate and adjust the threshold to a setting both above the volume of the buzz when the guitar is not playing and below that of the volume when it is, you should be able to get rid of all the noise, which can be most annoying between takes and during non-playing sections.

Speaking of noise gates, I once had an amp that could not be used without one. Slipping back in time, the popular guitar sounds of the late 1960s and early 1970s were pretty raunchy. They used a lot of fuzz box and overdriven amplifier distortion back then. This became a problem when recording basics because the high levels required from the amplifiers to get these sounds made instrument isolation just about impossible. Previous to all this, I had done a bit of messing around with old junk-store tape decks. One was a cassette deck, the kind often used to record speech dictation. First, I permanently disabled the nonrecord safety (cassette tab feeler) arm so that the deck could go into record mode without a tape inserted. I then wired the monitor speaker output connection back to the deck's internal speaker. When I attached an inexpensive mic to the input, and put the unit into record mode, it functioned like a small PA system. I noticed that when the record level control was set high, the speaker would distort like crazy, and too high a setting resulted in screaming feedback.

Metal-lined milk crates were common back then. A couple of hinges, a lid, and some padding made it a nice home for my little Frankenstein. I then glued an old, beat-up SM57 capsule (that was all I had) right to its speaker grill. With a guitar plugged into the line input, the record level could be adjusted for anything from a fairly clean (early transistor) sound to one that was substantially distorted. Like most of these devices, it incorporated an *automatic gain control* (AGC). This limited sounds with a quick attack and a slow release. When it "detected" excessive levels, it jumped on the input and clamped it down. As the level decreased, the limiting was automatically backed off until it reached the point where the automatic gain kicked in. It made this thing sound as if it was just looking for something to amplify. I mean the son of a gun made breathing sounds and when it reached the point of maximum gain and still found nothing, it would sit idle, right there at maximum gain. So when the sound returned, it would scream out for a brief moment until it was squashed down again by the limiter.

This would have been enough for most folks, but not me. I then bypassed the internal record-level potentiometer and ran wires out to my own pot. This was then mounted with an input jack to the outside of the crate with a two-way toggle switch. The switch? Well, this thing could sound pretty mean with a guitar plugged into the line input, but the sound was absolutely frightening when an instrument was connected to the mic input. When I also used a DI to restore a little of the guitar's definition, the sound was a touch cleaner but still massive, to say the least.

Fast-forward to 1985 and somebody (there's always a cynic in the crowd) thinks the guitar is not full enough sounding. Try panning the signal to the left, patch it out of the board into a digital delay unit, put a couple of milliseconds of delay on it, bring it back into a different channel, and pan that one right for a fat, stereo-like spread. In this case, the parts that are played must have some spacing involved or it can end up garbled, too thick, or muddy sounding and have the effect of killing any airy feel that may be required.

I personally prefer real (over automatic) doubling, that is, having the part played twice as opposed to just adding delay to the original. The two separate takes will never be exactly the same. The player will often be ahead of the first take at some points and behind at others. Mind you, the difference here will only be fractions of a second, but if mixed together in mono, it will sound like one take, just fuller. However, when the two are panned left and right, it's a whole different ball game. Even though the timing differences are so slight that they won't really be heard as such in mono, in stereo they can become pleasingly apparent. The guitar parts are now panned: the first take left and its double, right. At one point where the doubled one is behind, the guitar will appear to start on the left and then move more to the right. At another point, where the doubled part is ahead, it will start on the right and seem to move left.

This movement, for the most part, will be random. If you then take one of these signals and patch it into a digital delay unit, add a longer delay (say 20 milliseconds), and bring it up into the mix, panned dead center, the effect is marvelous. The guitar will now travel left to right and then back to center, and another time from right to left and then center. The overall movement honestly seems as if the guitar is moving in a circle or rotating in front of you. In headphones it is almost like it's spinning around your head.

With single tracks (no doubling) I will often use another effect. Maybe I'll put the signal through a harmonizer, change the pitch slightly sharp or flat (less than 1 percent), add a few milliseconds of delay, and pan it a little to the opposite side of the original. The sound will be fuller with an upper or lower harmonic content that's almost distant sounding.

For a Leslie sound on guitar, instead of an effect, try the real thing. Rip open the back of a Hammond organ. If you're lucky, the preamp will still be there and it will have an RCA-type jack on its side. That's an input, my friend. Be careful not to overload it, which means no boosters on the guitar or, if you're using the board's send, keep the level *way* down. Mic the Leslie and you're in phase-shift heaven. Better yet, hook up two Leslies, mic each, and pan them one left and one right. They'll never spin at the exact same interval and with the straight guitar up the middle, the sound is beautiful, especially with something like a 12-string Rickenbacker. Don't stop now!

Because of your close micing, the room acoustics can be avoided altogether, letting you wreak havoc with the growing multitude of variable acoustic dimension devices available. Try super-limiting for a bigger-than-life sound, or super-compression to make the guitar sound like it's breathing. Don't forget reverb and chambers. Send it through a speaker into a hallway, an elevator shaft, or an alley. You'll be surprised at the sounds you can come up with. But enough for now; there will be more on special effects and outboard gear when I get to the mixdown stage. But while we are on the subject, it may be best to finish the discussion on Leslies and how to go about micing these devices.

First off, I consider a Leslie to be an instrument in itself. They come in a fairly large variety of configurations and are probably one of the most hot-rodded (or customized) items found in recording studios. It is common to find the low-end speaker changed out for a better one, either to handle more power or to extend the low-frequency response. The high-end drivers can be changed, too, and at least one manufacturer used to make a special adapter just for this purpose. The amp can be replaced as well, and biamping the unit renders separate volume control over both the high and low end.

The motor speed control can be changed from its simple fast and slow switch to vari-speed and continuously variable settings. I know of one studio that put separate motors on the high and low end, both with their own vari-speed controls. The chief engineer kept a special slightly oversized set of belts handy that, with a single twist, would still fit between the pulleys of the spinning parts and the motors for reverse rotation. Needless to say, some very weird sounds can be had with this thing. I mention this because it is important to ask questions about Leslies you've had no previous experience with. Micing these units is usually straight ahead, but there are some variables. A single mic placed a few feet from the cabinet will give you a good sound, but I find micing both of the speakers individually to be much better. Placing the mics at the vents (either on the front or at the sides) will render a beautiful, wood sound from the cabinet. Obviously, this will be more effective on the bass speaker. Micing the back or opened part of the cabinet will give you more of a straight speaker sound. Here, the use of windscreens is required because the spinning parts of the unit will cause a good amount of wind to whoosh across the diaphragm of the mic, which is not very pleasant sounding. Almost any type of mic can be used on a Leslie cabinet and sound fine. Because of this, your choice should be made to aid in the placement of this sound in the mix. A vocal mic like an SM57 or 58 will boost the upper mid-frequencies, while an old tube mic like a Neumann U 47 will accent those low, rich sounds. A crisp-sounding dynamic, say a Sennheiser 4l6u, for example, will, if needed, help the high end to cut through the mix. On the other hand, a condenser, like a Shure SM81, will be flatter in response and not cut off any of the low end. I've taped a Crown PZM to the inside of the cabinet, which seems to increase that warm, wood sound even further. Some care should be taken, however, to check out the unit. Being a mechanical device of often dubious condition, you should see if it needs oiling, and if its belts and motor mounts are secure. The last thing you need to happen is for it to stop spinning, or for it to start squeaking during a take. The volume control, located inside the cabinet on the top of the amp, can also be used for different sounds. All the way up can be dirt city, while lower settings, while never clean sounding, will be somewhat less raunchy. This is the reason many keyboard players used to swap out the amp for a cleaner, and usually a more high-powered model.

Bass Guitar Overdubs Although generally routine, the bass guitar can be one of the toughest instruments to record. This is due to its wide dynamic range. A low sustained note may hardly register on the VU meter, and when played back you will find that to hear it at all, the faders must be raised all the way up. A note higher in pitch, played with more definition, however, will send your needle right through the other side. Add to this the playing style that makes use of slapping, plucking, and snapping percussive sounds (à la Larry Graham), and you could end up with levels capable of causing damage to subsequent equipment. The result here could easily be distortion. It can also mean bleed onto adjacent tracks, as well as the probability of not being able to fully erase this signal from analog tape (saturation).

We already went over setting up the bass guitar inputs during basic sessions, but since that was fairly brief, I'll spend a little more time on it here. The actual wiring chart won't be repeated as it is the same. Since the DI is getting its signal directly from the bass guitar's pickup, its sound should be true, but you will get more of the bottom end as opposed to an amp sound, which often has an inherent boost at higher frequencies. Therefore, for a rich, full, deep, bottom sound, a DI may be the best bet, but, on the other hand, they also deliver more string sounds. So, if the player starts to slap and pop, the levels are going to get pretty high. An amp-mic combination, on the other hand, will render more growl and mid-bass punch, and tends to mask out high-end string sounds.

At this point, the sound of the speaker itself may warrant a change in cabinets. Smaller speakers generally have higher resonant frequencies so that this choice can be used to complement the more percussive playing styles. Reggae players like tons of bottom end and will jack the bass EQ on the amp all the way up while rolling-off the treble both here and on their guitar to get that earth-moving sound. However, on tape this may come out as pure mud with no definition and therefore too drone-like sounding. Instead of using a 20-inch speaker in a bass reflex or folded horn cabinet, I would hook them into a front-loaded 15, or maybe even a couple of 12-inch speakers. With their style of play, a DI works perfectly, giving me all the low-end that they want, while these speakers give me plenty of definition and punch. Mixing a little of the speaker with the DI rounds out all the notes a bit while the desirable deep low-end remains.

On the other hand, I could get downright radical in the heavy bottom direction when it was called for (see Part 1, "The Basics").

Another example of a bass-playing style engineers can have problems recording is with funk. Many of these bassists like playing through guitar amps in order to accentuate the top end of their percussive-type sounds. I've seen this drive some engineers crazy. "There's not enough bottom, and every time you start slapping the strings my meters go through the top." The player's retort is usually, "But that's the way I want it to sound." Here the engineer is in the control room listening to high-quality monitors and watching his meters go from barely registering, to slamming into the red, while the musician is in the studio, listening through cans and feeling the vibrations of low-frequency content from the amp while playing exactly the way he likes.

The problem is that although both are listening to the same thing, the sounds each is hearing are completely different. The engineer often ends up rolling-off all the top end to reduce the problems. The result is a sound that is definitely not happening for anyone, with both people blaming each other.

I found the best method is to convince the bassist to play through larger speakers in a front-mounted cabinet (but *not* a folded-horn cabinet). This will not overly reduce the amount of high-end percussive sounds, but when a straight bass line is being laid down it comes through rich and full. Then I would add a DI to a separate track.

A high-pass filter can be used to make the high end more distinct. I would then mix this direct signal with the signal from the amp/speaker in the cue system until the high-end sound is dynamic enough to allow the musician to play in his own style *and* hear it the way he wants it in the cans. The bass is recorded to two separate tracks, and later mixed together to a single track with the bassist in the control room, when we are both listening to the same thing. Aside from amp settings, speakers, and DIs, the sound of a bass can also be varied by using different mics.

Often, I'll have a great sound when in the solo mode, and the player and the producer agree that it's the right sound for the project we're working on. I'll drop it into the mix and "Bam!" that great sound gets washed out. In this case, the bass and kick drum may be occupying the same frequency range so that they're stepping all over each other. Here, all the apparent definition of the bass can be completely lost when heard in context. What it may need is a little more low midrange. Using a dynamic mic like a Sennheiser 421 will give you back some of that definition. It may also help to cut the bass EQ at the board between 250–500 Hz, depending on what area of the lower spectrum the keyboard is occupying (as well as the key the tune is in) and the bassist's style of play. Boosting some of the lows between 100–150 Hz and bumping up some mid-highs may return that "great" sound by making the bass come through crystal clear. If more bottom is needed, try a dynamic mic with a large diaphragm like an RE-20 by *Electro-Voice* (EV), a Shure SM7, a ribbon mic like a Royer, or an old tube mic for added richness. If you need more high end, try a mic that you generally reserve for cymbals. If the amp sound is perfect as is, use a more high-quality transparent mic like an AKG 414.

Now that your overdubbing and isolation is not a problem, and mic placement isn't crucial, you can play around a little. Bass notes do not fully develop until some distance from the speaker, so try moving your mics farther out and you'll find the difference to be amazingly more natural.

The sound is now fine, but the levels are still not right. Sometimes they're too low and other times they're peaking out. *Now* is the time to add in a limiter-compressor.

If you set your attack time to be too fast, everything will be limited and you may lose too much punch. That is okay for some things, but definitely not happening for rock or funk. In these cases, try setting it somewhat slower. This will leave the first part of the note as is. In this way, the untouched first part of the note will give you the punch. The rest of the note will still be leveled off so that you won't peak out, or worse yet, have a track that is distorted. The release time must be fairly fast so that the next note will not be affected by the internal setting the device has computed for the first. But if it is too fast, it can cause a pumping distorted output.

No drawings are needed to explain this problem, but picturing a slow-rolling sine wave helps. If the attack is set for a very fast response, the wave begins to be attenuated almost as soon as it reaches the threshold setting. This can cut off too much of an accented note. With a fast release setting, the gain reduction stops almost as soon as the wave's level decreases back down to the threshold point. This doesn't seem to be all that bad, at least for the wide-rolling sine waves. Now imagine an increased number of complex wave forms as would be expected from a music instrument. In this case, the limiter would be switching between nonattenuating and attenuating modes more rapidly. Even though better compressors/limiters produce little or no noise, this opening up and closing down of the attenuation affects the input being processed in such a way that the background noise that's part of this signal is also turned on and off. The resulting sound is often called breathing, for this reason. Now erase this image and replace it with the low-frequency waveforms of a bass guitar and you'll understand why this distortion is also called pumping. It is not pleasant sounding at all.

So play around with the attack and release settings to add a taste of gutsy sound to the bass, but keep these settings a little conservative with this instrument. The ratio and output settings should be adjusted to preserve some of the inherent dynamic range of the bass, which happens to be one of the most pleasing aspects of this instrument. The difference between the quietest and loudest notes should be about 6 dB. The output control can then be adjusted to establish where this range sits, in terms of the level sent to the recorder. I like to keep the console's VU meter reading between −4 and +2 dB (re: 0 dB = +4 dBu). Within this range a bass line can "walk along" at −3 dBu. This is enough level so that when the fader is raised it will be easily heard with little to no added background noise. At the same time, when the playing style changes to one that is more percussive in nature, the +2 to 3 dBu levels will not be distorted.

This sound is exactly what the recordist should strive for. If this is too much of an effort, then get an Aphex Compellor and Dominator dynamics processors and let them do the work for you.

I used to think it was a good idea to carry a set of bass strings with me because sometimes you would run into a bass with old strings that were dull sounding. However, new bass strings require a fairly lengthy break-in period before their tuning becomes stable. I found this out when I witnessed how a bass player, who had just replaced his strings, had to repeatedly stop playing and retune. As far as bass guitars go, your best bet is to talk to two groups of professionals: session players and repairmen.

I do not do guitar intonation settings nor fret reshaping, but I know how it is done, and I can recognize the buzzes and slightly off-pitch sounds they can cause when out of whack. Therefore, I can explain to someone what they need to have done and why. I can also point them to quality repair people or let them use my equipment. No, I don't have a guitar that has been professionally set up and ready to go, but many of the top, world-class studios do. Yet even though I do not play guitar, I own three amps that are absolutely killers in the studio and a bass guitar as well. It is an early 1970s Swedish-made Hangstrom with a short scale neck that is all set up and decked out with 1960s-style nylon-coated strings. What's *that* all about? Well, check it out. The weight of this bass is closer to that of a guitar than a normal bass. Its short scale neck is conducive to very fast playing and its nylon strings are squeak-free. So?

So, I would just hand this Hangstrom to the bassist and ask him to warm up on it. Within a few minutes he would be flying up and down the neck, happily enjoying the unhampered ease of playing a lightweight instrument, while his fingers slid from note to note without making any squeaking noises. Many times I was told at the end of the session that he thought that was the best he had ever played. To me, it usually sounded like happy enthusiastic playing, but the only problem was the occasional difficulty I would have in getting it back from him. Oh yeah, I also have a 10-pound transformer that is specifically used on bass guitars.

Mics, DIs, amps, effects, and, yes, even instruments are all part of the recordist's arsenal of recording tools. But sometimes the recordist's preferences, as far as what "sounds best," are meaningless.

Rock bass players don't like nylon-coated strings because they are not as bright as round-wound strings. Hey, they don't even want to hear about using flat-wounds, let alone a pick. At one time, all bass players used picks. The sound had all the great low-end fundamentals while at the same time being bright and clear. But playing styles change. Today, rock players don't have to worry about finger-on-string squeaks, as these sounds are masked out by everything else.

When a six-string bass with a built-in preamp that includes an active equalizer shows up at a session, it may be best to treat it like a rhythm guitar. I wouldn't necessarily put it through a Fender twin, but it's definitely going to need some top-end reproduction.

Recording Acoustic Guitars Okay, first it's best to look at the instrument itself. Like all other instruments, its sound is based on a certain frequency content as well as harmonic overtones. All things considered, its basic range is somewhere between just below 90 Hz and up to around 3 to 4 kHz.

Here all similarities end, as no two acoustic guitars sound the same. The varieties are endless with different materials; man-made and all kinds of wood; different neck widths and lengths; flat or arched tops; with and without holes; F or round holes; with and without tailpieces; different body sizes; gut, nylon, or steel strings; strings of differing thickness; various bracing patterns; and on and on. So how can I tell you how to record every acoustic guitar that walks in the door? I can't and I'm not going to try. What I will try to do is give you a basic understanding of the instrument and a starting point from which, with as little trial and error as possible, you can achieve the desired sound.

The sound of the guitar is dependent on which particular type is being used. For example, if the strings are attached to the bridge itself, you'll get less midrange, whereas if a tailpiece is employed there will be more of that chunky midrange type of sound. I won't bother getting into bracing, since you can't see what you've got unless you rip the guitar apart. An arched top, F-hole guitar generally has more high end than the ones with flat tops and round holes. These generalities can go on forever. The only real measure of what a particular guitar sounds like is to actually hear it in the environment in which it will be recorded.

Many acoustic guitars that show up at a session are borrowed. Someone says "Hey, let's put an acoustic guitar on such and such a track." "Great idea, you got one?" "Nope." "Well, let's borrow one." "Hey Tommy, you still got that old acoustic?" "Sure."

Tommy would love to be able to say that his guitar is being played on someone's album. The problem is that the axe has either been sitting in his closet for five years, or he's been strummin' the livin' hell out of it while panhandling change on the street. Either way, it's not a good recording instrument to say the least. I can't count how many bad guitars I have seen that were untunable, let alone able to hold tuning, with ancient strings, sometimes even rusty ones. Some have weird buzzes and occasionally one will fall apart in the middle of a take.

Number one, no matter what instrument you are attempting to record, you can save everyone a lot of time, money, and serious aggravation by first making sure you are dealing with one that is in good working condition. Next, you have to know where it is going to sit within the tune in terms of its frequency content: lots of high-end string sound if you want it to cut through rock 'n' roll, more mids for a rhythm part, or some deep bottom for an emotional ballad. Whatever it is they want, you have to first listen to the instrument to see what you have to work with. Acoustic guitars were obviously designed to sound best for both the person playing and the people that are listening to them. Therefore, micing the instrument from either the player's or the listener's point of view should work out. Unfortunately, because of both leakage from other instruments during basic or live recording sessions, as well as the probability of feedback when using an amp or a monitor source other than headphones, closer micing positions are often necessary.

Even with the best mics available, the tone will be quite different than what is heard by both yourself and the player in the room. This is because, like all instruments, the sound is projected from an acoustic guitar in all directions and different frequencies are projected more strongly from individual areas of the instrument.

All these sounds come together at the player's or listeners' ears to make up the whole. Close micing techniques can defeat this. However, you can overcome this problem with the correct choice of microphones, their placement, and, if need be, a little EQ. Even in a no-holds-barred overdub situation, you have to consider mic polar patterns, phase cancellation, standing waves, and the room's reverberation time.

The mic you choose can cancel out a particular part of the guitar's frequency range due to the positioning or because of its frequency response. This can be either intentional and advantageous or unanticipated and detrimental. Therefore, it will be very obvious to all whether you've done your homework or not.

Don't forget the peculiarities of the acoustical environment. Reflective and absorptive surfaces can also be critical variables.

But for argument's sake, let's say you're in the perfect situation: fine guitar, good player, nice room, and an overdub session. Everything sounds great just as you hear it in the room and you want to record the sound just as it is. If you're lucky, you have a binaural head with a good set of flat-response mics in it. You position this where your own head is, and when it sounds best, you record it and move on to the next instrument. Or you can use two identical mics, angled 110 degrees from each other with their diaphragms 17 centimeters apart, and get it all down as is, with at most a tasteful touch of EQ. Rare indeed!

More often, the sound you're going to want to end up with in the mix will need more presence or intimacy. This means that you've got to move the mics in closer. Here is where the trouble begins. You point the mic at the hole and the sound can be so bottom heavy that you have to use an EQ to roll-off most of the bass. Move it to the bridge, and all your detail goes out the window, while at the neck, the presence you're after is nowhere to be found.

Remember that an acoustic guitar is designed to sound best to the player and the listener. Try placing the mic about 8 inches to 1 foot above the guitar pointed downward directly at the bridge (think of the player's ear location). You'll be surprised at how natural sounding it is. To avoid some of the excess bottom blast of some guitars, use a quality mic like an AKG 414 or a Neumann U 87, and back it up a bit to let the sound develop.

Point it straight in at the resonator section of the guitar's body (in between the hole and the bridge), and about 1.5 to 2 feet back. Here, an old tube mic or a good condenser can warm up the sound.

You may still need to roll-off a bit of the bottom end below 100 Hz and add a little boost between 1 and 4 kHz for clarity. To put it more succinctly, if the guitar lacks bottom, aim the mic at the hole. If it needs highs, aim it more toward the neck. For increased mids and presence, the bridge may be a better target.

This is all fine for a guitar that sounds good, but if you're not so lucky, you'll have to know the characteristics of your mics. A dynamic will clip off some of those extra-low bass harmonics, while a vocal mic will accent the high mids. A mic with a limited high-frequency response may help control tinny string noise.

Usually, multiple micing will be needed. Try two mics in close to accent frequencies that may be lacking, say, one at the bridge for mids and one at the neck for highs with a guitar that is mainly bottom heavy. I know of some engineers who go at the guitar from both the front and the back. The front picks up the high end, while the one on the body gives as much warmth as you need. Mixed together, the two render a full sound that can be adjusted for just about any desired tonality. A mic positioned out farther, say, 4 to 6 feet, and mixed in with a close one will help when you need the instrument to be a little less "up-front" sounding in the mix.

My personal favorite micing technique is one that I discovered by accident. I was experimenting with double-micing a guitar that sounded okay, but it was a bit too bottom sounding and because it lacked a bit in brightness and true warmth the result was dull sounding transients. I placed one mic at the bridge to see if that would clear up the sound and left to check it out before setting up the second mic. In the control room, I was happily surprised to find that my single mic had done the trick, and we immediately cut the track. Later I discovered that I had confused the two mic inputs and the one I recorded with wasn't the one at the bridge, but rather the one that I never got around to setting up. Being no fool, I investigated immediately. The second mic was positioned a couple of feet in front of the guitar. It was pointed at and almost touching the floor, which was hardwood. This seemed so bizarre that I had the guitarist play some more while I checked it out. Sure enough, that was the mic I had recorded with and it works great as long as the player doesn't start tapping his foot. Since that time I've mentioned it to several engineers, a few of which told me that it is an old technique and, in fact, quite common.

So, I didn't actually discover it, big deal. It still works fine as will a PZM when placed in this location. Foot tapping can be controlled with a small piece of carpet. By the way, the original mic that this had occurred with was an Audio-Technica ATM 11, but any flat-response nonball-end mic can be used for this purpose. It should be close enough to the reflective surface (floor, control room window, or wall) so that a dollar bill can just slide between the two.

Another favorite micing technique is one I used to get an extremely big and fat sound. Here, I placed a pair of mics like Shure SM81 condensers with their diaphragms touching in an L-type fashion, that is, at a 45-degree angle. This configuration is then placed in real tight, almost touching the strings and close to the neck, above the player's strumming arch. Sometimes, this can be altered to somewhere between the hole and the neck if the musician's style of play permits. The mic that is pointed up toward the bass strings, and represented by the bottom line of the L, is panned to the left speaker. The other mic, which is pointed more toward the high strings and represented by the side of the L, is panned to the right. Now, say that the mic's diaphragms are about one-half inch in diameter, as with the SM81. This is about one-fifth the size of your ear. With the mic in that close, the farthest string will also be about 2 to 3 inches away. The overall effect, when panned hard left and right, is such that the guitar honestly sounds as though it is 8 feet wide! Big, to say the least.

Acoustic Guitar During Basics?   All this is fine and dandy for overdubs, but what if someone wants to use an acoustic guitar on a rhythm date. Punt? Tell them that they are out of their mind? Quit? Not really. Oftentimes the song will have been written using an acoustic guitar. The writer needs to play this instrument during basics in order to convey the intended feel of the tune to the drummer, bass player, and anyone else who is laying down tracks during this fundamental groundwork. In this case, I would go for a technique that may, or may not, give me a keeper track as far as the guitar is concerned, but will definitely enable the feeling to be communicated to the other players through their cans.

Here, setting up the acoustic guitar in an isolation booth will work fine. If it's a good booth and the mics don't pick up too much reflection off the glass windows, I may get a "keeper" track. If for some reason it's not that good, the other players will still hear it fine in their headphones and I can redo it as an overdub. If you do not have an iso booth at your disposal, baffling will at least get the guitar into the cue system without too much leakage. The taller partitions with windows will work more favorably. Pickups will also do the trick nicely and are especially helpful in live recording. Many acoustic guitars are manufactured with some sort of pickup already built in. Some even come equipped with a balanced mic and line outputs. These pickups may not have an ideal sound, but are more than adequate when complemented with an additional mic.

Whenever you put a contact-type transducer on someone's guitar, it is better to use hand-softened wax, or velcro instead of some unknown sticky stuff, as sometimes, when removing the pickup, they can cause the guitar's finish to come off.

During overdubbing, choose your mic according to what's lacking in the pickup. For live or basic session work, choose a mic with a tight polar pattern to avoid leakage. The best guitar built-in pickup combinations that I know of are the Ovation and Takamine. They can be very real sounding. Many studios and sound companies keep some sort of pickup around in case an instrument is not equipped with one.

So far, I haven't found a bridge-mounted pickup that I was totally happy with, but they all can be helpful in difficult situations. The experimentation that I've done with polymer piezo-film pickups was aimed more at capturing low frequencies, but they performed admirably.

Pickups can be placed pretty much anywhere around the body as this whole area of the guitar resonates. The position that I have had the most luck with when using pickups (that are actually tiny microphones) is inside the guitar right behind the bridge.

Lavaliere mics, the kind that newscasters clip to their ties, are also very helpful and again, although not perfect, they are sometimes exactly what is needed to augment the sound of pickups. These can be placed inside the guitar as well, and a position that is just below the hole toward the bridge seems to be the best. More commonly, however, you'll find these clipped to the edge of the hole, facing in toward the center. This can also be okay, as long as the guitar is not overly endowed with bottom end. In either case, a thin ($1/16$ to $1/8$ inch) piece of foam rubber underneath the mic will aid in getting better sound quality (less vibration), as well as help prevent scratching the finish. Of these lavaliere types, the Sony ECM-50 is one of the best and is used by many of the top players. The Sanken OSS-11s, because of their omnidirectional polar pattern, works well inside guitars that are not overly boomy sounding.

These attachable mics are very important when recording live performances as they give the player more freedom of movement since there is less worry about drop-outs in the signal when the guitar is backed off from a stationary mic by as little as a few inches. Whenever possible, I'll also hook up a wireless transmission system since a lavaliere's wires are generally very thin and inclined to breakage (usually during a raucous solo). As far as live sound recording goes, it is also important to watch for excessive low-end content as this will be accentuated by the PA system and the acoustics of the hall. This often produces a low-end droning sound that can wreak havoc on recordings.

All in all, a well-recorded acoustic guitar is one of the most pleasant sounds there is. It can round out the bottom end of a tune, add needed punch to the rhythm, or give you some real nice high-end accents that will easily cut through the din. Once recorded, the nature of its sound is such that it is very applicable to triggering various outboard effects. A little time modulation and the notes start to fall all over each other. Delay will make it sound very fat and, by gating out the attacks, chords will swell out of nowhere, giving you a very ethereal type of sound.

One last point about all stringed instruments played by hand (as opposed to hammer strike) and particularly electric and acoustic guitars. They are shipped from the factory in an unfinished state. That is, they are not completely set up. This is due to the fact that the manufacturers have no idea what any particular instrument will be used for. The setting of a guitar's action will certainly differ from, say, bluegrass finger-picking to blues, and also differ from acoustic jazz as compared to classical playing.

These adjustments, along with the choice of strings, slight neck corrections, and improvements in intonation, are left up to the owner to have done as per their own particular needs and style of playing. Even when set up properly, parts wear and move out of adjustment. Buzzing guitar sounds used to drive me crazy until I learned that frets that have been flattened from wear can initiate string vibrations.

Eyeball how the strings sit in height, horizontally across the guitar's body and neck. They should all be equal in height from the bridge all the way to the head stock. If they are not, intonation could be a problem.

When engineering, I make an effort to take down the name and number of any player that I work with who had an acoustic guitar that recorded well. If another band came in with a dud, I would give this guitarist a call. This often saved the spirits of all concerned because it kept the creative flow going. Sometimes, the person who is supposed to play the acoustic may strictly be an electric player and have difficulty cutting it, especially if it is a 12-string. In these situations, the guitar's owner may be called upon to play the part. This possibility, while far from being guaranteed (along with a small rental fee), is often enough incentive for a cat to come running over with his axe. It is services like this that increase your own, as well as the studio's, value.

Bouncing Tracks   Depending on the microphone techniques used, the instruments that have been laid down, and the number of overdubs needed, so far we could have used as many as 18 tracks. No matter what format you're working in 24, 16, and certainly 8, it's getting to the point where we're running out of open tracks. When there is a need to open up some tracks, or to just clean up any of the loose ends like partial takes, bouncing is necessary. I can't be explicit about actual bounce or ping-pong procedures because they will vary according to the different types of equipment used. Simply stated, all you're doing is taking two or more tracks and combining them onto another single track. The original tracks can then be erased to give you more room. A few pointers about bouncing are necessary. Going from one track to an adjacent one can cause oscillation, so leave at least one track in between the tracks that you're bouncing to and from. Make sure the bounce was done correctly and the mix or combination is acceptable to all before erasing the originals. Finally, if you're working with analog tape, since you bounce in the synch mode, the signal is coming off the record head and is therefore losing some of its fidelity, so to preserve the integrity of the sound, try to keep your movement of tracks to a minimum.

Recording Vocals   Since most or all of the basics are finished, it's time to deal with the lead vocal. I prefer to get this done before doing any of the embellishment instruments such as strings, horns, and percussion as these will be tailored to "fit" around the lead vocal. As you'll see, the methods I use to record vocals often require the use of at least a couple of tracks, and occasionally several. This is because the most important part of any popular recording is the lead vocal.

Of primary importance when laying down vocal tracks is the singer's comfort. It is important to their performance for them to feel at ease. The accomplishment of this entails aspects of both common consideration and diplomacy. Little things like a glass of water nearby, a music stand already set up, and a comfortable chair for between takes are important in making someone feel special.

I don't particularly like the sound of vocals in an isolation booth. They tend to fill up with sound very quickly and the reverberant effects of this can be unpredictable so I'll seldom use a vocal that's been done in one during basics except to use as a reference. On the other hand, the main studio area, while sounding better, can have the effect of making a solitary individual feel very much alone and on the spot. Some engineers eliminate this with the use of baffles around the singer, but for my money I want the openness of a large-sounding room. I find simply turning off a few lights is often sufficient.

The cue or foldback mix, and how it is heard by the vocalist, will definitely affect the performance. For this reason, I'd normally let the singer set up their own cue mix in the control room prior to recording.

Obviously, since there's only one person to deal with, the best and most comfortable headphones should be used. Occasionally, a vocalist will have some trouble listening to themselves through headphones. Usually, lifting one side off the ear and onto the side of their head will help. Today the single-sided headphones utilized for field broadcasting are extremely comfortable. If these do not help, here's a way to avoid the use of cans altogether. Take a matched-pair of speakers and place them both the same distance in front of the mic being used and aimed in at it so that the three form the points of an equilateral triangle. Now, wire one of the speakers 180 degrees out of phase with the others. The signal you send to these speakers must also be mono. This can be done either by hitting a "mono" switch on the board, or panning everything to the center. Because the mic has only one "ear," the mono signal from the out-of-phase speakers will almost completely cancel each other out so there will be little to no leakage. The vocalists, on the other hand, having two ears, will still hear a sufficient amount of the cue mix. Some playing around with positioning may be required, but the end result not only gets you a better performance, but rectifies what otherwise could be a problematic situation.

One last point about your cue send. Any effects or embellishments such as reverberation, delay, pitch change, and so on must be included. These often work in a way that allows the vocalist to control them. Just as a guitarist would need to hear a flanging effect, the vocalist must hear reverberation in order to determine how long a note must be sustained.

Now, while I'm getting levels straight and checking out different effects, I will have the vocalist sing along with the track. This helps in two ways. It allows the performer's voice to warm up, and it gives me an idea of how loud he or she sings. Now is also a good time for the producer to check out different types of embellishments that may be added. These passes will be recorded just in case one is an out-and-out "keeper." We'll get more into effects, as well as things like limiters, later, but first let's back track to mic selection.

Knowledge of the characteristics of different microphones is more important concerning vocals than with any other instrument. This is due to the fact that no two voices are alike. Generally, with guitars, for example, throwing a Shure SM57 in front of a speaker will give you a predictable sound. However, I know of no such generality that applies to singers. So, what I'll do is give you examples of different types of mics used for various vocal sounds along with remedies to several common problems. This will give you a starting point in mic selection as opposed to a set formula.

One trick I've used concerning vocal mic selection is that when you're checking out different studios, instead of simply listening to tapes that have been done there, either bring along a singer whose vocal characteristics you know or use your own voice. Aside from evaluating the room's acoustics, you can use this opportunity to test any mics the studio has that you are not familiar with and make a mental note as to how they sound. Even this is a far cry from anything definitive since two identical mics may have different histories and at this point may no longer sound alike.

Most vocal sessions begin with high-quality, expensive mics like Neumann U 87s, AKG 414s, or rare old tube or ribbon mics. These work fine most of the time but if they don't for any reason, or they're not within your budget, other mics like Audio Technica's AT813 condenser can be comparable in sound quality and very cost effective. Other somewhat standard vocal mics include Sennheiser MKH 80, Sanken CU41s, RCA 44s, Shure SM7s, Neumann U 47s, Milab's VIP 50, RE20s, and various old tube mics. Every manufacturer's top-of-the-line model will usually be a great choice for a vocal mic. All have distinct sounds that may or may not be what you're looking for. You may want to accent the low-, mid-, or high-frequency content of the vocal. Just as a handheld vocal mic will give you more high-mids and those with a good deal of proximity effect will render more lows, all mics have their own characteristics as far as frequency is concerned. For more lows, I will try some dynamic mics like an EV RE20, a Sennheiser 421, or maybe a ribbon mic. If more mids are required for a punchier sound, a mic with a frequency response curve comparable to that of a handheld vocal mic will be the ticket. Many mics have the effect of beefing up the high end. Examples of these would be Shure's SM 85/87 and Sennheiser's 412u, and various other mics you generally find used on cymbals. If the vocalist is somewhat soft-spoken, I go for a powered mic for gain or maybe a more sensitive ribbon like a Beyer or a Shure SM33. Every mic has its uses, and these examples could go on forever. There are cases where no one mic may fit the bill. Here, two or more mics will be used, and we'll discuss this later.

Let's say you've chosen the mic that sounds the best, but there are still some problems. Because the mic is in close to the mouth, it may pick up and amplify sounds that are not heard even by the person singing, let alone someone a few feet away. These include breath sounds that impinge on the mic's diaphragm, adding a popping noise to the signal, "ticks," "fracks," and other sounds from the lips and mouth, as well as sibilant sounds from "Esses," "Zs," and "Chs." All of these have their remedies. Using a windscreen, moving the mic back a bit, or some combination of the two, can generally eliminate pops. Handheld vocal mics are designed with this in mind and before you start to comment on quality, there are several of these mics available that are easily good enough sounding to be used in world-class recording situations and many have been. A couple that come to mind are the Shure SM87, CAD's Model 95, Milab LC-25, Sennheiser's 416 or 441, AKG D-330BT, Pearl's TLC 90, Audio Technica's ATM41, Neumann's KMS 105, and the Beyer M 500.

Other problems may be due to excessive mid- and high end, so using a mic or a positioning that not only does not accentuate these frequencies, but actually attenuates them may be helpful. A mic placed above the head will not pick up the lip smacks as intensely and using an old tube mic can warm up the signal to a point where these sounds will be negligible. If one of these is not available, the more common Shure SM7 or RE20 should do the trick. There's a reason why these mics have been so popular for so long. Finally, several outboard devices are generally available for solving these problems.

Compressors will not stop the breath sounds, but they can get rid of mouth sounds, as will noise gates. Many de-essers are available to eliminate any sibilant sounds. One of the most common of these is the Orban Dynamic Sibilance Controller, which can work wonders. I've also used the FM de-esser incorporated into Valley People's Dyna-Mite with very good results. These days it seems like "vocal processors" are a dime a dozen and I guess that's okay if you want to have everything in a single device, but I prefer the studios I work in to have an assortment of different equipment, as this kind of variety lends itself to a wider range of applications.

So, as you can see, mic choice, while very important, is not the only consideration. The placement of mics to avoid mouth sounds is also important. The most standard vocal mic placement is directly in front of the vocalist, 4 inches to 1 foot away and at eye level. The vocalist should stand straight with his or her head tilted slightly back. This posture is best for opening up the throat, allows for proper breathing, and enables the diaphragm muscles to be used to project the sound. You'll find a singer in this position with one of the above-mentioned quality vocal mics during most vocal sessions. Generally, these mics have or are set to a cardioid or heart-shaped pickup pattern, but if the performer has a tendency to move with the rhythm, an omnidirectional mic or setting may provide a smoother output signal, that is, one that has fewer dropouts or dips in level.

This is all fine, but what if it is not comfortable for the person singing? I can recall one occasion where the performer could not use a stationary mic. He had no previous studio experience, but did have years of live work under his belt. In order for him to "put out," he had to use a handheld mic. On this session I used an AKG D-330BT, which not only has great isolation from handling noise, but also incorporates switches for both low-frequency roll-off to compensate for the proximity effect as well as high-frequency boost capabilities right on the mic. In the end, I also had to shine several hundred watts of light on him and use his two stage monitors wired out of phase for cue. A bit extreme, but the resultant tracks were worth it.

Vocal sounds not only come from the nose and mouth, but from the head and chest as well. A mic aimed at the head, between the top and the forehead, will give you a deeper sound that's difficult to describe except to say that it's "bony." Mics aimed at the chest area also give a deeper sound. I know of one engineer who utilizes both of these mic positions almost exclusively. Here, the vocalist sings straight out between these two and this placement picks up all the sounds while avoiding pops and mouth noises. The mics he uses are an RE20 on the chest and a U 47, set for cardioid, on the head. It works, but I'd add another mic positioned between these two because it gives the vocalist something to sing into other than open space. The output from this mic should be checked as it may provide something beneficial, such as added presence, to the overall vocal sound.

Because of their nature, Crown PZMs readily lend themselves to experimentation. I've taped one facing a vocalist's chest for a real deep intimate sound and if placed on a wall, the whole area acts like a pickup. The sound is more distant, but when added in with the main vocal mic it lends a bit of air to the sound.

As with other instruments, multiple vocal micing also gives the ability to play with phase relationships, allowing you to cancel or accentuate different frequencies at the push of a button. I was told of a session by one engineer where he placed a single mic right in front of a female vocalist, with a cluster of one-half dozen or so other mics a couple of feet back. She thought that she was singing into that one mic, but was in fact being picked up on all of them, to some degree. When mixed together the sound was great. A method I personally enjoy is putting up two or three mics and having the vocalist perform the verse on one, the bridge on the second, and the chorus on the third. The chorus mic will have the most bite for cutting through, the verse will be mellower, and the one for the bridge will be chosen for some other characteristic. Always add a room mic to pick up any ad libs that may come from the singer, as an enthusiastic comment made right into a vocal mic can sound phony.

As far as choice of microphone and placement is concerned, we have only scratched the surface. But with all this in mind, you should be able to get a quality sound with any vocalist. The actual recording of vocals is pretty straightforward. As with any other instrument, you want to capture as much of the dynamic range as is feasible. However, the louder parts shouldn't cause distortion, while those lower in volume must be clearly heard without interference from background noise. No matter which mic or mics (and placements) you choose, some form of compression or limiting may be necessary to accomplish this. Here, the peak or maximum level as set by the threshold, should be processed "softly" so as not to have a rising level slammed to a stop and sound unnatural. Once the peaks have been brought down, everything can be raised back up via the output control to within a safe volume while retaining as much of the original dynamic range as possible.

If a singer makes a mistake while recording, do not stop the recorder. Keep rolling right to the end. Some vocalists can stop and start indefinitely and still continue with the exact same emotion, but most can't. They have to build up to it and the last thing you want to do is blow their feel. Furthermore, after a mistake a performer will generally loosen up, thinking that you're rolling for them to practice. This continuation, along with any verbal encouragement from you, can be very important to the vocalist. He or she often experiments more and takes a "go-for-it" attitude that often yields some great ideas as well as some of your best tracks. If this happens, I'll keep the track with the mistake at the front and quickly patch my input chain as is (for continuity) onto a different track and run it down again. A composite of sections of the two takes may be right on the money, or you can easily drop in words or phrases from the second take over the mistakes on the original track.

When using two mics on a vocalist you can take each to a separate track. This allows you to mix them later while listening to the finished song or, better yet, you'll be able to pan them out. Since they're from the same performance, it will appear thicker, fuller, and have no definite placement within the mix, which is real nice for lead vocals. When dealing with a limited number of tracks, however, your different mics and takes will have to be mixed down to a single track. Here, as with all other recording, continuity (as far as sound and level) is extremely important. It is embarrassing for a recordist to get to the point where the part is seemingly completed, but when mixed or combined the two takes' sounds or levels don't match up.

Occasionally, you'll be lucky enough to work with a vocalist capable of many tonalities. Once the main track is completed, doubling or tripling can be done in order to add these tonal qualities to the original, thereby broadening the color of the sound. For vocalists without this gift there are plenty of outboard effects to help. Automatic double-tracking devices abound. With these, a small amount of delay can be added to make it appear as if two people are singing or you can use even less delay for a tighter thickening. Pitch-shifting devices can give you both upper and lower harmonics. When listened to separately they can make a female sound like a baritone or a male like Mickey Mouse, but when mixed behind the main vocal it will sound like a natural upper or lower harmonic of their voice. We'll get into other effects like chorusing, phasing, super-limiting, and so on, during the section on mixdown.

Recording multiple background vocalists pretty much follows the same guidelines with the exception that you can't use out-of-phase speakers for cueing even if you are using a single mic, as someone will end up blocking a speaker to some degree. When recording *professional* sessions with vocalists divided into two sections, a mic in a figure eight polar pattern may be used, thus bringing everyone down to a single track. In this case, as well as when the group is gathered around an omni, the individual vocalists will have to position themselves the appropriate distance from the mic, or raise and lower their voices for a proper mix. This is more easily accomplished when only one ear is covered by a headphone. You will find that professional musicians will move without a word from you. With smaller ensembles, each person can be given his or her own mic and you can mix these down to one track. A large choir can be covered by dividing it up into several sections, giving each a mic, and blending them together. Vocal micing is something that readily lends itself to variation and experimentation, perhaps more than with other instrument micing techniques. Take advantage of this fact, and don't be afraid to try something new.

Overdubbing Percussion   Although we will be covering the recording methods used to capture the sounds of all kinds of percussion instruments, the best place to start is with the drum kit. The type of mics used and their placement was handled in the section on recording basic sessions; however, with overdubs there are some changes.

First, let's look at some of the reasons for drum overdubs. Often, the drummer has listened to the song's progression from the day the basics were recorded, right up to the present where we've added in the rest of the rhythm section, lead guitar, and vocal parts. At this point, because of everything that's been added, along with any changes production-wise that may have taken place during subsequent recording, the drummer and the producer may feel that the original interpretation, while originally fine, no longer makes it as far as feel. Other times, the drummer hears a slight mistake or feels that they could do a part better. Having heard their performance many times now, they may not be able to live with a certain section. Whatever the case, it is the recordist's job to put in the fix.

Since other parts of the kit may resonate, ring, or cause other unwanted noise, these overdubs should be done using only a section of the kit. This was not possible during the basic sessions, but now it will be no problem to set up only the section of the kit that is to be overdubbed. In other words, if you're recording only the bass drum, get it away from the rest of the kit. The same goes for the snare, toms, hi-hat, and cymbals. Each should be tuned to the old track and miced the same as before so the new sound will blend in with that of the original take. Punch in the part and you are done.

If the original recording of the snare lacks that high-end snare sound, or if the snare tensioning itself was too tight during basics so that the decay does not last as long as wanted, a repair method that works great is to put a speaker into the studio facing down into the drum's skin. Now send the original snare's signal to that speaker. The sound pressure of the original snare coming out of the speaker will cause the snares under the drum to vibrate in perfect time with the original. Put a mic on them and record it in synch. It works.

Since the snare drum is so important to popular music, I'll add a few more tips. Monofilament string, the stuff used on Weed Eater® and other lawn trimmers, makes a very durable and superior sounding replacement for the common nylon, waxed, or other cord (I've seen shoelaces no less) that is used to affix the snares to the bottom of the drum. You need to use the thinner variety and since it is impossible to tie, the drum must have clampdown string mounts. It is worth the effort. The stuff doesn't break easily and lasts "forever," but the sound is the reason I went through the trouble. Monofilament is a little stretchy so it causes the snares to almost recoil into the bottom skin. The result is a sound that really pops.

Another method improves the matchup between the overdub and an original track that has some leakage on it. The new part may come in too clean and therefore stick out as an overdub. Here, use a speaker but, instead of the snare track, feed it *with* any instruments that had leaked onto the original snare mic during the basic session.

Sometimes people have the wrong impression of drum overdubs. They see a single drum, like a solo kick or just the hi-hat setup and think that getting that single piece down should be a piece of cake. Far from it. It is often easier for a drummer to play the whole kit as opposed to only a part of it. Try it yourself. Listen to a song and tap your foot exactly on time, every time the kick drum is about to sound. It is not that easy, right? Now imagine what it's like without hearing the kick, or sitting there anticipating the arrival point of the triple double-tap you're supposed to overdub as the song goes into the final chorus.

I always thought it was part of the recording engineer's job to make things a little easier for the working musician. You have a drummer trying to overdub a kick, so raise it off the floor, mount it with the skin facing up, remove the pedal, take off the beater, and hand it to the drummer. Believe me, this way he'll have no difficulty nailing his part. Better yet, hand him your mallet. Yeah, I carried several varieties of sticks, including a mallet. I even made a set of sticks myself with Super Balls™ on the ends for an extra fast roll.

How about doing the kick drum overdub right in the control room? Rest a mic on top of a yellow pages phonebook sitting on the drummer's lap, hand him a beater, and hit record. What about a reggae tune? Put a face cloth on top of the book. What about a funk, soul, or old-time disco sound? Take the first five pages and fold them over the top by 1 inch. Close the cover and paper clip it at the top, side, and bottom to those pages that follow. The folded pages keep the cover slightly raised between hits so you get that nice "sticky" slap sound along with the deep thump.

I used to also carry my own snares with me until I mounted them on the custom snare drum I assembled. These snares were made for marching band use and measured 4 inches across. The snare drum I built was based on an old Ludwig 14 × 14-inch floor tom. All I did was add the snare mounting hardware (the clamping kind) and swap out the bottom rim with one made for a snare (so it had the snare retainer cutouts). Many people laughed at it *until* they heard it. Its sound said "back off!" A *real* thunder drum, tough-like.

My last drum tip *could be* that if you want to learn about drums, get a job in a rehearsal studio. But there is one more tip.

There is such a thing as competitive marching band tournaments. Winning a competition is a big deal for the schools entered. A lot of alumni money rides on stuff like this. Supplying these "teams" with the competitive edge is a big business. Many of the competitions are held indoors. Accessories that make a drum more directional and more dispersant, and that change its frequency response, not to mention the percussion instruments themselves, provide a whole new well of killer sound possibilities for the audio recordist. Check out the *National Association of Music Merchandisers* (NAM) for a listing of these companies, and better yet, drop by schools near you and check out the bands. They may need a roadie/sound man/recordist; you never know!

Percussion   The next type of drums we'll go over are those that have animal skin heads. While they don't all sound the same (different thicknesses and shell materials will vary their sound), they are similar in many ways. Here you have two main choices: a mic placed tight in with added effects, or a more live room sound. If it's the latter, and the studio you're working in sounds great, put up a bunch of mics such as unidirectionals in the corners and omni-directionals in the middle of the room. You'll end up with at least a couple that sound fine. These types of drums are often rare and I never felt obliged to muck about with them.

However, if you want to pan the different drums or the room's acoustics are not quite right for the sound you're after, you've got to mic in close. First, see how the player sets up his or her equipment. Look at the skins: There's usually a spot that's darker than the rest. Because this "sweet spot" gives the best sound it is therefore hit more often, which is why it is darker. Use very directional mics down close to the skin for separation. If the sound I want is transparent (natural), I'll use a Shure SM81. If that "snapping" sound (as with congas or bongos) is to be accentuated, stick to the standard, which is a Shure SM57. Almost all mics work fairly well here, though. An AKG 414 will be transparent, while the AKG 330 can be set to have more presence. Sennheiser's 421 will be deeper, while the Sennheiser 416 will accent the high mids. About the only type of mic I would not use, besides an omni, would be an old ribbon mic as they have a tendency to distort too easily when the player gets into it, but newer ribbons can give an exceptional sound.

All discussion concerning dealing with animal skin drums can be summed up with this example. I once engineered a session on which batas were used. These are very rare drums that were literally smuggled out of Cuba as they have deep religious importance. Shaped much like an hourglass with one end smaller, these wooden-bodied drums employ a skin on each end, which is tuned by melting wax onto the center. They are played while held in the lap and have both a very low, deep sound on one end, and a high, crisp sound on the other. This studio had a full complement of mics, but since the players spoke no English, I listened to the drums, put an SM81 on each of the low skins, and an SM57 on the highs and recorded both drums without a hitch. Once you've found the mics that work best for you, it's better to go with them in situations like this where time delays could kill the feel. These are not the times to experiment.

I don't care how close someone can program a synthesizer to sound like a steel drum, the original is still better. Here the notes come from all around the bell, so stay away from the temptation to mic it at the rim. Anything close in, from above (outside of a binaural set of mics mounted on a set of headphones, which work great), will either be blocked by the player or end up in his way. Therefore, if you don't want an airy, open live-room sound, go at this instrument from underneath. Use a mic with a wider polar pattern or an omni, anything but a figure eight as floor reflections are not helpful here. The mic should be backed off at least 3 to 4 feet in order to pick up all the notes equally. If you still get too much reflection off a hard wooden floor, place a piece of carpet behind the mic. PZM mics resting on top of a piece of carpet fit this bill to a tee.

The typical timbale player will play the skins, rims, and sides of the drums along with possibly a small cowbell mounted between them. Backing off the mic for more of a room sound may be the ticket, but if the producer wants more of a stereo-panned sound you will need to use multiple mics. I have had mics on the bodies, skins, and bell, giving me a total of five, which were later mixed down to stereo. Placement is pretty straightforward. Look where the stick hits the body and aim there. Make sure the mic on the bell is out of the way, as this instrument renders different sounds when struck all over (even the end). Treat timbales just like toms in that you set your skin mics at the rim and face them in where the stick normally contacts the head. The mics chosen should be as directional as possible and able to take a lot of level, as well as a few direct stick hits.

With wooden percussion instruments such as claves and temple blocks (hollow blocks of wood with slots), the most important part of the sound is their natural mellowness. Unless some other effect is wanted, I would go for a mic that doesn't contradict this deep sound by accentuating its frequencies at the high mids such as would happen if a handheld vocal mic were used. Condenser mics work great here, something like a Shure SM81 or an Audio Technica ATM 11. Tube mics are nice sounding as well. These percussive sounds also trigger digital delay and time modulation units to such a degree that I always immediately run them through one, so as to let everyone choose whether they want these sounds affected or not. It is very important that the player hear these effects so that he or she can play along with them. With the right spacing of multiple delay times, each note will be multiplied into many, and with modulation, the pitch of the notes will rise and fall with a waterfall effect that will cause the track to sound as if it were done by many players. Add automatic panning, and the listener will feel surrounded. The sound of delayed and modulated temple blocks is wonderful.

Marimbas are similar to the previous situation in that they are made of wood, so mics should be selected accordingly. However, like vibes, they are wide enough to make coverage by a single mic difficult unless it is hung about 6 feet above the instrument. With these, I prefer three mics evenly spaced and in close so that all the notes are even in level. By panning them left, right, and center, you can get as big a sound as you want. Most stereo micing techniques also work well as long as they do not get in the player's way.

Vibes, on the other hand, are different, as are triangles, cymbals, small cowbells, chimes, and a host of other percussion toys in that they are made of metal. These (for the most part) are more high-ended as far as frequency response goes. If you want to capture a sound that will cut through any mix, a transparent mic like an AKG 414, a Neumann U 87, an ATM 11, or a Shure SM81 will handle the job. Newer ribbon mics like Beyers and Royers should be good, too. Because sticks are used, irreplaceable older ribbons are out of the question. A handheld vocal mic will increase the high-mids and add more presence. On the other hand, if a mellower sound is what you're after, an old tube mic, an SM81 with no bass roll-off, or any other good, quality flat-response mic through an old Pultec tube limiter will do the trick. If none of these are available, try damping the instrument with tape. The thickness, amount, and kind of tape used will depend on the amount of damping needed (light/masking, medium/medical adhesive tape, heavy/duct tape). Metal chimes, on the other hand, are mellower sounding because they are laid out in a double row, one almost behind the other. I know of no great close micing technique. Instead, I'd put a couple of mics either behind or above this instrument and combine them onto a single track, or go for a wide stereo spread, depending on the situation.

Percussion instruments with multiple attacks, like cabasas and shakers, are easy to mic, as just about any mic anywhere will sound okay, but because they are used as accents or are sounded infrequently, I prefer the sound of a dynamic. What I would do is place this mic in real close, less then 6 inches away, and put it through a limiter, so that I can raise the overall volume without fear of overload distortion. The resulting sound can be bigger than life as if the shaker were the size of a 50-gallon drum. Another often-used percussion instrument is the tambourine. While never a problem to mic, it can still be tricky to get a good sound since, at least with the traditional tambourines, you're dealing with skin, wood, and metal. These are usually in dubious condition and I would often opt for a newer nonskinned one if the original isn't going to cut it. If it is okay, I would always use multiple mics. Place the mics around the player's arc of swing and maybe one above. Since (for the most part) I would pick up everything with the mics in front, I would go for something more exotic from above. An interesting choice is a sensitive ribbon mic. They can pick up the shake sound nicely but will distort like hell on the hits. Try an out-of-phase shotgun mic pointed down, somewhere along the arch of the swing. When mixed with a mic in front, strange things happen, such as certain frequencies dropping out just before the hit.

Quite often a percussionist will come to the studio with a whole trunk full of toys. They will set up a table and lay out as many as two to three dozen different noisemakers. These players have usually done extensive roadwork and have picked up different items in their travels. You find things like whistles, baby rattles, bunches of keys, children's frustration hammers, and who knows what. There is no way to know how they will sound. It is impossible to check each piece and even the musician doesn't know what they will actually play, how loud they will play it, or where they will play it. So what can you do? The best bet here is a pair of quality mics about 4 feet above the table. Again AKG 414s, Shure SM81s, Neumann U 87s, and AT 813s do fine. Aim them for an overall coverage of the table. Again, for stereo separation, try two PZMs back to back at the center of the table, or angle them out to direct their pickup. Some things like tiny bells, however, may not project enough and therefore won't make it in the mix at lower levels. What you need to do here is place a mic that is very directional, aimed in at the table where these instruments are played. The most important thing with percussion is dynamic range. Because it is part of the percussionist's sound, he generally understands it well. So you should not play any games with his headphone send. I will often send him the full stereo mix with his sounds added in at mix level. Then he will be able to hear for himself which pieces should be played closer or farther away from the mics and will generally take care of the input level for you.

You can shorten a triangle's decay time by gaffer taping a small stone to it. If the session entails a bell tree, wind chimes, a glockenspiel, a xylophone, or any other small instrument that has its notes laid out in a row, just for fun I will double mic it, in the coincident style with the two diaphragms at 45 degrees. When this mic configuration is centrally placed and hard panned on the monitors, these small instruments swell to gigantic proportions.

Temperature, Humidity, and Instrument Tuning   The next groups of instruments we'll deal with are more prone to the effects of temperature and humidity. It's true that electric guitars, acoustic grand pianos, drums, and certainly acoustic guitars are affected by the environmental conditions of the rooms they're in, but seldom to the extent where their tuning will noticeably change. It's common sense that the sound of a flute will be more easily affected by temperature changes than a 12-foot grand piano, and it will certainly heat up more rapidly when played. It should also be obvious that because an acoustic guitar is held close-in to the player's body, its temperature will be more stable than a violin that extends out over the player's shoulder with all its sides exposed to air circulation.

I've put in the time to study the acoustics of music instruments and you should, too. It can be pretty boring stuff, but physicists seem to get a kick out of the subject so there's a huge body of research available for audio recordists to take advantage of.

It shouldn't come as a surprise to anyone familiar with music that size counts in terms of frequency content. And when it comes to the violin family, the size range is considerable (from upright bass through cello, to viola and finally those small country "fiddles").

Size also plays a major role in the range difference between baritone and soprano woodwinds, and common sense also tells us that lower notes will be projected through the larger openings of these instruments. However, this is not true of the bell, and the clarinet and oboe's wooden bodies react to temperature change differently than horns with metal bodies. Metal-bodied woodwinds are more easily thrown out of pitch by varying temperatures than the wooden-body (single and double) reed instruments. These shifts in pitch are only slight and all wind instrument musicians quickly learn to adjust their embouchure to compensate for the flat pitch of a cold instrument.

Condensation, however, is another problem for these players. Saliva builds up on the inside walls of wooden horns more rapidly than on metal, but both suffer from difficult play when a mouthpiece gets waterlogged. When leather pads become wet, they can stick in the closed position. The smaller keys and lighter spring tensions of clarinets make them especially susceptible to this problem. Listen to a recording of "Peter and the Wolf" and imagine what it must be like for a clarinetist playing all those 32nd notes when their pads are sticking closed due to being exposed to the constant temperature changes brought about by HVAC units.

This is all common sense, all scientific facts, and information freely available to anyone who may possibly be interested. Yet every night, down in the orchestra pit of theaters all over the world, musicians are being subjected to the result of people ignoring these facts. How do I know this?

Whenever I had a horn or string recording date, I tried to get the instruments in the studio a half hour ahead of time. I *never* removed any instrument from its case, but would only open the lid and suggest the musician fully expose their instrument either by placing it on its stand or lying it on the carpet in a cordoned-off area of the room. I also spent time adjusting the air conditioner's temperature settings and the direction of air flow from the vents.

Studio temperatures should be no lower than 77 degrees for the sake of the instruments, and the control room's temperature should be no higher than 72 degrees for the sake of the electronic equipment. A clarinet player once stopped me and asked why I wasn't employed in theater design, "'cause they could sure use someone like you." When I asked him what he meant, he told me about life in the pits.

Theater HVAC systems must be heavy duty to regulate the temperature of an enclosed space that is often three stories in height with dimensions in the thousands of cubic feet. But do they put vents up in the balcony to chill out the rowdy crew up there? Or how about aiming it in from both sides toward the front of the floor area to keep the expensive ticket holders nice and cool? No chance; they locate them in the orchestra pit aimed directly on the musician's instruments. According to this gentleman, every time there's a break in the playing "everyone starts swabbing and wiping. It's so bad, some of the guys bring sweaters with them."

You'd expect that the thin hollow bodies of instruments in the violin family would react to temperature changes rather quickly. You may even have considered the fact that the strings may be affected too. But how about the bow? It's true. Once, while in a music instrument store (recordists *need* to spend a lot of time in music stores), I decided to buy a block of bow rosin. I always tried to do that kind of thing: A couple new guitar strings in their envelopes, a tube of cork grease, and a pair or two of drumsticks cost you practically nothing but usually make the musicians feel more "at home," and sometimes their availability can help keep the musical flow going. But I didn't buy any rosin that day as there were too many varieties to choose from, including soft, medium, and hard grades designed for cold, medium, and warm *climates!* So when you are recording horns or strings, besides setting up music stands, having chairs nearby, reading lights, and an unencumbered headphone and mic cable layout, think about the room's temperature and humidity, and *definitely* set aside enough time for the musicians to warm up their instruments.

Recording Horns and Strings

We now get to micing and recording strings and horns, first individually, and then as sections. As you'll see, both situations differ greatly. Strings and horns, as with most acoustical instruments, have very complex sound-radiation patterns and project their sounds multidirectionally. What's more, this directionality changes with the changing of the pitch or frequency of the note being played, with brass instruments being more symmetrical in their directional characteristics. To illustrate this more plainly, let's look at a couple of examples.

Bowed string instruments have numerous resonances at both the top and bottom plates at the front and back of the body, which reinforce the harmonics of their strings' vibrations and produce their distinct sound. The low end is pretty much omnidirectional. The low-mid frequency (400 to 800 Hz) is more prevalent at the back of the body and is important in rendering a deeper quality, much like that when you sing the vowel "O." Midrange is better at the right side of the top plate, with high mids at the left. In general, the top plate also yields most of the highs and therefore the instrument's brilliance or crispness. Cellos radiate their lows in an omnidirectional pattern as well, with low mids here stronger at the front, mids stronger toward the neck, and the highs are best picked up right at the neck/body area. Contrabasses radiate low frequencies toward the front plate with the higher frequencies found at both the front and the back.

Contrary to popular belief, most of the sound from woodwinds does not come from the bell, but from all the open tone holes. Only when all the holes are closed, as when playing the lowest notes, does most of the sound radiate from the bell. Aside from this, midrange frequencies come from a number of tone holes that run the length of the upper part of the body. The lower frequencies come from the tone holes at the instrument's lower right-hand side. The highs are projected from the tone holes closer to the mouthpiece as well as the bell. In fact, it's the bell that projects the raspy sound quality of woodwinds the most.

Brass instruments have the most unidirectional pattern, but their low end is more omnidirectional. Here the low frequencies will be trumpet below 600 Hz, trombones 450 Hz, and tuba 80 Hz. But the most distinctive frequency ranges of the trombone (around 700 Hz) and the trumpet (1200 Hz) are best radiated from their sides. Thus, examining the acoustics of these instruments shows that a single, close, unidirectional mic is not quite capable of picking up the full frequency content of these instruments.

However, even on a single instrument overdub date, half a dozen mics placed around, say, a sax is not only impractical, but also very uncomfortable, as it will overly limit the musician's ability to move. If you want a lively take, some compromise is needed and generally double-micing the instrument will be more than sufficient.

Mic Placement   For saxophones, I'd put a mic at the right-hand side, facing the tone holes to pick up the low end. Since I'll want some of its raspy sound, and highs as well as the mids, I point a mic at the top of the bell, but back off a bit and therefore also aim at the top of the horn.

With violins, I'd place one mic about a foot under for lows and another about a foot and a half above the instrument for the high end. Cellos get miced either at the front or the back for the low end (depending on the instrument's sound and separation of tones), with a second mic close to or right where the neck connects to the body for the highs. Contrabasses need a mic at the front for the low mids, while I mic the back for both presence and depth.

Brass instruments are obviously miced at the bell and I use an additional mic placed at the side to pick up more of their rich tonal qualities. It is important to be concerned with placement when using two mics on an instrument as phase cancellation can occur.

Often, I have to move a mic several times in order to avoid the two mics picking up too much of the same frequency content. An aid here is the choice of mics whose frequency response is such that they will automatically ignore either the low or the high end. If, and only if, no mics are available with these qualities, low- and high-pass filtering can be employed.

Once these two inputs are balanced properly, in terms of acquiring the full tonal quality of the instrument, you're ready to print. Because you are capturing the full range of the instrument's sound, you'll find that not only does the sound excel in the mix, but the addition of artificial reverberation is also enhanced due to the fuller frequency content. You can now mix these two mics down to a single track, again checking for any unwanted phase anomalies.

Make use of a good room when you have the opportunity. The previous micing method does involve a bit of patience and practice, but was the only one I found that was acceptable in the acoustically deadened climates of most 1980s recording studios. If, on the other hand, you're fortunate enough to work in a room that has good live characteristics, it becomes a lot simpler. In a reverberant environment, the direct sound contains more high end than does the room sound. A mic placed in close and on-axis with the direct sound will give you punch, definition, and crispness. But there's a big difference in the room sound. Only a small portion of the total sound of these instruments will be picked up by the direct mic. The full omnidirectional radiation of the instrument will be reflected off the floors, ceiling, and walls many times and build up in volume. This sound will contain large amounts of the lower-frequency energy radiated by the instrument in directions other than that which feeds the original direct mic (which again is mostly "presence"-dominated). The room will add fullness and warmth to the sound. Because these rooms are generally somewhat small, this mixture of reflections will reach a room mic quickly enough to follow the direct sound very closely, even during swiftly played passages and rhythmic figures. Thus, you seldom end up with a confusion of note timing or unnatural delay.

The placement of the second mic requires a bit of experimentation The amount of direct and reverberant sound the mic picks up will depend on its polar pattern or directional characteristics, the way sound radiates from the instrument, the position of the mic in relation to the instrument, and the acoustics of the area that surrounds it. While all of these variables come into play, there will still be a preferred mic positioning that provides a superior sound. Try changing the polar pattern of the mic, adding some absorptive or reflective screens, or changing the position of the instrument itself.

My favorite, and one that works most often, is a mic placed a fraction of an inch above and aimed down at a hardwood floor, a few feet in front of the instrument. Try taping a Crown PZM to the control room window's glass panel. You will be surprised at how well it works. As always, be careful of phase cancellations, and choose mics whose frequency responses will aid in capturing the instrument's full bandwidth.

Whatever the case may be, once you've put the time into figuring out the best mic and instrument placements for the room, you'll have it clocked forever. The next time the same instrument shows up, you'll look like a genius.

Recording a Full Section   All this esoteric experimentation with microphone placement is fine and dandy when you're dealing with a single player, but what happens when you're faced with a full horn or string section? You can't have anywhere from 3 to 12 musicians standing around getting paid scale while you play around with a handful of mics. The producer will be on your butt in a minute. So, even though you may not get the complete esoteric fullness of tonal quality, it is probably better for all involved if you go for a single micing technique. The sound quality the producer is after is generally going to be more punchy and attack-accented anyway. Here, after all the horns are lined up, music stands set up, and headphones run, start placing your mics on the individual players.

With brass, aim right at the bell, about 1 to 1.5 feet back. Always use a mic that can handle a high SPL on horns, as these can deliver a very hot signal. For the tuba, the mic should be above the bell about 2 to 3 feet and pointed down. For this reason, if you want separation, this instrument should be behind the line of fire of the other brass instruments. Trombones, as do tubas, have a lower-end response than the trumpets, so mics should be chosen accordingly. Their output is often less than that of a trumpet so the mic should be placed closer in on the bell, say about 6 inches or so. In all cases, check to see where the instrument is being held and ask the musician if this is the way it is generally positioned. These players definitely want their instruments to be picked up and will therefore be more than helpful when asked about their playing styles.

Now, move your woodwind and flute players off to the side. If no isolation booth is available, go-bo around them so that not too much of the brass is picked up through their mics. On saxes, place the mic facing into the bell but aimed up a bit to pick up the sounds out of the body's tone holes as well. Either end of the flute will render a more breathy sound, with some sounds coming from the tone holes, but it is mainly projected right at the embouchure or lip piece. In fact, many of the top players use wooden (lip and body) head joints because they darken or mellow out the instrument's sound.

We're all set up, so now what? Once everyone is miced up, you check the blend, reposition mics if necessary, and set levels. If any particular instrument or section is not as hot as you need it to be, add another mic to that area. All in all, in this way, with not too much effort and very little time wasted, a 12-piece horn section can be set up and its sound gotten down with a level of fidelity that is very good in terms of punch and presence. If the desired sound needs to have more warmth, mics can be repositioned for more proximity effect, which will somewhat compensate for this limited micing technique.

Another trick is to run the whole mix through an old, tube-leveling amp like a Techtronics LA-2A, adding even harmonics, which always warms things up.

So far, you've had to use no EQ. On some instruments (that is, other than the loudest horns), it is safe to use an old tube mic without worrying about damaging its irreplaceable capsule. This should give you a warmer sound.

After you've become better acquainted with the double-micing techniques previously described, and have had a chance to experiment with the room you're dealing with, smaller horn sections can be double-miced without taking up too much time. When attempting this, however, a certain amount of separation is needed between the players to isolate the different instruments from each other. Go-bos should be placed between the instruments in such a way as to not block any of the player's visibility, both between each other and the conductor. Half-height go-bos will be fine between the woodwinds, as they are held lower.

String Sections   When it comes to string sections, these people are more often than not classical players and therefore masters at ensemble playing. In other words, professionals at blending sounds. This makes the engineer's job a piece of cake. I've done string sections with a single omnidirectional mic in an acoustically well-balanced room with great results. If a stereo mix is desired, have them play a passage while you move about the room. When it sounds good to you, put up a binaural head mic like a Sony, Sennheiser, or a JVC (if you can find them). If no binaural heads are available, take a matched pair of mics, place their diaphragms 17 centimeters apart, and aim them at a 110-degree angle from each other. In either case, the placement should be exactly where the sound is best as heard by you in the room. A great mic in this situation is AKG's 422 stereo mic. Put this at that location in the room and the sound should be on the money as far as the blend and the stereo spread.

To help these musicians in self-blending, I'd give each one a single-sided headphone, or have them place a double-sided one on so as to leave one ear uncovered. Then I'd feed them the tracks they needed to hear, minus themselves. I then adjusted the level of this feed so that they could still hear their own playing (they'll tell you when) and hit record. It's that simple.

If for some reason the musicians you are working with are not studio pros, or one of the instruments is not coming through loud enough so it's getting lost in the mix, you may have to add another mic. That mic should be fairly unidirectional and aimed at the area of the instrument that will render the frequency content you need.

Back in the control room, bring this instrument up to a level that correctly fits into the mix of the rest of the section and you're set. Hit record and, aside from the possibility of needing to punch in, you're finished.

Pickups, as covered in the mic section, can be very helpful in capturing the low-frequency content of these instruments while still providing separation. However, be careful when adding these to contrabasses and cellos as their high outputs can damage control room monitors if they are cranked up.

Mixing

The Road to the Pre-Mix Stage   Close attention must be paid to capturing an instrument's full frequency response and dynamic range as distortion and high background noise levels will be impossible to correct later on. The instrument's timbre must blend with the overall conceptual tone of the song while remaining faithful to the musician's own personal sound. However, more effort must be put into capturing the sound of instruments that are recorded in a purist style, that is, dry, or without effects, since the recording engineer must now pay particular attention to including the instrument's complete frequency content and as much of the its dynamic range as possible. By a purist's style of recording I mean capturing an instrument's sound exactly as it appears in the studio.

The engineer may opt to augment the sound to improve its quality, which can be acceptable as long as it is accomplished without excessive processing. The main intent when recording any instrument is to keep the sound open and natural, and still have the maximum amount of information to work with so that it will be receptive to effects later on during the mixdown stage.

If the tracks are available, I'd record something like a lead guitar or synth track with the effects sounds that the musician wants, as well as a dry track without the effects, in case later on things change. I've seen a single overdub completely change the direction a tune was heading. At any point a musician may find themselves with a new toy and they feel its sound is just what the song needs. By having the dry track, the original sound can easily be changed even during the mixing process.

Therefore, the engineer must keep in mind that all recording should be done as preparation for the final mix. Equalization should be used along with mic selection and placement to realize a frequency response that blends the instrument's tone or timbre with the intended or conceptual "sound scape." Noise gates and compressors are used to reduce noise and distortion while at the same time augmenting the recording of the instrument's full dynamic range. While building up all the tracks for the mix, the recordist needs to make a habit of checking for phase anomalies that can cause quirks in the frequency response. When recording purist or classical music, not capturing every bit of an instrument's frequency content may be regrettable, but ending up with tracks that have high levels of background noise or *any* amount of distortion will be considered a crime. Here, an important part of the recording engineer's job is keeping vigil over the integrity and authenticity of the sounds so that they will better fit the intended aural picture when it comes time to mix.

Mixing for Edits   Don't forget that the mix is not the end result of your work. Mixing is actually another step in the buildup and preparation of the song for editing and mastering. The act of mixing for edits begins during the initial recording process and continues all during the subsequent overdubs. This is why it is so important that the recording engineer be able to at least follow along with a music score. This is especially true when long "dance versions" of songs are being done or are on the horizon. Here's an actual example: A drum figure during a bridge that was always used as an intro to the chorus that ended with a heavily hit crash cymbal. I pointed out to the producer that this may become distracting if it repeatedly sounded throughout the extended version that they planned on compiling. Furthermore, the length of the cymbal's decay would make many of the proposed edits difficult and some downright impossible. We decided to have the band play just the bridge section a couple of times. The drummer whacked his sticks together on the first one and the producer provided aural cues on the rest. I also insisted on recording a solo of that crash hit so that it could be sampled and triggered to sound over any edits that needed some help in achieving a smooth transition. This is the kind of thing I was referring to when I said that the recording engineer's job may possibly entail small suggestions concerning production.

These days the biggest concern as far as the final mastering process seems to be whether to use the highest sample rate or the same rate that the end product will be processed at. What's wrong with mixing to both?

The actual media, tape, CD, hard drive, or DVD is meaningless as far as cost is concerned when compared to the expense of remixing. During some of my mix sessions, I'd have the output buses feeding up to six recorders: an analog ¼-inch deck recording and feeding a repro of the mix to a DAT recording at its highest sample rate; two half-inch analog decks, one running at 15 ips aligned to operate at elevated levels using no Dolby noise reduction (NNR) and another running at 30 ips; two CD recorders (one just for backup); and a DAT set to a sampling rate of 44.1. If there was a possibility that a video would be made for a song, I'd also bus the mix to a video deck along with time code patched right from the multitrack. But that's just my way. Even during the days of Quad, when mixing through an encoder, I would simultaneously mix the four buses to an analog four-track deck.

When I'm called on to record a project, the first thing I want to know is if there is a possibility that it will be used for film or video. If the answer is yes, I'd call the post house and find out how they liked to receive their mixes. If it were to be broadcasted live, I'd call upon some of my contacts in broadcast or call a couple of program directors to see what levels are currently passing through their system without being whacked by limiters. As far as the album release, I would simply ask the intended mastering engineer or record company what they wanted.

Therefore, since you'll be mixing to prepare the material for the editing and the mastering stages I think it's best you start with an understanding of what these are all about before we get into the mixing process.

My hope is that it will help you to avoid making the same mistakes I made when I first started out in early 1970s.

Editing

In order to be a good editor, you have to have the attitude of a sharp-shootin' gunslinger. You hear the tape, know the place to edit, get to that place, and cut. Any hesitation or doubt, punt. This skill is both natural and developed. Natural in terms of having, if not perfect, at least the relative pitch perception needed in order to distinguish between sounds that are played at slower and faster speeds, as well as in reverse. Your brain/ear coordination can, however, be developed to make these distinctions along with those of tempo, feel, and musical timbre. I know of no truly adequate study guide or teaching process for editing, so what I'll do is relay to you the best way I know of how to learn this skill.

First, the tools: I'd suggest buying your own editing block, as often the one attached to the studio's two-track will be a bit worn (the size of its cutting slot could be up to three times wider than when new). You can easily tape yours right to the top of the deck. It's kind of like using a mouse or joystick you're familiar with in that it's more comfortable to work with and can thus speed the whole process up. Next, get demagnetized single-edged razor blades, lots of them; you'd be surprised how fast mylar can dull steel, so change blades often.

Next, grease pencils you should have at least two different colors (helpful in distinguishing between markings).

You'll need splicing tape. I'd suggest either Scotch or Alien Bradley, as both have good strength, consistency, and the right amount of glue. Pre-cut tabs are also good, but I was brought up using a roll of tape on a normal Scotch tape dispenser. Clean hands are a must. White gloves are not necessary, but nobody wants funk on their tape. Get into the habit of keeping these items in the exact same place as flow is important in editing.

The only types of tape that can be edited correctly are those with information running in only one direction. If your deck has the ability to record two pair of stereo tracks, one running in each direction (half-track decks), you can edit, but your second stereo track will never match up. Cassettes are even more problematic to edit. Even if all tracks are running in the same direction (as with porta studio-type decks), there is no way to cue or position the tape accurately before you mark it. So what we're dealing with in this article is your basic two-track deck. Multitrack tape (16 or 24 track) can be edited, but you'll need a lot of experience before you should attempt it.

In order to perform an edit, the tape deck you're using must fulfill two requirements; its playback amplifiers have to be on, even when the deck is not in the play mode, and its brakes must not be firmly locked in the stop mode. Consulting the owner's manual will clear this up, but if yours is lost, simply move the reels by hand in the stop mode. If they move without excessive effort and you hear the sound recorded on the tape or see the VU meters react, you're fine. All decks must first be laid down on their backs. If this causes a problem due to connectors plugged into the rear panel, the deck has to be supported in this position in such a way as to not rest on these connectors; wood blocks work well.

Let's try it.

Even though the following processes as described are for editing on an open-reel tape deck, you can try them out whether you're using a mouse, joy stick, or touch screen, "scrubbing" in forward and backward motion.

The first thing we'll be editing will be a click track, so borrow a metronome and record a few minutes of this sound via either micing it or using its direct output, if it has one. Set it for about one beat per second. This signal is simple to edit but will still teach you to cut before, after, and in the middle of a sound. You'll also gain experience with basic tempo as well as slow and reversed sound perception. Once you've laid down your track, go back to the beginning, find the first click, and stop the deck. What you now want to do is find the very beginning of that click.

With open-reel tape decks, this method requires a bit of coordination, as well as the use of both hands. While in the stop mode, place one hand on each reel and rotate both reels at the same time counterclockwise and then clockwise.

You rock the tape back and forth across the playback (repro) head. When you get to the point where you hear the beginning of the click, stop and the brakes will hold the tape in that exact place. To do this effectively, your monitoring environment should be relatively quiet and your volume levels adequate to hear low volume signals. Frequently watching your VU meters is very helpful.

Next you need to know where to mark your tape. Sound is produced at the gap of the playback (reproduce) head. When you hear something, it is on the section of the tape that is positioned at this gap. Finding this is easy; it's in the middle of the head farthest to the right. However, there are still some decks around with only one head having two gaps. Again, it's the right-hand gap and it can be easily seen with a magnifying glass. Most engineers, including myself, marked the tape right on the head. But this can easily leave undetected grease pencil residue on the head, hampering correct reproduction and even cause uneven head wear. Furthermore, you might knock your head out of alignment, causing incorrect reproduction.

So, where do you mark the tape? I like the edge of the head assembly casing just to the right of the repro head, but anywhere along the tape path of the deck is good as long as it's within 2 to 3 inches to the right of the repro head. You can even glue something onto the deck if you wish as long as it doesn't interfere with the tape path. Find the beginning of the first click and mark your tape at both the repro gap and your new marking position.

Removal of the tape from the heads can be done by simply pulling it out and away, but if a deck is faulty or the brake tensioning is adjusted too stiffly you can stretch the tape. If it's someone's master, you'll be out of a job. So the best habit is to rotate the right-hand reel clockwise until you've got enough slack to reach your edit block. The edit block's tape slot is not as wide as the tape itself, so you have to push the middle of the tape into it with your index finger. Now run both index fingers out from the center of the block.

The tape will now be held into position via the pressure exerted by the tape itself against the dovetail edge of the slot. However, it still can be easily slid from side to side for positioning. Line up your left-hand mark directly over the 45-degree angle blade slot. The right-hand mark should also be within the edit block; if it's not, find a new second marking spot. This mark's position is then permanently recorded on the block itself either by a scratch or indelible marker. From now on, anytime you mark tape at the new marking spot and place it at the newly recorded point of your block, the blade slot will match up perfectly with the section of tape that was at the repro head gap. Now cut the tape with a smooth slicing motion. It doesn't take a lot of pressure to accomplish this, so if you find it at all difficult you're probably bending the blade and cutting into the block. Practice until you have a smooth, even stroke. Once the cut is completed, slide the two pieces of tape away from the blade slot.

Look at them to see if you've accomplished a correct cut—an even, straight 45-degree angle with no rippled edges. Now place the front of a foot or so of white leader tape into the block and cut it to form another 45-degree angle. Slide the left-hand section of leader out of the block and butt the left-hand section of tape up to the remaining right-hand section of the leader, at some position other than right above the razor slot, and push them together until they butt up against each other. With tape-to-tape edits a small gap (no more than $1/32$ of an inch) should be left between the two pieces. Stick about a $3/4$-inch length of splicing tape on the edge of your razor blade.

Use the blade to guide the splicing tape to a position above where the tape and leader meet. Press down on the splicing tape with your left index finger and remove the blade. Use your fingertip to adhere the splicing tape (it should end up almost the same color of the recording tape). Now pull this composite tape from both sides and lift. It should pop out of the block. Never peel tape from a block from one side or you'll put ripples in the tape's edges, which could ruin it. Repeat this process for the right-hand sides of the tape and leader. If all does not go well the first time you try this, don't despair; take short cuts, or settle for shoddy work. Keep at it 'til you get it right.

Now, reload the tape onto the deck by rotating the right-hand reel counterclockwise, making sure that you don't deform the tape or rethread it incorrectly. Since you've leadered the beginning, it will be easy to find the first click. Position the splice to the left of the capstan, hit play, and count to the sixth click. Running the new splice between the capstan and pinch roller helps it to adhere. The sixth click represents the second beat in your second $\frac{4}{4}$ measure. Find the beginning of this click and mark it. Go to the seventh click and do the same. Edit out the section of tape that runs from the beginning of the sixth click to the beginning of the seventh. Rewind your tape and play it back. The tempo should remain the same, the loss of the sixth click should not be detectable. If it is, try again and keep doing it until you get it right.

Once you've accomplished this, pick out another click and mark the end of its sound and do the same for the next one. Edit out this section of tape and repeat the process until its loss is undetectable. With this done, it's time for you to try to edit from the middle of a click. As you rock the tape back and forth from the beginning to the end of the sound, note the distance by marking both and measuring, or eyeballing and marking the middle.

Do the same for the next click and edit out the section containing the second half of the first click to the second half of the second. Again, repeat this process until the edit is undetectable.

By now, you should have noticed that you're always cutting both clicks in the exact same place, beginning, end, or middle. This will also work with any other fraction, such as $\frac{1}{4}$ or $\frac{2}{3}$ of the way through a sound. This is because we are dealing with the same sound set at an exact tempo. Later when we get into musical "feel" where the tempo can be varying, elastic, or anticipated, you may have to cut with a bit of tape added or subtracted to retain the proper tempo, feel-wise.

A good introduction to this can be obtained now by varying the tempo of your click track. Add another section of leader after all your previous edits. Count off three clicks. Mark the end of the third and the beginning of the fourth. Edit out half the space between the end of the seventh and the beginning of the eighth. Repeat this for the ninth and tenth. Clicks 10 and 11 are left as is. Take out half the space between the end of the eleventh and the beginning of the twelfth and at the end of the twelfth and the beginning of the thirteenth. Now rewind and play this section back.

You've Got Rhythm

You have just created a rhythm. If it sounds out of time, try again until you get it right. At this point, you can see that you could actually compose a rhythmic structure for a complete song out of a single sound. Next you need to get into distinguishing between different sounds, understanding feel, and complex rhythms.

Determining whether a certain sound is that of a high-pitched guitar note around 1 kHz or that of a word sung at 800 Hz is no problem in real time. But when the tape is running at 15 ips, check the same two sounds out while you're rocking the tape back and forth across the repro head and it's a different story. Here, the tape is traveling at, say, 1 ips and the frequencies drop to very low levels (around 53.3 Hz. for the vocal and 66.6 for the guitar). Definitely not easy to distinguish between, yet it's still possible.

The brain/ear combination is capable of distinguishing between two sounds even closer in frequency than 10 Hz. You're also aided by the fact that the space between these sounds is, with manual movement, now 15 times greater than before. The helpfulness of this distance in editing has already been seen in our previous work. So now, beg, borrow, steal, or buy the simplest rhythm machine you can get and record a couple of minutes of a tempo you like. Now, hold it. Before we go any further, we have to discuss a sound's decay time.

Ride cymbals have a longer decay time than even a ringing tom, which will certainly hang out longer than a deadened bass drum. If you try to edit out a snare beat that occurs while a cymbal is decaying, it's not going to work. *This is why we're dealing with editing before we get to mixing.*

Let's say a producer wants to delete the second "B" section and go right to the chorus. If, just before the part to be edited out, there's a cymbal sustain that crosses into it and you make the cut, that cymbal is going to be noticeably chopped. The result is not going to pass; therefore, during the mix you'll have to mute or fade that cymbal. Even with simple rhythm machines, there will be beats occurring under decaying sounds and therefore not accessible to the editor, so a lot can be learned from this type of track. First, pull out of the mix a snare, a bass, or a crash. To do so, you'll really have to listen to those sounds at a much lower frequency, which is just the kind of practice that's needed. Your VU meter will help, but your ear is the key. You'll find that some beats can be taken out while some cannot.

Machines can be very boring to play along with, but with a bit of editing you can put together a complete song with more of the human feel you'd get from a drummer. Start by pulling out different beats and replacing them with unrecorded tape of the exact same size. Now, try removing blank spaces in the next measure and add in some of the notes you cut from the first. You may have to play around with the position of the preceding beat as well to keep the tempo uniform, but you can end up with, say, double snare beats very easily. Once you've completed a couple of measures of this syncopation, make a tape loop and add in a few unchanged measures. There's no need to edit every measure. Now transfer this looping track to another deck.

Next, work on your bridge or "B" section and chorus, until you've got a complete song.

This exercise will not only increase your editing ability, but also your appreciation of drummers.

Editing the Spoken Word

This is one of the most complex things to edit, in terms of distinguishing between different sounds is speech. In order to separate syllables you must be able to tell the difference between various phonic sounds or tones; not simple, but again not impossible. The vowels (A, E, I, O, and U) are the toughest, since, depending on the speaker, they can all sound exactly the same. The best ways to pick them out are by distinguishing the sounds around them, listening while rocking the reels backwards, and watching for a dip in your VU meter as these sounds are generally softer than those of consonants. Consonants are harder and louder since they're almost percussive in nature (say "B" as in boy, "C" as in cat, "D" as in dog, etc.). Take the word conquer. It starts off with a hard click-like sound. The "O" and "U" will be joined together, followed by another clicking sound of the "Q" ending with the "UER" also joined together. Therefore, the easiest edit points would be either before or after these click sounds. In a situation where the word was supposed to be conquest and the line was spoken as "Following the army's conquer-uh-conquest of the enemy's strategic position . . .," you may think that you can just edit out from the beginning of the word "conquer" to the end of the "uh," but it would not work that way. The most important aspect of speech editing is making the edited version believable.

Since parts of words can be accented in different ways for different situations, extreme care must be taken in editing speech. After a mistake like the previous one, the speaker would naturally accent the beginning of his own correction. By simply placing this accented beginning up against the preceding word (enemies), you'll throw the flow of the sentence off. This is precisely why you must be able to splice in the middle of words.

The correct edit points here would be from after the unaccented "con" of the word conquer (just before the second click sound) to after the accented "con" of the corrected word. The sentence will now flow correctly, and therefore your edit will not be detectable.

Sometimes mistakes will go undetected by the speaker, like saying "computer" instead of "commuter." Here the "P" will be a puff-like sound. Splicing out a bit of the "mm" and adding in the same amount of blank tape before the very beginning of the "U" will take care of this type of problem.

Why Edit Speech?

Why edit at all? Well, often you'll find yourself dealing with broadcast time, a very precious commodity. The amount of flubs (mistakes) and extra phrases (ums, ers, maybes, I thinks, etc.) can easily eat up valuable time and must be gotten rid of. It gets to the point where, especially in commercials. I'll end up speeding up the whole narration just to be able to fit it into the allotted time slot. This process is relatively simple, but requires two things: some way to control playback speed and a pitch transposer. If the commercial is a 10-second spot and the narration runs 12 to 13 seconds, speed up the replay so that the length will be 10 seconds. Now, since the pitch has been increased, run it through a harmonizer and drop it back down to the correct pitch. Some digital processors handle both of these steps. These devices bring the pitch of the speech back to the proper point, while shortening it's length time-wise.

Speech is also very complex, as far as tempo is concerned. Everyone has a distinct pattern for all their emotions. These patterns must be adhered to in order to make the edit not sound like an interruption. This is more important in editing longer programs, such as documentaries. Once all the flubs are fixed, you may find that whole phrases that are not important to the dialog must be eliminated. The way human speech works physically is very helpful here. You'll state a phrase, take a breath, and then continue. These pauses for breath make perfect edit points. Again, the trick is to keep the flow natural and continuous. For this reason, the breath you choose to keep or cut is important. Always keep the breath before the edit. It goes like this: sentence 1 (A), breath 1 (B), sentence 2 (C), breath 2 (D), sentence 3. To leave out sentence number two, you may be inclined to cut from point A to point C, but it probably won't work. Breath 2 may not have the proper rhythm to follow sentence 1. Additionally, if any reverberation existed in the acoustical environment that the speech was recorded in, you'll get the tag or decay of sentence 2 between sentence 1 and 3. The proper edit points would be from B to D. Here, the tempo of breath will flow better after sentence 1 and any decay due to reverberation of the first sentence will fit perfectly.

Besides reverberation, other background noises must be considered in speech editing. I once had to edit a section of tape out of a play. One of the props in the play was a running grandfather clock. I made the seemingly easy edit, played it back, and found that the disruption of the clock's tick in the background was very noticeable.

Since there was a natural pause in the dialogue before the edit, I was able to lengthen the space before the edit, record a tick onto another deck, and then splice it in to keep the clock's tempo consistent. If you're not able to keep the tempo consistent for one reason or another, there's another good trick to try. Simply by inserting a noise, such as a door slamming or a loud cough, you can startle the listener enough so that his or her brain/ear concentration on the previous tempo is interrupted enough so that a change will go unnoticed.

Most decks have an edit button that allows you to dump unwanted tape off the machine or simply erase that section of the file. While very handy for getting rid of long pauses, I generally don't do this with outtakes (sentences or phrases that will not be used). Often, edits do not work because of inconsistent speech inflection. Take the sentence "I knew I was dealing with a man having a vast imagination after he said that." Due to time factors you may have to dump the "after he said that" part of the sentence. Unfortunately, the "ation" of imagination will not have the proper inflection for an ending to this sentence. More often than not, you'll be able to pick up an "ation" with the needed inflection from your outtakes, so keep them on a separate reel, or a separate file.

As stated, different moves during mixdown such as muting a cymbal can make your job of editing easier. With speech there's a technique used during recording that can make your job almost routine. Set up an oscillator (or synthesizer with a matchbook stuck between the keys to hold one down) to output a signal at about 50 to 60 Hz. Bring this tone up to a fader, mute it, and pan it to the right-hand side of the stereo recorder. Now, record the narration on the left-hand side of the deck. As you read along with the script, mark it every time there's a mistake made and at the same time unmute the tone for a second or two.

When recording is completed, it will be very easy to find the mistakes on the tape since those low-frequency tones will be readily heard in fast forward or rewind. Additionally, since the script has been marked, you'll be able to make the corrections without getting lost or confused.

Why all this emphasis on speech editing? It's very much part of the engineer's job, sometimes more so than music tracks. The health of the music industry may vary from time to time, but the availability of commercial and broadcast work will almost always be consistent. Obviously, any engineer who can handle both is worth more to the studio he works for.

Editing a Complete Track

Now, we move on to complete tracks. Let's start off with Muzak or that syrupy type of background music that's often found on elevators in supermarkets, and in most of your parents' record collections. This is the type of music that's most often found in commercials. Since every listener should be concentrating on the narration, it's important that your edits not be distracting. To ensure this, fading in and out, or cross-fading (one section faded in at the same time the other fades out), is often used.

Tape a few pieces and edit them down to, say, 15- or 40-second spots. Now tape your speech sections and edit them down to under 10 or 30 seconds.

You can now make the peak or climax of the music match up with an important word or phrase in the narration. All that's needed now is to edit or fade the excess music and you've completed an honest-to-God commercial. Granted, you may need three analog recorders to accomplish this, but with the amount of musicians recording their own material today, it shouldn't be hard to borrow work station time. At this point you can literally go into business for yourself.

Restaurants and record or hi-fi shops are always advertising on the radio, Find out your local station's rates, come up with an idea (jingle, sale, contest, etc.), complete your tape, and present it to the store owner with all the facts on costs. You may get turned down a few times, but keep at it, as it's not only a good way to make money, but if you get more work and have to use a studio, it could land you a job there, and don't be surprised if the radio station is interested in hiring you.

Aside from all this, editing can be done strictly for fun and entertainment. Record a favorite album and start changing parts, cutting into different songs, and repeating sections. You're no longer obliged to listen to music just as it was laid down by the artist. This kind of editing is very good training for studio work: it'll help train your ears and enable you to know just where cuts can or cannot be made, as far as tempo, background sounds, and having enough spacing between notes, as well as key tuning and your relative pitch.

I wouldn't recommend the editing of 2-inch masters until you've gotten a lot of experience under your belt. If you must, here's a very simple way to avoid either cutting a client's master in the wrong place, or attempting an edit that can't be done. Record a rough mix to a stereo recorder and try and edit there first. If it works, great. If not, punt.

Production Cues

Finally, we come to production cues. Often an engineer will get his instructions for an edit in written form. In some cases, the producer may be involved with another project and can't make the date, but for some editors this is just the way business is done. Instructions such as "I want the 'B' section repeated, then delete the final chorus, and go right into the tag" are common. Understanding these terms along with tempo, measures, and beats is important to all engineers. During an overdub, the producer may want to lay down, say, a backup vocalist's part on the third chorus first, then go back to the first verse, and then do the bridge. If you don't know what he's talking about, you'll not only slow down the session, but could destroy the creative flow, and pretty much irritate everyone involved. That's why almost everyone interviewed for an engineering job is asked if they play an instrument, or at least have some knowledge of music.

There is, to my knowledge, no book on the meaning of these terms, but I'd imagine that an inexpensive volume on the writing of pop tunes would deal with terminology for the various sections of songs. Even though this terminology can change from producer to producer, a song's structure always remains within a certain framework so that you can easily find your spot by just asking one of the musicians what section is being called the "A" section. This is usually the first verse section and may follow an introduction or "Intro."

The "B," "C," or subsequent sections will generally be other verses that differ from the "A" verse in, say, key, structure, tempo, or some other way. Choruses are usually numbered as in first, second, third, or final chorus. There are also solo sections and/or instrumental sections. The tag is usually the ending of the song, which can be extended, vamped, or faded out. Yet vamps or variations on the melody and tempo can occur anywhere in the song as well.

With extended-play songs, many sections of a tune are repeated, shuffled, and edited together to achieve their longer length. By simply knowing what the "A" section is, you can easily determine the other parts of a song.

As an example, let's take a tune that everyone should be familiar with and break it down into its sections. "Not Fade Away" is a very simple Bo Diddley-type of tune. The only thing happening here is a repeated verse and chorus. The song goes: verse, chorus, verse, chorus, verse (instrumental section, meaning no lyric, but the same as the verse), verse, chorus, tag (ending or fade on the chorus).

In this tune, the verse would be the "A" part and the chorus would be the "B" section so that it could also be spoken as A1, B1, A2, B2, A3, B3, tag.

Some producers wouldn't call the chorus the "B" section, or the instrumental an "A" section, and you might be cued as to the ending by "tag" or "fade." A bridge in the song will generally be called just that. An intro can even be called the "A" section, but not very often. A vamp will be a variation on the melody, usually found in longer tunes, while a solo is a solo. The way it's stated by one producer may not necessarily be the same way as another, but a few questions (even over the phone) will straighten everything out.

The Mixing Stage Having gone through the processes of recording the various instruments and the effects that will be added to their tracks, we have now covered enough of the procedures of laying down tracks to enable us to move on to the mixing of a complete tune. Although I have not given a detailed step-by-step description of recording every single music instrument, you should be able to use some combination of the techniques described as a starting point for recording any instrument that comes in the door. I once recorded a trio of old-timers playing a marching band bass drum, an accordion, and a steel drum, all with shoulder harnesses. What a gas *that* was! But there's not enough room for *every* story.

No matter what kind of music you're mixing, it can suffer from a muddy, cluttered sound, comb filtering caused by phase anomalies, and distortion or high background noise if you did not pay close attention to the phase response and levels of each individual track, the frequency content, how the sounds blend together when combined, all the sends and returns, plus the main bus output and the mix recorder's input levels. How can one person possibly keep track of all of this at the same time?

Make sure you have the right tools and know how to use them. If a mason or carpenter showed up at your home to do some work and they did not have a plumb, a snap line, and a level, you would not have too much faith in the outcome of their work. I could not put much faith in the outcome of a mix done by an engineer who didn't have a pair of meters calibrated to a standard reference level, a spectrum analyzer, *and* some sort of phase meter.

One of the most important things you must deal with during a mix is separation. Let's start with frequency content. We already discussed recording different instruments that occupy the same frequency range.

Well, if this situation had not been handled correctly, now would be the time to "fix it in the mix." Many instruments do overlap frequency-wise, and even if they are not playing the same part, they may occasionally play the same note at the same time, and this will cause them to mask out each other. This is a common case with a bass line and a kick drum. It may seem like a production flaw, but very often a song's parts will develop on the spot and you may want to keep an excellently played overdub even though it may conflict with another previously recorded instrument.

You should become familiar with the common graph representation of the frequency range relationships between different instruments. While graphs such as these are very helpful aids to the developing audio engineer, they should not be considered rules of exact placement. For example, let's take that problem of the bass guitar and kick drum playing on the same beat. If the note played by the bass guitar is a low G, it translates to about 100 Hz. The kick drum's resonant frequency (especially if you use an AKG D-12 mic) can also be around 100 Hz. Every time the two occur simultaneously, you may have trouble distinguishing between them. Now the graph of relationships may show the kick drum's frequency range extending between 50 Hz and 6 kHz. This may be the case for some bass drums, but adding a bit of EQ within this range can cause conflicts with additional instruments.

However, the sound of a beater hitting a skin (especially if the beater is made of wood and the skin is plastic) actually exceeds this range. So, by boosting the EQ a few decibels, in the neighborhood of 8 to 10 kHz, you can add a little more of the high-end ticking sound produced when these two materials collide.

When soloed, this may not be too appealing, but when brought up in the mix an amazing thing happens. Our brain/ear combination may hear a tick, yet it does not focus in on it as such; instead it accentuates our perception of the kick's low end. In other words, you may hear the high end, but it causes you to feel the low end. Try it yourself while using your modified transistor radio monitor. There is no way the speakers in these little guys can reproduce much low-frequency content, but when they reproduce the beater ticking sound, you'll swear that you can hear the bass note. You don't have a transistor radio reference monitor? I carried that, walkman-type headphones, and a automobile speaker with me because I felt it may be important to hear what my work sounded like when reproduced on those systems. But let's get back to frequency content.

The bass note is full of harmonic content. This fact can also help. By boosting the bass guitar's EQ a few dB around 60 Hz, you will enable it to duck below the kick resonant range. The end result of these two simple adjustments can be a complete clearing up of the problem.

As far as applying this to, say, a piano and electric guitar's rhythm parts, the same technique holds but obviously at different frequencies. Actually, the ranges of these two instruments are such that, depending on the notes played and considering the content of the rest of the instruments being played with them, you will have various options as to where they can be placed within the spectrum. A couple of decibels of boost around 2 kHz will make that guitar cut through anything. Try boosting a piano track up at 7 kHz for added sparkle. Maybe there is space for you to add EQ at the low-mid to mid area of one of these instruments, but be careful not to clutter up the vocalist's range.

All in all, we could spend a great deal of time on frequency ranges and placement, but since instruments are not only recorded, but also played differently, there can't be any real absolutes for all mixes. But do get a graphic representation and also start to train your ears to match up sounds to their respective frequencies. It helps when you have the aid of a spectrum analyzer.

Soon you'll be able to listen to a complete mix and hear holes in the frequency range that can be used in order to unclutter instruments (via boosting one of the instrument's content at that frequency), increasing your track's separation in the mix, and adding definition to all parts played.

Another tool used for separation is depth. By this, I mean how far back, or up front, a track is placed. The simplest form of this is obtained with your level controls. A guitar lick laid in very quietly under a louder lead vocalist's line will make it appear to be some distance behind the singer. Time-processing devices further enhance this effect. If you add reverberation to the guitar and bring it up a bit louder than the dry guitar signal, it will appear that the guitar is farther back in a large room while the dry vocal is right in front of your face. Echoes can make an instrument's environment cavernous, while with the advanced digital reverberation units available today, you have the ability to place instruments an an almost infinite selection of acoustic spaces.

In addition, I'd always set up a couple of chambers (bathrooms, hallways, and even the studio itself) before a mix. Depending on the tempo of the tune, as well as the time settings of my other processors, at least one or two will work out well. Obviously, the more places you can put things, the better. For example, I've done mixes using three digital reverbs, two springs, two plates, three chambers, tape slap-back echo, and digital delay. The average home recordist is not going to have anywhere near this amount of flexibility, but an inexpensive spring, a three-head tape deck, and speakers placed in a couple of miced-up rooms can often do the trick.

Finally, if you're doing a stereo instead of a mono mix, your panning positions will help with separation. For example, panning the piano a little to the left and the rhythm guitar a bit to the right can effectively separate the two.

Most engineers today don't think of panning as being significant, let alone as an effect. That's because they never had to deal with a mono mix. I had already mixed the tune to mono when I did my first stereo mix. I'll never forget how easy panning made things. In fact, whenever I ran into a particularly tough mix, I'd first get it to sound good in mono and *then* widen the spread out to stereo.

Also, don't forget that you can use a little delay along with panning in situations where an instrument's frequency content is such that attempting to add separation by changing its sound with equalization would jeopardize its fidelity. An example of this would be a finely tuned and played classical instrument.

As you can see, adjusting the pan, level, and frequency content of a signal not only provides the mix engineer with three independent tools, but they also can interact with each other in order to achieve the desired result: an uncluttered mix. Furthermore, they can be used to enhance tracks, whether to correct or embellish a sound that is less than adequate, to bring out a particular instrument or vocal within the mix, to round out an entire mix, or to add a touch of excitement to the piece. The best way to exemplify all this is to go through each of these methods by using examples of actual mixes.

An Eight-Track Demo Mix   I once mixed an eight-track demo for a top-40 club band that was never intended to be mastered, just used as an example of their playing to be sent to club owners to help them get bookings. Therefore, this mix took on a rather novel approach. All recording was done in a small studio and ended up having a very tight sound. However, the band wanted to give the listener the feeling of the excitement of the band's live performance. While actually recording a live set in a club could have easily been accomplished, the advantages gained by the ability to overdub and punch in over mistakes was deemed more desirable.

The drum kit was recorded with the bass and was bounced down to a fairly tight stereo spread across two tracks, much the same as you would hear at a live show. During that recording date all instruments were played and everyone sang, even though nothing but the bass and drums were recorded. This provided the realistic leakage of the other instruments onto the drum's mics. When all the rest of the parts were completed, the song sounded as though it was performed all together, as the vocalist and guitarist wisely refrained from excessive use of studio effects outside of doubling. Next we would take this taped performance to a live club date.

A couple of weeks later, the band was booked at a club that had fairly live acoustics. I had already once duplicated this particular environment's reverberation artificially. Now, I fed a mix of the four recorded tracks into the stage monitor system and these musicians were good enough to play along in tune and stay in tempo with this feed. We recorded the live set to two tracks along with the audience noises and applause to the final two tracks and were now ready to mix them together.

Back in the studio, I duplicated the club's acoustics by setting up an effects chain through an EQ, into a spring reverb, and then to an amp/speaker placed out in the studio and used a mic to pick up the sound that was reflected off the control room window. Like I said, that club was fairly live. During the mix, one of the board's effects sends was used to run all the instruments that were recorded in the studio through this reverb setup.

With this done, I now had a live and a studio performance that was synched up both position- or time-wise on the same tape and everything had a matching ambient sound. It was now child's play to add in some of the ambient audience sounds before, after, and even during the songs. We then mixed our "live" performance and later even added in some additional applause. This worked so well that at one point I was actually able to mute the final chord of a song that had been recorded in the studio and use only the song's ending as done live with all the natural live applause. Nobody could ever tell the difference. This example, while being a very simplistic mix as compared to the mega-track computer-assisted mixes, shows the kind of knowledge that is sometimes needed by a recording engineer in order for them to be flexible enough to achieve the desired result.

When it comes to mixing an actual live recording, things are different, to say the least. Here, everything is dependent on your ability to set up your mics so as to isolate them from other instruments. There will always be a certain amount of bleed and later on this could make overdubbing difficult to impossible. Occasionally, you will need to mute a certain track in order to have the overdubbed part not conflict with the original. Once this is accomplished, you can use the above-mentioned method to fit this overdub into the same performance.

To accurately duplicate an acoustical environment, a good understanding of reverb, decay, delay, and echo is needed in order to avoid a hit-or-miss approach. Listen carefully to the room's sound, the frequencies that are accentuated, how long it takes a sound to die away (decay), any doubling or flaming of the snare, slap-back off the rear wall, and so on. I was constantly analyzing acoustics, both on the street and indoors, and when I heard something I liked I would check it out, even if by simply clapping my hands. Later, when I had some free studio time, I would attempt to duplicate that particular environment using any time-processing device I could get hold of. Don't quit if it doesn't work out. Use your free time to develop your own understanding of these tools. The result will be the difference between an engineer who can get a few particular sounds and one with a bag full of tricks.

Changing a Pop Tune to a Rock Song During a Mix   Before we continue with mixing, I want to describe the circumstances that led to my getting a particular mixing job. I can tell you a million things about mixing, but it is also important to understand how an engineer "picks up" clients. So, although the history of this particular project may seem unimportant, the knowledge to be gained from it may actually be very helpful.

Almost all studio engineers, including myself, do some live mixing. The reasons are various. You may want to help a band get a record deal when there are important people in the audience; it may be to instruct their live engineer as to how to achieve the same kind of sounds you got on their album, or it may be just for the fun and excitement of being involved with a live show.

Some friends of mine were doing a club date and asked me to help out with their sound. They were headlining a show for the first time and it was important to them, so I agreed. The headlining band always does their sound check first so that the next band can leave all their gear set up as is for the show and save a lot of needless work.

As always, I had brought along a few effects and mics, and was able to get a hot mix up very easily, which impressed the other band a lot. They were from somewhere down south, making their first appearance in New York and were a bit naive about the club situation here. Bands make little or no money playing some clubs here, they are often charged money to use the club's PA, and sometimes they even must pay an extra fee for the club's soundman's services. This reality shocked the members of this band big time. They figured that they wouldn't make much but hoped to at least break even. As it turned out, the $25 engineering charge would have made it impossible for them to even get back home. We had talked a bit during the sound check, they had complimented my work, and I felt a little sorry for them. So I offered to handle the board for them without charge. Both shows went down without a hitch and we all had a great time and that was, I thought, the end of it.

Several weeks later, I got a call from the leader of that band, saying that they had made a 16-track demo of one of their songs. At the time of the recording they were shooting for a strictly pop market, but were now a little more rock-oriented. Not exactly heavy metal, but they were still after a meaner sound. He wanted to know if he should re-record the song or if it could be salvaged at the mix stage. After explaining to him that there was no way of telling without hearing the tracks, it was decided that he would bring the tape to New York and we would see what could be done. So, by being a nice guy, I not only landed myself a session, but also got what ended up to be a perfect example of what can be done in a mixdown session for this book. So now let's delve into this mix.

First, I should say that the engineer who had recorded the tracks had done an excellent job. Every instrument was clean, clear, and retained its full bandwidth. Without this I could not have accomplished what I was able to. Yet he was recording a light pop tune and, as you will see, in order to turn this into rock, some pretty drastic measures had to be taken.

Step One: The Rundown   The first thing I did was to bring up all the faders and run the tape down a couple of times to get both a feel for the tune and understand its structure. It was definitely Pop City and ran like this: intro, A-part, chorus, A-part, chorus, solo/bridge, B-part, bridge, chorus, A-part, chorus. The A-part was the verse while the first bridge was only a couple of notes at the end of the solo. The B-part was an up-tempo verse in a different key. The second bridge was a tempo-and-key change around a tom fill, which led back into the chorus. It was a good song, but as I said, it was Pop City and would need a lot of work to rock it out.

Step Two: Track by Track   Now, I brought up each individual track to see exactly what I had to work with. The kick was solid, with a resonant frequency around 150 Hz, and there was some high-end beater tick. The snare had a nice snap but lacked low-end as well as high-end sizzle. The hi-hat was constantly closed. The toms occurred only during the second bridge, while the only cymbal was a heavy ride with a really long decay. There was no way I could give this a "Led Zeppelin" drum-kit sound. The bass guitar was clean as a whistle, a bit high-ended by rock standards and, as I expected, was recorded only through a direct box. The guitar part was a repeated simple melodic two-measure riff, doubled for thickness. The acoustic piano had been recorded excellently, having its full frequency content intact. However, the player's left hand was doing the same thing as the bass guitar and his right hand was playing exactly the same part as the guitar. The synthesizer was good, but quite common sounding, and his use of the pitch-modulation wheel at the end of his phrases seemed a bit conservative and was therefore almost completely unnoticeable. The tambourine (hereafter, tamb) was consistent throughout the tune, an endless two and four, except that it wasn't played during the second bridge. Its sound was good, but had a decay time that was a bit too long for what we were shooting for. The guitarist's background vocal was well done, but his voice was a bit thin and obviously some embellishment would be needed. The lead vocal had been doubled to enhance the female vocalist's performance; however, the two were not matched up closely enough to be heard as a single vocal. Listening to both tracks separately, each version had some lines that were better than those on the other. Even though this seems like a complete rank-out of the whole tape, remember that I was listening from strictly a rock point of view and was looking for all the things that were lacking as far as this type of sound and feel were concerned. So, things were not all that bad.

Initially, I wanted to make a single lead vocal track comprised of the best parts of the two original tracks. While this could be done during the mix, the strain of constant muting without flubs could be a little difficult to handle. I felt it would be much better to bounce the best lines down to a single track first, and then just let it run during the mix. Since all 16 tracks were loaded up, it would have meant erasing something. Luckily, the toms and tamb were never played at the same time, so I could bounce the tamb to the tom track and stack them. This meant losing some of the ambient kit signal that those mics were picking up, but as you will see, I was able to get it back artificially. The tom fill itself was fairly weak by rock standards, so I knew that I would need to have their playback levels way up there. With a bit of trial and error, I found the proper level at which to bounce the tamb track over so that once the tom level was set, it would be right for the tamb, too.

Next, we listened down to the whole track to be sure it was okay before erasing the original tamb track. This now opened a track, which gave me a place to compile a single lead vocal. After comparing and listening to the best lines from each take, I patched up the bounce, matched up the levels, and (via muting one vocal or the other) made my composite track. Now I was ready to take on the task at hand, which was rocking out a purely pop tune.

The Drums As I said earlier, the kick was fine, but it needed more bottom end. I set up the board's (an MCI 500 series) sine wave oscillator for a 60 Hz output and put it through a gate (Valley People's Dyna-Mite). The original kick was patched to the gate's external keying input, and its threshold was set to just under the kick's lowest level. Fortunately, there was very little bleed from the snare, so I did not have trouble with that sound keying the gate. The release control was set to make the sound resonate about as long as it normally would on an actual drum. This 60-cycle pulse, when added to the original signal, gave us a tough bass drum. To make up for my erasing some of the tom mic's ambience, I sent this signal to a small chamber I had set up in a hallway, which had about a one-half second of decay time.

The snare's signal was sent through the board's headphone cue system to a speaker (JBL 431I B) in the studio, placed close to and aimed at another snare, which was miced (AKG D-58E) from the bottom, real close to the snares themselves, which were the ones that were 4 inches across in size. After a bit of playing around with their tensioning, I got the right amount of sizzle decay, which was brought up to another channel on the console. The original snare had been equalized (boosted a couple of dB at 200 Hz) for punch, and once the two sounds were mixed properly (level-wise), they were multed together at the patch bay and put through a three-head two-track tape deck for a very short slap-back echo. This, in turn, was brought up to a third snare channel. This signal was then compressed for a ratio of just under 20:1 and then put through a digital reverb (Lexicon 224X) that was set for an almost two-second decay. Three channels for just the snare may seem a bit much, but in order to make an almost jazz-like snare sound, like it's being hit with a baseball bat in the Grand Canyon, you've got to get a little aggressive.

The hi-hat wasn't adding any feel to the tune, just keeping time. I put it through a phaser, setting the waveform control to envelope, and set the depth so that I ended with an almost a splash-cymbal sound. It was a really nice elastic effect. Also, since the phaser (an Eventide Instant Phaser) had a noiseless bypass switch, I could (and did) switch back and forth between the two sounds. This signal was also sent to the same small chamber as the kick.

The tambourine was next. Since its loudest point was at the initial hit, I was able to gate it and set the threshold to just under this level. Now I could set the release for a reduced length of decay. The shortened signal was now put through Tascam 122 three-head cassette deck for a short slap-back, and then to a Biamp MR/140 spring reverb set fairly long (but under two seconds), which gave me a ricochet-like sound. At this point, I told the bandleader that it was his job to mute and unmute this track to avoid it constantly sounding on the two and four.

The toms were a different matter altogether. I could mute a couple of tamb beats, which preceded the second bridge, in order to switch over our effects chain. However, the EQ I needed for the mid and high toms wouldn't do for the tamb that now shared the same track.

A Loft 401 parametric was inserted into the chain before the cassette deck. Its low-level out was fed to the deck, while its high level came back to the same channel of the board via the tamb's spring return. In this way, while the tamb was sounding I put the Loft in bypass, and while the toms played, the tape deck was put in pause mode. This reduction in changeover steps made it an easier move.

The sound of the toms was pretty good but, as stated, a bit weak. By boosting each one up as much as 6 dB at different points in the mid-frequencies and sending them to the same large concert-hall digital setup as the snare, they became convincingly rock, to say the least.

The ride cymbal was probably my biggest problem. The nature of a ride is both a long decay and a volume change that is almost like a sine wave in form. This makes it hard to shorten its decay via gating, as the processing keeps opening up and closing down. I wanted to have the track re-recorded with a different ride and even a crash beat or two, but the bandleader liked it the way it was, so I just put it into the chamber and kept it down in the mix. At this point, I set the kit's levels and grouped all these tracks, including the chamber return, to one fader for overall level changes.

The bass was not too much of a problem, even though it was a bit too clean and high-endish. I put it through a tube limiter (LA2A), which helped some, but I wanted a more earth-shaking rock sound. So the output of the limiter was put into an old Altec "green monster" tube amp and the output was run to an iso booth into (if I remember correctly) a folded-horn bass speaker cabinet with an 18-inch speaker. The output of this was picked up by a Sony C 37 tube mic. The EQ had to be adjusted only slightly to get rid of some string noise, but the result was a fatter, gutsy sound.

As stated, the guitar part was a repeated melodic line that lasted two measures. The playing was very clean and the sound was full frequency. He had used a pedal-type chorusing device and doubled his part for thickening. Whenever you have someone double their part, they are never exactly right on, time-wise. Sometimes the first track will be ahead of the second while at other times it will be behind. We're talking about milliseconds here, yet when the two tracks are panned, one hard left and the other hard right, the lick will appear to move from one side to the other.

The starting point of this movement is as random as the difference in the player's starting times, so it can move from left to right one time and right to left another. By putting the first guitar track through a digital delay (DeltaLab Effectron), and setting it to around 50 milliseconds (longer than any of the doubling delays), and putting the output to a third channel panned dead center, the guitar line now started at one side, moved to the other, and always ended up in the middle. Sitting in between the stereo spread, the guitar seemed to be moving in a melodic circle in front of you. I wanted a little more pizzazz, so the delayed signal (the last one to sound) was given some modulation to add a rippling sound to its ending. To this I added a little spring reverb to further lengthen its decay, 2 dB of boost at 2 kHz for bite, and then grouped the three tracks to one fader.

Our stereo piano track, although recorded beautifully, was bad news as it masked both bass and guitar due to the playing of the exact same notes. In rock, the bass and guitar are more prominent instruments; this meant that the piano needed to be adjusted. Via EQ, I rolled-off most of the low end that was also occupied by the bass guitar and boosted up the low mids (around 400 Hz) to add fullness to the track. The high end or right-hand side was right there with the guitar, so I cut it from 800 Hz to 3 kHz. Using an outboard EQ (Orban), I also boosted it a few dB at 7 kHz for some sparkle. A bit of this signal was sent to the plate to add to its melodic sound, and the two tracks were then grouped to a single fader.

The synthesizer solo was not drastic enough for my taste so I tried a couple of outboard effects even though I knew that I would end up using a flanger (in this case a favorite, the Loft 450). I was able to get the solo to just jump out at you from the speakers. The player's use of his modulation wheel was hardly noticeable so I also sent this signal through the plate. The trick here is to catch the beginning of the modulation and turn up the send to the point where you're overdriving the plate, then back it off gradually, and get ready to catch the next one. The resulting sound is like dropping a tuned steel beam onto concrete. It is attention getting, to say the least.

Now it was time to get our vocals together. I wanted to give the lead vocalist her own reverberation, but the other two chambers I had set up were not happening. First, the bathroom had the wrong decay time and rather than spend a lot of time trying to adjust it acoustically by adding absorption materials and repositioning everything numerous times, I tried my last space, a large closet. However, it was very close to the iso booth, which I was using for the bass speaker, and there ended up being too much leakage. I didn't want to put her vocal through a spring, and the digital was needed for the drums, so that only left the plate. This ended up being not as much of a problem as it could have been, since there were only two other things going through it. Furthermore, she did not sing during the solo, where I was overdriving the plate, and the input from the piano was set at a very low level.

I first had to put the vocal through a de-esser (Orban) due to some sibilance, and I added a touch of EQ (about 2 dB) around 800 Hz, to warm up the sound. This was patched through a digital delay (Eventide) set at 35 milliseconds, which was patched to the plate (its setting was about one second). The overall effect was very close to an early Elvis Presley record. The voice was up front yet with reverb somewhere not too far in the background.

One background vocal, even with the lead vocalist's accompaniment, is not my idea of a chorus. In this tune, the chorus was also the hook and it needed to be a bit more addictive. Since my composite lead vocal was singing the chorus as well, I kept it a little up front level-wise. What I wanted was a bunch of people to help out. The guitarist's background vocal was panned slightly left, put through a digital delay (Lexicon PCM-41) set at 20 milliseconds, brought up to another channel, and panned slightly right. Both of these were sent to the same stereo spring as the guitar (AKG BX-20), but the left vocal was sent to the spring's right side, while the right vocal was sent to the spring's left side. Since the spring's returns were panned hard left and right, it created a kind of wash of reverb signal. The two original lead vocals were also needed. These were each panned hard left and hard right, then put through two chorus effects (the guitarist's Boss pedal and an ADA multi-effects unit), brought back up to the board, and sent to the BX-20 in the same way as the background vocals.

Further Adjustments The reverb became a bit much as the vocals were starting to swim around (become undefined), so I used different sends for the vocals than those that were being used for the guitar. I limited the vocal send (Valley People), which had no effect on the guitar. I then gated (Keypex) the spring's returns, which also affected the guitars. This opened up the sound quite a bit while retaining most of the guitar's reverb and keeping the background vocals sounding full. The vocals were now grouped to a single fader. That was about it for signal processing.

Plotting It Out If you get the kind of music instrument vs. frequency range chart that I mentioned earlier, you can check the content of this tune along with the equalization settings given. You will see that there was very little overlapping or cluttering frequency-wise. Since most of the pan positions are also given, we must now discuss depth. The kick, hi-hat, and low tom track with ambience and cymbal were put through a small chamber (with a one-half-second decay). This effectively separated them from everything else and made up for some of the erased ambient sounds. The snare and toms were big, to say the least, with the toms in a large hall due to the 224 and the snare in something like the Grand Canyon, due to the added slap-back echo. The plate gave me a full second of decay for the piano, much more when overdriven for the synthesizer and with the 35-millisecond delay before the lead vocal, it gave a completely different acoustical space. The stereo spring was set up for over two seconds of decay, but by gating the returns, the guitar's reverb was down to around 1½ seconds. Due to gates put on the background vocalist's sends, their reverb, although from the same device, sounded quite different.

Everything sounded good, distinguishable, and uncluttered. Because of the grouping, levels were easily balanced and we were ready to mix. Now we discussed and practiced our moves. It was decided that the tambourine would only be used on the choruses, so this channel along with the original lead vocal's two channels had to be muted and unmuted and it was decided that the bandleader would handle those moves and we'd both deal with the hi-hat's muting and phasing. My job was to handle the synthesizer send to the plate and the changeover required for the tambourine/tom transition at the second bridge.

We ran the tune down and everything went perfectly until we reached the chorus after that second bridge, where the toms end and the tambourine comes back in. Due to the tom's reverb, they hung over into the chorus. That was okay, but all the moves required in order to get the first tambourine in on the second beat of the chorus were a bit much.

Getting into the bridge was no problem as the tom/tambourine track was already muted. I had plenty of time to adjust the tom's pans, switch in the outboard EQ, pause the cassette deck, bring up the sends to the digital, and unmute the tracks before the bridge.

However, in order to keep that first tambourine ricochet, I had to mute the tamb/tom track after the last tom of the roll sounded, re-pan the two tom tracks up the center, kill the send to the digital reverb, switch the outboard EQ into bypass, hit the deck into record, and then unmute the tamb/tom track. The amount of time I had to do all this was less than a second and thus impossible. We rounded up another set of hands, backed up the tape to before the bridge (the easiest splicing point), practiced the moves a few times, got it right, returned all controls to where they had been before the bridge, recorded the last half of the tune, and later spliced the two halves together.

If you think this is complicated, wait until you get into 24-track, computerized mixdown!

End of Mixing That is about it for mixing as further documentation of mixdowns should not be needed since between the descriptions that were given and the material in the "tour" sections, there is enough information to provide a good handle on not only how processing is utilized, but also why. It does not make any sense to continue listing specific EQ or effects settings because sounds change with each instrument, player, session, and even song. The idea is to be prepared so that you can calmly take on the problems you run into using a logical approach based on your prior experiences and experimentation.

For example, I once overdubbed a vocal onto a finished instrumental. It was to be a demo tape for the vocalist. After listening to her sing alone with the tape I decided that she could benefit from an added touch of reverb. However, the backing tracks were mixed dry. If I washed the whole mix with reverb it would have sounded phony. If I added reverb to just her vocal, the embellishment would have stuck out like a sore thumb. What was I to do?

I sent the whole backing track to a parametric EQ and jacked up the midband by +8 dB, narrowed the Q to about a fifth of an octave, rolled off all the other bands, and swept the midbands's center-frequency control until the snare jumped out of the mix. This was sent to a gate's keying input and the gate's controls were set so that it would open for a brief period every time the snare sounded. I used another send to feed the full backing track to the gate's normal input and the gate's output was then patched to the input of a digital reverb. The reverb's decay was set fairly short at about one-half second and its output was fed to a three-band EQ where the mids and lows were rolled off and the highs accentuated. Now with every snare hit the whole backing track was briefly washed with a bright reverb. To the listener it honesty sounded like the reverb was only on the snare.

But let's get back to the vocal. Since it needed a little of the smoothing out that the decay of reverb provides, I set up a plate for a full second of decay as I felt this would fit the bill. However, I not only sent the vocal to the plate, but also a touch of the full backing track was fed to this unit as well. Now the vocal had reverb. Because the full backing track's reverb, while lower in level, had the same sound and decay slope (and due to the attention-grabbing snare's reverb), nothing about the vocal sounded unnatural.

Mission accomplished? I guess so, because a couple of days later this young female vocalist returned to the studio with a pie she baked *for me!* Granted, it was no great shakes as far as big pay, but for an assistant engineer doing one of his first solo recording sessions, that pie tasted and *felt* mighty good.

As always, your approach to a project will depend on the particular situation. There are so many variables: the format used, where the project is being recorded, the particular talent involved, and, if it is some kind of an event, everything else that may be going on. All of these will dictate which is the best approach to take. If you're dealing with an event, you may be expected to record it as purely as possible. If you're going down to stereo, it means mixing on the fly where there is no room for "mis-takes."

There are things that people like in certain types of music that they cannot take in others. For example, rock distortion may or may not play well with jazz musicians, but it will definitely not win you any admirers among the classical crowd.

With purist classical recording the musicians, expect the audience to hear (live or on tape) exactly what is being played. Remember, they are very close to their instruments, so if you only rely on an XY configuration at audience distance the sound will most likely not cut it as far as they're concerned. Mics placed closer in, say, a couple of feet from the instruments, will not only add more presence and fullness to their sound, but it will let the sound develop by also picking up some natural reflections from the floor, walls, and ceiling. Additionally, should you run into balance problems between instruments, you can just raise the mic on the instrument that's being overpowered.

The same kind of live situation also occurs during many studio sessions as when music for commercials is being tracked. These can be difficult for the beginning engineer because here time is of the essence. You've got to mic things as fast as you can, put baffles where you can, and start recording, because the whole project has to be completed in one hour. Things can get hectic, but you need to hold back the panic and keep the flow going; always keep in mind that the musicians out in the studio probably have just enough time to make their next session. You do not want these folks getting tense. You need them calm and happy in order for things to sound good. This is especially important here as you do not have the time to spend on getting exacting sounds.

More than likely you would be feeding two stereo mix recorders so that you can edit one while the other is still recording. Yes, it's pressure, but you must be willing to handle it. In fact, you'll actually get to where you enjoy the challenge. Once you reach this point, your education will come in floods.

Live Recording

Recording music live is more demanding than studio recording. When you are doing live music you should be just as concerned with quality. You may be dealing with the acoustics, and adding a bit of equalization and compression in order to maximize the audio quality, but the program material is also flooding the stage at high levels. This area will be full of all kinds of instruments so you often find yourself right on the edge of wailing feedback all night long. This can be very demanding and it takes a good deal of experience and knowledge to pull it off. So, do not go into live situations lightly. Do your homework beforehand.

In every recording situation you walk a fine line. Everybody expects you to handle it all. They don't want to know about any of the difficulties you may be having, nor do they want to know how you do it. They may understand the process and know about everything that is going on, but it's up to you to get a good sound as quickly as you can.

As much as you may fear it, somewhere along the way you've got to do some live to stereo recording. This is the highlight, where it *all* comes together. I'm not talking about a piano recital, or even a duet, but at least three pieces (if not a quartet), because it's the degree of interplay, the spontaneity, the "right here and now or never" aspects of live performances that make it an event. And when you've done your homework and you're plugged into those performers, you're part of that event. What do I mean by homework?

You attend rehearsals, check out the equipment being used, and clock how the players interrelate. Now you know that the acoustic guitar being used has a resonant frequency right around 200 Hz, just the same as the drummer's floor tom. If you convince the guitarist to stand to the drummer's left and to use a smaller spot monitor that is mounted on a stand instead of the stage monitors with 15-inch low-frequency drivers, you will have saved everyone a lot of trouble. And if you have an EQ handy that is all set up for a roll-off at 200 Hz, you will be in a much more comfortable position. With this comes confidence. After you've done it a couple of times, and begin to know how to get the instruments to sound good, and can balance it all in the stereo mix, it's almost repeatable. Sit back and enjoy it, because it doesn't happen every day.

With live recording, you do not have overdubs or multiple takes, and there is a lot less separation than in the studio. So start off with what you know best; in other words, go with your standard, close-mic traditional layout. If you can, add a bit of go-boing (unused speaker cabinets work well) around loud instruments. Even an upright bass will not be the end of the world if you know beforehand that it's used and you've outfitted and checked it with a pickup. Add a directional mic and a tad of EQ and the bleed you dreaded having to deal with becomes hardly noticeable. *Just don't lose your cool!* The live event is a place where they generally do not have accurate scores as there is no way of telling exactly how a part (let alone a whole piece) will be played in every situation. Here is a *tip*: Get a bunch of unfamiliar classical albums. Start playing one in the middle of a piece. When the coronet takes a solo, listen to it closely, but also pay attention to the other instruments. See if you can anticipate when and how the rest of the instruments will come back in. Believe me, this can become a fun game and it is incredible ear training. A change in key, a percussion beat, an increase in level, soon become eye-opening clues to what is about to happen. If your fingers are already on the faders when it does, you may not be a member of the band, but you are certainly part of that event.

You may not be able to convince musicians to put up with go-bos on stage, or to use small, low-level monitors, but if their instruments' levels are so loud that they are bleeding all over each other, close micing, in-ear monitors, and basing the mix on a pulled-back coincident pair of mics may be the only way to go. This will be the situation you need to deal with most of the time as those nice, jazz or ballad recording dates just do not come up that often. This is where all your homework, studying, and experimentation starts to pay off because from here on out you may have to go it alone.

Well, you will never be totally alone because those who get involved in audio engineering as a career can always review sections of this book whenever needed. I hope you have learned some of the basics of studio recording and I know this information can be a big help when you get your chance to lay down tracks.

I have refrained from stating definitive rules because there are none. While there are rules, it will always be a question of taste. People may believe that *the* microphone for vocals is a high-priced multipattern condenser and that it is the only mic you should ever use. Well, according to the sound of some of the vocal tracks I have laid down with handheld vocal mics like Shure Beta 58s, Sennheiser 441s, AKG 330 BTs, and Beyer M500s, that's simply not true. This is the reason I wholeheartedly encourage experimentation as you never really know what works and what doesn't until you try it yourself.

So Where Do You Go from Here?

Improve Your Understanding of Music   The importance of both ear training and music knowledge cannot be overemphasized. The recording engineer needs to learn to speak in musical terms and know the frequency ranges of musical instruments and how they work. For example, it is not a question of just micing the bell of a saxophone; you mic the area from where the sound you want radiates. Musicians don't want to hear about "frequency ratios" when you're setting up a harmonizer. To them, a 2:1 frequency ratio may be meaningless. It'll make more sense when it's stated as an octave. You certainly don't have to memorize all the correlation points between ratio percentages and intervals; just make an effort to read the musical annotation off the unit's display and relay *that* to them.

For audio engineering, it's not necessary to be able to read music, but it's critical that you be able to follow along with charts. Producers often give engineers instructions like "go to the third measure after the bridge," and if you can at least get to the bridge, it won't be a chore for them to then point out which is the third measure. Soon you'll be counting along and thinking in terms of edit points.

Start playing an instrument. Anything, even a thumb piano, is good for training your ears to pick up relative pitch. Listen to a lot of different kinds of music, yes, but listen hard to the sounds of the individual instruments. I believe a competent recordist should be able to duplicate any sound they hear. If you hear it and figure out how to make it happen, it becomes *your* sound.

Sitting down at a piano can seem daunting, but it's all right there laid out for you. As Oscar Peterson said, "If you've got one finger you can express yourself." But get a book of chords and start dabbling around. You'll soon get to the point where every time you walk by the piano you can sound a chord. Hey, it may be no great shakes as far as performance, or composition, but you'll know when it's time to call the piano tuner. Additionally, when you have charts to work with, you can play chords in the keys you will be working in beforehand and thus tune up your critical listening capabilities.

The next thing you know, you are following along using the charts, counting along with the tempo, hearing the pitch, and picking up on specific turning points in the piece. At this point, you're only a short step away from responding to a production cue like "we need to punch over the third note in the fourth measure of the solo" by moving to just before that point, starting playback, ready to nail that one note. In and out, no waiting around, the creative flow uninterrupted, no getting lost or losing an extra note because of a bad punch. It hurts the session's flow when the engineer has to rehearse the move three or four times. When a recordist is totally locked into what's going on, how valuable is that?

At least get yourself a book on beginning music theory. That way you will learn to operate the console like it's a music instrument, making your mutes, fades, and punches right along with the tune's meter with drummer-like anticipation, feel, and tempo. When you are tapped into what's going on, everyone will know it *and* they'll appreciate it, too.

Learn About Music Instruments   The recordist must know the frequency ranges of the music instruments being recorded. It's important to know that trumpets bottom out at about 180 Hz, while trombones reach down close to 80 Hz, and *both* reach out close to 10 kHz as this could easily affect your choice of mics.

You should know that the lowest note that is produced by a C-flute is not any lower than 250 Hz. If that flute is being played in its upper register, its lowest frequency content is going to range somewhere between 600 Hz and 1 kHz. So if you want to mellow out a bright sound by equalizing up the bottom end, you will know enough not to start cranking things up at 100 Hz.

All woodwinds have several notes that have a slight dip in their output levels. They are right around the point where the octave key changes registers. Some of these notes are at the horn's lowest extremity, while others are at the top, so it is not a point of adding a mic. However, you can use an EQ with a slightly narrow band bump to even things out. But listen first, as most players are used to this problem and will automatically compensate for it. If that is the case, any EQ could spoil a good sound. I'm afraid I could never go over all the instruments I'd like to, but as I previously stated, physicists love music instrument acoustics so most every library is full of this information.

Ear Training   I also encourage you to do a lot of ear training. Study the sounds when you listen to different styles of music. A classical concert grand piano sounds different than a honky-tonk upright. Why? What can you do to make each one sound more like the other?

Follow through when you are in the studio. When you place two mics on an amp, listen to the two sounds and make a written note about the differences. It is not only about which one is better sounding, it's just as important to note which produces more high end or if one is boomy sounding. Now you have a tool that you can call upon in the future.

Everyone knows that kick drums sound different in reggae and jazz music. The audio engineer has to listen to what those two kicks sound like and think about how to go about getting those same sounds. What kind of mic: a dynamic, condenser, or ribbon? What about the polar patterns: highly directional or omni? Any attenuation or low-frequency roll-off? A placement that's in real tight, right at the point where the beater hits the skin, or pulled back to get more of a room sound?

Ear training is also about good relative pitch. The engineer has to know when things are not sounding right or if something like a synthesizer is drifting out of tune. Sometimes the producer will be too occupied to pay close attention to every sound. Musicians may be so engaged in reading and playing along with complicated charts that they only hear themselves and may not be matching up with the others. This is no amateur mistake; it happens to the best session players, too. The engineer is really the only one who's getting the audio equivalent of a bird's eye view of the whole musical picture. An engineer discretely mentioning the possibility of a slight accidental wavering of intonation will not cause any strife or embarrassment, and it is one of their responsibilities.

Furthering Your Education   There are several ways to learn more about audio recording and engineering, including both books and courses on the subject, but hey, how about starting off by first educating yourself? You don't have to think about it all the time, but you can't afford to waste time sitting at home watching TV. Spend your evenings at home by diagramming a couple of variations on standard micing techniques. You have a lot of reading and listening to do, and I'm not referring to audio trade magazines, or your favorite rock groups.

The books found at your local library will be more informative than industry trades as these monthly magazines simply cannot cover topics to any great depth. The general rule is that they need one page of advertising for every page of editorial, and these pages get eaten up pretty quickly. Say you've got a 100-page publication with 50 pages of text. You lose the first page to the editor's opening comment and the publication's masthead. Letters from readers cost you three, and with the advertising index, corrections to last month's issue, and industry announcements, you're down to 40. Every month there's a "Special Focus" that includes a listing of available equipment, so you're down to 30.

At $2,000 a page, the most that can be devoted to a specific topic will be three pages and with photos, drawings, or graphs and tables; so the text gets whittled down to 3,000 words. This does not often add up to what you can call in-depth tutorial coverage.

If you think these publications should change their priorities, try putting out a 150-page magazine every month with a newsstand price of $2.50 and *no* advertising. It will have a very short lifespan, and not just because of the money. In order for advertisers to shed the long green, they must be confident in the publication's viability. Therefore, a large number of ads ends up being the validation for the publication.

So if what you need are detailed explanations that "get to the bottom of things," your best bet is the library at a local technical school.

Open up your ears by broadening your musical horizons. You may not be able to handle the almost dissonant sound of near eastern music, but when you listen to classical and big band orchestrations, you begin to understand how to blend sounds. Drop by music stores, and ask musicians about their instruments and other equipment. While you obviously need to check out your local club scene, don't forget the high school marching band.

Learn by Doing   As anyone who knows anything about audio engineering will tell you, the best training is experience. So get out there and do it. Work with local bands, even if it is just roadie work; at least you will learn about equipment. That's how I started. Get yourself a cheap tape recorder, anything. Quality is not the most important factor. Try different microphone positioning, and listen to the difference they make in the sound. I would record church choirs and local bands for free, just for the experience. Later, I started to make a little money for this service.

It is never too soon to start acting like a professional so take the time to plan your strategy before you show up to engineer a live or recording date. Lay out the way you are going to set up the stage or the studio. Think about those positions in terms of both bleed and phase anomalies from reflections and, if it is a live venue, plan everything to avoid feedback. Yes, you have to be willing to work very hard.

Start with just the good, old-fashioned standard micing technique setups, then experiment and try different mics and different placements. Just keep progressing. How are you going to mic a piano? Well, first you have to know if it is an upright, a baby grand, or a keyboard-triggering device that produces digitally recorded sounds. If you are in luck and it is a playable acoustic grand, you then have to decide what kind of sound you want from it. Do you want a purist classical sound or something with more high-end cut like a honky-tonk rock sound? The answers to these questions will determine whether your mic placement is close to the hammers, farther back, or if it would be best to place them in the sound holes and drop the lid to the short stick to increase isolation.

By documenting everything and comparing how the sounds differ, you will soon begin to learn how to get two completely different sounds just by re-aiming a mic. This is big-time ear/brain training, and it is very important because it helps you to develop your own style.

Obviously, the best thing that you can do is to get a job in a recording studio as a trainee, but openings for these positions are rare. The economic problems most studios face today require that they be almost constantly booked just to stay in business. There just is not the same amount of free time as there was years ago. Still, just being around the studio is a learning experience. You may start off sweeping floors and running out for coffee for little or no pay, but it's worth it.

How do you go about getting a gig in a recording facility? Well, the job is not going to come to you. You've got to get out there, hit the pavement, knock on doors, and generally make yourself visible. However, do not pester a studio with repeated visits and phone calls: Always call a studio before you drop in. When invited to stop by, leave when your interview is over. Do not hang around and make a nuisance of yourself. But don't give up! This is probably the most difficult field to get a start in. Impatience and a lack of self-composure have ended more careers than anything else.

It is very easy to forget that the recording studio is a place of business. One studio owner told me that a kid got a start in his studio by promoting it before he worked there. He stopped by and talked to the studio manager, found out about the listed hourly rate and equipment that was available, and then asked for and received about a dozen of the manager's business cards. He wrote his name on the back and then started talking up the studio to his musician friends and handing out the cards. He did not discuss discount pricing or try to cut any deals on his own. He simply got them to check out the studio. His deal with the studio was that if someone came in with one of his cards and booked time, he would be allowed to sit in on the session, provided he would not interfere. Within a few months, this individual not only had sat in on a number of sessions, but had been given a trainee job.

No matter what approach you take to get yourself into a studio, be intelligent about it, have some common consideration for the studio as a business, and stay out of people's hair.

You will get endless refusals, and in some cases you'll be turned down for a job because they had just found somebody else. This is where your perseverance is really tested. But, hey, face the facts. As I used to tell my students (usually after laying a particular boring and lengthy homework assignment on them), "If I were to go to any street corner in the world and start asking people if they wanted a job in a recording studio, how many do you think would turn me down?" That's your competition: every kid in the world! So keep hitting the streets visiting a large number of studios. Just kept going at it, even after you're continually refused. If you're very, very lucky, you may get the privilege to answer phones, sweep up, and move equipment around for little or no pay. However, from this lowly position, you will get to observe the whole process of studio operations and meet different musicians and engineers. Some day you may get an opportunity to work with an engineer as an assistant. Then you'll get the chance to watch a real engineer at work in a studio.

Next, you may have to hop to another studio for advancement. Since you already understand assisting, you may get a chance to actually engineer at a smaller studio. Are you worried that you don't know enough to cut it and that you'll mess up and never get another chance? Then do what I did. I hired an engineer to assist me. Using the engineering fee I received and the small amount allotted for an assistant (and because the engineer gave me a break on his price), it only cost me a little more than twice what I was making. But I got through my first basics session without a hitch and I'm proud to say, that engineer had the easiest gig of his life. I only had one question all night so he was off the studio couch for a total of maybe three minutes.

It's all about homework, but you must have a passion to succeed and find yourself a niche. From then on, the whole engineering thing will just really break open. Pretty soon you will be handling demos for the studio. Because of the knowledge you gained when you worked with other engineers, you begin to comprehend the differences in their styles and what they are able to accomplish, and begin to interweave those approaches with your own.

Now is the time to arrive at the studio early. Get in there with the second engineers, and help them set things up. When doing basics, you should have a hand in making sure everything is positioned properly. You want your sessions to run as smooth as possible, so the musicians upon arrival at the studio will not be faced with a huge technological wiring and interface maze that's humming, buzzing, and intermittent. If you want everything all neatly set up and properly working when the musicians arrive and the drummer sits down at the drums, you will not only have to take charge, but actually do most of the setup work yourself. But that's okay, because your seconds want to learn too, so each time around it will go a little easier.

Now the real work begins. The separation of every mic has to be checked out before you start to get drum sounds and get rid of rings and other unwanted noises. Separation is very important, so you must work very hard on your microphone placement.

Often the first thing an inexperienced engineer will do after placing a mic on an instrument is go into the control room and give it a listen. When they're not happy with the sound they start EQing before they even think about the sound they have. Is there a problem? What is it? Does it have something to do with the positioning? Why not try to change it? Don't get lazy on yourself! How about trying a different mic? Any change may make it sound worse or better, but you won't know until you try. Now the resulting information is logged for future reference in that filing cabinet on your shoulders.

Part of developing your own style is finding ways to record that make everything easier later on. This is especially true when capturing an instrument's full frequency response and dynamic range without any distortion or background noise. Avoiding mistakes here makes mixing a lot easier. You'll really begin to understand this when you start mixing another engineer's tracks.

Be extra vigilant about watching out for phase cancellation and the effects of comb filtering from your mic placement. Keep your eyes on the level and phase meters when getting drum sounds. After all the positioning is set for all the mics, keep checking the phase meter. Better yet, get a shop scope set up in the control room so you can actually watch the phase very closely while having all the musicians play everything so that you can see what's really going on.

As a recordist, you must learn to work with all the different recording mediums and equipment. It does not really matter if the project is done on a Neve, an API, an SSL, or a live mixing board. Really. If the song is good and the band performs it well, you are in a very lucky situation and all your personal preferences will not be a factor. But engineers have to know the limitations of the equipment they are dealing with. Because of this, one of the best things you can do as a second engineer is hang around when the maintenance people come into the studio to work. Watch what they're doing and ask them about it. You may never need to get involved in it yourself, but it's a good idea to keep in touch with all the electronic nuances of the devices that you are dealing with. Some people play instruments, and others do intonation work and repairs on them. It is not necessary for every engineer to be able to tear a console apart and start soldering in replacement components or do modifications, but every engineer should know how to check for proper alignments. Keep a very close eye on these things. It is extremely important. There is a weird trend lately to refer to a technician as a geek. Geeks are sad individuals that are part of a circus sideshow. Most of the techs I know are also pretty good at handing out fat lips. In other words, calling a tech a geek could be hazardous.

You have to know about music if you want to sit behind a console and run a session. And it's the recording engineer's job *to* run the session. Even when the musicians are top-flight pros, and the producer and arranger are right there on top of things, during a session it's still up to the engineer to set the pace and mood. If you don't know what they're talking about when they say "go back to the second measure of the first B part," you're going to throw the flow out of whack. Your job is to help keep everything in the pocket, happily groovin' along, with no operational or technical problems, even if you have to fake it.

Getting the job? For the very rare few it's finding a producer who likes you. This is the greatest thing that can happen to any engineer because it means steady work. In my 40 years, I have seen it happen to a few, but I was never the lucky one. In the meantime, there's also all kinds of training available outside the studio. These types of training vary, so it is difficult to offer a definitive foolproof plan. For example, many engineers have technical (electronic) backgrounds, while others start out as musicians and come from the music side. Others come from a field unrelated to recording. In terms of increasing your knowledge, such areas as electronic repair, acoustics, music theory, and business (recording is a business, you know) are very helpful.

I have a friend who never regretted using the time off between working live tours to study for his FCC technician's license. Broadcast companies and other outfits require that their technicians have this license. Technically, the test is not difficult, as far as state-of-the-art electronics go, but it does require the command of a lot of information. Many schools offer courses to prepare applicants for the license, but in this case my friend just locked himself away for two months to study on his own. Either way, it is not impossible to pull it off. If the whole idea of being a tech is not right for you, go for the FCC operator's license. Then you'll be allowed to take to the airwaves, should that opportunity arise. The health and wealth of the recording industry is constantly varying, so being able to wear a couple of different hats is a necessity.

Studio Etiquette   How you act makes *all* the difference. Let's take me, for example. A little self-criticism is good for the head, if not the soul. I had to work extremely hard to learn the technological side of recording engineering. This subject wasn't being taught anywhere when I came up, and books that described the meanings of specifications were cryptic at best. But I knew it was very important to being able to perform the job of recording engineering well, so I stuck to it. I never held it against anyone who did not understand the technological side of recording. But at times I could take it a little too seriously. I never let it interfere with a session, but occasionally, while doing tech work, I was not the most easy-going person to be around. While it is okay to be concerned with the integrity of the project's fidelity, people just naturally want to be around people who are having fun, and if you are wringing your hands because the S/N ratio is only down 70 dB, "Happy" ain't going to be your nickname. It would sometimes surprise people when I would crack a joke! Take it from one who is experienced in this, lighten up, for your own sake. You've got the best job in the world, so enjoy it!

In terms of personality, you should realize the importance of studio etiquette, as there is nothing more important than common consideration and keeping your mouth shut when it should be. Inner-studio political backstabbing will come back on you fast (I've seen it happen), but you also need to be aware of the fact that this is a very competitive business and people will try to block you out. Stick with doing your job and stay away from the politics involved as much as possible.

As stated, you can be the greatest engineer in the world, but if you are no fun to be around, nobody will want to work with you. I would say that attitude is one of the most important things audio engineers should concern themselves with. Knowing every micing technique is not as important as knowing how to deal with interpersonal relationships.

You do the job because you love it, and it feels good when you do it right. Having some knowledge of why you are there and why everybody else is there, whether from a musical or technical perspective, helps a lot. You will find that it makes things go a lot easier for everyone involved. So take pride in what you do and enjoy it.

Always be ready to share information. You should consider it your duty. Did you learn everything you know totally on your own? I didn't, and if you are reading this book neither did you. So when it comes time to pay back into the "handing down the audio knowledge" fund, step up to the plate, and make it concise and exacting, yet easily understood while remaining humble. It is a tall order, I know, but if you get so set in your ways that you are afraid to let anyone else in on your methods, you will get stuck on them and your growth as an audio recordist will stagnate.

It is important to keep an open forum so that you can pick up new ideas. I was never just a straight technical engineer; I also played a little bit, too. You have to think about the musical side of the process and that has a lot to do with listening and learning about production. Don't hide away from the music side just because you are surrounded by some of the greatest musicians, arrangers, and producers in the world. You will often find that they are very much interested in what you know. I was always interested in what they had to say.

Personality is the most important aspect of audio engineering. There, I said it and it is true. Your willingness to work with people, to be flexible, and helpful is the key. It will undoubtedly get you more opportunities than someone who appears egotistical. People will prefer to work with you even if you don't know everything there is to know simply because you have a personality that they can count on. They know that you're there for them.

When it comes to creative ideas and commenting on the music, you should draw a fine line there. My policy was to shut up unless asked, and it kept me out of a lot of the trouble that I saw others get into. People resent it when you start critiquing their work. It is not the recording engineer's job to go running out into the studio suggesting ideas, even if they seem appropriate. I actually once saw an engineer tell a producer that *he* could play the part better than the musician in the studio! I was so flabbergasted that I couldn't speak for about five minutes. I can tell you, this guy's tenure at that studio did not last very long.

The engineer's main function is to keep the creativity flowing. You are watching everything, even if it just means that you can be right there to lend a hand when something needs to be moved. Sometimes making people laugh is part of it when it breaks the tension. I would often blame myself for a "technical problem" to decrease a musician's embarrassment over a flub they made, and I'd never bring it up to them.

If you're enjoying yourself, people will enjoy being with you, and they will want to have you around. It also has a lot to do with the way you handle yourself in difficult situations. Stay clear of studio politics. Do not talk about the studio, its owners, the acts booked, or any problems. You will just be stirring up bad blood and it has nothing to do with your job.

When I finished working, I found it best just to leave and stay away from all that negative stuff. Besides, after a long session I would usually need a break if for nothing else than to rest my ears.

It is extremely important for anyone working in any service-oriented business not to have an attitude problem. In the rarified world of the professional audio recording studios, where egos can bounce from one extremity to another in the blink of an eye, no one should have to deal with any attitude problems at all. That is something that is unbelievably important.

Some people can be a little difficult to work with. But check out their side of things. They're laying it all on the line. Every time they play, their reputations are at stake, not just in the eye of the public, but more importantly in terms of their own feeling of musical self-worth. Every engineer should be put on the other side of the glass just to see what it feels like. It is not easy. In fact, it can be very hard. Instruments, charts, headphones, production cues, and self-consciousness can wreak havoc on a musician's ability to concentrate. That is why you must try to make things go a bit easier. I can't tell you how much musicians appreciated it when I took the time and effort on little things like making sure their instruments would not be subjected to tuning anomalies caused by HVAC temperature changes. You must be aware of how you handle yourself and, at times, you must be willing to bend the rules a bit for people.

It is rare for anyone to find themselves in a completely intolerable working situation, but studios are also known for overworking their trainees. You have rights too, so simply explain that you cannot work in the way that they are demanding (i.e., lifting heavy equipment, long hours, no breaks, abusive engineers or clients). It is always better to handle these problems maturely instead of bad-mouthing it all over town. Would you want to hire somebody who was known for that kind of behavior?

Taking care of yourself is very important. It is obvious that you should not risk blowing out your ears, but I've seen long hours, lack of sleep, and bad eating habits (ordering in cheap take-out food) take their toll on many recording engineers. I wish I could tell you about an organization of audio engineers that made a health plan available to all the folks working independently in this field, but I can't. So, in this respect, you *are* on your own. I never had any kind of a safety net or a large financial portfolio. When I was hit with ill health, it completely wiped me out. So while a steady job may not seem as glamorous as being a free-wheeling independent audio engineer, it is something I would urge you to reconsider.

As I wrote in the preface, there is a lot more I would have added to this volume if had I the time, but at least you have the makings of a great start. Is this everything you need to know? Maybe 25 percent. The rest is up to you.

INDEX

ABOUT THE AUTHOR

Mike Shea has been a leader in music recording for over 20 years. Staff Engineer at the Hit Factory in New York City in the mid-eighties when it was the leading, state-of-the-art recording facility in the world, he has built (and recorded in) numerous studios around the country. A writer for *Pro Sound News, International Musician, Recording World,* and *The Record,* he is former Technical Editor of *Recording World.* He teaches audio engineering at the Institute of Audio Research and lives in New York City.